Microorganisms for Sustainability

Volume 54

Microorganisms have been in existence since the origin of life on earth and can survive the most extreme habitats or conditions on earth. Microorganisms are involved in regulating biogeochemical cycles, maintaining plant and animal health, and sustaining the global food chain. Moreover, they play crucial roles in addressing the challenges of climate change and achieving the targets of Sustainable Development Goals (SDGs).

This multidisciplinary book series captures the role of microbes towards building a sustainable world, while encompassing cutting-edge technologies and current needs across various fields such as agriculture sustainability, bioremediation, restoration of degraded habitats and wastelands, and food security. Additionally, this series explores microbial applications in industries, and building their utilization in clean and green energy solutions. Furthermore, themes like microbial secondary metabolites, extremophilic microbes and modern omics, including next generation sequencing and metagenomics, are also covered in this series.

With contributions from researchers across the globe, this series addresses the important call of 'One Planet-One Health-One Future'. It comprises a collection of diverse volumes that provides insights for scientists, young researchers, educators and decision-makers in the government, private sector, and non-governmental organizations, empowering their efforts to achieve the global goals.

The series invites, evaluates, and accepts book proposals to ensure a diverse, inclusive, and evolving program. The final decision regarding acceptance rests with the series editor.

Peer-review

This book series follows a stratified review process. Proposals for individual volumes are reviewed by the series editor and then the editorial board members. On a case-to-case basis, external reviewers are also invited for further evaluation of the book proposal. Review of the chapters is the responsibility of the volume editor(s). A manuscript submission platform has been recently made available to the authors, volume editors and the series editor.

Ajay Kumar • Manoj Kumar Solanki
Editors

Microbial Biocontrol Techniques

Importance in Ensuring Food Security

Editors
Ajay Kumar
Amity Institute of Biotechnology
Amity University
Noida, Uttar Pradesh, India

Manoj Kumar Solanki
Department of Life Science and Biological Sciences
I.E.S. University
Bhopal, Madhya Pradesh, India

ISSN 2512-1901 ISSN 2512-1898 (electronic)
Microorganisms for Sustainability
ISBN 978-981-97-8738-8 ISBN 978-981-97-8739-5 (eBook)
https://doi.org/10.1007/978-981-97-8739-5

This Springer imprint is published by the registered company Springer Nature Singapore Pte Ltd.
The registered company address is: 152 Beach Road, #21-01/04 Gateway East, Singapore 189721, Singapore

Preface

The increasing demand for sustainable agricultural practices has led to a renewed focus on biological control methods to manage pests and pathogens. Among these methods, microbial biocontrol agents (BCAs) have emerged as a promising alternative to chemical pesticides, offering environmentally friendly and sustainable approach. This book, *Microbial Biocontrol Techniques*, aims to provide a comprehensive overview of the latest developments, methodologies, and applications of microbial BCAs in the sustainable agriculture.

The book provides a fundamental concept of microbial biocontrol and explores the significance of beneficial microbes in maintaining plant health and productivity. The chapters of the book cover a diverse range of microbial agents, including bacteria, fungi, and viruses, detailing their modes of action, target specificity, and mechanisms of pathogen suppression. Emphasis is placed on the importance of screening techniques, viz. in vivo, in vitro, and in silico methods, which are essential for identifying and optimizing potent BCAs.

Microbial Biocontrol Techniques serves as an essential resource for researchers, agronomists, and practitioners seeking to enhance their understanding of microbial biocontrol strategies and their role in sustainable agriculture. By fostering a deeper appreciation of the potential and challenges associated with microbial BCAs, this book aims to contribute to the advancement of eco-friendly pest management solutions for a healthier and more resilient agricultural ecosystem.

Noida, Uttar Pradesh, India — Ajay Kumar
Bhopal, Madhya Pradesh, India — Manoj Kumar Solanki

Contents

1 **Microbial Consortia: An Approach in Plant Growth Promotion and Plant Diseases Management** ... 1
Mario Alberto Soltero-Rios, Alondra Yazmin Medina-Orozco, Marco A. Valenzuela-Yerena, Naivy Montserrat Andrade-Robles, Ashley Andrea Mendoza-Flores, Jesús A. Y. Morelia-Jiménez, Milton Yahir Ochoa-Castañeda, Alberto Robles-Hernández, and Ramsés Ramón González-Estrada

2 **Microbes-Associated Activation of Plant Defense Against Phytopathogens** ... 19
M. C. Bañuelos-González, E. A. Cuéllar-Torres, S. Aguilera-Aguirre, U. M. López-García, R. I. Ortiz-Basurto, and M. A. Chacón-López

3 **Impact of Environmental Factors on Microbial Association with Plant** ... 41
Gema Cárdenas-Flores, Ana Y. Flores-Ramírez, Adriana G. Aranda-Vázquez, Martina A. Chacón-López, José V. Valencia-Zavala, Rosa I. Ortiz-Basurto, Ramsés R. González-Estrada, and Ulises M. López-García

4 **Role of Endophytic Microbes for Postharvest Disease Management** ... 63
Monika Singh, Aamna Hassan, Shanvi Nautiyal, Rakesh, and Sonika Kalia

5 **Utilization of Rice Husk-Derived Silica in Fertilizer and Pesticide Formulation** ... 79
Masoumeh Kordi, Robab Salami, Parisa Bolouri, Zeynab Borzouyi, Behnam Asgari Lajayer, Weixi Shu, and G. W. Price

6 **An Application of Biopesticides in Control of Pest and Crop Protection: An Eco-Friendly Management** ... 95
Narendra Kumar and S. M. Paul Khurana

7 Beneficial Microbes in Agriculture: Recent Development and Outlook 125
Bushra Rasool, L. V. Ravishankar, Nisar Ahmad Dar, Amrish Vaid, Zaffar Bashir, Bisma Rashid, Sana Surma, Baby Summuna, M. D. Thabrez, and Sakshi Sharma

8 CRISPR/Cas9-Based Genome Engineering in Plants for Enhancing Disease Resistance 143
Haris Butt and Kubilay Kurtulus Bastas

9 Biocontrol Mechanisms by Root-Associated *Bacillus* Species 155
Muhammad Hamza Ashraf, Nazim Hussain, Muhammad Zafar Saleem, and Rimsha Haqqi

10 Cyanobacteria as Mediators of Systematic Resistance for Sustainable Disease Control in Agriculture 179
Muhammad Hamza Ashraf, Nazim Hussain, Rimsha Haqqi, Aramish Naz, and Syed Wajahat Ullah

11 Microbial Inoculants for the Management of Pesticide Toxicity in Plants 199
Dharmendra Kumar, Som Dutt, Dinesh Kumar, Kailash Chand Naga, Vikas Mangal, and Brajesh Singh

12 Screening of Microbial Biocontrol Agents Using Traditional and *In Silico* Approaches 227
Vishal, Navaneet Chaturvedi, Nischal Pradhan, Aamna Hassan, Kshitij Parmar, Monika Singh, Rachana Singh, and Ajay Kumar

13 *Talaromyces* spp. Are Promising Biocontrol Agents for Sustainable Agriculture 245
Aqleem Abbas, Sajjad Ali, Mustansar Mubeen, Abid Hussain, Kainat Ahmed Gutumsary, Babar Hussain, and Taixiang Chen

14 Metagenomics: Unveiling Microbial Communities Theme 281
Ravishankar L. V., Bushra Rasool, R. Nilesh Kumar, Amit Kumar Maurya, Ayushi Sharma, Ranbir Singh, and Mir Zaffar Bashir

15 Antimicrobial Resistance: Introduction and Challenges 293
Arunima Singh, Yogesh Kumar Vishwakarma, Mayank, Neelmani Bhardwaj, and R. S. Singh

16 Multifunctional Nanocomposites Enhance Biotherapeutics Sensitivity in Replication-Associated Diseases 311
Tanya Kapil, Palak Gupta, Arathi Radhakrishnan, and Rajpal Srivastav

17 Blockchain Technology for Environmental Conservation 323
Susmita Shukla, Umme Abiha, Ritambhara Bhutani, Naman Kapoor, and Shiv Kant Shukla

18 Exploring Host Immunity and Virulence in Mucorales 363
Aastha Chauhan, Pooja Sen, Aman Singh, Krishna Prajapati, Vanshika Jain, and Pooja Vijayaraghavan

Editors and Contributors

About the Editors

Ajay Kumar is currently working as an Assistant Professor (Scale II) at the Amity Institute of Biotechnology, Amity University, Noida, India. Dr. Kumar recently completed his postdoctoral research at the Agriculture Research Organization, Volcani Center, Rishon LeZion, Israel, on the theme "Postharvest Management of Fruits Using Beneficial Microbes" or "Biopesticide Against the Postharvest Pathogens of Fruits." Prior to this, Dr. Kumar obtained his doctoral degree from the Department of Botany, Institute of Science, Banaras Hindu University, Varanasi, India, on the theme "Plant–Microbe Interaction." In his research career, Dr. Kumar has published more than 250 scientific contributions, including research and review articles, books, and book chapters, with leading international journals and publishers. Dr. Kumar also serves as an Associate Editor for *Frontiers in Microbiology and BMC Microbiology* and as a Special Guest Editor for Microorganisms, Plants, sustainability MDPI Journals. In addition, he is a Review Editor for several leading international journals, such as *Antioxidants, Scientific Reports, Agronomy, Environmental Pollution, PLOS ONE, Agriculture, Ecosystems & Environment, Biological Control, Plant Biochemistry, Biotechnology*, and *3Biotech*.

Manoj Kumar Solanki is an Associate Professor at IES University, Bhopal, India. He previously served as a scientist at the University of Silesia in Katowice, Poland, from March 2021 to March 2023. He earned his Master's degree in Microbiology from Barkatullah University (2006) and a PhD in Microbiology from Rani Durgavati University, India (2013). Dr. Solanki's dedication to scientific progress has been recognized with prestigious awards, including visiting scientist fellowships at the Guangxi Academy of Agriculture Sciences in China (2013–2015) and the Volcani Center, Agricultural Research Organization in Israel (2016–2020). Dr. Solanki's research includes but is not limited to plant–microbe interactions, soil microbiology, plant disease management, and enzymology. His work is widely published in leading international journals. He is passionate about using agriculturally important microorganisms, including bacteria, actinomycetes, fungi, and yeast, to improve soil and crop health. Dr. Solanki's strong commitment to scientific discovery and practical solutions continues to drive his significant contributions to the field.

Contributors

Aqleem Abbas Department of Agriculture and Food Technology, Karakoram International University, Gilgit, Pakistan

Umme Abiha IDRP, Smart Healthcare, IIT Jodhpur, Rajasthan and All India Institute of Medical Science, Jodhpur, Rajasthan, India

S. Aguilera-Aguirre Integral Food Research Laboratory, National Technological Institute of Mexico/Technological Institute of Tepic, Tepic, Nayarit, Mexico

Sajjad Ali Department of Chemistry, Karakoram International University, Gilgit, Pakistan

Naivy Montserrat Andrade-Robles Tecnológico Nacional de México/I. T. Tepic, Tepic, Nayarit, Mexico

Adriana G. Aranda-Vázquez Agrobiotechnology and Electrochemistry Laboratory-Integral Food Research Laboratory, National Technological Institute of Mexico/Technological Institute of Tepic, Tepic, Nayarit, Mexico

Muhammad Hamza Ashraf Centre for Applied Molecular Biology (CAMB), University of the Punjab, Lahore, Pakistan

M. C. Bañuelos-González Integral Food Research Laboratory, National Technological Institute of Mexico/Technological Institute of Tepic, Tepic, Nayarit, Mexico

Mir Zaffar Bashir Division of Basic Sciences and Humanities, FoA, SKUAST-Kashmir, Wadura, Srinagar, Jammu and Kashmir, India

Kubilay Kurtulus Bastas Department of Plant Protection, Faculty of Agriculture, Selcuk University, Konya, Turkey

Neelmani Bhardwaj Institute of Science, Banaras Hindu University, Varanasi, India

Ritambhara Bhutani Applied Plant Biotechnology Research Lab, Centre for Plant and Environmental Biotechnology, Amity Institute of Biotechnology, Amity University, Noida, India

Parisa Bolouri Department of Genetics and Bioengineering, Yeditepe University, Istanbul, Turkey

Zeynab Borzouyi Agriculture Department, Islamic Azad University of Sabzevar, Sabzevar, Khorasan Razavi, Iran

Haris Butt Department of Plant Protection, Faculty of Agriculture, Selcuk University, Konya, Turkey

Gema Cárdenas-Flores Agrobiotechnology and Electrochemistry Laboratory-Integral Food Research Laboratory, National Technological Institute of Mexico/Technological Institute of Tepic, Tepic, Nayarit, Mexico

Martina A. Chacón-López Agrobiotechnology and Electrochemistry Laboratory-Integral Food Research Laboratory, National Technological Institute of Mexico/Technological Institute of Tepic, Tepic, Nayarit, Mexico

Navaneet Chaturvedi Amity Institute of Biotechnology, Amity University, Noida, Uttar Pradesh, India

Aastha Chauhan Amity Institute of Biotechnology, Amity University Uttar Pradesh, Noida, Uttar Pradesh, India

Taixiang Chen State Key Laboratory of Herbage Improvement and Grassland Agro-Ecosystems, Gansu Tech Innovation Centre of Western China Grassland Industry, College of Pastoral Agriculture Science and Technology, Lanzhou University, Lanzhou, People's Republic of China

E. A. Cuéllar-Torres Integral Food Research Laboratory, National Technological Institute of Mexico/Technological Institute of Tepic, Tepic, Nayarit, Mexico

Nisar Ahmad Dar Maize Research Station, SKUAST-Jammu, Udhampur, Jammu and Kashmir, India

Som Dutt Division of Crop Physiology, Biochemistry and Postharvest Technology, ICAR-Central Potato Research Institute, Shimla, Himachal Pradesh, India

Ana Y. Flores-Ramírez Agrobiotechnology and Electrochemistry Laboratory-Integral Food Research Laboratory, National Technological Institute of Mexico/Technological Institute of Tepic, Tepic, Nayarit, Mexico

Ramsés R. González-Estrada Agrobiotechnology and Electrochemistry Laboratory-Integral Food Research Laboratory, National Technological Institute of Mexico/Tecnológico Nacional de México/I. T. Tepic, Tepic, Nayarit, Mexico

Palak Gupta Amity Institute of Biotechnology, Amity University, Noida, Uttar Pradesh, India

Kainat Ahmed Gutumsary Department of Agriculture and Food Technology, Karakoram International University, Gilgit, Pakistan

Rimsha Haqqi Centre for Applied Molecular Biology (CAMB), University of the Punjab, Lahore, Pakistan

Aamna Hassan Jamia Hamdard University, New Delhi, India

Abid Hussain Department of Agriculture and Food Technology, Karakoram International University, Gilgit, Pakistan

Babar Hussain State Key Laboratory of Geohazard Prevention and Geo Environment Protection, Chengdu University of Technology, Chengdu, People's Republic of China

Nazim Hussain Centre for Applied Molecular Biology (CAMB), University of the Punjab, Lahore, Pakistan

Vanshika Jain Amity Institute of Biotechnology, Amity University Uttar Pradesh, Noida, Uttar Pradesh, India

Sonika Kalia Department of Biotechnology, School of Applied and Life Sciences, Uttaranchal University, Dehradun, Uttarakhand, India

Tanya Kapil Amity Institute of Biotechnology, Amity University, Noida, Uttar Pradesh, India

Naman Kapoor Applied Plant Biotechnology Research Lab, Centre for Plant and Environmental Biotechnology, Amity Institute of Biotechnology, Amity University, Noida, India

S. M. Paul Khurana Amity Institute of Biotechnology, Amity University Haryana, Manesar, Gurgaon, Haryana, India

Masoumeh Kordi Department of Cell & Molecular Biology, Faculty of Life Sciences and Biotechnology, Shahid Beheshti University, Tehran, Iran

Ajay Kumar Amity Institute of Biotechnology, Amity University, Noida, Uttar Pradesh, India

Dharmendra Kumar Division of Crop Physiology, Biochemistry and Postharvest Technology, ICAR-Central Potato Research Institute, Shimla, Himachal Pradesh, India

Dinesh Kumar Division of Crop Physiology, Biochemistry and Postharvest Technology, ICAR-Central Potato Research Institute, Shimla, Himachal Pradesh, India

Narendra Kumar Department of Botany, Guru Ghasidas Vishwavidyalaya (a Central University), Bilaspur, Chhattisgarh, India

R. Nilesh Kumar Division of Plant Pathology, Faculty of Agriculture, SKUAST-Jammu, Jammu, Jammu and Kashmir, India

Behnam Asgari Lajayer Faculty of Agriculture, Dalhousie University, Truro, NS, Canada

Ulises M. López-García Agrobiotechnology and Electrochemistry Laboratory-Integral Food Research Laboratory, National Technological Institute of Mexico/ Technological Institute of Tepic, Tepic, Nayarit, Mexico

Vikas Mangal Division of Crop Physiology, Biochemistry and Postharvest Technology, ICAR-Central Potato Research Institute, Shimla, Himachal Pradesh, India

Amit Kumar Maurya Division of Plant Pathology, Teerthanker Mahaveer University (TMU), College of Agricultural Sciences, Moradabad, Uttar Pradesh, India

Mayank Institute of Science, Banaras Hindu University, Varanasi, India

Alondra Yazmin Medina-Orozco Tecnológico Nacional de México/I. T. Tepic, Tepic, Nayarit, Mexico

Ashley Andrea Mendoza-Flores Tecnológico Nacional de México/I. T. Tepic, Tepic, Nayarit, Mexico

Jesús A. Y. Morelia-Jiménez Tecnológico Nacional de México/I. T. Tepic, Tepic, Nayarit, Mexico

Mustansar Mubeen Department of Plant Pathology, College of Agriculture, University of Sargodha, Sargodha, Pakistan

Kailash Chand Naga Division of Crop Physiology, Biochemistry and Postharvest Technology, ICAR-Central Potato Research Institute, Shimla, Himachal Pradesh, India

Shanvi Nautiyal Department of Biotechnology, School of Applied and Life Sciences, Uttaranchal University, Dehradun, Uttarakhand, India

Aramish Naz Centre for Applied Molecular Biology (CAMB), University of the Punjab, Lahore, Pakistan

Milton Yahir Ochoa-Castañeda Tecnológico Nacional de México/I. T. Tepic, Tepic, Nayarit, Mexico

Rosa I. Ortiz-Basurto Agrobiotechnology and Electrochemistry Laboratory-Integral Food Research Laboratory, National Technological Institute of Mexico/ Technological Institute of Tepic, Tepic, Nayarit, Mexico

Kshitij Parmar Amity Institute of Organic Agriculture, Amity University, Noida, Uttar Pradesh, India

Nischal Pradhan Amity Institute of Biotechnology, Amity University, Noida, Uttar Pradesh, India

Krishna Prajapati Amity Institute of Biotechnology, Amity University Uttar Pradesh, Noida, Uttar Pradesh, India

G. W. Price Faculty of Agriculture, Dalhousie University, Truro, NS, Canada

Arathi Radhakrishnan Amity Institute of Biotechnology, Amity University, Noida, Uttar Pradesh, India

Rakesh Department of Biotechnology, School of Applied and Life Sciences, Uttaranchal University, Dehradun, Uttarakhand, India

Bisma Rashid Division of Basic Sciences and Humanities, FoA, SKUAST-Kashmir, Wadura, Jammu, Jammu and Kashmir, India

Bushra Rasool Division of Plant Pathology, Faculty of Agriculture, SKUAST-Jammu, Jammu, Jammu and Kashmir, India

L. V. Ravishankar Division of Plant Pathology, Faculty of Agriculture, SKUAST-Jammu, Jammu, Jammu and Kashmir, India

Alberto Robles-Hernández Tecnológico Nacional de México/I. T. Tepic, Tepic, Nayarit, Mexico

Robab Salami Department of Cell & Molecular Biology, Faculty of Life Sciences and Biotechnology, Shahid Beheshti University, Tehran, Iran

Muhammad Zafar Saleem Centre for Applied Molecular Biology (CAMB), University of the Punjab, Lahore, Pakistan

Pooja Sen Amity Institute of Biotechnology, Amity University Uttar Pradesh, Noida, Uttar Pradesh, India

Ayushi Sharma Division of Plant Pathology, Faculty of Agriculture, SKUAST-Jammu, Jammu, Jammu and Kashmir, India

Sakshi Sharma Division of Plant Pathology, Faculty of Agriculture, SKUAST-Jammu, Jammu, Jammu and Kashmir, India

Shiv Kant Shukla School of Services, Kaushalaya—The Skill University, Ahmedabad, Gujarat, India

Susmita Shukla Applied Plant Biotechnology Research Lab, Centre for Plant and Environmental Biotechnology, Amity Institute of Biotechnology, Amity University, Noida, India

Weixi Shu Faculty of Agriculture, Dalhousie University, Truro, NS, Canada

Aman Singh Amity Institute of Biotechnology, Amity University Uttar Pradesh, Noida, Uttar Pradesh, India

Arunima Singh Department of Chemical Engineering & Technology, IIT (BHU), Varanasi, India

Brajesh Singh Division of Crop Physiology, Biochemistry and Postharvest Technology, ICAR-Central Potato Research Institute, Shimla, Himachal Pradesh, India

Monika Singh Department of Biotechnology, School of Applied and Life Sciences, Uttaranchal University, Dehradun, Uttarakhand, India

R. S. Singh Department of Chemical Engineering & Technology, IIT (BHU), Varanasi, India

Rachana Singh Amity Institute of Biotechnology, Amity University, Noida, Uttar Pradesh, India

Ranbir Singh Division of Plant Pathology, Faculty of Agriculture, SKUAST-Jammu, Jammu, Jammu and Kashmir, India

Mario Alberto Soltero-Rios Tecnológico Nacional de México/I. T. Tepic, Tepic, Nayarit, Mexico

Rajpal Srivastav Amity Institute of Biotechnology, Amity University, Noida, Uttar Pradesh, India

Baby Summuna Division of Plant Pathology, FoH, SKUAST-Kashmir, Jammu, Jammu and Kashmir, India

Sana Surma Division of Plant Pathology, FoH, SKUAST-Kashmir, Jammu, Jammu and Kashmir, India

M. D. Thabrez Division of Plant Pathology, Faculty of Agriculture, SKUAST-Jammu, Jammu, Jammu and Kashmir, India

Syed Wajahat Ullah Centre for Applied Molecular Biology (CAMB), University of the Punjab, Lahore, Pakistan

Amrish Vaid Division of Plant Pathology, Faculty of Agriculture, SKUAST-Jammu, Jammu, Jammu and Kashmir, India

José V. Valencia-Zavala Agrobiotechnology and Electrochemistry Laboratory-Integral Food Research Laboratory, National Technological Institute of Mexico/Technological Institute of Tepic, Tepic, Nayarit, Mexico

Marco A. Valenzuela-Yerena Tecnológico Nacional de México/I. T. Tepic, Tepic, Nayarit, Mexico

Pooja Vijayaraghavan Amity Institute of Biotechnology, Amity University Uttar Pradesh, Noida, Uttar Pradesh, India

Vishal Amity Institute of Biotechnology, Amity University, Noida, Uttar Pradesh, India

Yogesh Kumar Vishwakarma Department of Chemical Engineering & Technology, IIT (BHU), Varanasi, India

1 Microbial Consortia: An Approach in Plant Growth Promotion and Plant Diseases Management

Mario Alberto Soltero-Rios, Alondra Yazmin Medina-Orozco, Marco A. Valenzuela-Yerena, Naivy Montserrat Andrade-Robles, Ashley Andrea Mendoza-Flores, Jesús A. Y. Morelia-Jiménez, Milton Yahir Ochoa-Castañeda, Alberto Robles-Hernández, and Ramsés Ramón González-Estrada

Abstract

At present, the use of fertilizers plays a key role in maintaining soil in good conditions and improving the quality of harvested produce. However, the agricultural costs can increase because of a significant number of fertilizers applied is lost due to environmental factors, and these compounds cause environmental pollution. On the other hand, plant-associated microorganisms play an important role in agriculture production. The application of microbial communities in order to increase the productivity and maintain soil fertility offers better beneficial effects than the single microorganism application. Besides, the use of microorganism offers the benefit of controlled plant pathogens, which traditionally are controlled by the application of chemical pesticides that can affect other microorganisms and are residual and pollute the environment. Thus, researchers need to develop environmentally friendly options instead of compounds from chemical origin. The present chapter will summarize recent investigations in the use of microorganisms for plant growth promotion and for controlling diseases.

M. A. Soltero-Rios · A. Y. Medina-Orozco · M. A. Valenzuela-Yerena · N. M. Andrade-Robles · A. A. Mendoza-Flores · J. A. Y. Morelia-Jiménez · M. Y. Ochoa-Castañeda · A. Robles-Hernández · R. R. González-Estrada (✉)
Tecnológico Nacional de México/I. T. Tepic, Tepic, Nayarit, Mexico
e-mail: maalsolterori@ittepic.edu.mx; alyamedinaor@ittepic.edu.mx; maanvalenzuelaye@ittepic.edu.mx; namoandradero@ittepic.edu.mx; asanmendozafl@ittepic.edu.mx; jeaimoreliaji@ittepic.edu.mx; miyaochoaca@ittepic.edu.mx; alrobleshe@ittepic.edu.mx; ramgonzalez@ittepic.edu.mx

A. Kumar, M. K. Solanki (eds.), *Microbial Biocontrol Techniques*, Microorganisms for Sustainability 54,
https://doi.org/10.1007/978-981-97-8739-5_1

Keywords
Agriculture · Production · Plant disease · Management · Eco-friendly

1.1 Introduction

In recent years, optimizing crop production has become crucial to meet the food demands of the growing global population (Kinnunen et al. 2020). To tackle this significant challenge, the use of agrochemicals, such as synthetic pesticides, in agricultural practices has risen dramatically (Curl et al. 2020; Ngegba et al. 2022; Schleiffer and Speiser 2022). The term "pesticides" encompasses all compounds used to destroy or regulate pests, including insecticides (insects), herbicides (weeds), and fungicides (fungi) (Hassaan and El Nemr 2020). However, the excessive and indiscriminate application of pesticides to boost food production leads to their excessive accumulation in soils (Schleiffer and Speiser 2022). Once they exceed threshold limits, pesticides negatively impact the abundance, diversity, composition, and functions of the rhizosphere microbiome (Geiger et al. 2010; Mahmood et al. 2016). Additionally, while most current chemicals have low toxicity for mammals, chronic or acute exposure can cause adverse health effects, such as neurodevelopmental disorders, reproductive problems, cancer, and endocrine disruption (Kim et al. 2017; Tago et al. 2014). Furthermore, the cost of pesticides and the emergence of resistance among insect pests to these chemicals are additional concerns that require attention (Jørgensen et al. 2020; Khan et al. 2020).

Several issues affect agricultural production, highlighting the need for new environmentally friendly technologies for sustainable crop production (Kour et al. 2020; Kumar et al. 2022; MacLaren et al. 2020; Nosheen et al. 2021). This chapter summarizes recent investigations into environmentally friendly options for disease control and plant growth promotion.

1.2 Fertilizers in Agriculture

Fertilizers are essential for food production in agriculture. With limited resources, shrinking arable land, and a growing population, fertilizers have become increasingly vital in boosting crop yields (Zheng et al. 2020). However, the overuse of chemical fertilizers poses a significant problem. Farmers are worried about the need to meet rising food demands and maintain soil fertility (Tian et al. 2016). This concern results in increased waste, higher agricultural and production costs, and a decline in soil quality, leading to acidification and non-point source pollution (Xia et al. 2020).

For instance, while synthetic fertilizers can enhance crop fertility, their misuse disrupts soil microbiota stability, leading to significant environmental pollution and health risks (Savci 2012). As a result, toxic substances can accumulate in plants, negatively impacting humans and animals. Consequently, managing the application of chemical fertilizers to ensure high and stable crop yields is a crucial focus of agricultural research (Hao et al. 2013).

Traditional fertilization practices can degrade soil structure and lead to significant erosion (Liu et al. 2022). Irresponsible use of agrochemicals and their limited biodegradation capacity have caused them to accumulate underground, leading to detrimental alterations in soil parameters such as structure, fertility, and water retention capacity (Savci 2012). Additionally, prolonged use of mineral fertilizers can deplete soil nutrient content and increase crop susceptibility to various diseases (Aktar et al. 2009).

Some nitrogen and phosphorus-based fertilizers contain a minimal amount of cadmium, but applying these compounds to agricultural land can cause an increase in the concentration of this compound, along with elevated levels of arsenic and chromium resulting from the leaching of heavy metals from the soil matrix (Atafar et al. 2010).

The application of fertilizers can also influence soil microbes, crucial for processes like organic matter decomposition and the cycling of carbon (C), nitrogen (N), phosphorus (P), and sulfur (S) in biogeochemical systems (Morris and Blackwood 2015). Moreover, the growth and activity of bacteria and fungi exhibit adaptability in response to crop yield and the chemical, physical, and biological attributes of soil (König et al. 2020; Liu et al. 2007; Verma and Jayakumar 2012). Numerous studies have documented changes in soil microbiota following fertilizer application, impacting rhizosphere communities (Zhao et al. 2020), their soil colonization capacity (Bzdyk et al. 2018), as well as bacterial and fungal diversity (Ma et al. 2021), among others factors (Ding et al. 2019; Li et al. 2014; Zhang et al. 2017). Zhong and Cai (2007) noted that mineral fertilizers enhance microbial biomass and diversity. Excessive use of fertilizers, particularly nitrogen-based ones like ammonium or urea, exacerbates soil acidity, influencing microbial diversity and activity, disrupting nutrient cycling, and impeding water infiltration and plant growth (Rasmussen et al. 1998).

1.3 Pesticides in Agriculture

The increasing use of pesticides is essential for boosting agricultural productivity and meeting food demands, but it adversely affects all living organisms, including plants and animals, both terrestrial and aquatic. Soil microorganisms are particularly vulnerable to pesticides (Mohamed et al. 2021). Despite pesticide applications being initially limited to small-scale and residential areas, homeowners often overlook proper dosage, inadvertently leading to high residues that pose risks across urban environments (Meftaul et al. 2020). The European Food Safety Authority (EFSA) has highlighted the absence of standardized methods for assessing the ecotoxicological impacts of pesticides on soil microorganisms (Thiour-Mauprivez et al. 2019).

Under ideal circumstances, a pesticide would be lethal exclusively to its target pest without harming any other species, including humans. However, achieving such specificity is rare and often unattainable (Kumar et al. 2024). Agricultural pesticides, including glyphosate (GLP), can persist as environmental pollutants, with GLP specifically drawing scrutiny due to its widespread use and purported risks to ecosystems and human health (Maggi et al. 2020).

The extensive and indiscriminate use of pesticides over the long term has severe repercussions for soil microbes, nutrient cycling, decomposition processes, and atmospheric dynamics, ultimately posing significant threats to food security, human health, and the environment (Majumder et al. 2021). While many agrochemicals can biodegrade with appropriate microbial action or consortia, some remain highly persistent, leading to persistent residual issues. Consequently, several national environmental protection regulations prohibit the use and application of such potent pesticides (Kumar et al. 2021).

In this context, there is a clear imperative to develop novel microbial indicators that are sensitive to pesticide exposure. Biomarkers capable of indicating direct impacts of pesticides on microorganisms are often seen as a solution. These biomarkers can be developed specifically for pesticides that inhibit enzymes found not only in target organisms but also in microorganisms classified as "non-target organisms" under current regulatory frameworks (Thiour-Mauprivez et al. 2019) (Table 1.1).

The application of pesticides, which affects microbial diversity (Table 1.2), can harm plant growth and development by reducing nutrient availability or disrupting the nutrient cycle (Majumder et al. 2021).

1.4 Plant Growth Promotion by Microorganisms

The implementation of microorganisms that promote plant growth in agriculture has been implemented as an innovative alternative that allows reducing the negative impact caused by the application of chemical-based fertilizers (Elnahal et al. 2022). Many studies have shown improvement in agricultural production by the inoculation of crops with microorganisms, representing a suitable alternative that reduces the dependence on chemical agents (Frezarin et al. 2023; Hakim et al. 2021; Naamala and Smith 2020; Santoyo et al. 2021; Urgiles-Gómez et al. 2021).

A set of plant growth-promoting microorganisms have been found in the microbiota, such as *Pseudomonas*, *Enterobacter*, *Bacillus*, *Variovorax*, *Klebsiella*, *Burkholderia*, *Azospirillum*, *Serratia*, *Azotobacter*, and *Trichoderma*. These microorganisms have been studied and stand out for their beneficial effects that help plant growth indirectly or directly, such as avoiding abiotic and biotic stress on plants, promoting plant nutrition and antagonizing phytopathogens in host plants (Tables 1.3 and 1.4).

Table 1.1 Effect of pesticides on environment

Type	Chemical compound	Environmental negative effects		References
		Soil	Water	
Pesticide	*Dichlorvos (organophosphate)*	Concentration indicated some increase at lower pH	Contamination of ground and surface waters with dichlorvos, due to its high solubility in water, with the presence of residues with average concentrations ranging between 0.2 and 1.3 μg/L	Okdeh et al. (2024)
	Endosulfan	Low concentrations of the chemical substance in soil aloof of farmland	High concentrations in the water during fall	Kim et al. (2020)
		Besides, it was reported a high concentration in the air, the compounds can be transported at higher temperatures and continuously circulates within the environment		
	Mevinphos (organophosphate)	Found low residues in the topsoil (0.14 ± 0.08 ppm). It is rapidly metabolized in soil	High detection frequencies in the aquatic system (60.4%). It is readily soluble in water and hydrolyzes at pH 7, with a half-life of 15–39 days	Chen et al. (2020b) and Faraj et al. (2024)
Herbicide	*2,4-Dichlorophenoxyacetic acid*	It can make its way into water bodies via runoff across surfaces or through infiltration, leaching, and soil percolation		Zuanazzi et al. (2020)
	Glyphosate	Low biodegradation and soil residue accumulation in the top soil and root zone	Runoff and leaching to aquifers, and persistence	Maggi et al. (2020)
	Paraquat	The compounds binds to the soil rapidly	The compounds can diffuse into the water network	Rashidipour et al. (2019)

(continued)

Table 1.1 (continued)

		Environmental negative effects		
Type	Chemical compound	Soil	Water	References
Fungicide	*Chlorothalonil*	High sorption in soils with organic matter, carbonyl and C–H groups, silt, clay, aluminum and iron oxides	It has a great potential for leaching into the groundwater from the soil surface	Meftaul et al. (2021)
	Mancozeb	Low persistence. Single first-order and biphasic kinetics DT_{50} 0.017–0.159 days	Exhibiting low persistence in groundwater. Medium to slight mobility K_{Foc} 363–2334 mL/g. in surface water and sediment exhibiting low risk to aquatic organisms with use of mitigation measures. High risk to aquatic organisms for all other representative uses	EFSA et al. (2020)

However, nowadays, a mixture of different microbial consortia including bacteria–bacteria or bacteria–fungi, or fungi–fungi have been applied in the field to enhance the agricultural productivity and plant disease management (Table 1.4).

1.5 Plant Disease Management by Microorganisms

Plant diseases are considered a topic of interest today due to their impact on the decrease in crop yields (mainly). Something alarming in this area is the large number of entities capable of generating plant diseases and highlighting bacteria, viruses, and fungi (Köhl et al. 2019). In the search to improve this situation, current techniques for its control involve the use of agrochemicals that, in addition to being less effective day by day, leads to contamination problems in soil and water. And if that were not enough, it causes poisoning in various individuals as part of poor management of agrochemical waste (Boro et al. 2022). As one of the alternatives to those mentioned, the use of microorganisms focused on plant diseases control has been introduced.

Table 1.2 Negative effects on microorganisms exposed to different pesticides

Type	Chemical substance	Relevant results	References
Herbicide	*Glyphosate*	Under glyphosate stress, genes involved in numerous metabolic pathways showed increased expression, with the microbial community boosting gene activity in pathways linked to translation, biosynthesis of secondary metabolites, transport, and catabolism. This response may help the community adapt to glyphosate contamination	Lu et al. (2020a)
	Monoisopropylamine salt of glyphosate	Reduced the population of phycocyanin-rich picocyanobacteria (PC-rich Pcy) and caused additional changes in the composition of picocyanobacteria	García et al. (2022)
	Paraquat	Lowered the nitrogen levels in *Bituminaria bituminosa* and reduced the symbiotic effectiveness of nitrogen-fixing bacteria such as *Pantoea agglomerans*, *Rhizobium nepotum*, *Rhizobium radiobacter*, and *Rhizobium tibeticum*	Mohamed et al. (2021)
Fungicide	*Mancozeb*	Outdoor use poses a high risk to soil microorganisms but represents a low risk to earthworms	EFSA et al. (2020)
	Triadimefon	The results showed significant reductions in the relative abundances of *Syntrophobacteraceae* (*Deltaproteobacteria*) and *Enterobacteriaceae* (*Gammaproteobacteria*)	Fan et al. (2021)
Insecticide	*Chlorpyrifos*	Colonization of the rhizosphere and biofilm formation by *Pseudomonas stutzeri* A1501 were suppressed, resulting in reduced nitrogenase activity in the rhizosphere. This inhibition occurred through the induction of reactive oxygen species, which hindered nitrogenase function	Lu et al. (2020b)
		Enzymes activities, viz. dehydrogenase and urease exhibited significant inhibition. Biomass activity and cell viability decreased significantly. Organics removal efficiency was reduced drastically under CPS stress	Kumar et al. (2024)
Pesticides	*Organophosphate*	Pesticides can induce oxidative stress in microbes, potentially impacting phosphorus (P) mineralization. However, they may also enhance activities of P-solubilizing bacteria, thereby increasing P availability. Conversely, pesticides can decrease biological nitrogen fixation and nitrogen mineralization by reducing populations of diazotrophs and saprophytic bacteria, due to cellular hyperkinesis and functional impairments, respectively	Virk et al. (2024)

Table 1.3 Efficacy of the application of microbial inoculants on plants

Microorganism	Benefical effect	Reference
Bacillus subtilis	The bacteria enhance the nitrogen fixation and the phosphate solubilization. In addition, the phytohormones produced by this bacillus enhance plant growth and increase fruit production	González-León et al. (2023)
Bacillus cereus	*Bacillus cereus* strains have shown an effect that promotes plant growth, improves the absorption of nutrients from the soil, reduces stress in case of water deficit, and increases photosynthesis	Jiménez (2024)
Pseudomonas fluorescens	Inhibits the attack of phytopathogens and improves photosynthetic activity	Nur Mawaddah et al. (2023)
Trichoderma spp.	It improves plant and root growth, in addition to promoting nutrient uptake and generating systemic resistance against abiotic stress and pathogenic agents	González-León et al. (2023)
Trichoderma spp.	Production of the enzyme 1-aminocyclopropane-1-carboxylate (ACC) deaminase and some phytohormones, besides acts as plant growth stimulator and protection against pathogen attack.	Urgiles-Gómez et al. (2023)
Serratia marcescens	Chitinase production related with biocontrol activity and plant protection. Besides, this microorganism can produce biosurfactants, volatile compounds, and other groups of bioactive compounds with antifungal properties	Moreno-Castillo et al. (2022)
Azospirillum spp.	Ability to fix atmospheric nitrogen, produce auxins and other growth regulators such as cytokinin, gibberellin, nitrous oxide, abscisic acid, and ethylene, also they induce germination and initial growth of seedlings	Cervera Achur and Velasco (2023)

Biocontrol is a fundamental part of ecological balance in ecosystems, which is no longer present in plant species belonging to agricultural areas or the other places inhabited by humans (Myers and Cory 2017; Kashyap et al. 2023). For this reason, emerging technologies propose to identify, isolate, and reproduce the microorganisms responsible for exercising biocontrol in different plants in wild environments and subsequently, apply them in areas where they can inhibit the development of specific diseases. These microorganisms are mainly bacteria and fungi (O'Brien 2017; Bastas and Kumar 2023; Bastas et al. 2023) (Table 1.5).

These entities are usually called microbial biological control agents. Its function is to prevent the establishment or infection of pathogens that would later lead to the development of a disease. To do this, these agents inhibit the development and reproduction of pathogens (mainly fungi) by affecting the integrity of their cellular structures, their metabolic pathways, or their access to nutrients. To do this, they use mechanisms such as antibiosis, parasitism, or competition for nutrients and space (Köhl et al. 2019).

Although the main interest in using biocontrol agents is to control diseases, additional beneficial effects have been discovered such as promoting plant growth of the species on which it is applied because they increase the availability of nutrients or synthesize hormones of interest to plants, thus causing these microorganisms to be

Table 1.4 Microbial consortia applied on plants

Treatment	Relevant results	Reference
Probiotic bacterial consortia (*Pseudomonas* strains)	Positive effect in the diversity and composition of plant rhizome microbiome	Hu et al. (2021)
Bacterial consortium (*Azospirillum brasilense* Sp7, *Pseudomonas putida* KT2440, *Acinetobacter* sp. EMM02, and *Sphingomonas* sp. OF178A)	The use of this consortium reduces the chemical fertilization by half, besides the same crop yield was obtained	Molina-Romero et al. (2021)
Microbial consortia (*Bacillus amyloliquefaciens*, *Bacillus subtilis*, *Pseudomonas chlororaphis*, and *Pseudomonas azotoformans*	The combination of a variety of biological control agents results positive, offering a wide variety of targets and direct or indirect mechanisms of action against pathogens	Minchev et al. (2021)
Halotolerant microbial consortia (*Bacillus* sp., *Delftia* sp., *Enterobacter* sp., *Achromobacter* sp.)	The application of the microbial consortia enhanced the production of secondary roots, favored the minerals accumulation and the transport of substances to the different parts of the plant, thus high biomass and growth were evidenced, besides under salinity stress tomato plants grew in a better way	Kapadia et al. (2021)
Microbial consortia (*Serendipita indica*, *Rhizophagus intraradices*, and *Azotobacter chroococcum*)	The application of this consortium notably enhanced the physical and physico-biochemical traits of maize. Treated maize exhibited significantly higher antioxidant enzyme activities, including peroxidase, catalase, polyphenol oxidase, and superoxide dismutase, compared to non-inoculated control plants	Tyagi et al. (2023)
Microbial consortia mixed with mineral fertilizers	Treatments had a positive effect in roots and biomass production, and nutrient uptake was improved	Islam et al. (2023)
Indigenous bacterial community (*Enterobacter*, *Pseudomonas*, and *Xanthomonas*)	Positive effects were noted on seedling radicle length, hypocotyl length, and total dry weight, promoting overall plant growth	May et al. (2023)
Native bacterial consortium (*Bacillus subtilis* TSO9, *B. cabrialesii* subsp. *tritici* TSO2T, *B. subtilis* TSO22, *B. paralicheniformis* TRQ65, and *Priestia megaterium* TRQ8)	The application of this bacterial consortium substantially improved wheat growth, yield, and plant quality	Ibarra-Villarreal et al. (2023)

used both as pesticides and fertilizers (El-Sayed et al. 2014; Devi et al. 2023). The impacts of the discovery, isolation, and application of these agents have allowed the development of new agricultural additives that could replace pesticides and fertilizers of synthetic origin, having beneficial effects in different areas (El-Saadony et al. 2022). Firstly, it would help improve the production of foods of plant origin by avoiding losses of 20–40% in crops due to pests, which also means that there are fewer economic losses. Added to this, the environmental impact that agricultural

Table 1.5 Efficacy of biological control agents against plant pathogens

Agent	Type	Tested against	Disease	Effectiveness	References
Trichoderma spp.	Fungi	Genera *Botrytis*, *Fusarium*, and *Rhizoctonia*	Gray mold in tomato, dry rot in potato, tobacco root rot, etc.	54.81–80%	Yao et al. (2023)
Gliocladium spp.	Fungi	*Colletotrichum coccodes*	Anthracnose in tomato	59–85%	Hassine et al. (2022)
Penicillium spp.	Fungi	*Colletotrichum coccodes*	Anthracnose in tomato	16–84%	Hassine et al. (2022)
Paecilomyces lilacinus	Fungi	*Meloidogyne javanica*	Causes abnormal leaf color and forms swollen roots and senescence in potato, tomato, wheat, etc.	47%	Ayaz et al. (2023)
Aspergillus candidus and *Aspergillus montenegro*	Fungi	*Clavibacter michiganensis and Erwinia amylovora*	Tomato late blight and wheat leaf rust	95%	Ngo et al. (2021)
Paecilomyces variotii	Fungi	*Mycosphaerella melonis and Podosphaera xanthii*	Gummy stem blight on the family Cucurbitaceae	76–78%	Moreno-Gavíra et al. (2021)
Wickerhamomyces anomalus	Yeast	*Rhizoctonia solani*	Rice sheath blight disease	55.2–65.1%	Khunnamwong et al. (2020)
Debaryomyces hansenii	Yeast	*Colletotrichum gloeosporioides*	Anthracnose in papaya	16–36%	Hernandez-Montiel et al. (2018)
Candida saopaulonensis	Yeast	*Rhizoctonia solani*	Damping-off and stem rot on cowpea plants	57.4%	de Tenório et al. (2019)
Cryptococcus laurentii	Yeast	*Rhizoctonia solani*	Damping-off and stem rot on cowpea plants	48.5%	de Tenório et al. (2019)
Bullera sinensis	Yeast	*Rhizoctonia solani*	Damping-off and stem rot on cowpea plants	66.3%	de Tenório et al. (2019)
Pythium oligandrum	Oomycete	*Pythium* spp.	Root rot	40–50%	Boro et al. (2022)
Pseudomonas spp.	Bacteria	*Ralstonia solanacearum*, *Erwinia amylovora*, and *Botrytis cinerea*	Tomato wilt disease, fire blight and bunch rot in fruit crops	40–74%	Boro et al. (2022), Sun et al. (2017)

Bacillus spp.	Bacteria	*Xanthomonas oryzae*, *Fusarium* spp., *Colletotrichum gloeosporioides*, and *Rhizoctonia solani*	Blight disease of rice, root rot in potato and *citrus*	65%	Chen et al. (2020a)
Lactobacillus plantarum	Bacteria	*Erwinia amylovora*	Fire blight disease or apple and pear	67%	Roselló et al. (2013)
Pantoea agglomerans	Bacteria	*Rhizopus stolonifer*, *Botrytis cinerea*, *Penicillium digitatum*, and *Penicillium italicum*	Fungal rot apple, pear, cherry and strawberry fruit	50%	Francés et al. (2006)
Weissella cibaria	Bacteria	*Penicillium expansum*	Fungal rot in apple	50%	Trias Mansilla et al. (2008)
Streptomyces spp.	Bacteria	*Dactylonectria* spp., *Ilyonectria* spp., *Phaeomoniella chlamydospora*, and *P. minimum*	Grapevine trunk diseases	33–61%	Álvarez-Pérez et al. (2017)
Stenotrophomonas rhizophila	Bacteria	*Alternaria alternata* and *Botrytis cinerea*	Gray mold in grapevine. Leaf spots, rots, and blights in different fruits like pears	20–65%	Raio et al. (2023)
Phages associated with different bacterial species	Bacteriophage	*Ralstonia solanacearum*	Bacterial wilt in tomato	80%	Wang et al. (2019)

chemicals cause on the soil, water, and living organisms would be mitigated and therefore the ecological imbalance and health problems in human beings would be reduced (Ons et al. 2020).

However, although its effects on food production are the most significant, it also has impacts on other productive areas such as those focused on ornamental plants, thus diversifying the economic impacts of these technologies (Lecomte et al. 2016). Finally, the fact of discovering species like this in various wild environments has highlighted the need to raise awareness among societies about the preservation of these areas due to their biological richness and its potential applications (Vargas et al. 2023).

1.6 Conclusion

In this chapter, we discuss the role of fertilizers and conventional methods for disease control, highlighting the adverse environmental impacts of chemical compounds. Additionally, we explore the potential of microbial consortia for disease management and plant growth enhancement. Utilizing effective microbial strains in agroecosystems demonstrates the potential to transform current agronomic practices, positively impacting soil health. Moreover, integrating diverse microorganisms with complementary capabilities presents an eco-friendly approach to boosting agricultural productivity without environmental harm.

References

Aktar W, Sengupta D, Chowdhury A (2009) Impact of pesticides use in agriculture: their benefits and hazards. Interdiscip Toxicol 2(1):1–12

Álvarez-Pérez JM, González-García S, Cobos R, Olego MÁ, Ibañez A, Díez-Galán A, Garzón-Jimeno E, Coque JJR (2017) Use of endophytic and rhizosphere actinobacteria from grapevine plants to reduce nursery fungal graft infections that lead to young grapevine decline. Appl Environ Microbiol 83(24):e01564-17

Atafar Z, Mesdaghinia A, Nouri J, Homaee M, Yunesian M, Ahmadimoghaddam M, Mahvi AH (2010) Effect of fertilizer application on soil heavy metal concentration. Environ Monit Assess 160:83–89

Ayaz M, Li C-H, Ali Q, Zhao W, Chi Y-K, Shafiq M, Ali F, Yu X-Y, Yu Q, Zhao J-T (2023) Bacterial and fungal biocontrol agents for plant disease protection: journey from lab to field, current status, challenges, and global perspectives. Molecules 28(18):6735

Bastas KK, Kumar A (2023) Omics technologies in the plant–microbe interactions. In: Bastas KK, Kumar A, Sivakumar U (eds) Microbial biocontrol: molecular perspective in plant disease management. Springer Nature, Singapore, pp 257–282. https://doi.org/10.1007/978-981-99-3947-3_13

Bastas KK, Kumar A, Sivakumar U, (eds). (2023) Microbial biocontrol: molecular perspective in plant disease management, vol 49. Springer Nature, Singapore. https://doi.org/10.1007/978-981-99-3947-3

Boro M, Sannyasi S, Chettri D, Verma AK (2022) Microorganisms in biological control strategies to manage microbial plant pathogens: a review. Arch Microbiol 204(11):666

Bzdyk RM, Olchowik J, Studnicki M, Oszako T, Sikora K, Szmidla H, Hilszczańska D (2018) The impact of effective microorganisms (EM) and organic and mineral fertilizers on the growth and mycorrhizal colonization of Fagus sylvatica and Quercus robur seedlings in a bare-root nursery experiment. Forests 9(10):597

Cervera Achur JC, Velasco A (2023) Formación del rendimiento de trigo tratado con microorganismos promotores de crecimiento

Chen K, Tian Z, He H, Long C, Jiang F (2020a) Bacillus species as potential biocontrol agents against citrus diseases. Biol Control 151:104419

Chen C, Zou W, Cui G, Tian J, Wang Y, Ma L (2020b) Ecological risk assessment of current-use pesticides in an aquatic system of Shanghai, China. Chemosphere 257:127222

Curl CL, Spivak M, Phinney R, Montrose L (2020) Synthetic pesticides and health in vulnerable populations: agricultural workers. Curr Environ Health Rep 7:13–29

de Tenório DA, de Medeiros EV, Lima CS, da Silva JM, de Barros JA, Neves RP, Laranjeira D (2019) Biological control of Rhizoctonia solani in cowpea plants using yeast. Trop Plant Pathol 44:113–119

Devi S, Sharma S, Tiwari A, Bhatt AK, Singh NK, Singh M, Kaushalendra, Kumar A (2023) Screening for multifarious plant growth promoting and biocontrol attributes in Bacillus strains isolated from Indo-Gangetic soil for enhancing growth of rice crops. Microorganisms 11:1085. https://doi.org/10.3390/microorganisms11041085

Ding J, Zhu D, Chen Q-L, Zheng F, Wang H-T, Zhu Y-G (2019) Effects of long-term fertilization on the associated microbiota of soil collembolan. Soil Biol Biochem 130:141–149

EFSA, Abdourahime H, Anastassiadou M, Arena M, Auteri D, Barmaz S, Brancato A, Bura L, Carrasco Cabrera L, Chaideftou E (2020) Peer review of the pesticide risk assessment of the active substance mancozeb. EFSA J 18(12):e05755

Elnahal ASM, El-Saadony MT, Saad AM, Desoky E-SM, El-Tahan AM, Rady MM, AbuQamar SF, El-Tarabily KA (2022) The use of microbial inoculants for biological control, plant growth promotion, and sustainable agriculture: a review. Eur J Plant Pathol 162(4):759–792

El-Saadony MT, Saad AM, Soliman SM, Salem HM, Ahmed AI, Mahmood M, El-Tahan AM, Ebrahim AAM, Abd El-Mageed TA, Negm SH (2022) Plant growth-promoting microorganisms as biocontrol agents of plant diseases: mechanisms, challenges and future perspectives. Front Plant Sci 13:923880

El-Sayed WS, Akhkha A, El-Naggar MY, Elbadry M (2014) In vitro antagonistic activity, plant growth promoting traits and phylogenetic affiliation of rhizobacteria associated with wild plants grown in arid soil. Front Microbiol 5:118165

Fan X, Fu Y, Nie Y, Matsumoto H, Wang Y, Hu T, Pan Q, Lv T, Fang H, Xu H (2021) Keystone taxa-mediated bacteriome response shapes the resilience of the paddy ecosystem to fungicide triadimefon contamination. J Hazard Mater 417:126061

Faraj TK, EL-Saeid MH, Najim MMM, Chieb M (2024) The impact of pesticide residues on soil health for sustainable vegetable production in arid areas. Separations 11(2):46

Francés J, Bonaterra A, Moreno MC, Cabrefiga J, Badosa E, Montesinos E (2006) Pathogen aggressiveness and postharvest biocontrol efficiency in Pantoea agglomerans. Postharvest Biol Technol 39(3):299–307

Frezarin ET, Santos CHB, Sales LR, dos Santos RM, de Carvalho LAL, Rigobelo EC (2023) Promotion of peanut (Arachis hypogaea L.) growth by plant growth-promoting microorganisms. Microbiol Res 14(1):316–332

García CAS, Vera MS, Vinocur A, Graziano M, Miranda C, Pizarro HN (2022) Rethinking the term "glyphosate effect" through the evaluation of different glyphosate-based herbicide effects over aquatic microbial communities. Environ Pollut 292:118382

Geiger F, Bengtsson J, Berendse F, Weisser WW, Emmerson M, Morales MB, Ceryngier P, Liira J, Tscharntke T, Winqvist C (2010) Persistent negative effects of pesticides on biodiversity and biological control potential on European farmland. Basic Appl Ecol 11(2):97–105

González-León Y, Ortega-Bernal J, Anducho-Reyes MA, Mercado-Flores Y (2023) Bacillus subtilis y trichoderma: características generales y su aplicación en la agricultura. TIP Revista Especializada En Ciencias Químico-Biológicas 25(1):1–14

Hakim S, Naqqash T, Nawaz MS, Laraib I, Siddique MJ, Zia R, Mirza MS, Imran A (2021) Rhizosphere engineering with plant growth-promoting microorganisms for agriculture and ecological sustainability. Front Sustain Food Syst 5:617157

Hao Q, Cheng B, Jiang C (2013) Long-term tillage effects on soil organic carbon and dissolved organic carbon in a purple paddy soil of Southwest China. Acta Ecol Sin 33(5):260–265

Hassaan MA, El Nemr A (2020) Pesticides pollution: classifications, human health impact, extraction and treatment techniques. Egypt J Aquat Res 46(3):207–220

Hassine M, Aydi-Ben-Abdallah R, Jabnoun-Khireddine H, Daami-Remadi M (2022) Soil-borne and compost-borne Penicillium sp. and Gliocladium spp. as potential microbial biocontrol agents for the suppression of anthracnose-induced decay on tomato fruits. Egypt J Biol Pest Control 32(1):20

Hernandez-Montiel LG, Gutierrez-Perez ED, Murillo-Amador B, Vero S, Chiquito-Contreras RG, Rincon-Enriquez G (2018) Mechanisms employed by Debaryomyces hansenii in biological control of anthracnose disease on papaya fruit. Postharvest Biol Technol 139:31–37

Hu J, Yang T, Friman V-P, Kowalchuk GA, Hautier Y, Li M, Wei Z, Xu Y, Shen Q, Jousset A (2021) Introduction of probiotic bacterial consortia promotes plant growth via impacts on the resident rhizosphere microbiome. Proc R Soc B 288(1960):20211396

Ibarra-Villarreal AL, Villarreal-Delgado MF, Parra-Cota FI, Yepez EA, Guzmán C, Gutierrez-Coronado MA, Valdez LC, Saint-Pierre C, de Los Santos-Villalobos S (2023) Effect of a native bacterial consortium on growth, yield, and grain quality of durum wheat (Triticum turgidum L. subsp. durum) under different nitrogen rates in the Yaqui Valley, Mexico. Plant Signal Behav 18(1):2219837

Islam MM, Rengel Z, Storer P, Siddique KHM, Solaiman ZM (2023) Microbial consortium inoculant and rock mineral fertiliser differentially improved yield and nutrient uptake of industrial hemp (Cannabis sativa L.) varieties. Ind Crop Prod 197:116599

Jiménez FS (2024) Efecto elicitor del uso de ADN propio en Bacillus cereus-Amazcala (Bc-A) como herramienta para aumentar su capacidad bioestimulante

Jørgensen PS, Folke C, Henriksson PJG, Malmros K, Troell M, Zorzet A (2020) Coevolutionary governance of antibiotic and pesticide resistance. Trends Ecol Evol 35(6):484–494

Kapadia C, Sayyed RZ, El Enshasy HA, Vaidya H, Sharma D, Patel N, Malek RA, Syed A, Elgorban AM, Ahmad K (2021) Halotolerant microbial consortia for sustainable mitigation of salinity stress, growth promotion, and mineral uptake in tomato plants and soil nutrient enrichment. Sustainability 13(15):8369

Kashyap N, Singh SK, Yadav N, Singh VK, Kumari M, Kumar D, Shukla L, Kaushalendra, Bhardwaj N, Kumar A (2023) Biocontrol screening of endophytes: applications and limitations. Plants 12:2480. https://doi.org/10.3390/plants12132480

Khan S, Uddin MN, Rizwan M, Khan W, Farooq M, Shah AS, Subhan F, Aziz F, Rahman KU, Khan A (2020) Mechanism of insecticide resistance in insects/pests. Pol J Environ Stud 29(3):2023

Khunnamwong P, Lertwattanasakul N, Jindamorakot S, Suwannarach N, Matsui K, Limtong S (2020) Evaluation of antagonistic activity and mechanisms of endophytic yeasts against pathogenic fungi causing economic crop diseases. Folia Microbiol 65:573–590

Kim K-H, Kabir E, Jahan SA (2017) Exposure to pesticides and the associated human health effects. Sci Total Environ 575:525–535

Kim L, Jeon J-W, Son J-Y, Kim C-S, Ye J, Kim H-J, Lee C-H, Hwang S-M, Choi S-D (2020) Nationwide levels and distribution of endosulfan in air, soil, water, and sediment in South Korea. Environ Pollut 265:115035

Kinnunen P, Guillaume JHA, Taka M, D'odorico P, Siebert S, Puma MJ, Jalava M, Kummu M (2020) Local food crop production can fulfil demand for less than one-third of the population. Nat Food 1(4):229–237
Köhl J, Kolnaar R, Ravensberg WJ (2019) Mode of action of microbial biological control agents against plant diseases: relevance beyond efficacy. Front Plant Sci 10:454982
König S, Vogel H-J, Harms H, Worrich A (2020) Physical, chemical and biological effects on soil bacterial dynamics in microscale models. Front Ecol Evol 8:53
Kour D, Rana KL, Yadav AN, Yadav N, Kumar M, Kumar V, Vyas P, Dhaliwal HS, Saxena AK (2020) Microbial biofertilizers: bioresources and eco-friendly technologies for agricultural and environmental sustainability. Biocatal Agric Biotechnol 23:101487
Kumar M, Yadav AN, Saxena R, Paul D, Tomar RS (2021) Biodiversity of pesticides degrading microbial communities and their environmental impact. Biocatal Agric Biotechnol 31:101883
Kumar S, Sindhu SS, Kumar R (2022) Biofertilizers: an ecofriendly technology for nutrient recycling and environmental sustainability. Curr Res Microb Sci 3:100094
Kumar R, Saini GK, Jawed M (2024) Resilience of aerobic sludge biomass under chlorpyrifos stress and its recovery potential. Chemosphere 352:141324
Lecomte C, Alabouvette C, Edel-Hermann V, Robert F, Steinberg C (2016) Biological control of ornamental plant diseases caused by Fusarium oxysporum: a review. Biol Control 101:17–30
Li C, Yan K, Tang L, Jia Z, Li Y (2014) Change in deep soil microbial communities due to long-term fertilization. Soil Biol Biochem 75:264–272
Liu B, Tu C, Hu S, Gumpertz M, Ristaino JB (2007) Effect of organic, sustainable, and conventional management strategies in grower fields on soil physical, chemical, and biological factors and the incidence of southern blight. Appl Soil Ecol 37(3):202–214
Liu B, Xia H, Jiang C, Riaz M, Yang L, Chen Y, Fan X, Xia X (2022) 14 year applications of chemical fertilizers and crop straw effects on soil labile organic carbon fractions, enzyme activities and microbial community in rice-wheat rotation of middle China. Sci Total Environ 841:156608
Lu T, Xu N, Zhang Q, Zhang Z, Debognies A, Zhou Z, Sun L, Qian H (2020a) Understanding the influence of glyphosate on the structure and function of freshwater microbial community in a microcosm. Environ Pollut 260:114012
Lu C, Yang Z, Liu J, Liao Q, Ling W, Waigi MG, Odinga ES (2020b) Chlorpyrifos inhibits nitrogen fixation in rice-vegetated soil containing Pseudomonas stutzeri A1501. Chemosphere 256:127098
Ma G, Gao X, Nan J, Zhang T, Xie X, Cai Q (2021) Fungicides alter the distribution and diversity of bacterial and fungal communities in ginseng fields. Bioengineered 12(1):8043–8056
MacLaren C, Storkey J, Menegat A, Metcalfe H, Dehnen-Schmutz K (2020) An ecological future for weed science to sustain crop production and the environment. A review. Agron Sustain Dev 40:1–29
Maggi F, la Cecilia D, Tang FHM, McBratney A (2020) The global environmental hazard of glyphosate use. Sci Total Environ 717:137167
Mahmood I, Imadi SR, Shazadi K, Gul A, Hakeem KR (2016) Effects of pesticides on environment. In: Plant, soil and microbes, Implications in crop science, vol 1. Springer, pp 253–269
Majumder S, Paul A, Kumar A, Verma CK, Divekar PA, Rani V, Rani AT, Halder J, Pandey KK, Singh J (2021) Impact of pesticides on microbial population. In: Microbial technology for sustainable environment. Springer, pp 467–481
May A, Coelho LF, Pedrinho A, Batista BD, Mendes LW, Mendes R, Morandi MAB, Barth G, Viana RS, Vilela ESD (2023) The use of indigenous bacterial community as inoculant for plant growth promotion in soybean cultivation. Arch Agron Soil Sci 69(1):135–150
Meftaul IM, Venkateswarlu K, Dharmarajan R, Annamalai P, Megharaj M (2020) Pesticides in the urban environment: a potential threat that knocks at the door. Sci Total Environ 711:134612
Meftaul IM, Venkateswarlu K, Annamalai P, Parven A, Sobhani Z, Megharaj M (2021) Behavior and fate of fungicide chlorothalonil in urban landscape soils and associated environmental concern. J Environ Sci Health B 56(12):1066–1077

Minchev Z, Kostenko O, Soler R, Pozo MJ (2021) Microbial consortia for effective biocontrol of root and foliar diseases in tomato. Front Plant Sci 12:756368

Mohamed M, Aliyat FZ, Ben Messaoud B, Marina M, Filippo G, Laila N, Jamal I (2021) Effects of pesticides use (glyphosate & paraquat) on biological nitrogen fixation. Water Air Soil Pollut 232:1–15

Molina-Romero D, Juárez-Sánchez S, Venegas B, Ortíz-González CS, Baez A, Morales-García YE, Muñoz-Rojas J (2021) A bacterial consortium interacts with different varieties of maize, promotes the plant growth, and reduces the application of chemical fertilizer under field conditions. Front Sustain Food Syst 4:616757

Moreno-Castillo B, Ortiz-Barrios H, Salvador-Figueroa M, de Lourdes Adriano-Anaya M (2022) Serratia marcescens (Enterobacteriaceae): promotora de la salud vegetal. IBCIENCIAS1 5(1):9

Moreno-Gavíra A, Diánez F, Sánchez-Montesinos B, Santos M (2021) Biocontrol effects of Paecilomyces variotii against fungal plant diseases. J Fungi 7(6):415

Morris S, Blackwood CB (2015) The ecology of the soil biota and their function. In: Paul EA (ed) Soil microbiology, ecology and biochemistry. Academic Press, Boston

Myers JH, Cory JS (2017) Biological control agents: invasive species or valuable solutions? In: Impact of biological invasions on ecosystem services. Springer, pp 191–202

Naamala J, Smith DL (2020) Relevance of plant growth promoting microorganisms and their derived compounds, in the face of climate change. Agronomy 10(8):1179

Ngegba PM, Cui G, Khalid MZ, Zhong G (2022) Use of botanical pesticides in agriculture as an alternative to synthetic pesticides. Agriculture 12(5):600

Ngo MT, Van Nguyen M, Han JW, Kim B, Kim YK, Park MS, Kim H, Choi GJ (2021) Biocontrol potential of Aspergillus species producing antimicrobial metabolites. Front Microbiol 12:804333

Nosheen S, Ajmal I, Song Y (2021) Microbes as biofertilizers, a potential approach for sustainable crop production. Sustainability 13(4):1868

Nur Mawaddah S, Ab Wahab MZ, Sapak Z (2023) The potential of Pseudomonas fluorescens as biological control agent against sheath blight disease in rice: a systematic review. Food Res 7(2):46–56

O'Brien PA (2017) Biological control of plant diseases. Australas Plant Pathol 46:293–304

Okdeh DMA, Sakr IA, Arraj T (2024) Determination of dichlorvos pesticide residues in the groundwater of the environment surrounding Al-Sin Lake using a high-performance liquid chromatography. Environ Qual Manag 34. https://doi.org/10.1002/tqem.22175

Ons L, Bylemans D, Thevissen K, Cammue BPA (2020) Combining biocontrol agents with chemical fungicides for integrated plant fungal disease control. Microorganisms 8(12):1930

Raio A, Brilli F, Neri L, Baraldi R, Orlando F, Pugliesi C, Chen X, Baccelli I (2023) Stenotrophomonas rhizophila Ep2. 2 inhibits growth of Botrytis cinerea through the emission of volatile organic compounds, restricts leaf infection and primes defense genes. Front Plant Sci 14:1235669

Rashidipour M, Maleki A, Kordi S, Birjandi M, Pajouhi N, Mohammadi E, Heydari R, Rezaee R, Rasoulian B, Davari B (2019) Pectin/chitosan/tripolyphosphate nanoparticles: efficient carriers for reducing soil sorption, cytotoxicity, and mutagenicity of paraquat and enhancing its herbicide activity. J Agric Food Chem 67(20):5736–5745

Rasmussen PE, Goulding KWT, Brown JR, Grace PR, Janzen HH, Korschens M (1998) Long-term agroecosystem experiments: assessing agricultural sustainability and global change. Science 282(5390):893–896

Roselló G, Bonaterra A, Francés J, Montesinos L, Badosa E, Montesinos E (2013) Biological control of fire blight of apple and pear with antagonistic Lactobacillus plantarum. Eur J Plant Pathol 137:621–633

Santoyo G, Guzmán-Guzmán P, Parra-Cota FI, de los Santos-Villalobos S, del Carmen Orozco-Mosqueda M, Glick BR (2021) Plant growth stimulation by microbial consortia. Agronomy 11(2):219

Savci S (2012) An agricultural pollutant: chemical fertilizer. Int J Environ Sci Dev 3(1):73

Schleiffer M, Speiser B (2022) Presence of pesticides in the environment, transition into organic food, and implications for quality assurance along the European organic food chain—a review. Environ Pollut 313:120116

Sun D, Zhuo T, Hu X, Fan X, Zou H (2017) Identification of a Pseudomonas putida as biocontrol agent for tomato bacterial wilt disease. Biol Control 114:45–50

Tago D, Andersson H, Treich N (2014) Pesticides and health: a review of evidence on health effects, valuation of risks, and benefit-cost analysis. In: Preference measurement in health. Emerald Publishing, pp 203–295

Thiour-Mauprivez C, Martin-Laurent F, Calvayrac C, Barthelmebs L (2019) Effects of herbicide on non-target microorganisms: towards a new class of biomarkers? Sci Total Environ 684:314–325

Tian S, Ning T, Wang Y, Liu Z, Li G, Li Z, Lal R (2016) Crop yield and soil carbon responses to tillage method changes in North China. Soil Tillage Res 163:207–213

Trias Mansilla R, Bañeras Vives L, Montesinos Seguí E, Badosa Romañó E (2008) Lactic acid bacteria from fresh fruit and vegetables as biocontrol agents of phytopathogenic bacteria and fungi. Int Microbiol 11:231–236

Tyagi J, Mishra A, Kumari S, Singh S, Agarwal H, Pudake RN, Varma A, Joshi NC (2023) Deploying a microbial consortium of Serendipita indica, Rhizophagus intraradices, and Azotobacter chroococcum to boost drought tolerance in maize. Environ Exp Bot 206:105142

Urgiles-Gómez N, Avila-Salem ME, Loján P, Encalada M, Hurtado L, Araujo S, Collahuazo Y, Guachanamá J, Poma N, Granda K (2021) Plant growth-promoting microorganisms in coffee production: from isolation to field application. Agronomy 11(8):1531

Urgiles-Gómez N, Loján P, Ávila-Salem ME, Benavidez-Silva C, Hurtado L, Livisaca F, Guaya P, Villamagua M, Poma L, Quichimbo L (2023) Microorganismos benéficos con potencial agrícola: una alternativa sostenible para la producción de café y calidad del suelo. Cedamaz 13(1):103–113

Vargas G, Rivera-Pedroza LF, García LF, Jahnke SM (2023) Conservation biological control as an important tool in the neotropical region. Neotrop Entomol 52(2):134–151

Verma S, Jayakumar S (2012) Impact of forest fire on physical, chemical and biological properties of soil: a review. Proc Int Acad Ecol Environ Sci 2(3):168

Virk AL, Shakoor A, Abdullah A, Chang SX, Cai Y (2024) Pesticide effects on crop physiology, production and soil biological functions

Wang X, Wei Z, Yang K, Wang J, Jousset A, Xu Y, Shen Q, Friman V-P (2019) Phage combination therapies for bacterial wilt disease in tomato. Nat Biotechnol 37(12):1513–1520

Xia H, Riaz M, Zhang M, Liu B, El-Desouki Z, Jiang C (2020) Biochar increases nitrogen use efficiency of maize by relieving aluminum toxicity and improving soil quality in acidic soil. Ecotoxicol Environ Saf 196:110531

Yao X, Guo H, Zhang K, Zhao M, Ruan J, Chen J (2023) Trichoderma and its role in biological control of plant fungal and nematode disease. Front Microbiol 14:1160551

Zhang Y, Shen H, He X, Thomas BW, Lupwayi NZ, Hao X, Thomas MC, Shi X (2017) Fertilization shapes bacterial community structure by alteration of soil pH. Front Microbiol 8:1325

Zhao C, Ni H, Zhao L, Zhou L, Borrás-Hidalgo O, Cui R (2020) High nitrogen concentration alter microbial community in Allium fistulosum rhizosphere. PLoS One 15(11):e0241371

Zheng W, Luo B, Hu X (2020) The determinants of farmers' fertilizers and pesticides use behavior in China: an explanation based on label effect. J Clean Prod 272:123054

Zhong WH, Cai ZC (2007) Long-term effects of inorganic fertilizers on microbial biomass and community functional diversity in a paddy soil derived from quaternary red clay. Appl Soil Ecol 36(2–3):84–91

Zuanazzi NR, de Castilhos Ghisi N, Oliveira EC (2020) Analysis of global trends and gaps for studies about 2,4-D herbicide toxicity: a scientometric review. Chemosphere 241:125016

Microbes-Associated Activation of Plant Defense Against Phytopathogens

2

M. C. Bañuelos-González, E. A. Cuéllar-Torres, S. Aguilera-Aguirre, U. M. López-García, R. I. Ortiz-Basurto, and M. A. Chacón-López

Abstract

Modern agriculture aims to produce safe, high-quality, low-cost food to meet the growing global demand. However, it faces various challenges, including losses caused by phytopathogens. Controlling these pathogens is often carried out using synthetic agrochemicals, which, although practical and effective, have disadvantages such as environmental damage, developing resistant strains, and human health risks. An alternative is the use of plant defense system inducers. Inducers can be physical, chemical, or biological factors that stimulate physiological or morphological responses in plants to cope with adverse conditions, including phytopathogen attacks. Inducible plant defense is activated in response to pathogen detection by producing antimicrobial compounds and activating defense signaling pathways. PGPR microorganisms can produce microbial inducers that can induce systemic resistance. These inducers activate the defense system in plants in response to some stresses, leading to the production of pathogenesis-related proteins (PRs) and antimicrobial compounds. These defense mechanisms help plants resist infections and ensure sustainable agricultural production.

Keywords

Elicitor · Plant defense · Biotic stress · Sustainability · Crop production

M. C. Bañuelos-González · E. A. Cuéllar-Torres · S. Aguilera-Aguirre · U. M. López-García · R. I. Ortiz-Basurto · M. A. Chacón-López (✉)
Integral Food Research Laboratory, National Technological Institute of Mexico/Technological Institute of Tepic, Tepic, Nayarit, Mexico
e-mail: midebanuelosgo@ittepic.edu.mx; esancuellarto@ittepic.edu.mx; saguilera@tepic.tecnm.mx; ulopez@tepic.tecnm.mx; riobasurt@ittepic.edu.mx; mchacon@tepic.tecnm.mx

A. Kumar, M. K. Solanki (eds.), *Microbial Biocontrol Techniques*, Microorganisms for Sustainability 54,
https://doi.org/10.1007/978-981-97-8739-5_2

2.1 Introduction

The main objective of agriculture is to produce high-quality, safe, and economical food products to supply in the constantly growing global population. However, this practice faces multiple adversities that reduce agricultural production, as losses and food waste, such as fruits and vegetables, and are caused by inadequate handling, short shelf life, and primarily damage caused by pathogenic microorganisms. In horticultural products, diseases caused by phytopathogens are responsible for significant economic losses, ranging from 5% to 25% in developed countries and 20% to 50% in developing countries (Gutiérrez-Martínez et al. 2018).

The control of phytopathogens is commonly carried out using synthetic agrochemicals, which are efficient in controlling pathogens but have various disadvantages, such as environmental damage, fostering the development of pathogen strains resistant to the active ingredients of these products, and posing a risk to human health (Torres-Calzada et al. 2013; Bhardwaj et al. 2024). Therefore, attention has been focused on applying alternatives that help maintain high-yield food production, are environmentally friendly, and are free of toxic residues. One attractive alternative is using elicitors of the plant defense system. Elicitors are physical, chemical, or biological factors that stimulate plants' physiological or morphological responses to cope with various conditions, including phytopathogen attacks. Several elicitor molecules capable of activating the plant defense system have been reported, such as carbohydrates, proteins, or secondary metabolites produced by microorganisms; these have been attributed to the ability to induce the plant defense system (Yu et al. 2022). Biological control agents use microorganisms such as bacteria, yeast, fungi, or some products generated in their metabolism (Köhl et al. 2019; Ali et al. 2024). It has been reported that antagonistic microorganisms such as Trichoderma and Clonostachys can produce antimicrobial compounds like 6-PAP, gliotoxin, viridin, and many more compounds with antimicrobial activity (Ghorbanpour et al. 2018).

The induction of plant defense by some biocontrol microorganisms has been considered a mode of action to control diseases in agricultural products (Jamalizadeh et al. 2011; Raaijmakers and Mazzola 2012). Defense induction by microorganisms originates from recognizing their structural compounds or metabolites (MAMPs) through the pattern recognition receptors (PRRs) of the plant's innate immune system (Singh et al. 2019). Defense response induction involves the production of different enzymes and metabolites, particularly pathogenesis-related proteins (PR proteins), including glucanases, chitinases, peroxidases, superoxide dismutase, and catalase, among others, as well as the production of compounds with antimicrobial activity such as phytoalexins or phenolic compounds (Garcia-Brugger et al. 2006).

This chapter focuses on the importance and application of microbial-origin elicitors as an alternative for controlling plant phytopathogens by inducing their defense system. Elicitors for plant defense activation represent a more sustainable option for agricultural production.

2.2 Plant Defense

The plant defense system is complex and sophisticated, activating responses to counteract the damage caused by pathogenic microorganisms or insects (Swapnil et al. 2023). Natural plant resistance to pathogens is based on the combined effects of inducible mechanisms and preformed barriers. A complex network of hormonal signals has been shown to control plant response to pathogen attack (Borges et al. 2014). Plants activate their defense mechanisms to protect themselves against attacks by pathogens (Shine et al. 2019). These defense mechanisms include developing physical barriers like thorns and tough tissue, and the production of toxins (Andersen et al. 2018).

Additionally, plants release volatile organic compounds that serve as signals to attract the natural enemies of herbivores, providing indirect defense against insect attack (Lazazzara et al. 2022). Besides, plants have developed specific defense mechanisms to protect themselves against bacteria, fungi, insects, and viruses. These defense mechanisms can be characterized into two main types: preformed defenses and inducible defenses (Sun et al. 2023). A basic level of protection is performed by constitutive defenses, meaning they are always present in the plant and provide protection. Structural barriers such as rigid cell walls and thorns and chemical compounds that are toxic to pathogens are examples of preformed defenses. Physical barriers such as the cuticle are crucial in the plant's preformed defenses; this structure is formed by polysaccharides and waxes in its internal layer and epicuticular waxes in the external layer. The cuticle is impermeable and is a physical barrier to protect the cells from invasion by pathogens (Andersen et al. 2018; Sun et al. 2023).

Essential components of the plant defense system are phytochemical compounds with antimicrobial effects. These include phenolic compounds and flavonoids, such as coumarins, lignins, terpenoids, alkaloids, glucosinolates, and stilbenes (Andersen et al. 2018; Lonakadi et al. 2020). Phenolic and flavonoid compounds can induce lipid peroxidation in the cell membranes of pathogens. This process generates reactive oxygen species (ROS) that damage the lipid components of membranes, increasing their permeability and causing the pathogen's cell death (Sun et al. 2023).

On the other hand, inducible defenses are activated in response to plants detecting the presence of pathogens, and they can initiate a rapid immune response that involves the production of antimicrobial compounds, enzymes that break down pathogen cell walls, and the activation of defense signaling pathways (Sun et al. 2023). These inducible defenses can be triggered by the recognition of pathogen-associated molecular patterns or by the activation of specific plant resistance genes. Overall, plant defense mechanisms against pathogens involve a combination of physical barriers, chemical compounds, and immune responses that work together to prevent pathogen invasion and limit its development (Francis et al. 2023). Biotic agents constantly attack plants, decreasing crop production and directly impacting global food security (Blakeney 2019). Therefore, it is essential to describe the primary defense mechanisms in plants, which are detailed below.

2.3 Plant System Immunity

The immunity of the plant system is a determining aspect of protecting plants against pathogens and other biotic stresses. A complex network of signaling pathways and guarding responses activated upon recognizing pathogens or their associated molecular patterns is involved (Kumar et al. 2021b). Both biotic and abiotic stresses trigger the activation of defense mechanisms that plants have developed to ensure their survival (Wang et al. 2023; Anand et al. 2023). These mechanisms involve the convergence of signaling pathways in response to different stress components and specific responses to combined stresses controlled by other signaling pathways (Kumar et al. 2021b). During defense system activation, hormones such as jasmonic acid (JA), salicylic acid (SA), ethylene (ET), nitric oxide (NO), cytokinins (CK), gibberellin (GA), auxin, brassinosteroids (BR), and abscisic acid (ABA) are involved in the signaling cascade. For instance, defense and growth mechanisms are interconnected through the metabolic activity of ABA, NO, auxin, CK, GA, and BR, which function in immunity and plant development. The ABA hormone is crucial in the physiological response to biotic stress, regulating stomatal closure, which influences the pathogen's access to plant tissues. Also, deficiencies in GA can lead to increased resistance to pathogens (Andersen et al. 2018; Wang and Bi 2021), which suggests that lower GA activity may favor the activation of systemic defense mechanisms, such as systemic acquired resistance (SAR) and induced systemic resistance (ISR).

Plant system immunity is a dynamic, multifaceted protection plan that enables plants to withstand various stressors and maintain their health and productivity (Shine et al. 2019). Plants, being sessile organisms, require effective defense mechanisms against biotic and abiotic stresses; among the most described are the following: systemic acquired resistance (SAR), herbivore-induced resistance (HIR), and induced systemic resistance (ISR) (Shine et al. 2019). SAR signaling is primarily mediated by SA and activated by pathogens, while ISR is mainly regulated by JA and ET and induced by plant growth-promoting microorganisms. Moreover, other molecules can activate the defense system. It has been reported that polysaccharides such as ulvans, carrageenans isolated from plant sources and fucans isolated from algae can induce the ISR or SAR response in plants (Shine et al. 2019; Zheng et al. 2020). Both SAR and ISR enhance the ability of plants to defend against infection; they differ in their triggers and the molecular processes that mediate these responses, which are explained below.

2.4 Plant Defense Classification

2.4.1 Systemic Acquired Resistance

One of the important defense mechanisms in plants is SAR. SAR is a response that plants trigger after exposure to specific pathogens. Commonly, when a biotrophic pathogen attacks a plant, the immune response begins with activating pathogenesis-related genes. Activation of SAR in uninfected tissues requires transmitting signals from the infected tissue to the whole plant; these signals travel through the vascular system (Dempsey and Klessig 2012). SAR is considered a broad-spectrum resistance to plant disease following a localized infection that initiates the protective/immunity signal due to a secondary infection of the same or a different microbial pathogen in an uninfected part of the plant (Malik et al. 2020). This response allows the plant to defend itself at the site of infection and throughout its system, providing a more effective and comprehensive defense. This defense mechanism is mediated by signaling networks controlled by phytohormones, particularly SA. SA is central in local and systemic resistance responses to biotrophic and hemibiotrophic pathogens (Malik et al. 2020) and is a crucial component in initiating SAR. When plants are exposed to these pathogens, SA is required to initiate a hypersensitive response and trigger SAR. When a pathogen is present, signaling activates the transcription factors expressing defense-related genes (Kumar et al. 2021a). It involves activating the expression of PR genes involved in defense responses, whose expression is regulated by transcription factors (TFs), such as WRKY, NAC, and MYB. SAR activation includes a complex regulatory network of interactions and MAPK cascades involved in signaling pathways to produce specific defense molecules, such as antimicrobial metabolites or peptides, to inhibit pathogen growth and limit the spread of infection within the plant.

2.4.2 Herbivore-Induced Resistance

HIR includes the ability of plants to dynamically and adaptively respond to herbivore attacks, ultimately leading to improved resistance against future herbivores. Plants sense and respond to herbivore-specific signals, such as triggers of insect saliva or mechanical damage, leading to the activation of defense genes and biosynthesis of various secondary metabolites (Malik et al. 2020). These metabolites may directly affect herbivore physiology, behavior, and fitness or indirectly attract predators and parasitoids that feed on herbivores. The production of volatile organic compounds is a particularly noteworthy aspect of herbivore-induced resistance, as these compounds may serve as important signals to attract natural enemies of herbivores, a process known as indirect defense. In addition, plants can also detect specific components of insect saliva during herbivory. When herbivorous insects damage plants, they respond with a wound response, identifying specific insect-derived molecules. These molecules, known as herbivore-associated molecular patterns (HAMPs), trigger signaling pathways that lead to defense mechanisms in other parts of the plant to prepare for further damage. Insect oral secretions, including saliva and oviposition fluid components, can trigger plant defense responses (Paniagua Voirol et al. 2020).

Molecules in oral secretions can activate plant defense responses, involving enzymes like glucose oxidase and β-glucosidase, peptides such as inception, and fatty acid complexes like volicitin. Some herbivorous insects indirectly damage plants by producing DAMPs from cellular damage during infection. Plants recognize insect elicitors through receptors such as AtLecRK-I.8, initiating defense responses, which were reported to mediate immune response to lipid extracts from the insect's eggs (Albert et al. 2020). Next, the activation of kinases y and phytohormone networks, such as JA, SA, and ET, are the main players in regulating the defense response against insects (Malik et al. 2020).

2.4.3 Induced Systemic Resistance

The ISR is associated with the action of PGPR microorganisms by eliciting plant defense (Sharma et al. 2019b). During the ISR, plants can activate their innate immune system and produce a wide range of secondary metabolites that act as defense mechanisms against potential threats. JA and ET play a key role in plants' responses associated with ISR (Andersen et al. 2018).

In ISR, the inducer activates resistance in the plant from distant cells, sending biochemical signals that systemically alert all plant tissues, a process known as "priming" (Goellner and Conrath 2008). The tissue where ISR develops must have some defense predisposition for induction to rapidly become widespread, evidenced by the systemic expression of PR proteins (Heil and Bostock 2002). The perception of the elicitor molecule by the plant triggers signal transduction pathways lead to ROS production, phytoalexins, cell wall reinforcement with phenylpropanoid compounds, callose deposition, synthesis of defense enzymes, and the accumulation of PRs, which exhibit antimicrobial activity (Thakur and Sohal 2013). A key component in signaling in both SAR and ISR is the non-expressor of pathogenesis (NPR1), which regulates specific genes associated with induced resistance against plant pathogens.

2.5 Defense Mechanism in Plants

The innate immune system is activated when plants recognize pathogens (Shine et al. 2019). Pathogens primarily activate the defense system via pathogen-associated molecular patterns (PAMPs). The perception of these patterns triggers pathogen-triggered immunity (PTI) followed by effector-triggered immunity (ETI) to protect plants from pathogen attacks (Malik et al. 2020; Kumar et al. 2021a). Plants sense pathogens through pattern recognition receptors (PRRs) with conserved domains recognizing PAMPs and DAMPs, activating PTI or ETI (Singh et al. 2019).

Plant receptors detect biotic stress threats for both SAR and ISR. To perceive some molecules from pathogens, the plant's cells have receptors named PRRs; these are receptor-like kinases (RLKs) or receptor-like proteins (RLPs) on the plasma membrane. Both receptors have similar chemist structures, the difference is that RLKs display a cytoplasmic kinase domain, and RLPs do not (Malik et al. 2020; Kumar et al. 2021a). Specific PRRs recognize PAMPs/DAMPs (molecules from pathogens) such as chitin, polygalacturonate, flagellin, and others, thus initiating a signaling cascade involving several molecular actors such as mitogen-activated protein kinases (MAPKs), G-proteins, ubiquitin, calcium signaling, hormone signaling, transcription factors (TFs) (Fig. 2.1), and epigenetic modifications, that modulate the expression of PR genes and others involved in the plant defense system (Albert et al. 2020). Also, the molecules from pathogens act as virulence factors, triggering ETI and speeding up immune system activation (Andersen et al. 2018). Initially, ETI induces a hypersensitive response (HR) due to the rapid production of ROS at the site of infection, leading to programmed cell death (PCD), which helps limit the spread of the pathogen and acts as a signal to activate other

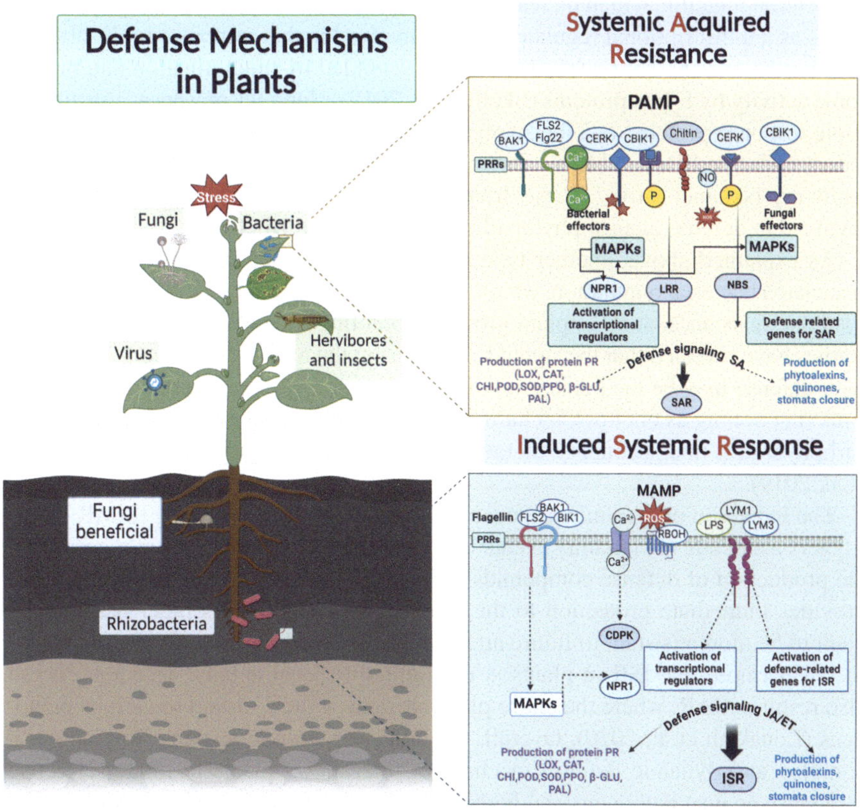

Fig. 2.1 Schematic representation of plant defense mechanisms by pathogenic/non-pathogenic microorganisms. Plants activate systemic acquired resistance (SAR) when bacteria, fungi, viruses, and insects attack them. In SAR response, the perception of organism pathogenic in the plant is by specific receptors (PRRs) such as BAK1, FLS2, Flg22, CERK1, CBIK1 and chitin; these specific receptors activate MAPKs and start the activation of transcriptional regulators and defense-related genes for SAR, which is mediated by hormones such as salicylic acid (SA), regulating the production of PR proteins, phytoalexins, and quinones. Concerning induced systemic response (ISR) and different receptors such as LPS, LYM1, LYM3, and FLS2, among others, activate the signaling cascade by MAPKs influencing NPR1 gene expression. The defense signaling by jasmonic acid (JA) and ethylene (ET) activates the production of PR proteins such as LOX, CAT, CHI, POD, SOD, PPO, β-GLU, and PAL, furthermore the physiological responses such as the production of phytoalexins, quinone, and stomata closure. (Figure created with Biorender.com)

defense mechanisms (Kumar et al. 2021a; Yang and Fernando 2021). If the plant responds successfully to the first pathogen attack, it will develop greater resistance to future attacks. This resistance is activated when the plant perceives specific biotic and abiotic stresses, initiating a response through signaling cascades where R proteins recognize effector molecules in plants through nucleotide-binding sites or leucine-rich repeats (NB-LRRs or NLRs) (Albert et al. 2020) for initiating a response through signaling molecules that activate the resistance mechanism in the plant, either SAR or ISR (Fig. 2.1).

An example of the activation signaling cascade to plant defense is the perception of bacterial flagellin, which increases ET biosynthesis, a key hormone in the ISR, as it acts as a transcriptional regulator of FTs involved in this response. In the absence of ET, ethylene-insensitive TF 3 (EIN3) is degraded by ubiquitination and proteasome activity by F-box proteins (Skelly et al. 2019). Upon ET presence, constitutive triple response protein1 (CTR1), a repressor of the ET signaling pathway, is inactivated, such that EIN2 and EIN3, downstream components of the ethylene signaling pathway, become active. These transcription factors positively regulate genes involved in defense against phytopathogens (Wang and Bi 2021).

As explained above, another type of induced resistance is caused by the root-associated bacterial population, which is non-pathogenic, and rhizobacteria-induced systemic resistance, such as plant growth-promoting rhizobacteria (PGPR). PGPR studies have revealed that the signal for triggering the metabolic network associated with defense may be the perception of some components from PGPR microorganisms functioning as elicitors, including proteins, lipids, cyclic lipopeptides (CLPs), carbohydrates, and secondary metabolites associated with ISR activation (Héloir et al. 2019).

The induction of systemic response in plants involves recognizing PAMP, which triggers a cascade of signaling events leading to the activation of defense genes and the production of defense compounds (Albert et al. 2020). This systemic response provides immediate protection to the infected area and primes other parts of the plant to be more resistant to future attacks (Khanna et al. 2020).

Furthermore, the ISR in plants is not limited to local defense reactions. It can also result in SAR, where the entire plant becomes more resistant to various pathogens (Lonakadi et al. 2020). Overall, the induced systemic response in plants is a complex and dynamic defense mechanism that allows plants to mount effective resistance against pathogens and herbivores. Various biochemical components and signaling pathways mediate this response, which produces defensive molecules to activate plant defense machinery (Kour et al. 2024). The activation of plant defense against phytopathogens involved in microbial inducers is more related to ISR, and the process of this type of defense induced by microorganisms will be detailed below.

2.6 Signaling Pathway Through ISR and Elicitors Microbial

When plants encounter biotic stress situations, they can perceive specific conserved signals associated with beneficial microbes, which are known as MAMP (microbial-associated molecular patterns). The immunity triggered upon recognition of MAMPs is known as MTI. These signals are recognized in the plant cell membrane by receptors on the plant cell membrane, called PRRs, which are part of the signaling system for triggering the plant's defenses and leading to resistance generation. These receptors act as the initial signaling molecules to activate the defense system in plants when identifying beneficial or pathogenic microorganisms.

PRRs can be kinase-like receptors (RLKs) or protein-like receptors (PRRs) which have an extracellular domain (Boutrot and Zipfel 2017; Trda et al. 2015; Bittel and Robatzek 2007; Jones and Dangl 2006; Zipfel 2008; Boller and He 2009). After MAMPs are recognized, a signaling cascade is triggered, such as the influx of calcium ions (Ca^{2+}), the production of reactive oxygen species (ROS) and nitric oxide (NO), and the activation of MAPKs, leading to transcriptional induction of primary metabolism and secondary metabolism in plants (Garcia-Brugger et al. 2006). This induction involves the expression of genes which involved in the production of PR proteins such as phenylalanine ammonialyase (PAL), chitinases or β-1,3-glucanases, polyphenol oxidase (PPO), and peroxidase (POD), the production of compounds involved in cell wall reinforcement such as lignin which can act as a barrier to inhibit pathogen entry, the production of antimicrobial compounds, and the production of proteins related to the hypersensitive response or HR whose production leads to cell death at the site of pathogen infection as a way to halt its advance (Garcia-Brugger et al. 2006; Ongena et al. 2000; Chen et al. 2000; Ramamoorthy et al. 2001; Boller and Felix 2009; Bukhat et al. 2020).

In the process of transcription or during post-transcriptional modifications, defence genes can be regulated through the addition of methyl groups to DNA, histone acetylation, changes made to chromatin, mRNA degradation, silencing, and inhibition of the translation process (Zhang et al. 2011; Latrasse et al. 2017).

Among the main bacterial genera that can induce Induced Systemic Resistance (ISR) are those belonging to *Bacillus*, such as *B. subtilis*, *B. amyloliquefaciens*, and other bacterial genera such as *Cladosporium*, *Ampelomyces*, *Phytophthora parasitica*, *P. fluorescens*, *P. aeruginosa*, *Trichoderma virens*, and others. These bacteria can act as bioelicitors when the plant recognizes components of their structure or the metabolites they produce, thus contributing to the improvement against different phytopathogens and diseases caused by them (Bakker et al. 2013; Segarra et al. 2009).

ISR-inducing microbial elicitors known as MAMPs (Microbe-Associated Molecular Patterns) enhance resistance to pathogens or pests. They can trigger the production of chemical compounds synthesized by plants as part of their secondary metabolism, which have been reported to protect plants from different types of biotic and abiotic stresses (Kesawat et al. 2023; Pathak et al. 2022). Some examples of these microbial elicitors include antibiotics, siderophores, volatiles, lipopolysaccharides, lipopeptides, exopolysaccharides, rhamnolipids, and *N*-acyl-homoserine which have been investigated and reported to have an effect against various plant pathogens by inhibiting their growth and thus conferring protection against diseases. These microbial elicitors can have potential applications in agriculture and plant protection, as they can enhance the plant immune system and reduce dependence of synthetic chemical compounds such as agrochemicals commonly used for plant pathogen control (Boller and He 2009; Höfte and Bakker 2007; Van Wees et al. 2008; Berendsen et al. 2015; Ongena and Jacques 2008; Conrath et al. 2006).

Among defence responses triggered in plants by ISR are those activated by the recognition of MAMPs. This type of resistance, mediated by rhizospheric bacteria such as PGPR and other signaling molecules, enhances the plant's immunity to improve its ability to respond to different types of stress. Activation of ISR mediated by PGPR bacteria and their metabolites acts as signallers that prepare the plant for an efficient response to environmental stresses, and this mechanism is known as priming. This mechanism in the plant is given by three different states such as naïve, priming, and post priming derived changes (Conrath et al. 2006; Ongena et al. 2007). Molecular mechanisms subsequent to this state confer a rapid defence response to the plant, inducing the production of different metabolites, compounds with antimicrobial activity, modifications at the chromatin level, in hormone synthesis, and in receptors responsible for the recognition of inducers are made (Dong 2004; Macho and Zipfel 2014; Boller and Felix 2009).

The signaling pathway for ISR mediated by PGPR bacteria is regulated by JA and ET, activating defense genes (Bari and Jones 2009; Mou et al. 2020; Glick 2012). The ISR signaling pathway involves plants' perception of elicitors, followed by the activation of signaling pathways that lead to the expression of defense genes. During the perception of ISR elicitors, signaling pathways and a consequent expression of genes involved in the defence system are triggered. Among the microbial elicitor molecules that are recognized by plants include amphiphilic lipopeptides that have been reported to inhibit plant pathogenic fungi, as is the case of cyclic lipopeptides (CLP) synthesized by PGPR bacteria such as *B. subtilis*, which synthesizes three different groups such as surfactins, iturins, and phengicins. Surfactin, a mycosubtilin belonging to the iturin group, and plipastatin belonging to the fengycins have been tested for their effect on different crops in the field and after harvest (Ongena and Jacques 2008; Héloir et al. 2019).

Various reports have studied the effect of different elicitors on activating ISR and conferring resistance in specific crops against some pathogens. Positive effects have been observed in bacterial genus such as *Bacillus* and *Pseudomonas*; some examples are compiled in Table 2.1.

Understanding molecular processes in response for plant improvement is essential in the development of tactics that contribute to the enhanced defence and productivity of crops due to the increased environmental challenges to which they can be exposed. Products have been formulated based on different microorganisms which have been able to contribute to the improvement of plant defence. Some of them have been commercialized and are applied on crops of agri-food interest, their use has reduced the application of more environmentally aggressive agrochemicals.

Table 2.1 Application of SRI-inducing microbial elicitors on different plant crops

Inducer microbial strain	Elicitor compounds	Crops treated	Target pathogens	References
Klebsiella spp. (MBE02)	Bacterial cells (MBE02)	Peanut (*Arachis hypogaea* L.)	*Aspergillus* sp.	Sharma et al. (2019a)
B. pumilus (S2-3-2)	Volatile compounds	Tobacco (*Nicotiana benthamiana*) Pepper (*Capsicum annuum* L. cv. Nok-Kwang)	Tobacco soft rot *X. axonopodis* pv. *vesicatoria*	Li et al. (2020)
B. subtilis (GLB191)	Cyclic lipopeptides (CLPs)	*Vitis vinifera* cv. Marselan	*Plasmopara viticola*	Li et al. (2019)
Bacillus subtilis KA9 and *Pseudomonas fluorescens* PDS1	Bacterial cells (KA9) (PDS1)	Pepper (*Capsicum annuum* L. cv. Pusa Jwala)	*Ralstonia solanacearum*	Kashyap et al. (2021)
Pseudomonas aeruginosa	Rhamnolipids (RLs)	Napus (*Brassica napus* cv. Darmor-bzh, v. Atenzo, cv. Gaspard)	*Botrytis cinerea* (B630)	Monnier et al. (2018)
Bacillus subtilis	Cyclic lipopeptide surfactin	Peanut (*Arachis hypogaea* var. Runner cultivar Granoleico)	*Sclerotium rolfsii*	Rodríguez et al. (2018)
Bacillus velezensis HN-Q-8	Bacterial cells HN-Q-8 (bacterial culture FLBS, bacterial supernatant without bacteria FL, bacterial suspension BCS)	Potato (cultivar Helan 15)	*Alternaria solani* (HWC-16)	Bai et al. (2023)
Trichoderma harzianum Th22	*Trichoderma* conidia suspensions	Corn (inbred line Huangzao 4)	*Curvularia lunata* CX3	Saravanakumar et al. (2016)
Pseudomonas spp.	Cyclic lipopeptides (CLPs)	Rice (*Oryza sativa* L.) indica cv. CO-39	*Magnaporthe oryzae* (WLIP)	Omoboye et al. (2019)
Pseudomonas sp. CMR12a	Cyclic lipopeptides (CLPs). Phenazines	Rice (*Oryza sativa* L.) Bean (*Phaseolus vulgaris*)	*Magnaporthe oryzae* *Rhizoctonia solani* AG2-2	Ma et al. (2016)

(continued)

Table 2.1 (continued)

Inducer microbial strain	Elicitor compounds	Crops treated	Target pathogens	References
Pseudomonas protegens	Orfamide A	Rice (*Oryza sativa* L.) indica cv. CO-39	*Cochliobolus miyabeanus* Cm988	Ma et al. (2017)
Bacillus subtilis BBG111	CLP-type biosurfactants	Rice (*Oryza sativa* subsp. japonica) cv. Taichung 65 (139 T65)	*Rhizoctonia solani*	Chandler et al. (2015)
Laccaria bicolor (monokaryotic strain CII-H-82, S238N)	*L. bicolor*	*Arabidopsis thaliana*	*Trichoplusia ni*	Vishwanathan et al. (2020)
Bacillus amyloliquefaciens Ba13	Bacterial cells (Ba13)	Tomato (*Lycopersicon esculentum* Mill. cv. Guofen 1)	Tomato yellow leaf curl virus (TYLCV)	Guo et al. (2019)
Pythium oligandrum, strain Po37	Oospores Po37	Grapevine cultivars (*Vitis vinifera L.*) Chardonnay and Cabernet Sauvignon	*Botrytis cinerea* CCB16 *Plasmopara vitícola* PV221	Yacoub et al. (2023)
Bacillus proteolyticus strain OSUB18	OSUB18 cells	*Arabidopsis thaliana*	*Pseudomonas syringae* pv. tomato (Pst DC3000) *Botrytis cinerea*	Yang et al. (2023)
Bacillus vallismortis EXTN-1	Iturin A	Chili pepper seedlings (*Capsicum annum* L.)	*Phytophthora capsici*	Park et al. (2016)
Bacillus subtilis GB03	Bacterial volatile compound (BVC)	(*Cucumis sativus* L. cv. *Backdadagi*)	*Pseudomonas syringae* pv. *lachrymans*	Song et al. (2019)
Bacillus amyloliquefaciens (AK1S)	Fengycin	Tomato	*Sclerotinia sclerotiorum*	Farzand et al. (2019)
Bacillus subtilis	Lipopeptides (mycosubtilin, surfactin, fengycin)	Wheat cv. Dinosor and Alixan	*Zymoseptoria tritici* T01193	Mejri et al. (2018)
Bacillus amiloliquefaciens NC6	Protein PeBA1	*Nicotiana tabacum*	*Tobacco mosaic virus* (TMV) *Botrytis cinerea*	Wang et al. (2016)

2.7 Bioelicitors Used in Agriculture

Due to the exponential increase in the world's population, estimated to be around 9.7 billion people by 2050, there has been an increase in the demand for food, resulting in improved and increased crop production as well as an increase in the agricultural surplus (UN DESA 2015; Kumar et al. 2021a). Different strategies have been implemented to improve and enhance crop yield in the face of various adversities. These include using chemical fertilizers and pesticides to protect crops from other stressors (dos Santos et al. 2012; Swapna 2019).

Chemical compounds or pesticides used on crops to protect them from various pathogens include those with antibacterial, fungicidal, insecticidal, herbicidal, and nematicidal activity (Anakwue 2019), these have been applied to crops such as corn, wheat, rice, sorghum, sugar cane, soybean, groundnut, and cotton, fruits, and various vegetables (Karamchandani et al. 2022).

Approximately 2 million tons of pesticides are globally used annually, including herbicides (50%), insecticides (18%), and rodenticides and nematicides (2%). Despite their efficacy in reducing various pathogens and pests for crop care and yield, different synthetic chemical compounds applied in agriculture have caused different negative impacts on soil composition, overall rhizosphere, and hence on crop health, external environment, e.g., on the health of growers and consumers which has had repercussions on the health of farmers and consumers (Sharma et al. 2019b; Sarkar et al. 2021). Therefore, awareness raising about the consequences of their use and the exploration of new pest control alternatives is necessary.

Scientists and commercial companies worldwide have turned their attention to the research and use of new alternatives to synthetic chemical compounds with the least possible impact on the environment (Droby et al. 2016).

Among the microorganisms investigated and used in different plants are the rhizosphere bacteria (PGPR) which promote plant growth, these bacteria do not cause negative effects on the health of crops, but provide different benefits by stimulating the defence system which causes a state of alertness that prepares plants to respond more immediately to different stress situations, contributing to phytostimulation (Verma et al. 2021). Other beneficial effects are their capacity for biofertilization, biopesticidal, and biocontrol activity through the production of compounds with antibiotic or inducing activity such as iturin, fengycin and surfactin lipopeptides, and other types of compounds such as volatile organic compounds (VOCs) and the production of defense enzymes (Goswami et al. 2016).

Different laboratory-level strategies have been investigated for utilizing microorganisms or their metabolic products. Various products have been semi-commercialized and commercialized by different companies (Table 2.2), for example, for the control of powdery mildew (*Uncinula necator*), AQ10® (*Ampelomyces quisqualis*) (Benuzzi and Baldoni 2000), and *Trichoderma atroviride*-based products (CS1 by Vintec®—Belchim Crop Production and I-1237 by Esquive®—Agrauxine by Lesaffre) (Pertot et al. 2015; Belchim 2022). *Trichoderma harzianum* (known as Trichodex®) acts against the oomycete by increasing lignin, callose, and hydrogen peroxide, in addition to upregulating the defense enzymes

Table 2.2 Commercial biological products that induce crop defense systems

Products	Biological agents	Company	Crops	Pathogens
Taegro®	*Bacillus subtilis* var. *amyloliquefaciens* cepa FZB24	Syngenta	Eggplant, Chile, tomato, green tomato, potato, bell pepper	(*Leveillula taurica*) (*Alternaria solani*)
			Zucchini, pumpkin, chayote, melon, cucumber, watermelon	(*Erysiphe cichoracearum*)
			Blueberry, raspberry, strawberry, currant, kiwi, grapevine, blackberry	(*Botrytis cinerea*)
			Banana	(*Mycosphaerella fijiensis*)
Sonata®	*Bacillus pumilus* QST 2808	Bayer CropScience	Grapevine, tomato	(*Uncinula necátor*) (*Oidium neolycopersici*)
Serenade® ASO	*Bacillus subtilis* QST 713	Bayer CropScience	Strawberry	(*Botrytis cinérea*)
Serenade® Opti	QST 713 strain of *Bacillus subtilis*	Bayer CropScience	Orange, grapefruit, lemon, tangerine, tangelo, pummelo, and other citrus crop	(*Xanthomonas citri* subsp. *citri*)
Romeo®	Cerevisane by *Saccharomyces cerevisiae*	Agrosur S.L.	Cucurbitaceae	(*Golovinomyces cichoracearum*) (*Podosphaera xanthii*)
			Lettuce	(*Bremia lactucae*)
			Tomatoes, eggplants, strawberries	(*Botrytis cinérea*)
			Vineyard, wine grape	(*Uncinula necátor*) (*Plasmopara vitícola*) (*Botrytis cinérea*)

(continued)

Table 2.2 (continued)

Products	Biological agents	Company	Crops	Pathogens
Duravel® WP	*Bacillus amyloliquefaciens* MBI600	BASF Agro Colombia	Tomato	(*Botrytis cinérea*)
BIO-IMUNE®	*Bacillus velezensis* BV02	Vittia Group	Coffee	(*Hemileia uvatrix*)
			Bean	(*Colletotrichum lindemuthianum*)
			Soja	(*Colletotrichum truncatum*)
			Apple tree	(*Sclerotinia sclerotiorum*)
Tricho-Turbo®	*Trichoderma asperellum* BV10	Vittia Group	Bean	(*Sclerotinia sclerotiorum*)
BACILO-X (MAPA sob n° 27021)	*Bacillus amyloliquefaciens cepa* D-747	Ballagro Agro Technology Ltd.	Onion	(*Alternaria porrí*)
			Banana	(*Mycosphaerella fijiensis*)
			Sweet potato	(*Rhizoctonia solani*)

phenylalanine ammonia-lyase (PAL), peroxidase (POD), and glucanase (Kamble et al. 2021). Serenade ASO 6 (bio-fungicide), foliar fungicide obtaining *B. subtilis* QST 713 is produced by Bayer CropScience Ltd. (Cambridge, UK). Rhamnolipids (RLs) are glycolipid biosurfactants being commercially produced by AGAE Technologies LLC, USA and Rhamnolipid Holdings Inc. (New York, USA) (Karamchandani et al. 2022).

There has been an estimated steady increase of approximately 10% each year in the application of biopesticides, which include microorganisms or their products obtained for plant protection. That is to say, around 3000 tons are produced annually (Olson 2015). In the United States (US), more than 200 biopesticide products are already on the market, compared to 60 products in the European Union (EU) (Damalas and Koutroubas 2018). Countries such as the United States, Canada, and Mexico contribute 45% of the sales of these products worldwide, while Asia accounts for 5% (Bailey et al. 2010).

The global market for pesticide products such as biofungicides, bioinsecticides, bioherbicides, and nematicides exceeded $3.3 million in 2017 and is expected to reach $9.5 million by 2025. Rapid increases totaling $3667.5 million are anticipated from North America, Europe, and Asia-Pacific. India and China are among the countries exploring the use of biopesticides in fruit production (GBMAIF 2017).

According to the Directorate of Plant Protection, Quarantine and Storage (DPPQS) of the Department of Agriculture and Farmers' Welfare of the Government of India, approximately 25 biopesticide products were registered in July 2021 in this country. Around 361 units and biocontrol laboratories operate in the country; however, few are dedicated to designing these products. Despite this, these numbers have significantly increased in recent decades (Mishra et al. 2020; Koul 2023).

Technological advances in the formulation of biopesticides allow the stabilization of different types of beneficial microorganisms and environmentally sensitive microbes, improving their handling and safety, increasing their stability and persistence under certain environmental conditions, and promoting their efficacy toward the pathogen (Lacey et al. 2015).

In Spain in 2016, more than two million ha were dedicated to organic agriculture, 8.5% more than in 2015, so from 2012 to 2016, it grew by 18.7% in the European Union (https://ec.europa.eu/eurostat).

In general, many laws and policies regulate the development and use of microbial biopesticides, complicating the registration of new microbial products because their regulation is based on a system designed for synthetic pesticides and their by-products. Adapting to the regulatory requirements is needed depending on the type of biopesticide used (Damalas and Koutroubas 2018).

The use of biological pesticides, such as those derived from various bacteria, is continuously growing, driven by market trends favoring healthy eating and the pursuit of minimizing negative environmental impacts. Bayer CropScience, within the bioinsecticide sector, has significantly impacted this field.

Industry and research institutes should collaborate more strongly to jointly develop new biopesticide products for biological disease control in different crops.

The results obtained and the various investigations conducted on biopesticides and their field application have contributed to their production on a larger scale, with the expectation that regulatory policies will be established to promote their better development at an industrial scale, where costs for market acceptance are higher for small businesses than for large companies, due to the large-scale production of the latter and the variety and timing of application needed to protect crops from various pathogens (Koul 2023).

2.8 Conclusions

Understanding plant defense mechanisms cannot be overstated in today's rapidly changing world. Given the increasingly devastating effects of climate change exacerbated by the significant amount of agrochemicals used to combat diseases in crops of agro-industrial interest in both pre- and post-harvest stages, it is urgent to take measures by shifting the focus toward more eco-friendly strategies. The current challenge is to produce food sustainably while controlling diseases without the excessive use of synthetic products that cause various harms. A promising alternative for agricultural sustainability is defense inducers, highlighting those of microbial origin due to the diversity of metabolites produced by microbial strains, representing different control mechanisms, and being accessible to apply. New inducers are constantly being discovered, and knowledge of the physiological response of plants is deepening; it will be especially relevant to demonstrate their practical application in agriculture and their feasibility of use in terms of energy and metabolic resistance of each plant type. Therefore, understanding the underlying mechanisms of the plant immune system at the molecular level is crucial for developing strategies using microorganism elicitors that enhance crop defense and agricultural productivity.

References

Albert I, Hua C, Nürnberger T, Pruitt RN, Zhang L (2020) Surface sensor systems in plant immunity. Plant Physiol 182:1582–1596. https://doi.org/10.1104/PP.19.01299

Ali A, Ölmez F, Zeshan MA, Mubeen M, Iftikhar Y, Sajid A, Abid M, Kumar A, Divvela PK, Solanki MK (2024) Yeast-based solutions in controlling plant pathogens. Biocatal Agric Biotechnol 58:103199. https://doi.org/10.1016/j.bcab.2024.103199

Anakwue R (2019) Cardiotoxicity of pesticides: are Africans at risk? Cardiovasc Toxicol 19(2):95–104. https://doi.org/10.1007/s12012-018-9486-7

Anand U, Pal T, Yadav N, Singh VK, Tripathi V, Choudhary KK, Shukla AK, Sunita K, Kumar A, Bontempi E, Ma Y (2023) Current scenario and future prospects of endophytic microbes: promising candidates for abiotic and biotic stress management for agricultural and environmental sustainability. Microb Ecol 86:1455–1486. https://doi.org/10.1007/s00248-023-02190-1

Andersen EJ, Ali S, Byamukama E et al (2018) Disease resistance mechanisms in plants. Genes (Basel) 9:1–30. https://doi.org/10.3390/genes9070339

Bai X, Li Q, Zhang D, Zhao Y, Zhao D, Pan Y, Wang J, Yang Z, Zhu J (2023) Bacillus velezensis strain HN-Q-8 induced resistance to Alternaria solani and stimulated growth of potato plant. Biology 12(6):856. https://doi.org/10.3390/biology12060856

Bailey KL, Boyetchko SM, Längle T (2010) Social and economic drivers shaping the future of biological control: a Canadian perspective on the factors affecting the development and use of microbial biopesticides. Biol Control 52(3):221–229. https://doi.org/10.1016/j.biocontrol.2009.05.003

Bakker PAHM, Berendsen RL, Doornbos RF, Wintermans PCA, Pieterse CMJ (2013) The rhizosphere revisited: root microbiomics. Front Plant Sci 4. https://doi.org/10.3389/fpls.2013.00165

Bari R, Jones JDG (2009) Role of plant hormones in plant defence responses. Plant Mol Biol 69(4):473–488. https://doi.org/10.1007/s11103-008-9435-0

Belchim (2022) Vintec: a natureza trabalhando por si. https://belchim.pt/producten/vintec/. Accessed 30 Jul 2022

Benuzzi M, Baldoni G (2000) AQ10, a new biofungicide based on Ampelomyces quisqualis for powdery mildew control on grapes. Inf Fitopatol 50(5):33–36

Berendsen RL, van Verk MC, Stringlis IA, Zamioudis C, Tommassen J, Pieterse CMJ, Bakker PAHM (2015) Unearthing the genomes of plant-beneficial Pseudomonas model strains WCS358, WCS374 and WCS417. BMC Genomics 16(1):539. https://doi.org/10.1186/s12864-015-1632-z

Bhardwaj A, Kaur K, Kumar A, Singh SK, Kharshiing B (2024) Development of microbial biotechnology products for phytopathogen management. In: The potential of microbes for a circular economy. Academic Press, pp 105–130. https://doi.org/10.1016/B978-0-443-15924-4.00013-8

Bittel P, Robatzek S (2007) Microbe-associated molecular patterns (MAMPs) probe plant immunity. Curr Opin Plant Biol 10(4):335–341. https://doi.org/10.1016/j.pbi.2007.04.02

Blakeney M (2019) Food loss and food waste: causes and solutions. Edward Elgar Publishing Limited. https://doi.org/10.4337/9781788975391

Boller T, Felix G (2009) A renaissance of elicitors: perception of microbe-associated molecular patterns and danger signals by pattern-recognition receptors. Annu Rev Plant Biol 60(1):379–406. https://doi.org/10.1146/annurev.arplant.57.032905.105346

Boller T, He SY (2009) Innate immunity in plants: an arms race between pattern recognition receptors in plants and effectors in microbial pathogens. Science 324(5928):742–744. https://doi.org/10.1126/science.1171647

Borges AA, Jiménez-Arias D, Expósito-Rodríguez M, Sandalio LM, Pérez JA (2014) Priming crops against biotic and abiotic stresses: MSB as a tool for studying mechanisms. Front Plant Sci 5:642. https://doi.org/10.3389/fpls.2014.00642

Boutrot F, Zipfel C (2017) Function, discovery, and exploitation of plant pattern recognition receptors for broad-Spectrum disease resistance. Annu Rev Phytopathol 55(1):257–286. https://doi.org/10.1146/annurev-phyto-080614-120106

Bukhat S, Imran A, Javaid S, Shahid M, Majeed A, Naqqash T (2020) Communication of plants with microbial world: exploring the regulatory networks for PGPR mediated defense signaling. Microbiol Res 238:126486. https://doi.org/10.1016/j.micres.2020.126486

Chandler S, Van Hese N, Coutte F, Jacques P, Höfte M, De Vleesschauwer D (2015) Role of cyclic lipopeptides produced by Bacillus subtilis in mounting induced immunity in rice (Oryza sativa L.). Physiol Mol Plant Pathol 91:20–30. https://doi.org/10.1016/j.pmpp.2015.05.010

Chen C, Bélanger RR, Benhamou N, Paulitz TC (2000) Defense enzymes induced in cucumber roots by treatment with plant growth-promoting rhizobacteria (PGPR) and Pythium aphanidermatum. Physiol Mol Plant Pathol 56(1):13–23. https://doi.org/10.1006/pmpp.1999.0243

Conrath U, Beckers GJM, Flors V, García-Agustín P, Jakab G, Mauch F, Newman M-A, Pieterse CMJ, Poinssot B, Pozo MJ, Pugin A, Schaffrath U, Ton J, Wendehenne D, Zimmerli L, Mauch-Mani B (2006) Priming: getting ready for battle. Mol Plant Microbe Interact 19(10):1062–1071. https://doi.org/10.1094/MPMI-19-1062

Damalas C, Koutroubas S (2018) Current status and recent developments in biopesticide use. Agriculture 8(1):13. https://doi.org/10.3390/agriculture8010013

Dempsey DA, Klessig DF (2012) SOS—too many signals for systemic acquired resistance? Trends Plant Sci 17(9):538–545. https://doi.org/10.1016/j.tplants.2012.05.011

Dong X (2004) NPR1, all things considered. Curr Opin Plant Biol 7(5):547–552. https://doi.org/10.1016/j.pbi.2004.07.005

dos Santos EA, Ferreira LR, Costa MD, dos Santos JB, da Silva MDCS, Aspiazú I (2012) The effects of soil fumigation on the growth and mineral nutrition of weeds and crops. Acta Sci Agron 34(2). https://doi.org/10.4025/actasciagron.v34i2.12971

Droby S, Wisniewski M, Teixidó N, Spadaro D, Jijakli MH (2016) The science, development, and commercialization of postharvest biocontrol products. Postharvest Biol Technol 122:22–29. https://doi.org/10.1016/j.postharvbio.2016.04.006

Farzand A, Moosa A, Zubair M, Khan AR, Massawe VC, Tahir HAS, Sheikh TMM, Ayaz M, Gao X (2019) Suppression of Sclerotinia sclerotiorum by the induction of systemic resistance and regulation of antioxidant pathways in tomato using Fengycin produced by Bacillus amyloliquefaciens FZB42. Biomolecules 9(10):613. https://doi.org/10.3390/biom9100613

Francis SP, Rene Christena L, Mishel PF (2023) Pathogen effectors: biochemical and structural targets during plant-microbe interactions

Garcia-Brugger A, Lamotte O, Vandelle E, Bourque S, Lecourieux D, Poinssot B, Wendehenne D, Pugin A (2006) Early signaling events induced by elicitors of plant defenses. Mol Plant Microbe Interact 19(7):711–724. https://doi.org/10.1094/MPMI-19-0711

GBMAIF, 2017, Global $7.6 Bn Opportunity Analysis And Industry Forecasts To 2022 - Research and Markets, News provided by Research and Markets, https://www.businesswire.com/news/home/20170912006127/en/Biopesticides-Market-2017---Global-7.6-Bn-Opportunity-Analysis-And-Industry-Forecasts-To-2022---Research-and-Markets (2017).

Ghorbanpour M, Omidvari M, Abbaszadeh-Dahaji P, Omidvar R, Kariman K (2018) Mechanisms underlying the protective effects of beneficial fungi against plant diseases. Biol Control 117:147–157. https://doi.org/10.1016/j.biocontrol.2017.11.006

Glick BR (2012) Plant growth-promoting bacteria: mechanisms and applications. Scientifica 2012:1–15. https://doi.org/10.6064/2012/963401

Goellner K, Conrath U (2008) Priming: it's all the world to induced disease resistance. Eur J Plant Pathol 121:233–242. https://doi.org/10.1007/s10658-007-9251-4

Goswami D, Thakker JN, Dhandhukia PC (2016) Portraying mechanics of plant growth promoting rhizobacteria (PGPR): a review. Cogent Food Agric 2(1). https://doi.org/10.1080/23311932.2015.1127500

Guo Q, Li Y, Lou Y, Shi M, Jiang Y, Zhou J, Sun Y, Xue Q, Lai H (2019) Bacillus amyloliquefaciens Ba13 induces plant systemic resistance and improves rhizosphere microecology against tomato yellow leaf curl virus disease. Appl Soil Ecol 137:154–166. https://doi.org/10.1016/j.apsoil.2019.01.015

Gutiérrez-Martínez P, Ledezma-Morales A, Romero-Islas L, Ramos-Guerrero A, Romero-Islas J, Rodríguez-Pereira C, Casas-Junco P, Coronado-Partida L, González-Estrada R (2018) Antifungal activity of chitosan against postharvest fungi of tropical and subtropical fruits. In: Dongre RS (ed) Chitin-chitosan myriad functionalities in science and technology. IntechOpen, London. https://doi.org/10.5772/intechopen.71146

Heil M, Bostock RM (2002) Induced systemic resistance (ISR) against pathogens in the context of induced plant defences. Ann Bot 89(5):503–512. https://doi.org/10.1093/aob/mcf076

Héloir M-C, Adrian M, Brulé D, Claverie J, Cordelier S, Daire X, Dorey S, Gauthier A, Lemaître-Guillier C, Negrel J, Trdá L, Trouvelot S, Vandelle E, Poinssot B (2019) Recognition of elicitors in grapevine: from MAMP and DAMP perception to induced resistance. Front Plant Sci 10. https://doi.org/10.3389/fpls.2019.01117

Höfte M, Bakker PAHM (2007) Competition for iron and induced systemic resistance by siderophores of plant growth promoting rhizobacteria. In: Microbial siderophores. Springer, Berlin, pp 121–133. https://doi.org/10.1007/978-3-540-71160-5_6

Jamalizadeh M, Etebarian HR, Aminian H, Alizadeh A (2011) A review of mechanisms of action of biological control organisms against post-harvest fruit spoilage. EPPO Bull 41(1):65–71. https://doi.org/10.1111/j.1365-2338.2011.02438.x

Jones JDG, Dangl JL (2006) The plant immune system. Nature 444(7117):323–329. https://doi.org/10.1038/nature05286

Kamble MV, Joshi SM, Hadimani S, Jogaiah S (2021) Biopriming with rhizosphere Trichoderma harzianum elicit protection against grapevine downy mildew disease by triggering histopathological and biochemical defense responses. Rhizosphere 19:100398. https://doi.org/10.1016/j.rhisph.2021.100398

Karamchandani BM, Pawar AA, Pawar SS, Syed S, Mone NS, Dalvi SG, Rahman PKSM, Banat IM, Satpute SK (2022) Biosurfactants' multifarious functional potential for sustainable agricultural practices. Front Bioeng Biotechnol 10. https://doi.org/10.3389/fbioe.2022.1047279

Kashyap AS, Manzar N, Rajawat MVS, Kesharwani AK, Singh RP, Dubey SC, Pattanayak D, Dhar S, Lal SK, Singh D (2021) Screening and biocontrol potential of rhizobacteria native to gangetic plains and hilly regions to induce systemic resistance and promote plant growth in Chilli against bacterial wilt disease. Plants 10(10):2125. https://doi.org/10.3390/plants10102125

Kesawat MS, Satheesh N, Kherawat BS, Kumar A, Kim HU, Chung SM, Kumar M (2023) Regulation of reactive oxygen species during salt stress in plants and their crosstalk with other signaling molecules—current perspectives and future directions. Plants 12(4):864. https://doi.org/10.3390/plants12040864

Khanna K, Kapoor D, Sharma P, Bakshi P, Sharma P, Saini P, Ohri P, Mir BA, Kaur R, Bhardwaj R (2020) Plant-microbe interactions under adverse environment. In: Plant ecophysiology and adaptation under climate change: mechanisms and perspectives I: general consequences and plant responses. Springer. https://doi.org/10.1007/978-981-15-2156-0_26

Köhl J, Kolnaar R, Ravensberg W (2019) Mode of action of microbial biological control agents against plant diseases: relevance beyond efficacy. Front Plant Sci 10(845). https://doi.org/10.3389/fpls.2019.00845

Koul O (2023) Biopesticides. In: Development and commercialization of biopesticides. Elsevier, pp 1–23. https://doi.org/10.1016/B978-0-323-95290-3.00009-1

Kour D, Negi R, Khan SS et al (2024) Microbes mediated induced systemic response in plants: a review. Plant Stress 11:100334. https://doi.org/10.1016/j.stress.2023.100334

Kumar J, Ramlal A, Mallick D, Mishra V (2021a) An overview of some biopesticides and their importance in plant protection for commercial acceptance. Plants 10(6):1185. https://doi.org/10.3390/plants10061185

Kumar J, Ramlal A, Kumar K et al (2021b) Signaling pathways and downstream effectors of host innate immunity in plants. Int J Mol Sci 22. https://doi.org/10.3390/ijms22169022

Lacey LA, Grzywacz D, Shapiro-Ilan DI, Frutos R, Brownbridge M, Goettel MS (2015) Insect pathogens as biological control agents: back to the future. J Invertebr Pathol 132:1–41. https://doi.org/10.1016/j.jip.2015.07.009

Latrasse D, Jégu T, Li H, de Zelicourt A, Raynaud C, Legras S, Gust A, Samajova O, Veluchamy A, Rayapuram N, Ramirez-Prado JS, Kulikova O, Colcombet J, Bigeard J, Genot B, Bisseling T, Benhamed M, Hirt H (2017) MAPK-triggered chromatin reprogramming by histone deacetylase in plant innate immunity. Genome Biol 18(1):131. https://doi.org/10.1186/s13059-017-1261-8

Lazazzara V, Avesani S, Robatscher P et al (2022) Biogenic volatile organic compounds in the grapevine response to pathogens, beneficial microorganisms, resistance inducers, and abiotic factors. J Exp Bot 73:529–554. https://doi.org/10.1093/jxb/erab367

Li Y, Héloir M, Zhang X, Geissler M, Trouvelot S, Jacquens L, Henkel M, Su X, Fang X, Wang Q, Adrian M (2019) Surfactin and fengycin contribute to the protection of a Bacillus subtilis strain against grape downy mildew by both direct effect and defence stimulation. Mol Plant Pathol 20(8):1037–1050. https://doi.org/10.1111/mpp.12809

Li W, Lee S-Y, Cho Y-J, Ghim S-Y, Jung H-Y (2020) Mediation of induced systemic resistance by the plant growth-promoting rhizobacteria Bacillus pumilus S2-3-2. Mol Biol Rep 47(11):8429–8438. https://doi.org/10.1007/s11033-020-05883-9

Lonakadi P, Jobby R, Desai N, Jha P (2020) Biochemical dynamics of plant-microbe interactions. In: Plant microbiome paradigm. Springer. https://doi.org/10.1007/978-3-030-50395-6_14

Ma Z, Hua GKH, Ongena M, Höfte M (2016) Role of phenazines and cyclic lipopeptides produced by pseudomonas sp. CMR12a in induced systemic resistance on rice and bean. Environ Microbiol Rep 8(5):896–904. https://doi.org/10.1111/1758-2229.12454

Ma Z, Ongena M, Höfte M (2017) The cyclic lipopeptide orfamide induces systemic resistance in rice to Cochliobolus miyabeanus but not to Magnaporthe oryzae. Plant Cell Rep 36(11):1731–1746. https://doi.org/10.1007/s00299-017-2187-z

Macho AP, Zipfel C (2014) Plant PRRs and the activation of innate immune signaling. Mol Cell 54(2):263–272. https://doi.org/10.1016/j.molcel.2014.03.028

Malik NAA, Kumar IS, Nadarajah K (2020) Elicitor and receptor molecules: orchestrators of plant defense and immunity. Int J Mol Sci 21. https://doi.org/10.3390/ijms21030963

Mejri S, Siah A, Coutte F, Magnin-Robert M, Randoux B, Tisserant B, Krier F, Jacques P, Reignault P, Halama P (2018) Biocontrol of the wheat pathogen Zymoseptoria tritici using cyclic lipopeptides from Bacillus subtilis. Environ Sci Pollut Res 25(30):29822–29833. https://doi.org/10.1007/s11356-017-9241-9

Mishra J, Dutta V, Arora NK (2020) Biopesticides in India: technology and sustainability linkages. 3 Biotech 10(5):210. https://doi.org/10.1007/s13205-020-02192-7

Monnier N, Furlan A, Botcazon C, Dahi A, Mongelard G, Cordelier S, Clément C, Dorey S, Sarazin C, Rippa S (2018) Rhamnolipids from Pseudomonas aeruginosa are elicitors triggering Brassica napus protection against Botrytis cinerea without physiological disorders. Front Plant Sci 9. https://doi.org/10.3389/fpls.2018.01170

Mou W, Kao Y-T, Michard E, Simon AA, Li D, Wudick MM, Lizzio MA, Feijó JA, Chang C (2020) Ethylene-independent signaling by the ethylene precursor ACC in Arabidopsis ovular pollen tube attraction. Nat Commun 11(1):4082. https://doi.org/10.1038/s41467-020-17819-9

Olson S (2015) An analysis of the biopesticide market now and where it is going. Outlooks Pest Manag 26(5):203–206. https://doi.org/10.1564/v26_oct_04

Omoboye OO, Oni FE, Batool H, Yimer HZ, De Mot R, Höfte M (2019) Pseudomonas cyclic lipopeptides suppress the rice blast fungus Magnaporthe oryzae by induced resistance and direct antagonism. Front Plant Sci 10. https://doi.org/10.3389/fpls.2019.00901

Ongena M, Jacques P (2008) Bacillus lipopeptides: versatile weapons for plant disease biocontrol. Trends Microbiol 16(3):115–125. https://doi.org/10.1016/j.tim.2007.12.009

Ongena M, Daayf F, Jacques P, Thonart P, Benhamou N, Paulitz TC, Bélanger RR (2000) Systemic induction of phytoalexins in cucumber in response to treatments with fluorescent pseudomonads. Plant Pathol 49(4):523–530. https://doi.org/10.1046/j.1365-3059.2000.00468.x

Ongena M, Jourdan E, Adam A, Paquot M, Brans A, Joris B, Arpigny J, Thonart P (2007) Surfactin and fengycin lipopeptides of Bacillus subtilis as elicitors of induced systemic resistance in plants. Environ Microbiol 9(4):1084–1090. https://doi.org/10.1111/j.1462-2920.2006.01202.x

Paniagua Voirol LR, Valsamakis G, Lortzing V, Weinhold A, Johnston PR, Fatouros NE, Kunze R, Hilker M (2020) Plant responses to insect eggs are not induced by egg-associated microbes, but by a secretion attached to the eggs. Plant Cell Environ 43:1815–1826. https://doi.org/10.1111/pce.13746

Park K, Park Y-S, Ahamed J, Dutta S, Ryu H, Lee S-H, Balaraju K, Manir M, Moon S-S (2016) Elicitation of induced systemic resistance of chili pepper by iturin A analogs derived from Bacillus vallismortis EXTN-1. Can J Plant Sci 96:564–570. https://doi.org/10.1139/cjps-2015-0199

Pathak P, Rai VK, Can H, Singh SK, Kumar D, Bhardwaj N, Roychowdhury R, de Azevedo LCB, Kaushalendra, Verma H, Kumar A (2022) Plant-endophyte interaction during biotic stress management. Plants 11(17):2203. https://doi.org/10.3390/plants11172203

Pertot I, Longa CMO, Savazzini F, Prodorutti D, Michelon L (2015) Trichoderma atroviride SC1 for biocontrol of fungal diseases in plants. International Patent Application US 8,431,120B2

Raaijmakers JM, Mazzola M (2012) Diversity and natural functions of antibiotics produced by beneficial and plant pathogenic bacteria. Annu Rev Phytopathol 50:403–424. https://doi.org/10.1146/annurev-phyto-081211-172908

Ramamoorthy V, Viswanathan R, Raguchander T, Prakasam V, Samiyappan R (2001) Induction of systemic resistance by plant growth promoting rhizobacteria in crop plants against pests and diseases. Crop Prot 20(1):1–11. https://doi.org/10.1016/S0261-2194(00)00056-9

Rodríguez J, Tonelli ML, Figueredo MS, Ibáñez F, Fabra A (2018) The lipopeptide surfactin triggers induced systemic resistance and priming state responses in Arachis hypogaea L. Eur J Plant Pathol 152(3):845–851. https://doi.org/10.1007/s10658-018-1524-6

Sansinenea E (2016) Regulatory issues in commercialization of *Bacillus thuringiensis*-based biopesticides. In: Agriculturally important microorganisms. Springer, Singapore, pp 69–80. https://doi.org/10.1007/978-981-10-2576-1_4

Saravanakumar K, Fan L, Fu K, Yu C, Wang M, Xia H, Sun J, Li Y, Chen J (2016) Cellulase from Trichoderma harzianum interacts with roots and triggers induced systemic resistance to foliar disease in maize. Sci Rep 6(1):35543. https://doi.org/10.1038/srep35543

Sarkar S, Gil JDB, Keeley J, Jansen K (2021) The use of pesticides in developing countries and their impact on health and the right to food. Policy Department for External Relations Directorate General for External Policies of the Union, Brussels. https://doi.org/10.2861/28995

Segarra G, Van der Ent S, Trillas I, Pieterse CMJ (2009) MYB72, a node of convergence in induced systemic resistance triggered by a fungal and a bacterial beneficial microbe. Plant Biol 11(1):90–96. https://doi.org/10.1111/j.1438-8677.2008.00162.x

Sharma A, Kumar V, Shahzad B, Tanveer M, Sidhu GPS, Handa N, Kohli SK, Yadav P, Bali AS, Parihar RD, Dar OI, Singh K, Jasrotia S, Bakshi P, Ramakrishnan M, Kumar S, Bhardwaj R, Thukral AK (2019a) Worldwide pesticide usage and its impacts on ecosystem. SN Appl Sci 1(11):1446. https://doi.org/10.1007/s42452-019-1485-1

Sharma S, Chen C, Navathe S, Chand R, Pandey SP (2019b) A halotolerant growth promoting rhizobacteria triggers induced systemic resistance in plants and defends against fungal infection. Sci Rep 9(1):4054. https://doi.org/10.1038/s41598-019-40930-x

Shine MB, Xiao X, Kachroo P, Kachroo A (2019) Signaling mechanisms underlying systemic acquired resistance to microbial pathogens. Plant Sci 279:81–86. https://doi.org/10.1016/j.plantsci.2018.01.001

Singh M, Srivastava M, Kumar A, Singh AK, Pandey KD (2019) Endophytic bacteria in plant disease management. In: Microbial endophytes: prospects for sustainable agriculture. Woodhead Publishing. https://doi.org/10.1016/B978-0-12-818734-0.00004-8

Skelly MJ, Furniss JJ, Grey HL et al (2019) Dynamic ubiquitination determines transcriptional activity of the plant immune coactivator NPR1. elife 8:1–26. https://doi.org/10.7554/eLife.47005

Song GC, Riu M, Ryu C-M (2019) Beyond the two compartments petri-dish: optimising growth promotion and induced resistance in cucumber exposed to gaseous bacterial volatiles in a miniature greenhouse system. Plant Methods 15(1):9. https://doi.org/10.1186/s13007-019-0395-y

Sun L, Wu X, Diao J, Zhang J (2023) Pathogenesis mechanisms of phytopathogen effectors. WIREs Mech Dis 15. https://doi.org/10.1002/wsbm.1592

Swapna AL (2019) Development of biofertilizers and its future perspective. J Pharm 2:327–332
Swapnil P, Meena M, Harish et al (2023) Plant-microbe interaction—recent advances in molecular and biochemical approaches, Agricultural aspects of microbiome leading to plant defence, vol 2. Academic Press
Thakur M, Sohal SB (2013) Role of elicitors in inducing resistance in plants against pathogen infection: a review. ISRN Biochem 2013:1–10. https://doi.org/10.1155/2013/762412
Torres-Calzada C, Tapia-Tussell R, Higuera-Ciapara I, Perez-Brito D (2013) Morphological, pathological and genetic diversity of *Colletotrichum* species responsible for anthracnose in papaya (*Carica papaya* L.). Eur J Plant Pathol 135(1):67–79. https://doi.org/10.1007/s10658-012-0065-7
Trda L, Boutrot F, Claverie J, Brule D, Dorey S, Poinssot B (2015) Perception of pathogenic or beneficial bacteria and their evasion of host immunity: pattern recognition receptors in the frontline. Front Plant Sci 6. https://doi.org/10.3389/fpls.2015.00219
Van Wees SC, Van der Ent S, Pieterse CM (2008) Plant immune responses triggered by beneficial microbes. Curr Opin Plant Biol 11(4):443–448. https://doi.org/10.1016/j.pbi.2008.05.005
Verma H, Kumar D, Kumar V, Kumari M, Singh SK, Sharma VK, Droby S, Santoyo G, White JF, Kumar A (2021) The potential application of endophytes in management of stress from drought and salinity in crop plants. Microorganisms 9(8):1729. https://doi.org/10.3390/microorganisms9081729
Vishwanathan K, Zienkiewicz K, Liu Y, Janz D, Feussner I, Polle A, Haney CH (2020) Ectomycorrhizal fungi induce systemic resistance against insects on a nonmycorrhizal plant in a CERK1-dependent manner. New Phytol 228(2):728–740. https://doi.org/10.1111/nph.16715
UN, DESA. United Nations Department of Economic and Social Affairs. World Population Prospects: The 2015 Revision, Key Findings Advance Tables; Working Paper No ESA/P/WP, United Nations Department of Economic and Social Affairs: New York, NY, USA, https://www.un.org/en/development/desa/publications/world-population-prospects-2015-revision.html (2015).
Wang B, Bi Y (2021) The role of signal production and transduction in induced resistance of harvested fruits and vegetables. Food Qual Saf 5. https://doi.org/10.1093/fqsafe/fyab011
Wang N, Liu M, Guo L, Yang X, Qiu D (2016) A novel protein elicitor (PeBA1) from Bacillus amyloliquefaciens NC6 induces systemic resistance in tobacco. Int J Biol Sci 12(6):757–767. https://doi.org/10.7150/ijbs.14333
Wang Z, Solanki MK, Kumar A, Solanki AC, Pang F, Ba ZX, Niu JQ, Ren ZX (2023) Promoting plant resilience against stress by engineering root micro environment with Streptomyces inoculants. Microbiol Res 277:127509. https://doi.org/10.1016/j.micres.2023.127509
Yacoub A, Haidar R, Mesguida O, Gerbore J, Hachicha M, Attard E, Guyoneaud R, Rey P (2023) Deciphering plant-induced responses toward Botrytis cinerea and Plasmopara viticola attacks in two grapevine cultivars colonized by the root biocontrol oomycete, Pythium oligandrum. J Fungi 9(5):511. https://doi.org/10.3390/jof9050511
Yang C, Fernando WGD (2021) Analysis of the oxidative burst and its relevant signaling pathways in leptosphaeria maculans—brassica napus pathosystem. Int J Mol Sci 22. https://doi.org/10.3390/ijms22094812
Yang P, Zhao Z, Fan J, Liang Y, Bernier MC, Gao Y, Zhao L, Opiyo SO, Xia Y (2023) Bacillus proteolyticus OSUB18 triggers induced systemic resistance against bacterial and fungal pathogens in Arabidopsis. Front Plant Sci 14. https://doi.org/10.3389/fpls.2023.1078100
Yu Y, Gui Y, Li Z, Jiang C, Guo J, Niu D (2022) Induced systemic resistance for improving plant immunity by beneficial microbes. Plants 11:386. https://doi.org/10.3390/plants11030386
Zhang W, Gao S, Zhou X, Chellappan P, Chen Z, Zhou X, Zhang X, Fromuth N, Coutino G, Coffey M, Jin H (2011) Bacteria-responsive microRNAs regulate plant innate immunity by modulating plant hormone networks. Plant Mol Biol 75(1–2):93–105. https://doi.org/10.1007/s11103-010-9710-8
Zheng F, Chen L, Zhang P et al (2020) Carbohydrate polymers exhibit great potential as effective elicitors in organic agriculture: a review. Carbohydr Polym 230:1–15. https://doi.org/10.1016/j.carbpol.2019.115637
Zipfel C (2008) Pattern-recognition receptors in plant innate immunity. Curr Opin Immunol 20(1):10–16. https://doi.org/10.1016/j.coi.2007.11.003

Impact of Environmental Factors on Microbial Association with Plant

3

Gema Cárdenas-Flores, Ana Y. Flores-Ramírez, Adriana G. Aranda-Vázquez, Martina A. Chacón-López, José V. Valencia-Zavala, Rosa I. Ortiz-Basurto, Ramsés R. González-Estrada, and Ulises M. López-García

Abstract

Plants and microorganisms are integral to ecosystems, engaging in diverse and complex interactions. These interactions are vital in plant-associated habitats such as the phyllosphere, endosphere, and especially the rhizosphere, which is crucial for plant health. Rhizosphere microbiome facilitates essential plant processes, including germination, water uptake, growth promotion, and hormone regulation. However, environmental factors like temperature, moisture, pH, salinity, and soil structure play a key role in shaping microbial communities and influencing plant–microbe interactions. Changes in soil microbial dynamics due to these factors impact nutrient cycling, plant growth, and stress tolerance. Understanding these complex plant–microbe dynamics is essential for developing climate-resilient agricultural practices and ensuring food security in the face of changing environmental conditions.

Keywords

Plant microbiome · Microbial communities · Plant–microbe interactions · Rhizosphere microbiome · Climate change · Environmental factors

---

G. Cárdenas-Flores · A. Y. Flores-Ramírez · A. G. Aranda-Vázquez · M. A. Chacón-López (✉) · J. V. Valencia-Zavala · R. I. Ortiz-Basurto · R. R. González-Estrada · U. M. López-García
Agrobiotechnology and Electrochemistry Laboratory-Integral Food Research Laboratory, National Technological Institute of Mexico/Technological Institute of Tepic, Tepic, Nayarit, Mexico
e-mail: gecardenasfl@ittepic.edu.mx; anyafloresra@ittepic.edu.mx; adguarandava@ittepic.edu.mx; mchacon@tepic.tecnm.mx; jovivalenciaza@ittepic.edu.mx; riobasurt@ittepic.edu.mx; ramgonzalez@ittepic.edu.mx; ulopez@ittepic.edu.mx

A. Kumar, M. K. Solanki (eds.), *Microbial Biocontrol Techniques*, Microorganisms for Sustainability 54,
https://doi.org/10.1007/978-981-97-8739-5_3

3.1 Introduction

Plants play a pivotal role in creating a myriad of ecological niches that foster the development and proliferation of diverse microorganisms. These niches, including the phyllosphere, endosphere, rhizosphere, and other parts like pollen and nectar, collectively form the plant microbiome (Liu et al. 2017). This microbiome, characterized by its high diversity, showcases varying bacterial community compositions across different plant-associated zones. For instance, the endosphere predominantly harbors Proteobacteria and Firmicutes, while the phyllosphere is rich in Proteobacteria, Actinomycetes, Bacteroidetes, and Firmicutes. Fungal communities in aerial and underground tissues are primarily composed of Ascomycota and Basidiomycota (Cregger et al. 2018; Trivedi et al. 2020). This symbiotic relationship between plants and microorganisms underscores the importance of understanding the plant microbiome's diversity. In the rhizosphere, microbial communities, similar to those found in the bulk soil, thrive. These communities include fungi, nematodes, protozoa, viruses, and bacteria, with the predominant groups being proteobacteria, acidobacteria, actinobacteria, bacteroidetes, and firmicutes. The composition of the rhizosphere, a dynamic and constantly fluctuating environment, can vary significantly depending on the plant species and soil type it interacts with (Bulgarelli et al. 2013; Dastogeer et al. 2020; Trivedi et al. 2020). The role of these microorganisms in the rhizosphere is of paramount importance, as they are instrumental in maintaining plant health. They facilitate germination, early establishment, water uptake, enhanced growth, hormonal regulation, and tolerance to biotic and abiotic stresses (Anand et al. 2023; Kumar et al. 2018). This role becomes even more pronounced in the face of our changing climatic conditions (Banerjee and van der Heijden 2023; Brevik et al. 2020). Understanding the role of the rhizosphere microbiome is crucial for enabling plants to adapt to diverse environmental conditions. Plant-associated microbial assemblages exhibit distinct physiological profiles, metabolic pathways, and temperature and humidity requirements (Trivedi et al. 2020). However, the impact of climate change on soil microbial dynamics, triggered by its sensitivity to alterations in nutrient content, pH, gas gradients, ultraviolet radiation, temperature, humidity, and CO_2 concentration, is a pressing concern (Gupta and Patil 2021; Tecon and Or 2017). Similarly, anthropogenic activities like agrochemical application and industrialization are significantly influencing these organisms, raising urgent questions about their long-term sustainability (Sharma et al. 2022). Plant-mediated responses, such as chemical signaling, play a significant role in these dynamics, particularly under environmental stress. For example, drought conditions induce the secretion of glycerol-3-phosphate in roots, promoting actinobacteria growth, which helps mitigate drought stress (Xu et al. 2018). This recruitment of specific microbial species aids plants in adapting to adverse conditions (Liu et al. 2020; Trivedi et al. 2022). Certain microorganisms significantly enhance crop resilience by improving nutrient availability, promoting growth, and increasing tolerance to environmental stressors. Understanding soil microbiota and plant interactions is essential for developing innovative practices supporting sustainable agriculture amidst climate variability (Altieri et al. 2015;

Singh et al. 2018). This chapter delves into the consequences of various environmental factors on plant–microbe relationships. It meticulously analyzes the effects of abiotic stressors like nutrient availability, soil pH, temperature variations, humidity, soil salinity, and other disturbances on these dynamics. Comprehending these interactions can provide crucial insights into how plants and microbes adapt to fluctuating environmental conditions, helping to develop more resilient and ecological agronomic practices.

3.2 Plant–Microbe Interactions in Agriculture

Plants provide a broad spectrum of ecological habitats, facilitating the growth and proliferation of various microorganisms. These niches include the phyllosphere (the surface of leaves), endosphere (inside leaves and roots), rhizosphere (soil region near roots), and other parts, such as pollen and nectar. All microorganisms present in these sites collectively form the plant microbiome (Liu et al. 2017).

The plant microbiome is highly diverse, with the composition of bacterial communities varying according to the specific plant-associated zone. For example, the main microorganisms in the endosphere are Proteobacteria and Firmicutes, with Bacteroidetes being present to a lesser extent. In contrast, the phyllosphere is enriched with bacteria from the phyla Proteobacteria and, to a lesser degree, Actinomycetes, Bacteroidetes, and Firmicutes. In terms of fungal species associated with plants, both in aerial and underground tissues, the main communities reported belong to the phyla Ascomycota and Basidiomycota (Cregger et al. 2018; Trivedi et al. 2020).

In the rhizosphere, microbial communities are similar to those found in the bulk soil environment and include a wide diversity of microbes such as fungi, nematodes, protozoa, viruses, and bacteria. The phyla Proteobacteria, Acidobacteria, Actinobacteria, Bacteroidetes, and Firmicutes are predominant among the bacteria. However, the composition of the rhizosphere can vary depending on plant species and soil type (Bulgarelli et al. 2013; Dastogeer et al. 2020; Trivedi et al. 2020). Rhizosphere microorganisms are widely recognized for their essential role in maintaining plant health. The functions of rhizosphere microbial communities include facilitating various plant processes such as germination and early establishment, water absorption, growth promotion, hormonal regulation, and tolerance to biotic and abiotic stresses (Wang et al. 2023; Pathak et al. 2022). The latter is particularly relevant in the global context of continuously changing climatic conditions (Banerjee and van der Heijden 2023; Brevik et al. 2020).

3.3 Climate Change Impacts on Soil Microbiota and Agriculture

In the face of a continuously changing climate, it is crucial to understand the role of the rhizosphere microbiome in enabling plants to adapt to diverse environmental conditions. Plant-associated microbial assemblages, with their distinct physiological profiles, metabolic pathways, and temperature and humidity requirements, are the key to plant survival (Trivedi et al. 2020). Because of climate change, variations in soil microbial dynamics may occur due to their sensitivity to changes, such as nutrient content in the soil, pH levels, gas gradients caused by soil compaction, ultraviolet radiation, temperature variations, humidity, and CO_2 concentration (Gupta and Patil 2021; Tecon and Or 2017). Moreover, anthropogenic activities such as the application of agrochemicals and industrialization can significantly impact these organisms (Sharma et al. 2022).

Furthermore, their dynamics are not only dictated by climatic variables but are also influenced by plant-mediated responses. Chemical signaling predominates in the plant–microbiome interface, facilitated by specific exudation events triggered, particularly under environmental stressors. For instance, when plants are subjected to drought conditions, the secretion of glycerol-3-phosphate is induced in the roots, favoring the growth of actinobacteria. In turn, these actinobacteria help alleviate drought stress's adverse impacts on plant development (Xu et al. 2018).

Through these processes, plants recruit certain microbial species into their rhizosphere to aid in their ability to acclimate to unfavorable environmental circumstances (Liu et al. 2020; Trivedi et al. 2022). Certain microorganisms can significantly contribute to crop resilience by enhancing nutrient availability, promoting plant growth, and increasing the tolerance to environmental stressors. Understanding the interactions between the soil microbiota and plants is essential for developing innovative practices supporting sustainable crop production in the face of climate variability (Altieri et al. 2015; Singh et al. 2018).

The chapter explores the potential consequences of diverse environmental factors on plant–microbe relationships. It scrutinizes the effects of abiotic stressors, including nutrient availability, soil pH, temperature variations, humidity levels, soil salinity, and other environmental disturbances, on the dynamics of these relationships. By comprehending these interactions, we can gain insights into how plants and microbes adapt to fluctuating environmental conditions, making significant strides toward developing more resilient and sustainable agricultural systems.

3.4 Environmental Factors Impacting Plant–Microbe Interactions

3.4.1 Temperature

Temperature is crucial in plant health, microbial dynamics, and their interactions (Sharma et al. 2022). Plants can be susceptible to drastic temperature changes. High temperatures are particularly stressful as they can alter membrane permeability, seed germination, and photosynthetic rates (Xu et al. 2014). In addition, elevated temperatures have often been reported to directly affect the ability of plants to defend themselves against pathogen attacks. Specifically, high temperatures suppress effector-triggered immunity (ETI), a crucial defense mechanism where plant resistance proteins recognize pathogen-derived effectors and trigger a robust immune response (Velásquez et al. 2018; Desaint et al. 2021).

Temperature changes also strongly affect microorganisms, leading to shifts in the composition of the rhizosphere microbial community. Studies have shown that high temperatures hurt Gram-positive bacteria while positively affecting Gram-negative bacteria. Additionally, higher rates of root colonization by endomycorrhizal fungi have been observed under these conditions (Jerbi et al. 2021). These alterations have essential implications for interactions with host plants, as temperature changes affect the proportions of different types of microorganisms, including pathogens, plant growth promoters, and extremophiles (Bai et al. 2022; Dastogeer et al. 2020; Zhang et al. 2023).

There are reports on how high temperatures affect the pathogenicity of soil microorganisms (Huot et al. 2017; Velásquez et al. 2018). These findings have significant implications for agriculture and microbiology. For example, high temperatures can increase virulence factors in the soft rot bacterium *Pectobacterium atrosepticum*, leading to enhanced production of plant cell wall-degrading enzymes and faster disease development (Baron et al. 2001). In contrast, they can also be negatively affected. For instance, heat stress can affect *Agrobacterium* virulence by inhibiting type IV secretion-associated pilus formation and virulence (vir) gene expression (Hasegawa et al. 2005). Furthermore, it has been revealed that both short-term and long-term increases in temperature increase microbial respiration rates, leading to changes in the proportion of certain nutrients, such as carbon and nitrogen, in the soil. Consequently, this increases nutrient competition between plants and microbes (Zhu et al. 2023; Li et al. 2024). Temperature fluctuations can stimulate adaptive responses in certain microorganisms to counteract the negative impact of such stressors, including morphological changes or the secretion of substances such as secondary metabolites, carbohydrates, polyols, etc. For instance, *Klebsiella* sp., a facultative thermophile, produces siderophores, indole-3-acetic acid, and ACC deaminase when grown at 60 °C (Mukherjee et al. 2020). In some cases, these adaptive responses of plant-associated microorganisms contribute to improving tolerance to high-temperature stress in plant species.

According to a report by Waqas et al. (2015), *Paecilomyces formosus*, an endophytic fungus, significantly contributes to plants' ability to tolerate heat stress by stimulating the production of secondary metabolites and phytohormones. Similarly, Khan et al. (2020) found that *B. cereus* SA1, a thermotolerant bacterium, synthesizes compounds such as hormones like gibberellin, indoleacetic acid, and salicylic acid, which increase the chlorophyll concentration and biomass in soybean under heat stress. Furthermore, *Glycine max* showed tolerance to stress caused by combined conditions of high temperature (45 °C), salinity and drought when a combination of *Paecilomyces formosus* and *Penicillium funiculosum* was applied. This inoculation enhanced plant development and photosynthetic efficiency, facilitated the absorption of micronutrients, minimized lipid peroxidation, and augmented the presence of antioxidants.

3.4.2 Humidity

Water is an essential requirement of life. Unfortunately, one of climate change's most identifiable impacts is precipitation fluctuation. This phenomenon alters atmospheric humidity levels and can cause floods and droughts (Gupta et al. 2022). Fluctuations in humidity levels can significantly impact on plants and the composition of the microbiome community, as well as on the interactions that may exist between them (Taktek et al. 2017). It is well known that microorganisms in the phyllosphere are susceptible to environmental humidity, as they depend on traces of available water and nutrients. Epiphytic microorganisms, which inhabit leaves' surfaces, thrive in high humidity conditions. In such circumstances, they have an increased amount of water available, which leads to enhanced bacterial motility or more significant proliferation and abundance of fungi and bacteria (Talley et al. 2002; Aung et al. 2018). For instance, certain bacteria, such as *Pseudomonas syringae*, require high humidity conditions to survive and successfully infect their hosts (Monier and Lindow 2005). In the case of fungi, spores require water to break dormancy (Serrano et al. 2014). High humidity is among the most critical factors influencing disease development among plant–microbe relationships. In addition to its effects on microbial proliferation, high humidity forces plants to open their stomata to regulate cellular homeostasis (Ruggiero et al. 2017). This allows phyllosphere microbes, particularly pathogens, to invade and infect plants successfully (Aung et al. 2018).

Flooding significantly affects the rhizosphere microbiome of plants. An anoxic environment is created as oxygen is depleted, and the soil becomes saturated with water. This leads to changes in the composition of the rhizosphere, favoring facultative and strict anaerobic microbes over aerobic organisms (Francioli et al. 2021). In addition, the microbiota can be indirectly affected by the impact of flooding on the plants to which they are associated. Flood stress can cause changes in root exudates, altering the rhizosphere's biochemical composition and metabolism of its microorganisms. For example, some authors have reported that plants under anoxic conditions caused by flooding reduce photosynthesis, limiting the amount of carbon exuded by the roots, which generally accounts for 25% of the carbon allocated below ground (Jones et al. 2009; Hamonts et al. 2013). Furthermore, under flood stress and reduced photosynthesis, plants induce glycolysis and fermentation, producing phytotoxic substances, such as ethanol and acetaldehyde, which are released through the roots and can affect the rhizosphere microbiome (Blom 1999; Bailey-Serres and Voesenek 2008).

In contrast, water scarcity caused by recurrent droughts profoundly affects the homeostasis of plant species' microbial diversity and soil properties (Manzanera 2021). The rhizosphere microbiome undergoes specific changes in the proportion of its bacterial communities under drought conditions, which are highly conserved among different plant species and soil types (Santos-Medellín et al. 2017). There is an enrichment of gram-positive bacteria, especially actinobacteria, and a depletion of gram-negative bacteria. This shift is mainly due to the structure of the cell walls of Gram-positive bacteria and their ability to form spores (Schimel 2018). In addition, drought-tolerant microbes use several direct and indirect mechanisms to manage water deficit, including the production of osmolytes such as sugars (trehalose, fructose), polyols (glycerol, mannitol), amino acids (proline, glycine), and the synthesis of hormones (abscisic acid, indoleacetic acid, and cytokinins) (Manzanera 2021).

Some authors suggest that the enrichment of certain microbial types under drought conditions is not only due to the intrinsic capabilities of the microbes but also to interactions with the host plant, which can excrete both positive and negative microbial growth regulators (Gupta et al. 2022; Xu et al. 2018). Similarly, the rhizosphere microbiome promotes plant adaptation and survival. The same metabolites produced by microbes to ensure survival also benefit plants by improving root-to-shoot ratios, increasing biomass, absorption of water and nutrients, and drought resistance (Shree et al. 2022; Tiwari et al. 2024).

3.4.3 Soil pH

One environmental factor that strongly influences the bioavailability of nutrients to plant roots and the microbial community is soil pH. Soil pH can be affected by water source, soil structure, texture, use, and composition (Abdul Rahman et al. 2021). Soil acidification can occur naturally due to rain and cation leaching, while alkaline soils result from a variety of natural processes. These include the content of certain minerals such as calcium carbonate and magnesium carbonate, which release alkaline ions into the soil. Changes in soil pH can also result from the interference of plant and microbial species inhabiting the soil. In the rhizosphere, the local soil pH can be acidified or alkalinized by up to two units through root exudates in response to the concentration of anions or cations in the surrounding soil (Hinsinger et al. 2003; Philippot et al. 2024).

Regarding the participation of the rhizosphere microbiome, microbial activities associated with the biogeochemical cycle can impact soil pH through proton and hydroxyl ion generation (Philippot et al. 2024). For instance, carbonic acid formed by microbial respiration through the carbon cycle is a source of protons that acidify the soil (Sánchez-Cañete et al. 2018). Another soil acidification process occurs through the nitrification of ammonia to nitrate by bacteria during the nitrogen cycle (Norton and Ouyang 2019). Furthermore, both bacteria and fungi can acidify the soil by producing and secreting organic acids, such as oxalate or citrate (Philippot et al. 2024). Similarly, soils can become acidified when bacteria and fungi oxidize inorganic sulfur compounds (Koch and Dahl 2018; Linder 2018). On the other hand, soil alkalinization is attributed to cyanobacteria due to their requirement for protons during carbon fixation (Mergelov et al. 2018). Soil alkalinization is also possible through fungal activity (Palmieri et al. 2020). Both bacteria and fungi possess the ability to actively modify soil pH, which consequently influences the composition of its soil communities (Dastogeer et al. 2020; Saad et al. 2020). This effect is more pronounced in the bacterial community, as it is more sensitive to changes in pH compared to fungi, explaining why the proportion of fungi to bacteria is higher in acidic soils (Lladó et al. 2017; Saad et al. 2020). Some microbes isolated from the rhizosphere include *Acidithiobacillus*, *Pseudomonas*, and *Flavobacterium*. These have been classified as plant growth-promoting bacteria because, under acidic soil conditions, they enhance plant iron absorption through siderophores by converting it into Fe^{2+} ions (Xu et al. 2018; Vansuyt et al. 2007). Similarly, plant growth-promoting microorganisms have been reported in alkaline soils, including *Curtobacterium*, *Paenibacillus*, and *Sphingomonas* (Mukhtar et al. 2019). Thus, these microbial communities not only influence soil pH but also, through their adaptive responses to proliferate in alkaline or acidic environments, can produce substances that benefit the growth of their associated plants.

3.4.4 Soil Nutrients

One of the primary causes and consequences of plant–microbe interaction in the rhizosphere is the exchange and acquisition of nutrients. Some nutrients are essential for plant growth and development, such as phosphorus and nitrogen. However, due to the physicochemical composition of the soil, as well as challenging environmental conditions, plants may need more access to these nutrients. One of the main pathways for nutrient absorption by plants is through the roots, which are absorbed through the epidermal cells via transporter channels in the plasma membrane. These epidermal cells can store, transport, or assimilate these nutrients. Another crucial pathway for nutrient absorption by plants is dependent on microbes. Some microorganisms in the rhizosphere form mutualistic relationships with plants as they enhance nutrient absorption by plants. In turn, plants release carbon derived from photosynthesis through root exudates, and this carbon promotes the growth of many microbial species near the root system (Oldroyd and Leyser 2020; Zhao et al. 2024).

For nitrogen (N), plants can incorporate nitrogen from the environment from sources such as nitrate and, to a lesser extent, nitrite and other organic nitrogen sources (Gent and Forde 2017). In the epidermal cells of their roots, plants can detect nitrogen sources as nitrate via intracellular nitrate sensors such as NLP7 and acquire it via specific transporters such as NPF6.3/NRT1.1, which regulate nitrate uptake depending on the external concentration (Ye et al. 2019; Zhao et al. 2024). Regarding microbial-mediated absorption, most of the fixed nitrogen is obtained through symbiosis between plants and bacteria from the families Rhizobiaceae and Actinorrhizae. Rhizobia establish a symbiotic relationship primarily with legumes, which, upon recognition, form nodules in the root cells that host the rhizobia and become cells called bacteroids. These cells fix atmospheric N_2 inside these nodules, transferring nitrogen to the plant as NH_4^+ (Udvardi and Poole 2013; Dellagi et al. 2020). Nitrogen fixation by actinobacteria can occur under symbiotic or non-symbiotic conditions. Species such as *Frankia*, which also form nodules, are less host-specific than rhizobia and associate with a wide variety of actinorhizal plants, including members of the Rosales and Cucurbitales. Another group of bacteria that can associate with plants and provide NH_4^+ from N_2 without forming specialized organs are cyanobacteria, mainly of the genus *Nostoc*. These bacteria perform nitrogen fixation through the differentiation of cells called heterocysts (Santi et al. 2013; Dellagi et al. 2020). Plants also form associations with fungi for the acquisition of nitrogen, of which the most abundant are the arbuscular mycorrhizal fungi (Gianinazzi et al. 2010).

Plants require various macronutrients for growth and development, and phosphorus (P) is one of them. P is found in the soil in both organic and inorganic forms. Organic P comes from biological tissues and is present in several biomolecules, such as nucleotides, sugar phosphates, phospholipids, and phosphoproteins (Billah et al. 2019). Inorganic phosphate, on the other hand, exists in different forms and amounts in soils, typically as highly stable and insoluble compounds such as apatite or variscite and as secondary minerals such as aluminum phosphate or calcium phosphate (Maltais-Landry et al. 2014). Despite the prevalence of phosphorus in soil, it is often not easily accessible to plants, because it diffuses very slowly, and plants mainly use inorganic P ions such as $H_2PO_4^-$ and HPO_4^{2-}, which react rapidly with other mineral oxides in the soil to form insoluble compounds (Zhao et al. 2024).

Plants acquire P from the soil via specific transporters responsible for capturing, translocating, and mobilizing this macronutrient within plant cells. These transporters, known as PHTs, belong to six families. The PHT1 transporters, situated at the cytoplasmic membrane, are expressed in root hairs and cortical cells of root tissues and take up inorganic phosphorus via a high-affinity Pi/H^+ symport mechanism (Daram et al. 1998; Guo et al. 2008). Other families of transporters are distributed in different organelles within plant cells, such as mitochondria, chloroplasts, and vacuoles (Versaw and Garcia 2017).

Despite the natural scarcity of available P, plants can also obtain this nutrient through the contribution of phosphate-solubilizing microorganisms (PSM) in the rhizosphere. These PSMs include mainly bacteria and fungi but also some actinomycetes and cyanobacteria. The most common phosphate-solubilizing bacteria include *Pseudomonas*, *Escherichia*, and *Bacillus*, while phosphate-solubilizing fungi include *Aspergillus*, *Rhizopus*, and *Penicillium* species. In addition, some arbuscular mycorrhizal fungi (AMFs), such as *Glomus aggregatum* and *Glomus mosseae*, also possess this ability (Zai et al. 2007; Jiang et al. 2020). The primary mechanism by which these microorganisms solubilize insoluble phosphate is through the secretion of organic or inorganic acids. These acids can chelate metal ions such as Al^{3+} or Ca^{2+}, thereby releasing P (Baumann et al. 2018). Although phosphate-solubilizing bacteria are more abundant in the rhizosphere, fungi have a greater capacity to solubilize P, as they can produce ten times more organic acids than their bacterial counterparts (Jiang et al. 2020).

Some phenomena can alter the natural availability of nutrients in the soil, both due to natural and anthropogenic causes. For example, global warming and increased CO_2 concentration raise plants' nitrogen requirements, which in long term lead to denitrification and causes stress in these organisms (Zaehle et al. 2015). Additionally, since P is often scarcely available in soils, natural phenomena such as excessive rainfall and flooding can leach P from the soil, resulting in nutrient deficiency conditions for plants (Jiang et al. 2020).

In situations of phosphorus and nitrogen limitation, plants have developed adaptive strategies such as changes in root morphology and the secretion of specific substances through the roots to enhance the uptake of N and P from the environment. This can occur directly or indirectly by recruiting microorganisms in the rhizosphere that fulfill this function (Zhao et al. 2024). For example, it has been reported that rice plants under N and P deficiency increase the concentration of metabolites such as organic acids in the roots, such as shikimic acid and gamma-aminobutyric acid (Tawaraya et al. 2018). These acids can act directly by improving the solubility of P or indirectly by lowering the pH of the environment and enriching specific bacterial and fungal taxa that, in turn, can solubilize P (Long and Yao 2020). Furthermore, the exudation of specific metabolites such as GABA, betaine, and glutarate has also been reported (Tawaraya et al. 2018). Rhizosphere microorganisms can use these metabolites to enhance their proliferation. For example, high concentrations of GABA exuded by roots are associated with increased bacteria in the Frankineae group, which are nitrogen-fixing bacteria (Badri et al. 2013).

3.4.5 Soil Salinity

Soil salinization results from the accumulation of salts in the soil owing to various factors. These include the deposition of high-salinity minerals, capillary upwelling of saline groundwater, unsustainable farming practices, and manufacturing activities (Eswar et al. 2021). There are two main types of saltwater intrusion: natural and human. The former is referred to as primary salinization, whereas the latter is known as secondary salinization (Zhang et al. 2023). Primary salinization is related to the regional weather conditions and the properties of the soil and groundwater. Lack of rainfall or rising groundwater levels can also increase salinity (Daliakopoulos et al. 2016). Human-caused or secondary salinization occurs when people misuse soil inappropriately. When saline irrigation is coupled with inadequate drainage conditions, it can lead to salt buildup on the soil surface. In future climate change scenarios, increasing temperatures and fast urbanization are expected to rise seawater intrusion in coastal areas (Eslami et al. 2021).

Microorganisms have the ability to modify their metabolism to adapt to salt stress within a certain range of soil salinity (Verma et al. 2021; Kesawat et al. 2023). However, the response of microbial diversity in soils under different salinity gradients is specific to the habitat or context under which it occurs (Zhang et al. 2023). Owing to advances in omics sciences, several studies have shown that salinity significantly affects the composition of microbial communities, more than factors such as temperature, pH, or other physicochemical factors (Bajay et al. 2023). Fungi are generally believed to be able to cope with osmotic stress more than bacteria (Karhu et al. 2014). This phenomenon of changes in microbial composition can negatively affect microorganism–plant interactions, as various biogeochemical processes, such as the soil nitrogen cycle, in which organic nitrogen is converted to inorganic nitrogen, can be affected (Kuypers et al. 2018). As reported by Bai et al. (2022), soil salinization can cause adverse effects on nitrogen mineralization because microbial activity is reduced with increasing salinity; however, according to a meta-analysis by Zhou et al. (2017), nitrogen mineralization may increase because the same saline increase in the soil may promote an increase in some bacterial communities called halophytes and halotolerant, which have evolved diverse adaptations to survive and thrive under high-salinity conditions, allowing them to play crucial roles in the decomposition of organic matter and solubilization of nutrients, thus contributing to the sustainability of these environments (Oren 2002). Halophilic and halotolerant microorganisms have evolved various adaptations to survive in high-salinity conditions, including the accumulation of compatible solutes such as amino acids (proline, glycine-betaine) and sugars (trehalose) in their cells to balance osmotic pressure. Additionally, these organisms modify their cell membrane structures and proteins to function efficiently in high salt concentrations (Gunde-Cimerman et al. 2018). In addition to decomposing organic matter, halophilic and halotolerant microorganisms solubilize nutrients such as inorganic phosphorus through organic acids and enzymes such as phosphatases and oxidoreductases (Rath and Rousk 2015), increasing their availability to plants in saline soils, where high salt concentrations limit nutrient availability (Mapelli et al. 2013).

The ability of halophilic and halotolerant microorganisms to adapt and perform critical functions in saline soils helps maintain soil health and supports plant growth in these adverse environments. However, complementary strategies must be implemented to mitigate the adverse effects of soil salinity. Some strategies include selecting salt-tolerant crops, using organic amendments, and appropriate irrigation practices. Behera et al. (2022) highlight the importance of using advanced breeding techniques to develop salinity-tolerant crops. Meanwhile, Lashari et al. (2018) show that organic amendments such as biochar and compost improve soil structure and fertility under saline conditions. In addition to the findings of Nagaz et al. (2012), the utilization of drip irrigation method was highlighted as an effective approach in minimizing the accumulation of salt in the root zone, thus contributing to the overall health and productivity of plants.

3.4.6 Solar Radiation

Solar radiation comprises wavelengths ranging from 150 to 4000 nm, encompassing the visible, infrared, and ultraviolet regions (Bhatia 2014; Vanhaelewyn et al. 2020). Among these, ultraviolet radiation is significant for its biological impact on organisms, which directly and indirectly influencing plant–pathogen and plant–rhizosphere relationships (Vanhaelewyn et al. 2020). Ultraviolet radiation is divided into UV-A, with wavelengths of 315–400 nm; UV-B, with wavelengths of 280–315 nm; and UV-C, with wavelengths of 200–280 nm (Carvalho and Castillo 2018; Robson et al. 2015). The effects of these ultraviolet radiation bands vary on plants and the microorganisms that inhabit them. UV-A rays affect plant immunity and stimulate both biomass and leaf size. Additionally, they can alter the composition of polyphenols, although to a lesser extent than UV-B rays (Verdaguer et al. 2017). UV-B rays induce immune responses in plants. UV-C rays can trigger defense responses and reduce symptoms caused by pathogenic bacteria, which participate in hormonal pathways (Vanhaelewyn et al. 2016; Aarrouf and Urban 2020). Therefore, UV-A and UV-B radiations affect plant morphogenesis and phototropism, while UV-B and UV-C radiations increase secondary metabolite production (Vanhaelewyn et al. 2020). Ultraviolet radiation with a wavelength of less than 350 nm has direct and indirect adverse effects on plant pathogens. Direct effects include DNA damage, protein polymerization, enzyme inactivation, and increased membrane permeability in bacteria. Indirect effects are related to plant defense responses, the accumulation of oxidative species, and the production of secondary metabolites, such as polyphenols (Vanhaelewyn et al. 2020). The effects of UV radiation on plants depend on the stage of development and the intensity of the radiation (Moreira-Rodríguez et al. 2017). Therefore, under normal conditions, solar radiation promotes photosynthesis in plants, which increases oxygen levels in the rhizosphere and the availability of root exudates (Bulgarelli et al. 2013; Sasse et al. 2018; Timmis and Ramos 2021). Through photosynthesis, exudates such as carbohydrates, amino acids, and organic acids are released into the soil, serving as a source of nutrients that stimulate growth and activity of certain beneficial bacteria in the rhizosphere, such as aerobic bacteria, which are crucial for processes such as nitrification and the decomposition of organic matter that improve soil health and plant resistance to pathogens (Bulgarelli et al. 2013; Sasse et al. 2018; Santoyo et al. 2021). Nonetheless, increased levels of solar radiation can have harmful effects on plants, including the production of reactive oxygen species that can damage DNA, proteins, and lipids in cell membranes (Akashi et al. 2008; Vanhaelewyn et al. 2020). To counter the stress, plants exhibit changes in leaf reflectance and orientation to reduce light absorption (Gibson 1996). Another mechanism for dealing with excess radiation is the presence of epicuticular waxes on the surface of leaves (Holmes and Keiller 2002). Additionally, plants synthesize polyphenols that act as photoprotectors, absorbing ultraviolet rays (Caldwell et al. 1998; Julkunen-Tiitto et al. 2015). Phenolic compounds, such as ascorbic acid, carotenoids, and glucosinolates, are UV-induced plant chemicals (Darré et al. 2022; Vanhaelewyn et al. 2020). Other UV-induced effects include photo repair and photoreactivation via UV-B, while UV-A absorption allows photolyase CPD and 6-4PP monoribozymes to join UV-B-induced dimers and repair DNA lesions, eliminating CPD and 6-4PP (Bray and West 2005; Vanhaelewyn et al. 2020). Similarly, fungi and bacteria also possess the ability to repair their DNA using photolyase activity and UV-A radiation.

The impact of radiation on the development and metabolism of fungi varies. Although not phototrophic, they have evolved mechanisms that enable them to sense and respond to ultraviolet radiation (Vanhaelewyn et al. 2020). UV radiation affects the growth of hyphae and the production of toxins, as well as having different effects on mushroom conidia, such as promoting conidiation by UV-A rays and reducing the germination of conidia by sublethal doses of UV-A and UV-B (Braga et al. 2015; Thind and Schilder 2018). One way that fungi and their spores protect themselves from UV radiation is through synthesizing pigment-causing compounds such as carotenoids and melanin (1,8-DHN or DOPA-melanin), which absorb a broad range of wavelengths.

In addition to these, fungi also produce other pigments, such as microsporines and amino acids like microsporines, which are also produced by bacteria such as cyanobacteria to absorb UV radiation, provide solar protection, and serve as antioxidant molecules. Microsporines and amino acids are considered specialized molecules that have a significant role in plant interactions, with increased production under UV-B radiation (Vanhaelewyn et al. 2020). Furthermore, some bacteria in the phylum, such as cyanobacteria, are tolerant to UV-B radiation and produce both enzymatic (such as catalase, superoxide dismutase, glutathione, and ascorbate peroxidase) and no enzymatic (such as carotenoids, α-tocopherol, ascorbic acid, and glutathione) antioxidants to avoid damage from UV radiation by producing photoprotective components that also counteract the synthesis of ROS (Rastogi et al. 2014). Other bacteria of interest to the phyllosphere are the actinobacteria, such as *Streptomyces*, *Saccharomonospora*, and *Nocardioides*, which can grow under extreme conditions of radiation and enhance plant development (Yandigeri et al. 2012; Nafis et al. 2019). UV radiation can significantly impact bacterial survival and growth, but the adverse effects depend on bacteria's various strategies to confront UV radiation stress (Rastogi et al. 2014).

This is because UV radiation can stimulate or inhibit bacterial activities, causing DNA damage in bacteria, reducing their viability, and altering the microbial composition on plant surfaces and in soil (Ruiz-González et al. 2013). Furthermore, the plant–microbe interaction is affected by increased exposure to UV radiation, which causes damaging effects on the biomolecules, and the induction of phytohormones and metabolites (Vanhaelewyn et al. 2020). Additionally, fungi, such as AMF, form a mutually beneficial relationship with other plants and assist in the uptake of nutrients that are indirectly influenced by UV-B radiation when the shoots of the host plants are exposed (Wahid et al. 2020). In this manner, solar radiation has a dual effect on bacterial activity in the soil and plants. While ultraviolet radiation can reduce bacterial viability through genetic damage, solar light promotes microbial activity by improving rhizosphere conditions and increasing root exudates, which serve as nutrients for bacteria. Maintaining the balance between the inhibitory and stimulatory effects is essential to ensure soil health and plant ecosystem productivity.

3.5 Conclusions and Future Perspectives

In changing environmental conditions, plants and their associated microbiota engage in a complex and specific communication to enhance adaptability and resilience. Plants recruit beneficial microorganisms into their rhizosphere by exuding certain metabolites that help counteract the adverse effects of biotic and abiotic stressors. These microorganisms, in turn, benefit from these exudates by finding improved growth conditions or utilizing them as carbon sources. To better understand these intricate plant–microbe interactions, it is crucial to identify the specific metabolites exuded by plants in response to environmental stressors such as drought, flooding, and extreme temperature fluctuations, as well as the combinations of these stresses. Research should focus on correlating these exudates with changes in the composition of the microbiome.

Identifying and understanding these adaptive plant–microbe relationships are essential for integrating this knowledge into agricultural practices. Such integration can help develop strategies to predict and manage the impacts of challenging conditions like climate change. Sustainable agricultural practices that leverage these insights will enhance crop resilience, ensure food security, and promote environmental sustainability. Continued research in this field will be vital for advancing our ability to address the complexities of climate change and its impact on agricultural systems.

References

Aarrouf J, Urban L (2020) Flashes of UV-C light: an innovative method for stimulating plant defences. PLoS One 15(7):e0235918. https://doi.org/10.1371/journal.pone.0235918

Abdul Rahman NSN, Abdul Hamid NW, Nadarajah K (2021) Effects of abiotic stress on soil microbiome. Int J Mol Sci 22(16):9036. https://doi.org/10.3390/ijms22169036

Akashi K, Yoshimura K, Nanasato Y, Takahara K, Munekage Y, Yokota A (2008) Wild plant resources for studying molecular mechanisms of drought/strong light stress tolerance. Plant Biotechnol 25(3):257–263. https://doi.org/10.5511/plantbiotechnology.25.257

Altieri MA, Nicholls CI, Henao A, Lana MA (2015) Agroecology and the design of climate change-resilient farming systems. Agron Sustain Dev 35(3):869–890

Anand U, Pal T, Yadav N, Singh VK, Tripathi V, Choudhary KK, Shukla AK, Sunita K, Kumar A, Bontempi E, Ma Y (2023) Current scenario and future prospects of endophytic microbes: promising candidates for abiotic and biotic stress management for agricultural and environmental sustainability. Microb Ecol 86:1455–1486. https://doi.org/10.1007/s00248-023-02190-1

Aung K, Jiang Y, He SY (2018) The role of water in plant–microbe interactions. Plant J 93(4):771–780. https://doi.org/10.1111/tpj.13795

Badri DV, Chaparro JM, Zhang R, Shen Q, Vivanco JM (2013) Application of natural blends of phytochemicals derived from the root exudates of arabidopsis to the soil reveal that phenolic-related compounds predominantly modulate the soil microbiome. J Biol Chem 288(7):4502–4512. https://doi.org/10.1074/jbc.M112.433300

Bai R, Wang J, Li N (2022) Climate change increases the suitable area and suitability degree of rubber tree powdery mildew in China. Ind Crop Prod 189:115888. https://doi.org/10.1016/j.indcrop.2022.115888

Bailey-Serres J, Voesenek LACJ (2008) Flooding stress: acclimations and genetic diversity. Annu Rev Plant Biol 59(1):313–339. https://doi.org/10.1146/annurev.arplant.59.032607.092752

Bajay SK, Ferreira RCU, Pimenta RJG, Mancini M, Aono AH, Niederauer GF, Horta MAC, De Souza AP (2023) Transcriptomics in agricultural sciences: capturing changes in gene regulation during abiotic or biotic stress. In: Transcriptome profiling. Elsevier, pp 257–283. https://doi.org/10.1016/B978-0-323-91810-7.00004-2

Banerjee S, van der Heijden MGA (2023) Soil microbiomes and one health. Nat Rev Microbiol 21(1):6–20. https://doi.org/10.1038/s41579-022-00779-w

Baron C, Domke N, Beinhofer M, Hapfelmeier S (2001) Elevated temperature differentially affects virulence, VirB protein accumulation, and T-pilus formation in different Agrobacterium tumefaciens and Agrobacterium vitis strains. J Bacteriol 183(23):6852–6861. https://doi.org/10.1128/JB.183.23.6852-6861.2001

Baumann K, Jung P, Samolov E, Lehnert LW, Büdel B, Karsten U, Bendix J, Achilles S, Schermer M, Matus F, Oses R, Osses P, Morshedizad M, Oehlschläger C, Hu Y, Klysubun W, Leinweber P (2018) Biological soil crusts along a climatic gradient in Chile: richness and imprints of phototrophic microorganisms in phosphorus biogeochemical cycling. Soil Biol Biochem 127:286–300. https://doi.org/10.1016/j.soilbio.2018.09.035

Behera TK, Krishna R, Ansari WA, Aamir M, Kumar P, Kashyap SP, Pandey S, Kole C (2022) Approaches involved in the vegetable crops salt stress tolerance improvement: present status and way ahead. Front Plant Sci 12:787292. https://doi.org/10.3389/fpls.2021.787292

Bhatia SC (2014) Solar radiations. In: Advanced renewable energy systems. Elsevier, pp 32–67. https://doi.org/10.1016/B978-1-78242-269-3.50002-4

Billah M, Khan M, Bano A, Hassan TU, Munir A, Gurmani AR (2019) Phosphorus and phosphate solubilizing bacteria: keys for sustainable agriculture. Geomicrobiol J 36(10):904–916. https://doi.org/10.1080/01490451.2019.1654043

Blom C (1999) Adaptations to flooding stress: from plant community to molecule. Plant Biol 1(3):261–273. https://doi.org/10.1055/s-2007-978515

Braga GUL, Rangel DEN, Fernandes ÉKK, Flint SD, Roberts DW (2015) Molecular and physiological effects of environmental UV radiation on fungal conidia. Curr Genet 61(3):405–425. https://doi.org/10.1007/s00294-015-0483-0

Bray CM, West CE (2005) DNA repair mechanisms in plants: crucial sensors and effectors for the maintenance of genome integrity. New Phytol 168(3):511–528. https://doi.org/10.1111/j.1469-8137.2005.01548.x

Brevik EC, Slaughter L, Singh BR, Steffan JJ, Collier D, Barnhart P, Pereira P (2020) Soil and human health: current status and future needs. Air Soil Water Res 13:117862212093444. https://doi.org/10.1177/1178622120934441

Bulgarelli D, Schlaeppi K, Spaepen S, van Themaat EVL, Schulze-Lefert P (2013) Structure and functions of the bacterial microbiota of plants. Annu Rev Plant Biol 64(1):807–838. https://doi.org/10.1146/annurev-arplant-050312-120106

Caldwell MM, Björn LO, Bornman JF, Flint SD, Kulandaivelu G, Teramura AH, Tevini M (1998) Effects of increased solar ultraviolet radiation on terrestrial ecosystems. J Photochem Photobiol B Biol 46(1–3):40–52. https://doi.org/10.1016/S1011-1344(98)00184-5

Carvalho SD, Castillo JA (2018) Influence of light on plant–phyllosphere interaction. Front Plant Sci 9:1482. https://doi.org/10.3389/fpls.2018.01482

Cregger MA, Veach AM, Yang ZK, Crouch MJ, Vilgalys R, Tuskan GA, Schadt CW (2018) The Populus holobiont: dissecting the effects of plant niches and genotype on the microbiome. Microbiome 6(1):31. https://doi.org/10.1186/s40168-018-0413-8

Daliakopoulos IN, Tsanis IK, Koutroulis A, Kourgialas NN, Varouchakis AE, Karatzas GP, Ritsema CJ (2016) The threat of soil salinity: a European scale review. Sci Total Environ 573:727–739. https://doi.org/10.1016/j.scitotenv.2016.08.177

Daram P, Brunner S, Persson BL, Amrhein N, Bucher M (1998) Functional analysis and cell-specific expression of a phosphate transporter from tomato. Planta 206(2):225–233. https://doi.org/10.1007/s004250050394

Darré M, Vicente AR, Cisneros-Zevallos L, Artés-Hernández F (2022) Postharvest ultraviolet radiation in fruit and vegetables: applications and factors modulating its efficacy on bioactive compounds and microbial growth. Foods 11(5):653. https://doi.org/10.3390/foods11050653

Dastogeer KMG, Tumpa FH, Sultana A, Akter MA, Chakraborty A (2020) Plant microbiome–an account of the factors that shape community composition and diversity. Curr Plant Biol 23:100161. https://doi.org/10.1016/j.cpb.2020.100161

Dellagi A, Quillere I, Hirel B (2020) Beneficial soil-borne bacteria and fungi: a promising way to improve plant nitrogen acquisition. J Exp Bot 71(15):4469–4479. https://doi.org/10.1093/jxb/eraa112

Desaint H, Aoun N, Deslandes L, Vailleau F, Roux F, Berthomé R (2021) Fight hard or die trying: when plants face pathogens under heat stress. New Phytol 229(2):712–734. https://doi.org/10.1111/nph.16965

Eslami S, Hoekstra P, Minderhoud PSJ, Trung NN, Hoch JM, Sutanudjaja EH, Dung DD, Tho TQ, Voepel HE, Woillez M-N, Van Der Vegt M (2021) Projections of salt intrusion in a mega-delta under climatic and anthropogenic stressors. Commun Earth Environ 2(1):142. https://doi.org/10.1038/s43247-021-00208-5

Eswar D, Karuppusamy R, Chellamuthu S (2021) Drivers of soil salinity and their correlation with climate change. Curr Opin Environ Sustain 50:310–318. https://doi.org/10.1016/j.cosust.2020.10.015

Francioli D, Cid G, Kanukollu S, Ulrich A, Hajirezaei M-R, Kolb S (2021) Flooding causes dramatic compositional shifts and depletion of putative beneficial bacteria on the spring wheat microbiota. Front Microbiol 12. https://doi.org/10.3389/fmicb.2021.773116

Gent L, Forde BG (2017) How do plants sense their nitrogen status? J Exp Bot 68(10):2531–2539. https://doi.org/10.1093/jxb/erx013

Gianinazzi S, Gollotte A, Binet M-N, Van Tuinen D, Redecker D, Wipf D (2010) Agroecology: the key role of arbuscular mycorrhizas in ecosystem services. Mycorrhiza 20(8):519–530. https://doi.org/10.1007/s00572-010-0333-3

Gibson AC (1996) Structure-function relations of warm desert plants. Springer, Berlin. https://doi.org/10.1007/978-3-642-60979-4

Gunde-Cimerman N, Plemenitaš A, Oren A (2018) Strategies of adaptation of microorganisms of the three domains of life to high salt concentrations. FEMS Microbiol Rev 42(3):353–375. https://doi.org/10.1093/femsre/fuy009

Guo B, Jin Y, Wussler C, Blancaflor EB, Motes CM, Versaw WK (2008) Functional analysis of the Arabidopsis PHT4 family of intracellular phosphate transporters. New Phytol 177(4):889–898. https://doi.org/10.1111/j.1469-8137.2007.02331.x

Gupta R, Patil R (2021) Phyllospheric microbes: diversity, functions, interaction, and applications in agriculture. In: Current trends in microbial biotechnology for sustainable agriculture. Springer, pp 301–323. https://doi.org/10.1007/978-981-15-6949-4_13

Gupta A, Bano A, Rai S, Mishra R, Singh M, Sharma S, Pathak N (2022) Mechanistic insights of plant-microbe interaction towards drought and salinity stress in plants for enhancing the agriculture productivity. Plant Stress 4:100073. https://doi.org/10.1016/j.stress.2022.100073

Hamonts K, Clough TJ, Stewart A, Clinton PW, Richardson AE, Wakelin SA, O'Callaghan M, Condron LM (2013) Effect of nitrogen and waterlogging on denitrifier gene abundance, community structure and activity in the rhizosphere of wheat. FEMS Microbiol Ecol 83(3):568–584. https://doi.org/10.1111/1574-6941.12015

Hasegawa H, Chatterjee A, Cui Y, Chatterjee AK (2005) Elevated temperature enhances virulence of Erwinia carotovora subsp. Carotovora strain EC153 to plants and stimulates production of the quorum sensing signal, N-acyl homoserine lactone, and extracellular proteins. Appl Environ Microbiol 71(8):4655–4663. https://doi.org/10.1128/AEM.71.8.4655-4663.2005

Hinsinger P, Plassard C, Tang C, Jaillard B (2003) Origins of root-mediated pH changes in the rhizosphere and their responses to environmental constraints: a review. Plant Soil 248(1/2):43–59. https://doi.org/10.1023/A:1022371130939

Holmes MG, Keiller DR (2002) Effects of pubescence and waxes on the reflectance of leaves in the ultraviolet and photosynthetic wavebands: a comparison of a range of species. Plant Cell Environ 25(1):85–93. https://doi.org/10.1046/j.1365-3040.2002.00779.x

Huot B, Castroverde CDM, Velásquez AC, Hubbard E, Pulman JA, Yao J, Childs KL, Tsuda K, Montgomery BL, He SY (2017) Dual impact of elevated temperature on plant defence and bacterial virulence in Arabidopsis. Nat Commun 8(1):1808. https://doi.org/10.1038/s41467-017-01674-2

Jerbi M, Labidi S, Bahri BA, Laruelle F, Tisserant B, Ben Jeddi F, Lounès-Hadj Sahraoui A (2021) Soil properties and climate affect arbuscular mycorrhizal fungi and soil microbial communities in Mediterranean rainfed cereal cropping systems. Pedobiologia 87-88:150748. https://doi.org/10.1016/j.pedobi.2021.150748

Jiang R, Wang M, Chen W, Li X, Balseiro-Romero M (2020) Changes in the integrated functional stability of microbial community under chemical stresses and the impacting factors in field soils. Ecol Indic 110:105919. https://doi.org/10.1016/j.ecolind.2019.105919

Jones DL, Nguyen C, Finlay RD (2009) Carbon flow in the rhizosphere: carbon trading at the soil–root interface. Plant Soil 321(1–2):5–33. https://doi.org/10.1007/s11104-009-9925-0

Julkunen-Tiitto R, Nenadis N, Neugart S, Robson M, Agati G, Vepsäläinen J, Zipoli G, Nybakken L, Winkler B, Jansen MAK (2015) Assessing the response of plant flavonoids to UV radiation: an overview of appropriate techniques. Phytochem Rev 14(2):273–297. https://doi.org/10.1007/s11101-014-9362-4

Karhu K, Auffret MD, Dungait JAJ, Hopkins DW, Prosser JI, Singh BK, Subke J-A, Wookey PA, Ågren GI, Sebastià M-T, Gouriveau F, Bergkvist G, Meir P, Nottingham AT, Salinas N, Hartley IP (2014) Temperature sensitivity of soil respiration rates enhanced by microbial community response. Nature 513(7516):81–84. https://doi.org/10.1038/nature13604

Kesawat MS, Satheesh N, Kherawat BS, Kumar A, Kim HU, Chung SM, Kumar M (2023) Regulation of reactive oxygen species during salt stress in plants and their crosstalk with other signaling molecules—current perspectives and future directions. Plants 12(4):864. https://doi.org/10.3390/plants12040864

Khan MA, Asaf S, Khan AL, Jan R, Kang S-M, Kim K-M, Lee I-J (2020) Thermotolerance effect of plant growth-promoting Bacillus cereus SA1 on soybean during heat stress. BMC Microbiol 20(1):175. https://doi.org/10.1186/s12866-020-01822-7

Koch T, Dahl C (2018) A novel bacterial sulfur oxidation pathway provides a new link between the cycles of organic and inorganic sulfur compounds. ISME J 12(10):2479–2491. https://doi.org/10.1038/s41396-018-0209-7

Kumar A, Singh VK, Tripathi V, Singh PP, Singh AK (2018) Plant growth-promoting rhizobacteria (PGPR): perspective in agriculture under biotic and abiotic stress. In: Crop improvement through microbial biotechnology. Elsevier, pp 333–342. https://doi.org/10.1016/B978-0-444-63987-5.00016-5

Kuypers MMM, Marchant HK, Kartal B (2018) The microbial nitrogen-cycling network. Nat Rev Microbiol 16(5):263–276. https://doi.org/10.1038/nrmicro.2018.9

Lashari MS, Bakht-un-Nisa Mangan IR, Ji H, Pan G, Lashari AA, Nan J (2018) Improvement of soil fertility and crop yield through biochar amendment from salt affected soil of central China. J Agric Sci Technol B 8(4). https://doi.org/10.17265/2161-6264/2018.04.002

Li S, Tang S, Chen H, Jin K (2024) Soil nitrogen availability drives the response of soil microbial biomass to warming. Sci Total Environ 917:170505. https://doi.org/10.1016/j.scitotenv.2024.170505

Linder T (2018) Assimilation of alternative sulfur sources in fungi. World J Microbiol Biotechnol 34(4):51. https://doi.org/10.1007/s11274-018-2435-6

Liu H, Carvalhais LC, Crawford M, Singh E, Dennis PG, Pieterse CMJ, Schenk PM (2017) Inner plant values: diversity, colonization and benefits from endophytic bacteria. Front Microbiol 8. https://doi.org/10.3389/fmicb.2017.02552

Liu H, Brettell LE, Qiu Z, Singh BK (2020) Microbiome-mediated stress resistance in plants. Trends Plant Sci 25(8):733–743. https://doi.org/10.1016/j.tplants.2020.03.014

Lladó S, López-Mondéjar R, Baldrian P (2017) Forest soil bacteria: diversity, involvement in ecosystem processes, and response to global change. Microbiol Mol Biol Rev 81(2):e00063-16. https://doi.org/10.1128/MMBR.00063-16

Long X-E, Yao H (2020) Phosphorus input alters the assembly of rice (Oryza sativa L.) root-associated communities. Microb Ecol 79(2):357–366. https://doi.org/10.1007/s00248-019-01407-6

Maltais-Landry G, Scow K, Brennan E (2014) Soil phosphorus mobilization in the rhizosphere of cover crops has little effect on phosphorus cycling in California agricultural soils. Soil Biol Biochem 78:255–262. https://doi.org/10.1016/j.soilbio.2014.08.013

Manzanera M (2021) Dealing with water stress and microbial preservation. Environ Microbiol 23(7):3351–3359. https://doi.org/10.1111/1462-2920.15096

Mapelli F, Marasco R, Rolli E, Barbato M, Cherif H, Guesmi A, Ouzari I, Daffonchio D, Borin S (2013) Potential for plant growth promotion of rhizobacteria associated with Salicornia growing in Tunisian hypersaline soils. Biomed Res Int 2013:1–13. https://doi.org/10.1155/2013/248078

Mergelov N, Mueller CW, Prater I, Shorkunov I, Dolgikh A, Zazovskaya E, Shishkov V, Krupskaya V, Abrosimov K, Cherkinsky A, Goryachkin S (2018) Alteration of rocks by endolithic organisms is one of the pathways for the beginning of soils on earth. Sci Rep 8(1):3367. https://doi.org/10.1038/s41598-018-21682-6

Monier J-M, Lindow SE (2005) Spatial organization of dual-species bacterial aggregates on leaf surfaces. Appl Environ Microbiol 71(9):5484–5493. https://doi.org/10.1128/AEM.71.9.5484-5493.2005

Moreira-Rodríguez M, Nair V, Benavides J, Cisneros-Zevallos L, Jacobo-Velázquez D (2017) UVA, UVB light, and methyl jasmonate, alone or combined, redirect the biosynthesis of glucosinolates, phenolics, carotenoids, and chlorophylls in broccoli sprouts. Int J Mol Sci 18(11):2330. https://doi.org/10.3390/ijms18112330

Mukherjee T, Banik A, Mukhopadhyay SK (2020) Plant growth-promoting traits of a thermophilic strain of the Klebsiella group with its effect on rice plant growth. Curr Microbiol 77(10):2613–2622. https://doi.org/10.1007/s00284-020-02032-0

Mukhtar S, Mehnaz S, Malik KA (2019) Microbial diversity in the rhizosphere of plants growing under extreme environments and its impact on crop improvement. Environ Sustain 2(3):329–338. https://doi.org/10.1007/s42398-019-00061-5

Nafis A, Raklami A, Bechtaoui N, El Khalloufi F, El Alaoui A, Glick BR, Hafidi M, Kouisni L, Ouhdouch Y, Hassani L (2019) Actinobacteria from extreme niches in Morocco and their plant growth-promoting potentials. Diversity 11(8):139. https://doi.org/10.3390/d11080139

Nagaz K, Masmoudi MM, Ben Mechlia N (2012) Effect of deficit drip-irrigation scheduling regimes with saline water on pepper yield, water productivity and soil salinity under arid conditions of Tunisia. J Appl Hortic 14(1):18–24. https://doi.org/10.37855/jah.2012.v14i01.03

Norton J, Ouyang Y (2019) Controls and adaptive management of nitrification in agricultural soils. Front Microbiol 10:1931. https://doi.org/10.3389/fmicb.2019.01931

Oldroyd GED, Leyser O (2020) A plant's diet, surviving in a variable nutrient environment. Science 368(6486):eaba0196. https://doi.org/10.1126/science.aba0196

Oren A (2002) Halophilic microorganisms and their environments, vol 5. Springer, Dordrecht. https://doi.org/10.1007/0-306-48053-0

Palmieri D, Vitale S, Lima G, Di Pietro A, Turrà D (2020) A bacterial endophyte exploits chemotropism of a fungal pathogen for plant colonization. Nat Commun 11(1):5264. https://doi.org/10.1038/s41467-020-18994-5

Pathak P, Rai VK, Can H, Singh SK, Kumar D, Bhardwaj N, Roychowdhury R, de Azevedo LCB, Kaushalendra, Verma H, Kumar A (2022) Plant-endophyte interaction during biotic stress management. Plants 11(17):2203. https://doi.org/10.3390/plants11172203

Philippot L, Chenu C, Kappler A, Rillig MC, Fierer N (2024) The interplay between microbial communities and soil properties. Nat Rev Microbiol 22(4):226–239. https://doi.org/10.1038/s41579-023-00980-5

Rastogi RP, Sinha RP, Moh SH, Lee TK, Kottuparambil S, Kim Y-J, Rhee J-S, Choi E-M, Brown MT, Häder D-P, Han T (2014) Ultraviolet radiation and cyanobacteria. J Photochem Photobiol B Biol 141:154–169. https://doi.org/10.1016/j.jphotobiol.2014.09.020

Rath KM, Rousk J (2015) Salt effects on the soil microbial decomposer community and their role in organic carbon cycling: a review. Soil Biol Biochem 81:108–123. https://doi.org/10.1016/j.soilbio.2014.11.001

Robson TM, Klem K, Urban O, Jansen MAK (2015) Re-interpreting plant morphological responses to UV-B radiation. Plant Cell Environ 38(5):856–866. https://doi.org/10.1111/pce.12374

Ruggiero A, Punzo P, Landi S, Costa A, Van Oosten M, Grillo S (2017) Improving plant water use efficiency through molecular genetics. Horticulturae 3(2):31. https://doi.org/10.3390/horticulturae3020031

Ruiz-González C, Simó R, Sommaruga R, Gasol JM (2013) Away from darkness: a review on the effects of solar radiation on heterotrophic bacterioplankton activity. Front Microbiol 4. https://doi.org/10.3389/fmicb.2013.00131

Saad MM, Eida AA, Hirt H (2020) Tailoring plant-associated microbial inoculants in agriculture: a roadmap for successful application. J Exp Bot 71(13):3878–3901. https://doi.org/10.1093/jxb/eraa111

Sánchez-Cañete EP, Barron-Gafford GA, Chorover J (2018) A considerable fraction of soil-respired CO_2 is not emitted directly to the atmosphere. Sci Rep 8(1):13518. https://doi.org/10.1038/s41598-018-29803-x

Santi C, Bogusz D, Franche C (2013) Biological nitrogen fixation in non-legume plants. Ann Bot 111(5):743–767. https://doi.org/10.1093/aob/mct048

Santos-Medellín C, Edwards J, Liechty Z, Nguyen B, Sundaresan V (2017) Drought stress results in a compartment-specific restructuring of the rice root-associated microbiomes. MBio 8(4). https://doi.org/10.1128/mBio.00764-17

Santoyo G, Urtis-Flores CA, Loeza-Lara PD, Del Carmen Orozco-Mosqueda M, Glick BR (2021) Rhizosphere colonization determinants by plant growth-promoting rhizobacteria (PGPR). Biology 10(6):475. https://doi.org/10.3390/biology10060475

Sasse J, Martinoia E, Northen T (2018) Feed your friends: do plant exudates shape the root microbiome? Trends Plant Sci 23(1):25–41. https://doi.org/10.1016/j.tplants.2017.09.003

Schimel JP (2018) Life in dry soils: effects of drought on soil microbial communities and processes. Annu Rev Ecol Evol Syst 49(1):409–432. https://doi.org/10.1146/annurev-ecolsys-110617-062614

Serrano M, Coluccia F, Torres M, L'Haridon F, Métraux J-P (2014) The cuticle and plant defense to pathogens. Front Plant Sci 5. https://doi.org/10.3389/fpls.2014.00274

Sharma B, Singh BN, Dwivedi P, Rajawat MVS (2022) Interference of climate change on plant-microbe interaction: present and future prospects. Front Agron 3. https://doi.org/10.3389/fagro.2021.725804

Shree B, Jayakrishnan U, Bhushan S (2022) Impact of key parameters involved with plant-microbe interaction in context to global climate change. Front Microbiol 13. https://doi.org/10.3389/fmicb.2022.1008451

Singh BK, Trivedi P, Egidi E, Macdonald CA, Delgado-Baquerizo M (2018) Crop microbiome and sustainable agriculture: depth, scope, and implications. Adv Agron 149:1–40

Taktek S, St-Arnaud M, Piché Y, Fortin JA, Antoun H (2017) Igneous phosphate rock solubilization by biofilm-forming mycorrhizobacteria and hyphobacteria associated with Rhizoglomus irregulare DAOM 197198. Mycorrhiza 27(1):13–22. https://doi.org/10.1007/s00572-016-0726-z

Talley SM, Coley PD, Kursar TA (2002) The effects of weather on fungal abundance and richness among 25 communities in the Intermountain West. BMC Ecol 2(1):7. https://doi.org/10.1186/1472-6785-2-7

Tawaraya K, Horie R, Wagatsuma T, Saito K, Oikawa A (2018) Metabolite profiling of shoot extract, root extract, and root exudate of rice under nitrogen and phosphorus deficiency. Soil Sci Plant Nutr 64(3):312–322. https://doi.org/10.1080/00380768.2018.1476828

Tecon R, Or D (2017) Biophysical processes supporting the diversity of microbial life in soil. FEMS Microbiol Rev 41(5):599–623. https://doi.org/10.1093/femsre/fux039

Thind TS, Schilder AC (2018) Understanding photoreception in fungi and its role in fungal development with focus on phytopathogenic fungi. Indian Phytopathol 71(2):169–182. https://doi.org/10.1007/s42360-018-0025-z

Timmis K, Ramos JL (2021) The soil crisis: the need to treat as a global health problem and the pivotal role of microbes in prophylaxis and therapy. Microb Biotechnol 14(3):769–797. https://doi.org/10.1111/1751-7915.13771

Tiwari P, Bose SK, Park K-I, Dufossé L, Fouillaud M (2024) Plant-microbe interactions under the extreme habitats and their potential applications. Microorganisms 12(3):448. https://doi.org/10.3390/microorganisms12030448

Trivedi P, Leach JE, Tringe SG, Sa T, Singh BK (2020) Plant–microbiome interactions: from community assembly to plant health. Nat Rev Microbiol 18(11):607–621. https://doi.org/10.1038/s41579-020-0412-1

Trivedi P, Batista BD, Bazany KE, Singh BK (2022) Plant–microbiome interactions under a changing world: responses, consequences and perspectives. New Phytol 234(6):1951–1959. https://doi.org/10.1111/nph.18016

Udvardi M, Poole PS (2013) Transport and metabolism in legume-rhizobia symbioses. Annu Rev Plant Biol 64(1):781–805. https://doi.org/10.1146/annurev-arplant-050312-120235

Vanhaelewyn L, Prinsen E, Van Der Straeten D, Vandenbussche F (2016) Hormone-controlled UV-B responses in plants. J Exp Bot 67(15):4469–4482. https://doi.org/10.1093/jxb/erw261

Vanhaelewyn L, Van Der Straeten D, De Coninck B, Vandenbussche F (2020) Ultraviolet radiation from a plant perspective: the plant-microorganism context. Front Plant Sci 11:597642. https://doi.org/10.3389/fpls.2020.597642

Vansuyt G, Robin A, Briat J-F, Curie C, Lemanceau P (2007) Iron acquisition from Fe-Pyoverdine by *Arabidopsis thaliana*. Mol Plant Microbe Interact 20(4):441–447. https://doi.org/10.1094/MPMI-20-4-0441

Velásquez AC, Castroverde CDM, He SY (2018) Plant–pathogen warfare under changing climate conditions. Curr Biol 28(10):R619–R634. https://doi.org/10.1016/j.cub.2018.03.054

Verdaguer D, Jansen MAK, Llorens L, Morales LO, Neugart S (2017) UV-A radiation effects on higher plants: exploring the known unknown. Plant Sci 255:72–81. https://doi.org/10.1016/j.plantsci.2016.11.014

Verma H, Kumar D, Kumar V, Kumari M, Singh SK, Sharma VK, Droby S, Santoyo G, White JF, Kumar A (2021) The potential application of endophytes in management of stress from drought and salinity in crop plants. Microorganisms 9(8):1729. https://doi.org/10.3390/microorganisms9081729

Versaw WK, Garcia LR (2017) Intracellular transport and compartmentation of phosphate in plants. Curr Opin Plant Biol 39:25–30. https://doi.org/10.1016/j.pbi.2017.04.015

Wahid F, Sharif M, Ali A, Fahad S, Adnan M, Noor M, Mian IA, Khan IA, Alam M, Saeed M, Ilyas M, Ullah R, Ilahi H, Azeem M (2020) Plant-microbes interactions and functions in changing climate. In: Fahad S, Hasanuzzaman M, Alam M, Ullah H, Saeed M, Ali Khan I, Adnan M (eds) Environment, climate, plant and vegetation growth. Springer International Publishing, pp 397–419. https://doi.org/10.1007/978-3-030-49732-3_16

Wang Z, Solanki MK, Kumar A, Solanki AC, Pang F, Ba ZX, Niu JQ, Ren ZX (2023) Promoting plant resilience against stress by engineering root micro environment with Streptomyces inoculants. Microbiol Res 277:127509. https://doi.org/10.1016/j.micres.2023.127509

Waqas M, Khan AL, Shahzad R, Ullah I, Khan AR, Lee I-J (2015) Mutualistic fungal endophytes produce phytohormones and organic acids that promote japonica rice plant growth under prolonged heat stress. J Zhejiang Univ Sci B 16(12):1011–1018. https://doi.org/10.1631/jzus.B1500081

Xu Z, Shimizu H, Ito S, Yagasaki Y, Zou C, Zhou G, Zheng Y (2014) Effects of elevated CO_2, warming and precipitation change on plant growth, photosynthesis and peroxidation in dominant species from North China grassland. Planta 239(2):421–435. https://doi.org/10.1007/s00425-013-1987-9

Xu L, Naylor D, Dong Z, Simmons T, Pierroz G, Hixson KK, Kim Y-M, Zink EM, Engbrecht KM, Wang Y, Gao C, DeGraaf S, Madera MA, Sievert JA, Hollingsworth J, Birdseye D, Scheller HV, Hutmacher R, Dahlberg J et al (2018) Drought delays development of the sorghum root microbiome and enriches for monoderm bacteria. Proc Natl Acad Sci 115(18):E4284. https://doi.org/10.1073/pnas.1717308115

Yandigeri MS, Meena KK, Singh D, Malviya N, Singh DP, Solanki MK, Yadav AK, Arora DK (2012) Drought-tolerant endophytic actinobacteria promote growth of wheat (Triticum aestivum) under water stress conditions. Plant Growth Regul 68(3):411–420. https://doi.org/10.1007/s10725-012-9730-2

Ye JY, Tian WH, Jin CW (2019) A reevaluation of the contribution of NRT 1.1 to nitrate uptake in Arabidopsis under low-nitrate supply. FEBS Lett 593(15):2051–2059. https://doi.org/10.1002/1873-3468.13473

Zaehle S, Jones CD, Houlton B, Lamarque J-F, Robertson E (2015) Nitrogen availability reduces CMIP5 projections of twenty-first-century land carbon uptake. J Clim 28(6):2494–2511. https://doi.org/10.1175/JCLI-D-13-00776.1

Zai X, Qin P, Wan S, Zhao F, Wang G, Yan D, Zhou J (2007) Effects of arbuscular mycorrhizal fungi on the rooting and growth of beach plum (Prunus maritima) cuttings. J Hortic Sci Biotechnol 82(6):863–866. https://doi.org/10.1080/14620316.2007.11512319

Zhang Z, Wang Q, Guan Q, Xiao X, Mi J, Lv S (2023) Research on the optimal allocation of agricultural water and soil resources in the Heihe River Basin based on SWAT and intelligent optimization. Agric Water Manag 279:108177. https://doi.org/10.1016/j.agwat.2023.108177

Zhao B, Jia X, Yu N, Murray JD, Yi K, Wang E (2024) Microbe-dependent and independent nitrogen and phosphate acquisition and regulation in plants. New Phytol 242(4):1507–1522. https://doi.org/10.1111/nph.19263

Zhou M, Butterbach-Bahl K, Vereecken H, Brüggemann N (2017) A meta-analysis of soil salinization effects on nitrogen pools, cycles and fluxes in coastal ecosystems. Glob Chang Biol 23(3):1338–1352. https://doi.org/10.1111/gcb.13430

Zhu Y, Yu K, Wu Q, Cheng X, Li Z, Wang Z, Zhao M, Wilkes A, Bisselling T, Han G, Ren H (2023) Seasonal precipitation and soil microbial community influence plant growth response to warming and N addition in a desert steppe. Plant Soil 482(1–2):245–259. https://doi.org/10.1007/s11104-022-05684-y

Role of Endophytic Microbes for Postharvest Disease Management

4

Monika Singh, Aamna Hassan, Shanvi Nautiyal, Rakesh, and Sonika Kalia

Abstract

Each plant organ has its own microbial assemblage, either on epiphytes or within the plant tissue, and these microbiomes perform a crucial part in encouraging development and as bio-controlling mean in response to number of phytopathogens during pre- and postharvest developmental stages of fruits. The endophytic microbiota of any plant organ interacts mutually and attracts one another in a number of action methods. The availability and spread of endophytic microbiota vary from place to place with each plant organ, even within the same plant, and are also influenced by genetics, cultivars, stages of development, growth conditions, biotic and abiotic factors, and specific interactions with one another. Endophytic microbiomes play an important role in disease management. In this particular chapter, attempt has been made to explore the carpophore of fruits, as well as the details of the application and management strategies of various fruit diseases through the use of beneficial endophytic microorganisms.

Keywords

Biological stress · Biocontrol · Carpophore · Endophytes · Fruits · Postharvest

M. Singh · S. Nautiyal · Rakesh · S. Kalia (✉)
Department of Biotechnology, School of Applied and Life Sciences, Uttaranchal University, Dehradun, Uttarakhand, India

A. Hassan
Jamia Hamdard University, New Delhi, India

A. Kumar, M. K. Solanki (eds.), *Microbial Biocontrol Techniques*, Microorganisms for Sustainability 54,
https://doi.org/10.1007/978-981-97-8739-5_4

4.1 Introduction

The carposphere of crops grown in agriculture includes distinctive bacterial strains and communities that are classified as epiphytes or at times as endophytes that play a significant role in the biological processes, physiological science, and resilience of vegetables and fruits to beforehand and after harvest degradation (Droby and Wisniewski 2018). The breakdown and mode of action of both epiphytic microbial communities in the carposphere, and especially their roles in fruit biological functioning, qualitative analysis, and disease management remain unclear. A wide range of research and analysis has been performed on the applications of epiphytic bacteria and other microbes such as yeast as a biocontrol agent for postharvest illnesses (Kumari et al. 2022; Droby et al. 2019; Solanki et al. 2024; Zhimo et al. 2021). However, there are only a few documents available on microbial endophytes in fruit and their potential roles and application in a favorable manner versus both abiotic and biotic stresses (Anand et al. 2023; Kumar et al. 2018; Pathak et al. 2022).

"Endophyte" term was first initially employed to describe fungi extracted from inside body cells or tissues of the target host plants, but later this concept was subsequently broadened to include populations of bacteria (Chanway 1996; Hardoim et al. 2015). De Bary (1866) was the first one to comment and define microorganisms that live within tissues of plants as "endophytes." The definition along with the types of endophytes was adjusted later in time based on researcher skills and observations. Endophytes have also been defined (Hallmann et al. 1997) as microorganisms or microbial communities that are collected from plant tissues after surface-disinfection and can persist throughout their host system without producing illness. Petrini (1991) defined endophytes as the minute organisms that occupy plant tissue at some stage in their life cycle. It concludes its life cycle without infecting the host plant. In accordance with Rosenblueth and Martinez-Romero (2006), endophytes can be categorized by kind (bacteria or fungus) and relationship to plants (facultative or obligatory). However, Hardoim et al. (2015) categorized endophytes based on their colonization niche rather than their function.

Postharvest storage is one of the global concern for maintaining the quality and longevity of fruits. Postharvest diseases cause much economic losses to harvested fruits during transportation and storage. Synthetic shelf life improving agents are commonly used to combat postharvest diseases. However, since the last decades, the trend is shifting toward safer and eco-friendly methods for the control of postharvest diseases fruits. Recently, the use of several endophytic microbes isolated from carposphere used for management of postharvest diseases of fruits (Singh et al. 2022a).

In this chapter, we endeavored to bring together previous research on endophytic bacteria in the carposphere and discuss the variety, distribution and availability pattern, and potential application of endophytic microbiota related with internal organ or tissues for regulation and control of pre- and postharvest diseases of crops.

4.2 The Entrance Point, Colonialism Along with Distribution of Endophytic Bacteria into Plant Tissues or Cells

The entrance of these microbial strains into the host tissue and their colonization is the result of multiple events involving the release of various secretory products, proteins, and signaling molecules. Some of the major secondary products and metabolites involved in this process are liposaccharides and expolysaccharides. These endophytic strains are mainly translocated toward the aerial plant parts such as leaves, wounds, roots, and cotyledons (Böhm et al. 2007; Suarez-Moreno et al. 2010; Singh et al. 2017a). The chemostatic interaction between the host and the microbe causes the microbial strain to move toward the host for attachment. Apart from playing a crucial role in microbial attachment, these secretory products are often known to have a defense mechanism against oxidative damage and free radicals that may damage the microbial strain. A successful adherence of the bacterial strain on to the host surface is crucial for successful colonization. This is accomplished by achieving a good interaction between structural components of the microbe, such as flagella, pilli, microbial appendages, and the secreted products, as mentioned earlier (Sauer and Camper 2001; Zheng et al. 2015).

Bacterial endophytes, after successfully attaching themselves to the host, follow different pathways for colonization patterns and sites, depending on the type of bacterial strain that causes infection. Penetration into the host cell is achieved either by passive means, i.e., through the cracks on roots, root tips, etc., or through active means involving activation of expolysaccharides and liposaccharides that facilitate the entry of endophytic bacteria into the host plant tissue or cells (Compant et al. 2010; Brader et al. 2014).

Once they enter the host plant, endophytes protect themselves from the host's defense mechanisms by maintaining a low-cell density and producing low-cell wall-degrading enzymes, unlike other regular phytopathogens (Ali et al. 2024; Bhardwaj et al. 2024). Ultimately, the success of colonization depends on the potential of plant–microbe interactions (Elbeltagy et al. 2000; Reinhold-Hurek et al. 2006).

The signaling molecules facilitate effective communication and cooperation among the endophytic bacteria and the targeted host plants (Keller and Surette 2006). Certain genes relating to nitrogen fixation and its specific metabolism mechanism, namely ntrY and ntrX along with cell fate control genes such as pleC and pleD are present dominantly in the endophytic bacteria as compared to the other bacteria such as rhizobacteria and phytopathogens (Hardoim et al. 2015). Several biological and biocontrol processes relating to regulation of energy and its effective utilization, such as carbon and nitrogen fixation, aerobic and anaerobic respiration, and transport of electron are regulated in effective response to cellular redox balance, which helps in effective endophytic colonization inside the host cell (Elsen et al. 2004; Hardoim et al. 2015; Devi et al. 2023). The presence of "NtrY" nitrogen sensor protein interacts and regulates NtrX for the induction and expression of genes (Pawlowski et al. 1991). The presence of a higher amount of ntrY and ntrX in the endophytic strains might be used more efficiently than rhizobacteria and phytopathogens for their survival under nitrogen-limiting conditions (Hardoim et al.

2015). Yi et al. (2017) reported that gene encoded for membrane proteins, regulators of transcription, metabolism of amino acids, and biosynthesis are upregulated in endophytic *Bacillus mycoides* in comparison to the soil isolated *B. mycoides*; however, sugar transport was down-regulated in the endophytic strain after the transcriptomic study.

Other variables contribute to effective endophytic colonization, such as host genotypes, microbial strain specificity, and nutrition availability (Compant et al. 2008; Hardoim et al. 2015) examined the colonization pathway of *Paraburkholderia phytofirmans* PsJN and revealed that it enters into the plant tissue through the skin layers, namely exodermis and cortical cell layers of the roots, then overcomes the endodermal layer physical barriers, eventually leading to the zone in the center. Endophytes spread from the zone in the middle to the plant's upper regions through xylem arteries. Endophytic microbes that enter the root system typically favor unsupervised endodermal cells in the apical roots (James et al. 2002; Gasser et al. 2011). Once microorganisms become endophytic, they feed on the energy and nutrients of the host. They use the minerals and nutrients of their host plant tissues as energy source and a source of carbon for survival and reproduction (Rasche et al. 2009; Malfanova et al. 2013). However, the main source of power differs in each case depending on the bacterial strain. This can be understood as one endophytic strain of *Pseudomonas* uses L-arabinose as an energy source in cucumber plants while some endophytic isolates of rice such as *Enterobacter dissolvens*, *Brevundimonas aurantiaca*, *Pantoea agglomerans*, *Pseudomonas* spp., and Enterobacteriaceae are reported to use L-arabinose and glucose to fulfill their carbon needs and energy source (Iwai et al. 2003; Prakamhang et al. 2009).

Endophytes can enter into the host horizontally that is through surroundings, or vertically that means inside the plant, most typically via seed, or both (Bright and Bulgheresi 2010; Frank et al. 2017). However, the richness of the microbial strains inside the seed that has developed under sterilized situations is smaller than those plants grown in natural environment that in turn suggest the preference of microorganisms for horizontal transmission (Compant et al. 2005). However, it is still possible that certain endophytes are transported both horizontally and vertically and become mixed. Endophytic bacteria that are regularly handed down and passed on from generation to generations must go from the seed stage to reproductive organ stage that is by formation of fruits, either via vessels or xylem or the shoot apical meristem, which cultivates into reproductive parts of the plant (Frank et al. 2017) (Table 4.1)

4.3 Evolution Process of Endophytes

The genetic process differed in both endophytes and pathogenic microbial strains during colonization. The attack of pathogenic strains differentially evokes the immune response and behavior of the host plants (Clay and Kover 1996). However, in the case of endophytes, no such type of immune response has been evoked

Table 4.1 List of various endophytic microbes used for postharvest disease management

S. no.	Endophytic bacterial strains	Domains of the strains	Disease or pathogens management	Plant/fruits	References
1	*Bacillus velezensis* QSE-21	Bacteria	Postharvest causes gray molds in fruits	Tomato	Frank et al. (2017)
2	*Bacillus subtilis*	Bacteria	*Penicillium digitatum*	Citrus fruits	Compant et al. (2005)
3	Endophytes	Bacteria	*Monilinia* spp. and *Rhizopus stolonifer*	Stone fruits	Rasche et al. (2009)
4	*Lactobacillus plantarum*	Bacteria	*Botrytis* spp.	Strawberry and other berries	Rasche et al. (2009)
5	*Paenibacillus polymyxa*	Bacteria	*Penicillium digitatum*	Citrus	Frank et al. (2017)
6	*Bacillus amyloliquefaciens*	Bacteria	*Botryosphaeria dothidea*	Kiwi fruit	Pang et al. (2021)
7	*Pseudomonas synxantha*	Bacteria	*Monilinia fructicola*	Stone fruit	Frank et al. (2017)

because of the symbiotic behavior of microbes to the host plants (Bergstrom and Lachmann 2003).

Endophytic bacteria are closely connected, which demonstrates their ecological and evolutionary relationship. Endophyte, life span, potential to develop within host plant, and plant sexual maturity age all have a significant impact on microbial growth patterns in plant hosts (Saikkonen et al. 2004). This accomplishes its life cycle of growth and reproduction in a single season, whereas perennial hosts may be able to develop the endophytic microbiota more effectively than the annual host. The genetic components of the cultivars also influence the makeup of the endophytic microbiome, and the hosts' genotypes are similar, suggesting evolutionary connection. In a study, Liu et al. (2018) found a genotype that influence on the endophytic bacteria of three apple cultivars, namely Golden Delicious, Royal Gala, and Honeycrisp, along with a combination of rootstock/scion. They revealed that the changes in the endophytic microbiota (rootstock and scion) of fungus for all the three above-mentioned cultivars, but no major differences as such were found in the respective endophytic microbiota. The contents of the respective fungal diversity differ across all the three cultivars. The diversity in "Golden Delicious" and "Royal Gala" is similar to each other with little differences, but they clearly differ from "Honey Crisp" in spite of rootstock genotypic attributes. Furthermore, the pedigree studies of all three cultivars revealed that Royal Gala and Gold Delightful are close relatives while Honeycrisp is distant.

The isolation and characterization of endophytic microorganisms have generally been carried out through traditional culture-dependent techniques or selective media plate methods. But this traditional technique has very limited efficacy because of the delay or lack of growth of most of the microorganisms on synthetic media in the laboratory (Pettersson and Baath 2003; Shi et al. 2010; Tholozan et al. 1999).

Table 4.2 Role of various strains of endophytes in plant growth promotion and development of plant

S. no.	Endophyte	Plant tissue	Bioprospecting activity	References
1	*Actinobacteria*	Potato plants	Antibiotic, antagonistic substances and inhibition activity	Cruz et al. (2019)
2	*Curtobacterium flaccumfaciens* found in oak tree	Against pathogen *Ceratocystis fagacearum*	Produces adenine residues and inhibition activity	Mocali et al. (2003), Frank et al. (2017), Pettersson and Baath (2003)
3	*Bacillus* sp.	Wheat, peas soybean	Promotes nodulation and PGP activity	Compant et al. (2011)
4	*Klebsiella* strain	Wheat plant	Nitrogen fixation	Compant et al. (2011)
5	*Arthrobacter* and *Mircobacterium*	Capsicum plant	1-Aminocyclopropane-1-carboxylate (ACC) deaminase activity and IAA production	Mocali et al. (2003), Frank et al. (2017), Pettersson and Baath (2003)
6	*Serratia plymuthica*	–	Fungal resistance	(Compant et al. 2011)

However, the latest next-generation sequencing techniques and metagenomics approaches provide a great opportunity to explore the hidden microbial communities inside plant tissues by using highly sensitive genetic detection and sequencing technologies. These latest technologies have the ability to detect both cultivable and non-cultivable microbiota present in very small quantities (Shi et al. 2010) (Table 4.2).

4.4 Endophyte Diversity

The composition and diversity of endophytic microorganisms depend on various factors such as host genotype or cultivar (Pettersson and Baath 2003; Frank et al. 2017), seasons, developmental stages, and environmental conditions (Bulgarelli et al. 2012; Miguel et al. 2013; Schlaeppi and Bulgarelli 2015; Ou et al. 2019). In this regard, various works have been published that describe the distribution pattern of the endophytic microbial community inside the host plant tissue and the associated limiting factors (Compant et al. 2011) recorded and compared the colonization habits of endophytic isolates of bacteria in grapevine flowers, berries, and seeds to those in other parts of the plants. Endophytic bacteria were found in almost in all sections of the plant; however, their density differs based on the plant organ. The total number of endophytic microbiota was considerably low in berries, flowers, and seeds than in other parts of the plant such as leaves and stems. It was observed that in the generative portions, the flowers had the highest diversity, along with the pulp, and the lowest was recorded in the grapeberry seeds.

The diversity and number of endophytes vary according to stage of development in the plant part. Miguel et al. (2013) found an assortment trend in the endophytes such as *Coffea canephora* in its three different developmental stages: green, green-yellow, and ripe. Among the endophytes, Proteobacteria, Firmicutes, Bacteroidetes, and Actinobacteria, were the most relevant community. *Kocuria turfanensis* spp. and *Pantoea* spp., *Bacillus thuringiensis*, *Bacillus licheniformis*, *Agrobacterium tubefaciens*, *Escherichia coli*, *Enterobacter hormaechei*, *Chryseobacterium* sp., and *Coffea ochrobactrum* sp., with bacterial strain diversity differing based on the fruit's developmental stage. The green stage showed the most variation, while the ripe stage had the least. *Bacillus subtilis* is the most frequent bacterial species in the initial phases of development, while *Klebsiella oxytoca* dominates at the ripening stage. Greater amount of sugar and caffeine have been reported to be present at the ripe stage which has been linked with less diversity. Luksa et al. (2020) have reported that ripening influences the endophytic fungus variety of the Sea Buckthorn (*Hippophae rhamnoides*) carposphere. They found that both riped and unriped phases had distinct communities of fungus. Unripe blueberries had dominant presence of endophytes such as of *Aureobasidium* and *Cladosporium*, while ripe berries were dominated by the yeast species such as *Aureobasidium* and *Metschnikowia*. The details of the endophytic microbial communities are listed in Table 4.3.

Recently, Cruz et al. (2019) reported that the effect of cultural practices on the inbuilt diversity of endophytic microbes in the coverings and peel waste of certain types of tropical fruits such as mango, papaya, guava, passion fruit, and banana that are available in both native or wild varieties. They found that diversity of endophytic bacterial communities was much more in the peels of fruits mentioned above in comparison to the other fungal neighbor hoods, regardless of method of production deployed. Further, it was discovered that type of fruit and its subsequent production had an important impact on the abundant groups of endophytes with the wild papaya sample has a far greater number of OTUs (operational taxonomic unit) than the conventional sample. There were not much variations present in the endophytic bacterial strains diversity between conventional and wild-cultivated peel samples. Fungal OTUs were higher in traditionally cultivated mangoes.

Mango fruit peels, irrespective of growing practice. All fruit types had comparable fungal varieties in both conventionally cultivated and non-traditional samples, with the exception of mango and guava, which showed a substantial difference between conventionally generated and non-cultivated samples of fruit peels. Certain group of scientists investigated that the relationship between watermelon cultivars (Red PI459074, Congo, and SDRose) and yellow fruit-flesh (cultivars PI227202, PI435990, and JBush cultivated in different geographically distinct parts), season variation, and developmental stage of ripening of fruit on endophytic bacteria. Metabarcoding tests revealed that every cultivar has a distinct microbial population which is consistent with previous research (Hardoim et al. 2015; Frank et al. 2017). Various groups of proteobacteria existed dominantly in the cultivars SDRose and PI227202, but cyanobacteria were the ruling community in the Congo.

Table 4.3 Endophytic strains isolated from different parts of fruits and vegetables used in disease management

S. no.	Bacterial species	Fruit or vegetables isolated from	References
1	*Bacillus weihenstephanensis*	Berries	Singh et al. (2017a)
2	*Pseudomonas* spp.	Lettuce Tomato Carrot Cucumber	Singh et al. (2017b)
3	*Salmonella enterica*	Carrot Cantaloupe Lettuce Carrot	Johnson et al. (1992), Galsurker et al. (2018)
4	*Helicobacter pylori*	Leek Basil leaves Coriander Cabbage	Singh et al. (2017a)
5	*Staphylococcus aureus*	Lettuce	Singh et al. (2017a)
6	*Enterobacter cloacae*	Tomatoes Lettuce Cucumber	Johnson et al. (1992)
7	*Klebsiella oxytoca*	Tomatoes Lettuce	Singh et al. (2017b), Nigris et al. (2018)
8	*Kluyvera ascorbata*	Carrot	Nan et al. (2019)
9	*Xanthomonas fuscans*	Onions Tomatoes	Singh et al. (2017b)
10	*Salmonella typhimurium*	Coriander leaves	Johnson et al. (1992)

Proteobacteria were found in large numbers in respective cultivars SDRose and PI227202, but cyanobacteria were dominant in the Congo and PI4559074 cultivars. They also concluded that the occurrence of completeness in SDRose. They suggested occurrence and availability of illness in this cultivar microbiota diversity. Glassner et al. (2015) studied the endophytes prevalent in fruits of (*Cucumis melo Reticulatus* group), a type of melon collected from two distinct locations: a commercial manufacturing farm and the wild. The most common endophytic bacterial classes at both sites were Alpha-, Beta-, and Gammaproteobacteria, Firmicutes, and Actinobacteria. DOPE-FISH microscopy was used to visualize and quantify endophytic bacteria, and significant variations were discovered between wild and farmed cucurbit species in terms of endophyte quantity and type. Seasonal variations occur during the cropping period.

It also affects the endophytic bacterial composition of fruits. Endophytic microbial diversity in mulberry fruit has been found to be very dynamic, with seasonal change acting as an influence in the composition of bacteria within the host parts, along with the impact the different cultivars' genetic origins. For example, endophyte colonization tends to increase during warmer, wet, and rainy seasons (Ou et al. 2019).

The diverse occurrence of endophytes in the carposphere is likewise affected by the fruits' health status. Wu et al. (2019) examined that the endophytic diversity in both healthy and ill version of kiwifruit. Ultimately they discovered that healthy fruits have greater endophytic diversity and distinct cosmetics than damaged fruits. In damaged fruit, the microbiota was predominant.

The fungus family Botryosphaeriaceae dominated in healthy fruit, while the Pleosporaceae family dominated in rotten ones. Bacterial community such as the Methylobacteriaceae, Sphingomonadaceae, and Nocardioidaceae categories were more common in the fresh and versatile fruit. On the basis of functionality, the microbiome of healthy fruits had a greater relative abundance and availability of ABC transporter genes, that are associated with bacterial secretion systems and possess two components and chemotaxis, but lack certain genes associated with polycyclic aromatic hydrocarbon degradation, sugar metabolism, and specific aminoacid metabolism such as glycine and serine. Al-Kharousi et al. (2016) investigated that the precise distribution pattern of the endophytic community. Enterobacteriaceae, *Enterococcus*, and *Staphylococcus aureus* were found in fresh vegetables and fruits from diverse countries and regions.

Sixty percent of fruits and more than 90% of vegetables in Oman have been reportedly infected with Enterobacteriaceae. About 20% of the fruits and 42% of the vegetables had Enterococus, 22% and 7% of them had *E. coli* and *S. aureus*, respectively. *Klebsiella pneumoniae*, *E. coli*, *Enterococcus casseliflavus*, and *Enterobacter cloacae* were the most common species in all of edibles.

The endophytic microbiota of Jingbai pear was studied for several plant parts, indicating that each tissue of the plant has its own population of bacteria (Ren et al. 2019a, b). The root tissue had the most diverse and abundant bacterial communities, followed by the flower, stem, and fruit. The least plentiful and varied were leaves, fruits, and seeds. The outcomes of the research revealed that the stems possessed the most diverse strains and richness of endophytic microbes followed by the flowers and fruit. Overall, the phyllosphere tissue had the lowest diversity and abundance. Except for tomato seeds and fruit jelly, proteobacteria were the most common phylum in all of the tissue specimens. The stem, leaf, and root tissues were dominated by the species *Acinetobacter*, *Enterobacter*, and *Pseudomonas*, although the fruit's endophyte diversity varied depending on the compartment. *Enterobacter* dominated the fruit, seeds, and pericarp; *Actinobacter* was predominantly present in the placenta; and *Weissella* dominated the fruit's jelly tissues.

The endophytes in the tomato fruit reported that fruit has distinct microbial communities from other parts of the plant. It was assumed that ovaries, which eventually became the fruit flesh, generate a distinct environment that supports certain microbial communities (Aleklett et al. 2014). Krishnan et al. (2012) strengthened the belief that different regions of the same fruit had distinct microbiomes, reporting varied and specialized endophytic bacterial populations in several fruit such as papaya. *Kocuria*, *Enterobacter*, *Acinetobacter* species were detected in both the stem and the fruit endocarp, whereas *Bacillus* spp. was prominent throughout the fruit. Singh et al. (2022b) reported that the endophytic bacterial strains such as *Rhizobium pusense*, *Bacillus* sp., *Methylophilus flavus*, and *Pseudomonas*

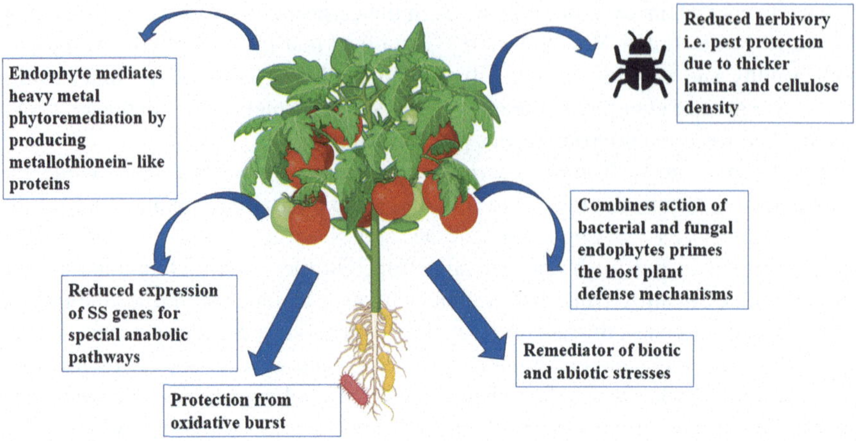

Fig. 4.1 Role of endophytic microbes in development of tomato plant under various adverse conditions

aeruginosa enhanced the growth and yield of tomato. In another study, the fungal endophytes *Ampelomyces* sp. and *Penicillium* sp. were provide a great tolerance against abiotic stress to the *Solanum lycopersicum* L. plant. Fungal endophytes also provide several benefits against common insects and non-insect pests to the tomato plant (Morsy et al. 2020; Grabka et al. 2022). Some other common roles of endophytes are shown in Fig. 4.1.

Hanin and Fitriasari (2019) identified five types of fungus in the pulp of the fruit and seeds of Jambolana. The endophytic strains such as *Neofusicoccum parvum* and *Pestalotiopsis* sp., were found in the pulp portion while *Colletotrichum fructicola* was found in the seeds. Isolated endophytic fungal strains from strawberries and their antioxidant capabilities were also checked. These endophytic fungal strains, viz. *Sporidiobolus* sp., *Rhodotorula* sp., *Pilidium concavum*, *Neodeightonia subglobosa*, *Aspergillus awamori*, and *Aspergillus* sp. *Aspergillus awamori* DT11 and DT13 showed excellent antioxidant properties and also suggested that these fungal strains had capability to enhance the nutritional value of fruits.

4.5 Endophytic Microbes in Disease Management

Plant endophytic microorganisms are currently used for promotion of growth, management of stress, and as of biocontrol agents against a variety of pest causing diseases during before and postharvest stages (Singh et al. 2017b, 2022c; Wang et al. 2023; Kashyap et al. 2023). These endophytes have been demonstrated to be capable of producing a wide number of primary and secondary metabolites, volatile chemicals, antibiotics and hormonal substances, exopolysaccharides, and cell wall-degrading enzymes (Nigris et al. 2018; Shameer and Prasad 2018). Endophytes play a role in signaling and regulating metabolic pathways in the fruit through enzyme synthesis along with ACC synthatase activity, quorum sensing, and other processes.

In general, most antagonistic microorganisms employed for treating postharvest fruits illness were extracted from the epiphytic community of the rhizosphere, with a few being focus on the possible use of endophytic isolates to manage fruit illnesses. Janisiewicz et al. 2013 proposed that endophytic microbial strains work as antagonistic microorganisms against postharvest fruit diseases of fruits. Endophytic microbiota also have great potential to synthesize a variety of secondary metabolites like phytohormones, which are involved in regulation of fruit physiological aspects, maturation, and ripening (Suhandono et al. 2016; Singh et al. 2019, 2020). Endophytes, as opposed to epiphytic microbiota, may be better colonizers because of their greater adaptability and acclimatize within the host tissue (Shi et al. 2010).A number of investigations have demonstrated that endophytic microbial strains can help manage an array of fruit illnesses (Sturz and Christie 1996). Istifadah et al. (2017) reported that the strains of *Lysinibacillus* sp., *Bacillus* spp., *Azotobacter* spp., and *Pseudomonas cepacia*, either alone or in combination, to reduce the occurrence bacterial wilting diseases in plants like chilli. Endophytic strains of *Pseudomonas putida* isolated from the papaya were used to reduce infection of *Phytophthora nicotianae* on the fruit after postharvest. This bacterial antagonist is able to capture multiple sections of papaya's fruit like, pericarp, leafstalk, and pulp while also inhibiting the growth of assortment fruit diseases. Anthracnose disease incidences were lower by the use of *Pseudomonas putida* by up to 65% when administered shortly after the florescence stage. Similarly, endophytic actinomycete strains of *Streptomyces* sp. were applied to control *Colletotrichum orbiculare* which is the causative agent of cucumber anthracnose disease. Lai et al. (2012) also used strain of *Paenibacillus polymyxa* extracted from *Sophora tonkinensis* root to prevent citrus green mold degradation after harvest.

Endophytic lactic acid bacteria (LAB) isolated from seeds of papaya have been shown to prevent papaya dieback disease outbreak caused by *Erwinia mallotivora* (Taha et al. 2019). Consortia application made from *Weissella cibaria* and *Lactococcus lactis* enhanced antibacterial activity against the pathogen and also decreasing disease severity. The citrus endophyte *Methylobacterium mesophilicum* was isolated, and this strain showed biocontrol activity against *Xylella fastidiosa* that causes citrus variegated chlorosis (CVC) in sweet orange, and it was suggested that the sweet orange's endophytic microbial community plays a crucial role to reduce the *Xylella fastidiosa*'s pathogenicity. Volatile organic substances released by the endophytic fungal strains *Nodulisporium* sp. and *Hypoxylon anthochroum* have been shown to hinder postharvest *Fusarium oxysporum* control in cherry tomatoes. Individual or combined application of fungal volatiles affects cell membrane permeability and impairs Fusarium hyphal shape of (Medina-Romero et al. 2017). The *Pseudomonas synxantha* isolated from the tree of kiwi fruit showed antagonistic activity after harvest against *Monilinia fructicola* and *Monilinia fructigena*, the causative agents of brown rot in stone fruits.

4.6 Future Prospects

Microbial endophytes from the carposphere have been shown to be an essential part of the microbial community. These have the potential to be excellent biocontrol agents and regulate fruit senescence and provide protection against various postharvest diseases. Endophytes are very capable of colonization various regions of tissue of fruit and interact with a variety of ways with one another and the host. Conclusively at this point, additional study is needed for a better understand the biology of fruit endophytes, as well as their role and interconnections with their target plants. Fruit endophytes would be a rich source of bioactive compounds, which have diverse biochemical functions for human welfare.

References

Aleklett K, Hart M, Shade A (2014) The microbial ecology of flowers: an emerging frontier in phyllosphere research. Botany 92(4):253–266

Ali A, Ölmez F, Zeshan MA, Mubeen M, Iftikhar Y, Sajid A, Abid M, Kumar A, Divvela PK, Solanki MK (2024) Yeast-based solutions in controlling plant pathogens. Biocatal Agric Biotechnol 58:103199. https://doi.org/10.1016/j.bcab.2024.103199

Al-Kharousi ZS, Guizani N, Al-Sadi AM, Al-Bulushi IM, Shaharoona B (2016) Hiding in fresh fruits and vegetables: opportunistic pathogens may cross geographical barriers. Int J Microbiol 2016:4292417

Anand U, Pal T, Yadav N, Singh VK, Tripathi V, Choudhary KK, Shukla AK, Sunita K, Kumar A, Bontempi E, Ma Y (2023) Current scenario and future prospects of endophytic microbes: promising candidates for abiotic and biotic stress management for agricultural and environmental sustainability. Microb Ecol 86:1455–1486. https://doi.org/10.1007/s00248-023-02190-1

Bergstrom CT, Lachmann M (2003) The Red King effect: when the slowest runner wins the co evolutionary race. Proc Natl Acad Sci USA 100:593–598

Bhardwaj A, Kaur K, Kumar A, Singh SK, Kharshiing B (2024) Development of microbial biotechnology products for phytopathogen management. In: The potential of microbes for a circular economy. Academic Press, pp 105–130. https://doi.org/10.1016/B978-0-443-15924-4.00013-8

Böhm M, Hurek T, Reinhold-Hurek B (2007) Twitching motility is essential for endophytic rice colonization by the N_2-fixing endophyte *Azoarcus* sp. strain BH72. Mol Plant Microbe Interact 20(5):526–533

Brader G, Compant S, Mitter B, Trognitz F, Sessitsch A (2014) Metabolic potential of endophytic bacteria. Curr Opin Biotechnol 27:30–37

Bright M, Bulgheresi S (2010) A complex journey: transmission of microbial symbionts. Nat Rev Microbiol 8(3):218–230

Bulgarelli D, Rott M, Schlaeppi K, van Themaat EVL, Ahmadinejad N, Assenza F, Rauf P, Huettel B, Reinhardt R, Schmelzer E, Peplies J (2012) Revealing structure and assembly cues for Arabidopsis root-inhabiting bacterial microbiota. Nature 488(7409):91–95

Chanway CP (1996) Endophytes: they're not just fungi. Can J Bot 74:321–322

Clay K, Kover PX (1996) The red queen hypothesis and plant/pathogen interactions. Annu Rev Phytopathol 34:29–50

Compant S, Duffy B, Nowak J, Clement C, Barka EA (2005) Use of plant growth-promoting bacteria for biocontrol of plant diseases: principles, mechanisms of action, and future prospects. Appl Environ Microbiol 71:4951–4959

Compant S, Kaplan H, Sessitsch A, Nowak J, Ait Barka E, Clement C (2008) Endophytic colonization of *Vitis vinifera* L. by *Burkholderia phytofirmans* strain PsJN: from the rhizosphere to inflorescence tissues. FEMS Microbiol Ecol 63(1):84–93

Compant S, Clément C, Sessitsch A (2010) Plant growth-promoting bacteria in the rhizo- and endosphere of plants: their role, colonization, mechanisms involved and prospects for utilization. Soil Biol Biochem 42(5):669–678

Compant S, Mitter B, Colli-Mull JG, Gangl H, Sessitsch A (2011) Endophytes of grapevine flowers, berries, and seeds: identification of cultivable bacteria, comparison with other plant parts, and visualization of niches of colonization. Microb Ecol 62(1):188–197

Cruz AF, Barka GD, Blum LEB, Tanaka T, Ono N, Kanaya S, Reineke A (2019) Evaluation of microbial communities in peels of Brazilian tropical fruits by amplicon sequence analysis. Braz J Microbiol 50(3):739–748

De Bary A (1866) Morphologie und Physiologie der Pilze, Flechten, und Myxomyceten. In: Hofmeister's handbook of physiological botany. W. Engelmann, Leipzig

Devi S, Sharma S, Tiwari A, Bhatt AK, Singh NK, Singh M, Kaushalendra, Kumar A (2023) Screening for multifarious plant growth promoting and biocontrol attributes in Bacillus strains isolated from Indo Gangetic soil for enhancing growth of rice crops. Microorganisms 11:1085. https://doi.org/10.3390/microorganisms11041085

Droby S, Wisniewski M (2018) The fruit microbiome: a new frontier for postharvest biocontrol and postharvest biology. Postharvest Biol Technol 140:107–112

Droby S, Wisniewski M, Teixido N, Spadaro D, Jijakli MH (2019) Biocontrol of postharvest diseases with antagonistic microorganisms. In: Palou L, Smilanick JL (eds) Postharvest pathology of fresh horticultural produce. CRC Press, p 842

Elbeltagy A, Nishioka K, Suzuki H, Sato T, Sato YI, Morisaki H, Mitsui H, Minamisawa K (2000) Isolation and characterization of endophytic bacteria from wild and traditionally cultivated rice varieties. Soil Sci Plant Nutr 46(3):617–629

Elsen S, Swem LR, Swem DL, Bauer CE (2004) RegB/RegA, a highly conserved redox-responding global two-component regulatory system. Microbiol Mol Biol Rev 68:263–279

Frank AC, Saldierna Guzman JP, Shay JE (2017) Transmission of bacterial endophytes. Microorganisms 5(4):70

Galsurker O, Diskin S, Maurer D, Feygenberg O, Alkan N (2018) Fruit stem-end rot. Horticulturae 4(4):50

Gasser I, Cardinale M, Muller H, Heller S, Eberl L, Lindenkamp N, Kaddor C, Steinbüchel A, Berg G (2011) Analysis of the endophytic lifestyle and plant growth promotion of *Burkholderia terricola* ZR2-12. Plant Soil 347:125–136

Glassner H, Zchori-Fein E, Compant S, Sessitsch A, Katzir N, Portnoy V, Yaron S (2015) Characterization of endophytic bacteria from cucurbit fruits with potential benefits to agriculture in melons (*Cucumis melo* L.). FEMS Microbiol Ecol 91(7):1–13

Grabka R, d'Entremont TW, Adams SJ, Walker AK, Tanney JB, Abbasi PA, Ali S (2022) Fungal endophytes and their role in agricultural plant protection against pests and pathogens. Plants 11(384). https://doi.org/10.3390/plants110303

Hallmann J, Quadt-Hallmann A, Mahaffee W, Kloepper J (1997) Bacterial endophytes in agricultural crops. Can J Microbiol 43:895–914

Hanin NA, Fitriasari PD (2019) Identification of endophytic fungi from fruits and seeds of jambolana (*Syzygium cumini* L.) skeels. IOP Conf Ser Earth Environ Sci 276(1):012060

Hardoim PR, Van Overbeek LS, Berg G, Pirttila AM, Compant S, Campisano A, Doring M, Sessitsch A (2015) The hidden world within plants: ecological and evolutionary considerations for defining functioning of microbial endophytes. Microbiol Mol Biol Rev 79(3):293–320

Istifadah N, Ningtyas DNY, Suryatmana P, Fitriatin BN (2017) The abilities of endophytic and biofertilizing bacteria and their combinations to suppress bacterial wilt disease (*Ralstonia solanacearum*) of Chili. In: 2nd international conference on sustainable agriculture and food security: a comprehensive approach. KnE Life Sciences, pp 296–304

Iwai H, Usui M, Hoshino H, Kamada H, Matsunaga T, Kakegawa K, Ishii T, Satoh S (2003) Analysis of sugars in squash xylem sap. Plant Cell Physiol 44(6):582–587

James EK, Gyaneshwar P, Mathan N, Barraquio WL, Reddy PM, Iannetta PP, Olivares FL, Ladha JK (2002) Infection and colonization of rice seedlings by the plant growth-promoting bacterium *Herbaspirillum seropedicae* Z67. Mol Plant Microbe Interact 15(9):894–906

Janisiewicz WJ, Jurick WM, Vico I, Peter KA, Buyer JS (2013) Culturable bacteria from plum fruit surfaces and their potential for controlling brown rot after harvest. Postharvest Biol Technol 76:145–151

Johnson GI, Mead AJ, Cooke AW, Dean JR (1992) Mango stem end rot pathogens-fruit infection by endophytic colonisation of the inflorescence and pedicel. Ann Appl Biol 120(2):225–234

Kashyap N, Singh SK, Yadav N, Singh VK, Kumari M, Kumar D, Shukla L, Kaushalendra, Bhardwaj N, Kumar A (2023) Biocontrol screening of endophytes: applications and limitations. Plants 12:2480. https://doi.org/10.3390/plants12132480

Keller L, Surette MG (2006) Communication in bacteria: an ecological and evolutionary perspective. Nat Rev Microbiol 4:249–258

Krishnan P, Bhat R, Kush A, Ravikumar P (2012) Isolation and functional characterization of bacterial endophytes from *Carica papaya* fruits. J Appl Microbiol 113(2):308–317

Kumar A, Singh VK, Tripathi V, Singh PP, Singh AK (2018) Plant growth-promoting rhizobacteria (PGPR): perspective in agriculture under biotic and abiotic stress. In: Crop improvement through microbial biotechnology. Elsevier, pp 333–342. https://doi.org/10.1016/B978-0-444-63987-5.00016-5

Kumari M, Qureshi KA, Jaremko M, White J, Singh SK, Sharma VK, Singh KK, Santoyo G, Puopolo G, Kumar A (2022) Deciphering the role of endophytic microbiome in postharvest diseases management of fruits: opportunity areas in commercial up-scale production. Front Plant Sci 13:1026575. https://doi.org/10.3389/fpls.2022.1026575

Lai K, Chen S, Hu M, Hu Q, Geng P, Weng Q, Jia J (2012) Control of postharvest green mold of citrus fruit by application of endophytic *Paenibacillus polymyxa* strain SG-6. Postharvest Biol Technol 69:40–48

Liu J, Abdelfattah A, Norelli J, Burchard E, Schena L, Droby S, Wisniewski M (2018) Apple endophytic microbiota of different rootstock/scion combinations suggests a genotype-specific influence. Microbiome 6:18

Luksa J, Vepstaitė-Monstavicė I, Apsegaitė V, Cereskienė LB, Stanevisiene R, Strazdaitė-Zielienė Z, Ravoitytė B, Aleknavičius D, Būda V, Mozuraitis R, Servienė E (2020) Fungal microbiota of sea buckthorn berries at two ripening stages and volatile profiling of potential biocontrol yeasts. Microorganisms 8:456. https://doi.org/10.3390/microorganisms8030456

Malfanova N, Lugtenberg BJ, Berg G (2013) Bacterial endophytes: who and where, and what are they doing there. In: Molecular microbial ecology of the rhizosphere. Wiley-Blackwell, pp 391–403. https://doi.org/10.1002/9781118297674.ch36

Medina-Romero YM, Roque-Flores G, Macías-Rubalcava ML (2017) Volatile organic compounds from endophytic fungi as innovative postharvest control of *Fusarium oxysporum* in cherry tomato fruits. Appl Microbiol Biotechnol 101(22):8209–8222

Miguel PSB, Delvaux JC, Oliveira MNV, Monteiro LCP, Freitas FDS, Costa MD, Totola MR, Moraes CA, Borges AC (2013) Diversity of endophytic bacteria in the fruits of *Coffea canephora*. Afr J Microbiol Res 7(7):586–594

Mocali S, Bertelli E, Cello FD, Mengoni A, Sfalanga A (2003) Fluctuation of bacteria isolated from elm tissues during different seasons and from different plant organs. Res Microbiol 154:105–114

Morsy M, Cleckler B, Armuelles-Millican H (2020) Fungal endophytes promote tomato growth and enhance drought and salt tolerance. Plants (Basel) 9(7):877. https://doi.org/10.3390/plants9070877

Nan L, Wang K, Pang B, Zhong Y, Zhang X, Li Y, Tao F, Yang U, Liu Y (2019) Isolation and identification of endophytes from carrots. AIP Conf Proc 2079:020018

Nigris S, Baldan E, Tondello A, Zanella F, Vitulo N, Favaro G et al (2018) Biocontrol traits of *Bacillus licheniformis* GL174, a culturable endophyte of *Vitis vinifera* cv. Glera. BMC Microbiol 18:133

Ou T, Xu WF, Wang F, Strobel G, Zhou ZY, Xiang ZH, Liu J, Xie J (2019) A microbiome study reveals seasonal variation in endophytic bacteria among different mulberry cultivars. Comput Struct Biotechnol J 17:1091–1100

Pang L, Xia B, Liu X, Yi Y, Jiang L, Chen C, Li P, Zhang M, Deng X, Wang R (2021) Improvement of antifungal activity of a culture filtrate of endophytic *Bacillus amyloliquefaciens* isolated from kiwifruit and its effect on postharvest quality of kiwifruit. J Food Biochem 45(1):e13551

Pathak P, Rai VK, Can H, Singh SK, Kumar D, Bhardwaj N, Roychowdhury R, de Azevedo LCB, Kaushalendra, Verma H, Kumar A (2022) Plant-endophyte interaction during biotic stress management. Plants 11:2203. https://doi.org/10.3390/plants11172203

Pawlowski K, Klosse U, Debruijn FJ (1991) Characterization of a novel *Azorhizobium caulinodans* ORS571 2-component regulatory system, NtrY/NtrX, involved in nitrogen-fixation and metabolism. Mol Gen Genet 231:124–138

Petrini O (1991) Fungal endophytes of tree leaves. In: Microbial ecology of leaves. Springer, New York, pp 179–197

Pettersson M, Baath E (2003) Temperature-dependent changes in the soil bacterial community in limed and unlimed soil. FEMS Microbiol Ecol 45:13–21

Prakamhang J, Minamisawa K, Teamtaisong K, Boonkerd N, Teaumroong N (2009) The communities of endophytic diazotrophic bacteria in cultivated rice (*Oryza sativa* L.). Appl Soil Ecol 42:141–149

Rasche F, Lueders T, Schloter M, Schaefer S, Buegger F, Gattinger A, Hood-Nowotny RC, Sessitsch A (2009) DNA-based stable isotope probing enables the identification of active bacterial endophytes in potatoes. New Phytol 181(4):802–807

Reinhold-Hurek B, Maes T, Gemmer S, Van Montagu M, Hurek T (2006) An endoglucanase is involved in infection of rice roots by the not-cellulose-metabolizing endophyte *Azoarcus* sp. strain BH72. Mol Plant Microbe Interact 19:181–188

Ren F, Dong W, Yan DH (2019a) Endophytic bacterial communities of Jingbai Pear trees in north China analyzed with Illumina sequencing of 16S rDNA. Arch Microbiol 201(2):199–208

Ren F, Dong W, Sun H, Yan DH (2019b) Endophytic mycobiota of jingbai pear trees in north China. Forests 10(3):260

Rosenblueth M, Martinez-Romero E (2006) Bacterial endophytes and their interaction with hosts. Mol Plant Microbe Interact 19:827–837

Saikkonen K, Wäli P, Helander M, Faeth SH (2004) Evolution of endophyte–plant symbioses. Trends Plant Sci 9(6):275–280

Sauer K, Camper AK (2001) Characterization of phenotypic changes in *Pseudomonas putida* in response to surface-associated growth. J Bacteriol 183(22):6579–6589

Schlaeppi K, Bulgarelli D (2015) The plant microbiome at work. Mol Plant Microbe Interact 28(3):212–217

Shameer S, Prasad TNVKV (2018) Plant growth promoting rhizobacteria for sustainable agricultural practices with special reference to biotic and abiotic stresses. Plant Growth Regul 84(3):603–615

Shi J, Liu A, Li X, Feng S, Chen W (2010) Identification of endophytic bacterial strain MGP1 selected from papaya and its biocontrol effects on pathogens infecting harvested papaya fruit. J Sci Food Agric 90(2):227–232

Singh M, Kumar A, Singh R, Pandey KD (2017a) Endophytic bacteria: a new source of bioactive compounds. 3 Biotech 7:1–14. https://doi.org/10.1007/s13205-017-0942-z

Singh VK, Singh AK, Kumar A (2017b) Disease management of tomato through PGPB: current trends and future perspective. 3 Biotech 7(4):255

Singh M, Singh D, Gupta A, Pandey KD, Singh PK, Kumar A (2019) Plant growth promoting rhizobacteria: application in biofertilizers and biocontrol of phytopathogens. In: PGPR amelioration in sustainable agriculture. Woodhead Publishing, pp 41–66

Singh M, Srivastava M, Kumar A, Singh AK, Pandey KD (2020) Endophytic bacteria in plant disease management. In: Microbial endophytes. Woodhead Publishing, pp 61–89

Singh M, Pandey KD, Rathore AC, Sharma SP, Kumar R (2022a) Bacterial antagonists: effective tools for the management of postharvest diseases in fruits, vegetables, and food grains. In: Kumar A (ed) Microbial biocontrol: food security and post harvest management. Springer, Cham, pp 295–309. https://doi.org/10.1007/978-3-030-87289-2_13

Singh R, Pandey KD, Singh M, Singh SK, Hashem A, Al-Arjani ABF, Abd-Allah EF, Singh PK, Kumar A (2022b) Isolation and characterization of endophytes bacterial strains of Momordica charantia L. and their possible approach in stress management. Microorganisms 10:290. https://doi.org/10.3390/microorganisms10020290

Singh M, Qureshi KA, Jaremko M, Rajput M, Singh SK, Kaushalendra, Pandey KD, Ferreira LFR, Kumar A (2022c) Bioprospects of endophytic bacteria in plant growth promotion and ag-nanoparticle biosynthesis. Plants 11:1787. https://doi.org/10.3390/plants11141787

Solanki MK, Wang Z, Kaushik A, Singh VK, Roychowdhury R, Kumar M, Kumar D, Singh J, Singh SK, Dixit B, Kumar A (2024) From orchard to table: significance of fruit microbiota in postharvest diseases management of citrus fruits. Food Control 165:110698. https://doi.org/10.1016/j.foodcont.2024.110698

Sturz AV, Christie BR (1996) Endophytic bacteria of red clover as causal agents of allelopathic clover–maize syndromes. Soil Biol Biochem 28:583–588

Suarez-Moreno ZR, Devescovi G, Myers M, Hallack L, Mendonça-Previato L, Caballero-Mellado J, Venturi V (2010) Commonalities and differences in regulation of N-acyl homoserine lactone quorum sensing in the beneficial plant-associated *Burkholderia* species cluster. Appl Environ Microbiol 76(13):4302–4317

Suhandono S, Kusumawardhani MK, Aditiawati P (2016) Isolation and molecular identification of endophytic bacteria from Rambutan fruits (*Nephelium lappaceum* L.) cultivar Binjai. HAYATI J Biosci 23(1):39–44

Taha MDM, Jaini MFM, Saidi NB, Rahim RA, Shah UKM, Hashim AM (2019) Biological control of *Erwinia mallotivora*, the causal agent of papaya dieback disease by indigenous seed-borne endophytic lactic acid bacteria consortium. PLoS One 14(12):e0224431

Tholozan JL, Cappelier JM, Tissier JP, Delattre G, Federighi M (1999) Physiological characterization of viable-but-nonculturable *Campylobacter jejuni* cell. Appl Environ Microbiol 65:1110–1116

Wang Z, Solanki MK, Kumar A, Solanki AC, Pang F, Ba ZX, Niu JQ, Ren ZX (2023) Promoting plant resilience against stress by engineering root microenvironment with Streptomyces inoculants. Microbiol Res 277:127509. https://doi.org/10.1016/j.micres.2023.127509

Wu W, Lei J, Hussain M, Cao S, Du B, Wang R (2019) Structure and function of the fruit microbiome in healthy and diseased kiwifruit. Pak J Agric Sci 56(3):577–585

Yi Y, de Jong A, Frenzel E, Kuipers OP (2017) Comparative transcriptomics of *Bacillus mycoides* strains in response to potato-root exudates reveals different genetic adaptation of endophytic and soil isolates. Front Microbiol 8:1487

Zheng H, Mao Y, Teng J, Zhu Q, Ling J, Zhong Z (2015) Flagellar-dependent motility in *Mesorhizobium tianshanense* is involved in the early stage of plant host interaction: study of an flgE mutant. Curr Microbiol 70(2):219–227

Zhimo VY, Kumar A, Biasi A, Salim S, Feygenberg O, Toamy MA, Abdelfattaah A, Medina S, Freilich S, Wisniewski M, Droby S (2021) Compositional shifts in the strawberry fruit microbiome in response to near-harvest application of Metschnikowia fructicola, a yeast biocontrol agent. Postharvest Biol Technol 175:111469. https://doi.org/10.1016/j.postharvbio.2021.111469

Utilization of Rice Husk-Derived Silica in Fertilizer and Pesticide Formulation

5

Masoumeh Kordi, Robab Salami, Parisa Bolouri, Zeynab Borzouyi, Behnam Asgari Lajayer, Weixi Shu, and G. W. Price

Abstract

Rice husk, a by-product of rice production, is generated in significant quantities across rice-producing nations. Annually, over 150 million tons of rice husks are produced from rice fields worldwide. Utilization of rice husk or its derived silica in agriculture offers an eco-friendly and cost-efficient alternative to traditional silica sources, promoting sustainable resource utilization and waste management. Silicon, a beneficial element for plant growth, exhibits growth-enhancing properties in plants. The application of nanoparticles in agriculture has emerged as an effective approach to crop production, addressing the escalating global demand for food. Rice husk, rice husk ash, and silica derived from rice husk have found extensive application in agriculture, serving as fertilizers, pesticides, and stress mitigators. This chapter explores various instances of their utilization in agricultural practices.

Keywords

Fertilizer · Pesticide · Rice husk · Silica · Silica nanoparticle

---

M. Kordi · R. Salami
Department of Cell & Molecular Biology, Faculty of Life Sciences and Biotechnology, Shahid Beheshti University, Tehran, Iran

P. Bolouri
Department of Genetics and Bioengineering, Yeditepe University, Istanbul, Turkey

Z. Borzouyi
Agriculture Department, Islamic Azad University of Sabzevar, Sabzevar, Khorasan Razavi, Iran

B. A. Lajayer (✉) · W. Shu · G. W. Price (✉)
Faculty of Agriculture, Dalhousie University, Truro, NS, Canada
e-mail: basgari@dal.ca; gprice@dal.ca

A. Kumar, M. K. Solanki (eds.), *Microbial Biocontrol Techniques*, Microorganisms for Sustainability 54,
https://doi.org/10.1007/978-981-97-8739-5_5

5.1 Introduction

In 2019, global paddy field production reached approximately 755 million tons, yielding an average of 20% rice husk as a by-product (Mistry et al. 2019; Silva et al. 2021). Rice husk stands as one of the most prevalent agricultural residues and by-products in rice-producing nations worldwide, including China, India, Bangladesh, Brazil, the USA, Cambodia, Vietnam, Myanmar, and Southeast Asia (Pode 2016; Mistry et al. 2019). Comprising cellulose (25–33%), lignin (25–31%), hemicellulose (18–21%), pentosans (15%), and minerals like silica (16%), alkalis, and trace elements, rice husk serves as a significant source of silica in agricultural applications (Mistry et al. 2019; Kordi et al. 2023).

In recent years, silica fertilizers have drawn attention for their potential to enhance plant growth, soil fertility, and resistance to diseases and pests (Tayade et al. 2022; Luyckx et al. 2017). Silica, the second most abundant element in the Earth's crust after oxygen, prevails as a dominant species in terrestrial environments (Vaculík et al. 2020).

Silica fertilizers can be derived from various sources, including silicate minerals like zeolite, quartz, and feldspar, as well as industrial waste materials such as fly ash, bottom ash, and slag (Tayade et al. 2022; Luyckx et al. 2017; Anggria and Masunaga 2021). Agriculture utilizes various sources of silicon fertilizers, including raw rice husk, rice husk ash, biochar from carbonized straw or husk, silicate slag, diatomite, wollastonite, sodium silicate, potassium silicate, and foliar silicon in soluble silicic acid forms (Gascho 2001; Sandhya et al. 2018).

Silica nanoparticles, ranging from 1 to 100 nm in size and featuring various structures such as spherical, hollow, and shaped (e.g., rod, cube), offer notable advantages over bulk silica (Takai-Yamashita and Fuji 2020; Pal et al. 2020; Khurana et al. 2023). The porous nature of silica nanoparticles has made them ideal carriers for delivering fertilizers, pesticides, plant growth regulators, DNA, and proteins in agricultural production and plant biotechnology (Resham et al. 2015; Goswami et al. 2022; Bueno and Ghoshal 2022). This chapter succinctly summarizes relevant information on the utilization of rice husk-derived silica in formulating fertilizers and pesticides.

5.2 Application of Rice Husk in Fields

The utilization of rice straw in soil has been extensively researched over the years, revealing its capacity to enhance soil productivity and improve physicochemical properties such as total nitrogen, phosphorus, potassium, and soil pH (Asadi et al. 2021). Rice husk serves as a significant carrier in fertilizer production owing to its distinctive properties (Ogbo and Odo 2011). It exhibits the potential for extended shelf life, rendering it an attractive option for preserving the efficacy of biofertilizers (Williams et al. 2023). In the study by Huang et al. (2019), the impact of rice husk biochar application in rice fields was investigated, with findings indicating that its effect on rice yield and functional attributes is dependent on the duration of application. Interestingly, the use of rice husk biochar, particularly after six seasons, led to an increase in rice grain yield, highlighting its potential to enhance soil fertility and the efficacy of fertilizers, especially in nutrient-deficient soils. These findings underscore the potential of rice husk in enhancing the quality and efficacy of fertilizers, thus positioning it as a valuable resource in the realm of agriculture (Williams et al. 2023). Additionally, rice husk serves as an organic fertilizer to improve water consumption efficiency in agricultural fields (Zou and Yang 2019). In another study by Frasetya et al. (2019), silica rice husk extracts were employed in rice fields, resulting in enhanced vegetative growth and rice yield compared to the control group. Similarly, Mahmad-Toher et al. (2022) compared two silica sources, calcium silicate and rice husk ash, in rice plants grown under greenhouse conditions, finding both to improve performance compared to the control. Consequently, these studies indicate the potential of rice husk ash as a renewable resource fertilizer in rice cultivation.

5.3 Rice Husk as a Source of Natural Silica

Rice husks contain approximately 20% hydrated amorphous silica, making them less susceptible to microbial degradation and an excellent carrier for biofertilizers (Chandrasekhar et al. 2003; Ogbo and Odo 2011). Various silicon fertilizers used in agriculture include raw rice husk, rice husk ash, biochar from carbonized straw or husk, silicate slag, diatomite, wollastonite, sodium silicate, potassium silicate, and foliar soluble silicic acid forms (Tayade et al. 2022). The significant silica content in rice husks highlights the importance of adequate silica levels in rice plants to support healthy growth. Silica deficiency in rice can lead to detrimental effects such as stunted stem growth, reduced rice production, and increased water evaporation (Tateda 2016). Recognizing this, Japanese scientists have investigated the feasibility of recycling rice husks to produce silica fertilizer for rice plants, aiming to reintroduce the silica present in husks back into paddy fields to sustain rice production (Sekifuji and Tateda 2019).

5.4 Silica

Silica, comprising 50–70% of soil mass, ranks as the second most abundant soil element (Ma and Yamaji 2006). Its diverse roles in plant processes are underscored by its accumulation in plants, ranging from 0.1% to 10% of dry weight, with variations observed among different plant organs (Luyckx et al. 2017). Plants are categorized into three groups—accumulators (e.g., rice), intermediates (e.g., cucumber), and excluders (e.g., tomato)—based on their tissue silica content (Luyckx et al. 2017).

Solid and liquid forms of silicon, such as potassium silicate, sodium silicate, and calcium silicate, are extensively utilized across diverse crops, horticulture, and turf management (Luyckx et al. 2017; Barão 2023). Silica fertilizers, also known as silicone fertilizers, serve as soil amendments supplying silicon to the soil-plant system (Hogan et al. 2019). Commercial silicone fertilizers, available in solid and liquid forms, include calcium silicate, magnesium silicate, and potassium silicate (El-Mehrat et al. 2017; Tayade et al. 2022).

Silica fertilizers offer several key benefits, including enhancing plant silicon nutrition and reinforcing plant resistance to biotic and abiotic stresses such as diseases, insects, and adverse environmental conditions (Tayade et al. 2022). Additionally, they contribute to augmenting soil fertility and improving soil pH (Tayade et al. 2022). By promoting sustainable agriculture and reducing reliance on chemical fertilizers and pesticides, these fertilizers help mitigate abiotic stresses like drought and nutrient deficiencies in crops (Tayade et al. 2022).

5.5 Nano-fertilizers and Nano-silica Fertilizers

Nanotechnology offers significant advancements in agriculture, particularly through the development of nano-fertilizers, which enhance plant nutrition by improving element bioavailability, increasing nutrient efficiency, and reducing leaching and wastage of nutrients in agricultural lands. These nano-fertilizers hold promise for boosting sustainable agricultural productivity and providing eco-friendly alternatives to traditional chemical fertilizers (Maghsoodi et al. 2020). Silica nanoparticles, characterized by their unique physicochemical properties such as small size, high surface-to-volume ratio, and solubility, have gained traction in agriculture for their ability to mitigate biotic and abiotic stresses, enhance nutrition, control pests, and improve plant product quality (Awad-Allah 2023). For instance, a study conducted by Amalya et al. (2020) investigated the combined effect of nitrogen (N), phosphorus (P), potassium (K), and nano-silica fertilizers on the yield of black rice. The study revealed that the optimal combination resulted in the highest yield per hectare.

5.6 Silica for Stress

Silica can enhance the expression of antioxidant enzymes, osmoprotectants, proteins, and secondary metabolites, thereby reducing both biotic and abiotic stresses. Silica nanoparticles further stimulate non-enzymatic defense reactions by promoting the synthesis of ascorbic acid, proline, glutathione, and phytohormones (Naaz et al. 2023). The use of nano-fertilizers offers numerous advantages over conventional fertilizers and enhances agricultural productivity (Snehal and Lohani 2018). Despite not being an essential nutrient for plant growth, silica plays a pivotal role in mitigating various stressors, including insect pests, pathogen diseases, metal stress, drought, salt stress, freezing, nutrient imbalance, and radiation damage (Snehal and Lohani 2018; Yan et al. 2024).

5.6.1 Silica for Metal Abiotic Stress

In recent decades, the impact of heavy metal stress on plants has gained considerable attention. These metals, characterized by their high relative density, exhibit toxicity even at low concentrations (Bhat et al. 2019). Cadmium (Cd), nickel (Ni), lead (Pb), iron (Fe), zinc (Zn), cobalt (Co), arsenic (As), chromium (Cr), silver (Ag), platinum (Pt), and others are among the elements grouped as heavy metals, with most not playing essential roles in plant physiology (Chibuike and Obiora 2014). They are further classified into non-essential (Ag, Cd, Pb, Hg, Cr, As) and essential micronutrient (Cu, Zn, Fe, Mn, Mo, Ni, and Co) categories (Kalaivanan and Ganeshamurthy 2016). Heavy metal toxicity disrupts various plant processes, including growth, photosynthesis, nutrient absorption, and water balance, leading to chlorosis, reduced biomass, and premature aging (Ali et al. 2011). These metals interfere with plant growth by disrupting essential cellular functions such as those of proteins, lipids, and thylakoid membranes (Kim et al. 2014). Silica has emerged as a promising tool for mitigating heavy metal toxicity in plants, offering potential relief from adverse effects (Liang et al. 2007; Kim et al. 2014). Abiotic stress, including heavy metal pollution, contributes significantly to global crop losses, accounting for 50% of agricultural yield reductions, thereby posing substantial threats to sustainable agriculture and food safety (Zargar et al. 2019; Singh et al. 2021; Qin et al. 2021).

Silica nanoparticles demonstrate protective properties for plants against various heavy metals, including Cd, As, Hg, Pb, Cu, Al, and Cr (Yan et al. 2024). Furthermore, rice husk proves to be an efficient material for heavy metal removal from soil, acting as a potent adsorbent for Cu, Fe, and Zn (Chockalingam and Subramanian 2006). Rice husk biochar and ash enhance soil conditions in metal-contaminated areas and facilitate plant growth by enhancing nutrient uptake and antioxidant enzymatic activities. Research indicates that soil application of rice husk ash can decrease Pb concentration in Ricinus communis roots by up to 87% compared to control plants (Kiran and Prasad 2019).

In another study conducted by Wang et al. (2019), biochar served as a cost-effective and efficient adsorbent for heavy metals in wheat. The findings revealed that biochar not only alleviated the impacts of heavy metals on seedling growth but also notably improved germination potential, germination speed, and root growth of seedlings. Silica nanoparticles have also been effective in reducing metal accumulation in plants experiencing metal stress (Table 5.1 and Fig. 5.1). Table 5.1 provides examples of silica applications aimed at mitigating heavy metal pollution in plants.

5.6.2 Silica for Salt and Drought Stress

Salt stress affects approximately 20% of all arable land worldwide (Pinedo-Guerrero et al. 2020), leading to diminished growth and yield of crops in saline environments. Incorporating macro silica elements into soil has been shown to improve soil conditions and enhance plant tolerance to stress (Abdel-Haliem et al. 2017; Zhu and

Table 5.1 Examples of the use of silica versus metal stress

Metal	Plant	Origin of silica nanoparticle	In vivo or in vitro	Kind of experimental	References
Cd	*Oryza sativa* L. japonica	SiNPs were prepared through sol-gel method	In vitro	Inductively coupled plasma mass spectroscopy	Cui et al. (2017)
Cd	*Brassica napus*	Rice straw extract	In vivo	Examined growth parameters in plant	Ahmed et al. (2023)
Cd	*Triticum aestivum*	SiNPs of Alfa Aesar with crystalline structure	In vivo	Examined growth parameters in plant	Ali et al. (2019)
Cd	*Triticum aestivum*	SiNPs of Alfa Aesar with crystalline structure	In vivo	Examined growth parameters in plant	Khan et al. (2020)
As	*Solanum lycopersicum* L.	Si NPs from SkySpring Nanomaterials Inc.	In vivo	Measured tomato growth and yield and microscopic analyses	González-Moscoso et al. (2022)
As	*Oryza sativa*	Chitosan-based silicon nanoparticles	In vivo	Measured growth parameters in plant	Jia-Yi et al. (2022)
Ag	*Glycine max*	Nano-SiO_2 from the J&K Scientific Ltd.	In vivo	Measured growth parameters in plant	Li et al. (2020)
Ag	*Vigna radiata*	Mesoporous silica nanoparticles	In vivo	Measured growth parameters in plant	Mitra et al. (2023)
Pb	*Oryza sativa*	Synergistic application of Si and Fe-NPs	In vivo	Measured growth parameters in plant	Ghouri et al. (2024)
Pb	*Coriandrum sativum*	The combined application of Pb resistant microbes and silicon nanoparticles	In vivo	Measured growth parameters in plant	Fatemi et al. (2020)
Cu	*Triticum aestivum*	Si-nanoparticles	In vivo	Measured growth parameters in plant	Riaz et al. (2022)
Al	*Zea mays*	Nano SiO_2 from Sigma Aldrich	In vivo	Measured growth parameters in plant	de Sousa et al. (2019)
Al	*Cicer arietinum*	SiNPs were synthesized chemical and biological (green) procedures	In vivo	Measured growth parameters in plant	Chandra et al. (2020)

(continued)

Table 5.1 (continued)

Metal	Plant	Origin of silica nanoparticle	In vivo or in vitro	Kind of experimental	References
Cr	*Brassica napus*	Si NPs were commercially purchased	In vivo	Measured growth parameters in plant	Huang et al. (2024)
Cr	*Triticum aestivum*	Sol-gel method was used for the synthesis of SiONPs	In vivo	Measured growth parameters in plant	Manzoor et al. (2022)

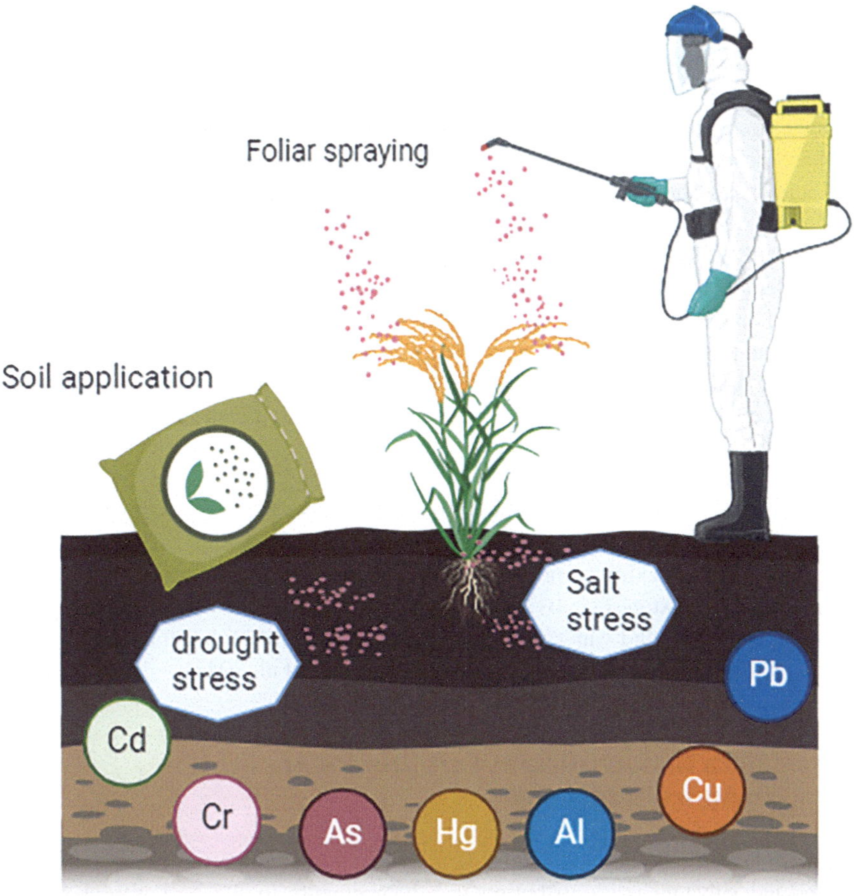

Fig. 5.1 The application of silica in the form of foliar application and soil application is effective against heavy metal stress and salinity and drought stress

Gong 2014). Salinity stress induces physiological drought in plants, resulting in the accumulation of reactive oxygen species (ROS) and disruption of normal metabolic processes, ultimately damaging the plasmalemma and membrane system (Zushi et al. 2009). Ionic toxicity, osmotic limitations, unbalanced nutrition (e.g., rapid sodium absorption compared to potassium), and oxidative stress collectively contribute to reduce plant growth and productivity under salinity stress conditions (Hussain et al. 2012; Ismail et al. 2022). In addition to ROS formation, salt stress triggers the production of free radicals, including hydroxyl radicals (OH), single oxygen ($^1O_2$), superoxide anions, and hydrogen peroxide (H_2O_2) (Shalata and Neumann 2001; Hossain et al. 2015; Ismail et al. 2022). Phuong et al. (2020) investigated the impact of rice husk biochar on reducing salinity and promoting rice growth under field conditions, revealing significant effects of biochar in soil amendment, particularly in mitigating salinity stress.

Enhanced root water absorption and anatomy are crucial for plants to withstand drought stress, with silica playing a pivotal role in these processes (Hattori et al. 2008). Silica application modulates polyamine and 1-aminocyclopropane-1-carboxylic acid levels under drought stress, promoting root growth over stem growth and enhancing water absorption through roots (Fleck et al. 2015; Shi et al. 2016). Safari et al. (2023) utilized rice husk biochar to alleviate water scarcity in perennial ryegrass, demonstrating improvements in soil physical and chemical properties, enhanced nutrient absorption, increased shoot and root growth, elevated chlorophyll content and photosynthesis efficiency, and enhanced drought tolerance in perennial ryegrass. Similarly, Shashi et al. (2018) investigated the effects of rice husk biochar on maize under drought conditions, revealing that the application of 20 tons/ha of rice husk biochar significantly enhanced maize growth, resulting in the highest yield under drought conditions.

Various studies have reported the mitigating effects of silica on salt and drought stress across different crops, as summarized in Table 5.2 and Fig. 5.1.

Silica fertilizer exhibits protective characteristics that bolster plants against biotic and abiotic stressors. Its incorporation into the soil not only enhances soil physical, chemical, and water properties but also augments nutrient availability for plants (Bocharnikova and Matichenkov 2014; Tayade et al. 2022).

5.7 Silica Nanoparticle as Fertilizer for Application in Agricultural Land

Nano-silica fertilizer, derived from silicon, serves as a nano-carrier for various molecules and substances, contributing to advancements in the agricultural sector, including the enhancement of biofertilizers. Due to its small size and higher surface-to-weight ratio, nano-silica fertilizer can function as a nano-pesticide and offers an alternative to mitigate the effects of soil salinity compared to the other silica fertilizers, nano-silica fertilizer boasts a higher concentration and can be utilized at a significantly lower consumption rate. It can be applied via foliar and soil applications (Rastogi et al. 2019; Goswami and Mathur 2022). In a study by Snehal

Table 5.2 Examples of the use of silica versus salinity and drought stress

Stress	Plant	Origin of silica nanoparticle	In vivo or in vitro	Kind of experimental	References
Salt	*Ocimum basilicum*	Silica nanoparticles	In vivo	Assessed morphological traits and physiological traits	Kalteh et al. (2014)
Salt	*Oryza sativa*	Si ions and silica nanoparticles (both extracted from rice straw)	In vivo	Assessed physiological and biochemical responses	Abdel-Haliem et al. (2017)
Salt	*Cucumis sativus*	Amorphous silica nanoparticles	In vivo	Assessed growth and productivity of cucumber	Alsaeedi et al. (2019)
Salt	*Pisum sativum*	Silicon and nano-silicon	In vivo	Assessed growth, yield, ions content	Ismail et al. (2022)
Drought	*Arabidopsis*	Mesoporous silica nanoparticles	In vitro	Assessed plant growth and drought stress tolerance	Tran et al. (2023)
Drought	*Pisum sativum*	Silica nanoparticles	In vivo	Assessed growth, oxidative stress, and antioxidative response of pea	Sutulienė et al. (2021)

and Lohani (2018), the combined application of NPK fertilizer and nano-silica was investigated for its effects on Ganyong grass plants, revealing a positive impact on increasing plant height, dry weight, and wet weight. Similarly, Wibowo et al. (2017) utilized nano-silica derived from rice straw in peanut, tomato, and soybean plants, observing an increase in plant height and leaf number with the application of this nano-fertilizer. However, in a study by Bakry et al. (2015), the effects of potassium silicate and magnesium silicate on two varieties of flax did not demonstrate improvements in nutrient content and yield.

5.8 Silica as a Pesticide and Carrier

Silica fertilizer enhances resistance against biological stresses induced by various pests, including Lepidoptera, *Hemiptera*, *Homoptera*, *Diptera*, *Thysanoptera*, *Oleoptera*, and non-insect attacks, in both monocotyledonous and dicotyledonous plants (Hou and Han 2010; Dias et al. 2014; Han et al. 2015).

In recent decades, nations reliant on agriculture have intensified research into pesticide delivery systems for pest population management (Rai and Ingle 2012; Kah and Hofmann 2014). Encapsulation and pesticide delivery via carriers offer manifold advantages, including shielding active substances from decomposition, environmental mitigation of active toxicity, enhanced bioavailability for hydrophobic actives, and sustainable delivery of substantial active substance quantities.

Consequently, maintaining effective poison concentrations for prolonged periods reduces the necessity for frequent applications (Wibowo et al. 2014). Although silica nanoparticles have primarily found use in the biomedical field, their application in agriculture has gained traction (Hussain et al. 2013; Rastogi et al. 2019; Shaban and Hasanzadeh 2020; Goswami and Mathur 2022). Thabet et al. (2021) demonstrated that silica nanoparticles possess the capability to repel or eliminate pests and predators upon application to plants. Moreover, they augment the attraction of natural enemies through herbivory-induced plant volatiles, underscoring their potential as potent pesticides against various pest insects and as components in novel insecticide formulations. The ensuing section delineates several applications of nano-silica as an insecticide and carrier in agricultural pesticide formulations.

In Wibowo et al.'s (2014) study, biocompatible oil core silica shell nano-capsules were utilized for the sustained release of fipronil insecticide. By adjusting the thickness of the silica shell, controlled release of fipronil in vitro was achieved, showcasing the efficacy and versatility of an environmentally friendly release system for insecticides. Saed et al. (2022) investigated the impact of nano-silica derived from sugarcane bagasse ash, alone and combined with diatomaceous earth, on adults *of Tribolium confusum* Jacquelin du Val. and *Rhyzopertha dominica*. Their findings indicated that nano-silica enhanced the insecticidal activity of diatomaceous earth. In another experiment, Thabet et al. (2021) explored the effects of silica nanoparticles on bean and soybean field, demonstrating a reduction in the population of *Aphis craccivora*, *Liriomyza trifolii*, and *Spodoptera littoralis* pests in the field following the application of silica nanoparticles. Debnath et al. (2012) investigated the efficacy of silica nanoparticles against second instar larvae of Spodoptera litura in a 2012 study, concluding that these nanoparticles effectively eliminate insect larvae. Similarly, Ayoub et al. (2017) explored the entomotoxic effects of silica nanostructures on *Spodoptera littoralis* in a 2017 study, observing that the surface characteristics and size of silica nanostructure particles significantly influence their entomotoxicity, suggesting silica nanostructures as valuable tools in pest management programs. Barik et al. (2012) evaluated the effects of nano-silica on malaria mosquitoes in a 2012 study, demonstrating the effectiveness of silica nanoparticles in controlling malaria vector mosquitoes. El-Naggar et al. (2020) investigated the effect of silica nanoparticles as a pesticide on four storage pests, *Sitophilus oryzae*, *Rhyzopertha dominica*, *Tribolium castaneum*, and *Oryzaephilus surinamensis*, in maize in a 2020 research. The results suggested that silica nanoparticles can serve as effective pesticides at very low and safe doses for products during storage. El-Bendary and El-Helaly (2013) utilized nano-silica on tomato plants against *Spodoptera littoralis*, revealing that the application of silica nanoparticles significantly enhances resistance in tomato plants against this pest.

5.9 Conclusions

The utilization of silica in agriculture extends across various applications. Rice husks and diverse agricultural residues contain substantial silica content, rendering them invaluable resources for agricultural purposes. By harnessing rice husks and their derivatives, not only can the expenses associated with agricultural waste disposal be mitigated, but they can also be repurposed as essential materials for the agricultural sector, thereby significantly contributing to the bio-economy. For example, rice husks or their silica derivatives can serve as key components in the production of premium-quality fertilizers and pesticides. Furthermore, silica and silica nanoparticles have emerged as potent fertilizers in agriculture, offering effective solutions against both biotic and abiotic stresses. Embracing the strategy of utilizing agricultural waste as a silica source proves to be both cost-effective and versatile, fostering sustainable agricultural practices.

References

Abdel-Haliem MEF, Hegazy HS, Hassan NS, Naguib DM (2017) Effect of silica ions and nano silica on rice plants under salinity stress. Ecol Eng 99:282–289. https://doi.org/10.1016/j.ecoleng.2016.11.060

Ahmed T, Masood HA, Noman M, Al-Huqail AA, Alghanem SM, Khan MM, Muhammad S, Manzoor N, Rizwan M, Qi X (2023) Biogenic silicon nanoparticles mitigate cadmium (Cd) toxicity in rapeseed (Brassica napus L.) by modulating the cellular oxidative stress metabolism and reducing Cd translocation. J Hazard Mater 459:132070

Ali S, Bai P, Zeng F, Cai S, Shamsi IH, Qiu B, Wu F, Zhang G (2011) The ecotoxicological and interactive effects of chromium and aluminum on growth, oxidative damage and antioxidant enzymes on two barley genotypes differing in Al tolerance. Environ Exp Bot 70(2–3):185–191

Ali S, Rizwan M, Hussain A, ur Rehman MZ, Ali B, Yousaf B, Wijaya L, Alyemeni MN, Ahmad P (2019) Silicon nanoparticles enhanced the growth and reduced the cadmium accumulation in grains of wheat (Triticum aestivum L.). Plant Physiol Biochem 140:1–8

Alsaeedi A, El-Ramady H, Alshaal T, El-Garawany M, Elhawat N, Al-Otaibi A (2019) Silica nanoparticles boost growth and productivity of cucumber under water deficit and salinity stresses by balancing nutrients uptake. Plant Physiol Biochem 139:1–10. https://doi.org/10.1016/j.plaphy.2019.03.008

Amalya N, Yuniarti A, Setiawan A, Machfud Y (2020) The effect of N, P, K fertilizer and nano silica fertilizer to total N content, N uptake, and black rice yield (Oryza sativa L. Indica) on inceptisols from jatinangor. Plant Sci 8:185–188

Anggria L, Masunaga T (2021) A method for production of pure silica as fertilizer from industrial waste material. In: IOP conference series: earth and environmental science, vol 1. IOP Publishing, p 012213

Asadi H, Ghorbani M, Rezaei-Rashti M, Abrishamkesh S, Amirahmadi E, Chengrong C, Gorji M (2021) Application of rice husk biochar for achieving sustainable agriculture and environment. Rice Sci 28(4):325–343. https://doi.org/10.1016/j.rsci.2021.05.004

Awad-Allah EF (2023) Effectiveness of silica nanoparticle application as plant nano-nutrition: a review. J Plant Nutr 46(11):2763–2776

Ayoub HA, Khairy M, Rashwan FA, Abdel-Hafez HF (2017) Synthesis and characterization of silica nanostructures for cotton leaf worm control. J Nanotechnol Chem 7:91–100

Bakry B, Shedeed S, Nofal O (2015) Production and quality traits of two flax varieties as affected by foliar application of silicon fertilizers under sandy soil conditions. Res J Pharm, Biol Chem Sci 6(5):181–188

Barão L (2023) The use of si-based fertilization to improve agricultural performance. J Soil Sci Plant Nutr 23(1):1096–1108

Barik TK, Kamaraju R, Gowswami A (2012) Silica nanoparticle: a potential new insecticide for mosquito vector control. Parasitol Res 111:1075–1083

Bhat JA, Shivaraj S, Singh P, Navadagi DB, Tripathi DK, Dash PK, Solanke AU, Sonah H, Deshmukh R (2019) Role of silicon in mitigation of heavy metal stresses in crop plants. J Plant Nutr 8(3):71

Bocharnikova E, Matichenkov V (2014) Silicon fertilizers: agricultural and environmental impacts. In: Fertilizers: components, uses in agriculture and environmental impacts. Nova Publishers, pp 183–198

Bueno V, Ghoshal S (2022) Inorganic porous nanoparticles as pesticide or nutrient carriers. In: Inorganic nanopesticides and nanofertilizers: a view from the mechanisms of action to field applications. Springer, pp 363–390

Chandra J, Chauhan R, Korram J, Satnami ML, Keshavkant S (2020) Silica nanoparticle minimizes aluminium imposed injuries by impeding cytotoxic agents and over expressing protective genes in Cicer arietinum. Sci Hortic 260:108885. https://doi.org/10.1016/j.scienta.2019.108885

Chandrasekhar S, Satyanarayana K, Pramada P, Raghavan P, Gupta T (2003) Review processing, properties and applications of reactive silica from rice husk—an overview. J Mater Sci 38:3159–3168

Chibuike GU, Obiora SC (2014) Heavy metal polluted soils: effect on plants and bioremediation methods. Appl Environ Soil Sci. https://doi.org/10.1155/2014/752708

Chockalingam E, Subramanian S (2006) Studies on removal of metal ions and sulphate reduction using rice husk and Desulfotomaculum nigrificans with reference to remediation of acid mine drainage. Chemosphere 62(5):699–708

Cui J, Liu T, Li F, Yi J, Liu C, Yu H (2017) Silica nanoparticles alleviate cadmium toxicity in rice cells: mechanisms and size effects. Environ Pollut 228:363–369

de Sousa A, Saleh AM, Habeeb TH, Hassan YM, Zrieq R, Wadaan MAM, Hozzein WN, Selim S, Matos M, AbdElgawad H (2019) Silicon dioxide nanoparticles ameliorate the phytotoxic hazards of aluminum in maize grown on acidic soil. Sci Total Environ 693:133636. https://doi.org/10.1016/j.scitotenv.2019.133636

Debnath N, Mitra S, Das S, Goswami A (2012) Synthesis of surface functionalized silica nanoparticles and their use as entomotoxic nanocides. Pestic Biochem Physiol 221:252–256

Dias P, Sampaio M, Rodrigues M, Korndörfer A, Oliveira R, Ferreira S, Korndörfer G (2014) Induction of resistance by silicon in wheat plants to alate and apterous morphs of Sitobion avenae (Hemiptera: Aphididae). Environ Entomol 43(4):949–956

El-Bendary H, El-Helaly A (2013) First record nanotechnology in agricultural: silica nano-particles a potential new insecticide for pest control. Adv Sci Eng Med 4(3):241–246

El-Mehrat H, Abdelaziz A, Ibrahim W, Hamed L (2017) Effectiveness of compost and potassium silicate under low rate of mineral fertilizers on production, quality, marketability and NPK contents of banana plants. J Soil Sci Agric Eng 8(12):787–795

El-Naggar ME, Abdelsalam NR, Fouda MM, Mackled MI, Al-Jaddadi MA, Ali HM, Siddiqui MH, Kandil EE (2020) Soil application of nano silica on maize yield and its insecticidal activity against some stored insects after the post-harvest. Nano 10(4):739

Fatemi H, Pour BE, Rizwan M (2020) Isolation and characterization of lead (Pb) resistant microbes and their combined use with silicon nanoparticles improved the growth, photosynthesis and antioxidant capacity of coriander (Coriandrum sativum L.) under Pb stress. Environ Pollut 266:114982

Fleck AT, Schulze S, Hinrichs M, Specht A, Waßmann F, Schreiber L, Schenk MK (2015) Silicon promotes exodermal Casparian band formation in Si-accumulating and Si-excluding species by forming phenol complexes. PLoS ONE 10(9):e0138555

Frasetya B, Harisman K, Sudrajat D, Subandi M (2019) Utilization of rice husk silicate extract to improve the productivity of paddy Ciherang cultivar. Bulg J Agric Sci 25(3):499

Gascho GJ (2001) Silicon sources for agriculture. In: Studies in plant science, vol 8. Elsevier, pp 197–207

Ghouri F, Sarwar S, Sun L, Riaz M, Haider FU, Ashraf H, Lai M, Imran M, Liu J, Ali S (2024) Silicon and iron nanoparticles protect rice against lead (Pb) stress by improving oxidative tolerance and minimizing Pb uptake. Environ Sci Pollut Res 14(1):5986

González-Moscoso M, Juárez-Maldonado A, Cadenas-Pliego G, Meza-Figueroa D, SenGupta B, Martínez-Villegas N (2022) Silicon nanoparticles decrease arsenic translocation and mitigate phytotoxicity in tomato plants. Environ Sci Pollut Res 29:1–17

Goswami P, Mathur J (2022) Application of agro-waste-mediated silica nanoparticles to sustainable agriculture. Heliyon 9(1):9

Goswami P, Mathur J, Srivastava N (2022) Silica nanoparticles as novel sustainable approach for plant growth and crop protection. J Bioprocess Biotech 8(7):e09908

Han Y, Lei W, Wen L, Hou M (2015) Silicon-mediated resistance in a susceptible rice variety to the rice leaf folder, Cnaphalocrocis medinalis Guenée (Lepidoptera: Pyralidae). PLoS One 10(4):e0120557

Hattori T, Sonobe K, Araki H, Inanaga S, An P, Morita S (2008) Silicon application by sorghum through the alleviation of stress-induced increase in hydraulic resistance. J Plant Nutr 31(8):1482–1495

Hogan B, McDermott F, Schmidt O (2019) Release of plant-available silicon from various silicon-rich amendments into soil solutions and leachates. J Soils Sediments 19:1272–1285

Hossain MA, Bhattacharjee S, Armin S-M, Qian P, Xin W, Li H-Y, Burritt DJ, Fujita M, Tran L-SP (2015) Hydrogen peroxide priming modulates abiotic oxidative stress tolerance: insights from ROS detoxification and scavenging. Front Plant Sci 6:141857

Hou M, Han Y (2010) Silicon-mediated rice plant resistance to the Asiatic rice borer (Lepidoptera: Crambidae): effects of silicon amendment and rice varietal resistance. J Econ Entomol 103(4):1412–1419

Huang M, Long F, Jiang L-G, Yang S-Y, Zou Y-B, Uphoff N (2019) Continuous applications of biochar to rice: effects on grain yield and yield attributes. J Integr Agric 18(3):563–570

Huang Q, Ayyaz A, Farooq MA, Zhang K, Chen W, Hannan F, Sun Y, Shahzad K, Ali B, Zhou W (2024) Silicon dioxide nanoparticles enhance plant growth, photosynthetic performance, and antioxidants defence machinery through suppressing chromium uptake in Brassica napus L. Environ Pollut 342:123013. https://doi.org/10.1016/j.envpol.2023.123013

Hussain M, Jang KH, Farooq M, Lee D-J (2012) Morphological and physiological evaluation of Korean rice genotypes for salt resistance. Sci Total Environ 14:970–974

Hussain HI, Yi Z, Rookes JE, Kong LX, Cahill DM (2013) Mesoporous silica nanoparticles as a biomolecule delivery vehicle in plants. J Nanopart Res 15:1–15

Ismail LM, Soliman MI, Abd El-Aziz MH, Abdel-Aziz HM (2022) Impact of silica ions and nano silica on growth and productivity of pea plants under salinity stress. Plants 11(4):494

Jia-Yi Y, Meng-Qiang S, Zhi-Liang C, Yu-Tang X, Hang W, Jian-Qiang Z, Ling H, Qi Z (2022) Effect of foliage applied chitosan-based silicon nanoparticles on arsenic uptake and translocation in rice (Oryza sativa L.). J Hazard Mater 433:128781

Kah M, Hofmann T (2014) Nanopesticide research: current trends and future priorities. Environ Int 63:224–235

Kalaivanan D, Ganeshamurthy AN (2016) Mechanisms of heavy metal toxicity in plants. Plants Agric Res:85–102

Kalteh M, Alipour ZT, Ashraf S, Marashi AM, Falah NA (2014) Effect of silica nanoparticles on basil (Ocimum basilicum) under salinity stress. Environ Nanotechnol Monit Manage 1:14–23

Khan ZS, Rizwan M, Hafeez M, Ali S, Adrees M, Qayyum MF, Khalid S, ur Rehman MZ, Sarwar MA (2020) Effects of silicon nanoparticles on growth and physiology of wheat in cadmium contaminated soil under different soil moisture levels. *Environ Sci Pollut Res* 27:4958–4968

Khurana P, Lingayat S, Thatai S (2023) Synthesis of silica spherical to Y-shape nanoparticles: a review with surface modifications and its characterization. Appl Chem Lett 13(2):106–126

Kim Y-H, Khan AL, Kim D-H, Lee S-Y, Kim K-M, Waqas M, Jung H-Y, Shin J-H, Kim J-G, Lee I-J (2014) Silicon mitigates heavy metal stress by regulating P-type heavy metal ATPases, Oryza sativa low silicon genes, and endogenous phytohormones. BMC Plant Biol 14:1–13

Kiran BR, Prasad MN (2019) Biochar and rice husk ash assisted phytoremediation potentials of Ricinus communis L. for lead-spiked soils. Ecotoxicol Environ Saf 183:109574

Kordi M, Farrokhi N, Pech-Canul MI, Ahmadikhah A (2023) Rice husk at a glance: from agro-industrial to modern applications. Rice Sci

Li Y, Zhu N, Liang X, Bai X, Zheng L, Zhao J, Li Y-F, Zhang Z, Gao Y (2020) Silica nanoparticles alleviate mercury toxicity via immobilization and inactivation of Hg (ii) in soybean (Glycine max). Environ Sci Nano 7(6):1807–1817

Liang Y, Sun W, Zhu Y-G, Christie P (2007) Mechanisms of silicon-mediated alleviation of abiotic stresses in higher plants: a review. Environ Pollut 147(2):422–428

Luyckx M, Hausman J-F, Lutts S, Guerriero G (2017) Silicon and plants: current knowledge and technological perspectives. Front Plant Sci 8:256009

Ma JF, Yamaji N (2006) Silicon uptake and accumulation in higher plants. Trends Plant Sci 11(8):392–397

Maghsoodi MR, Ghodszad L, Asgari Lajayer B (2020) Dilemma of hydroxyapatite nanoparticles as phosphorus fertilizer: potentials, challenges and effects on plants. Environ Technol Innov 1(19):100869

Mahmad-Toher A-S, Govender N, Dorairaj D, Wong M-Y (2022) Comparative evaluation on calcium silicate and rice husk ash amendment for silicon-based fertilization of Malaysian rice (Oryza sativa L.) varieties. J Plant Nutr 45(9):1336–1347

Manzoor N, Ali L, Ahmed T, Rizwan M, Ali S, Shahid MS, Schulin R, Liu Y, Wang G (2022) Silicon oxide nanoparticles alleviate chromium toxicity in wheat (*Triticum aestivum* L.). Environ Pollut 315:120391. https://doi.org/10.1016/j.envpol.2022.120391

Mistry R, Karmakar S, Kumar Roy T (2019) Experimental evaluation of rice husk ash and fly ash as alternative fillers in hot-mix asphalt. J Road Mater Pavement Des 20(4):979–990

Mitra S, Mukherjee S, Sil M, Adak S, Maitra P, Goswami A, Hessel V (2023) Role of mesoporous silica nanoparticles in combating mercury-induced stress in Vigna radiata (mung bean) and Bacillus coagulans (soil bacteria). Environ Sci Nano 30(50):109343–109353

Naaz H, Rawat K, Saffeullah P, Umar S (2023) Silica nanoparticles synthesis and applications in agriculture for plant fertilization and protection: a review. Environ Chem Lett 21(1):539–559

Ogbo F, Odo M (2011) Potential of rice husk and cassava peel as carriers for bio-fertilizer production. New J Biol 23:1–4

Pal N, Lee J-H, Cho E-B (2020) Recent trends in morphology-controlled synthesis and application of mesoporous silica nanoparticles. J Nanobiotechnol 10(11):2122

Phuong NTK, Khoi CM, Ritz K, Linh TB, Minh DD, Duc TA, Sinh NV, Linh TT, Toyota K (2020) Influence of rice husk biochar and compost amendments on salt contents and hydraulic properties of soil and rice yield in salt-affected fields. Agriculture 10(8):1101

Pinedo-Guerrero ZH, Cadenas-Pliego G, Ortega-Ortiz H, González-Morales S, Benavides-Mendoza A, Valdés-Reyna J, Juárez-Maldonado A (2020) Form of silica improves yield, fruit quality and antioxidant defense system of tomato plants under salt stress. Agronomy 10(9):367

Pode R (2016) Potential applications of rice husk ash waste from rice husk biomass power plant. Rev Environ Sci Biotechnol 53:1468–1485

Qin G, Niu Z, Yu J, Li Z, Ma J, Xiang P (2021) Soil heavy metal pollution and food safety in China: effects, sources and removing technology. J Hazard Mater 267:129205

Rai M, Ingle A (2012) Role of nanotechnology in agriculture with special reference to management of insect pests. Appl Microbiol Biotechnol 94:287–293

Rastogi A, Tripathi DK, Yadav S, Chauhan DK, Živčák M, Ghorbanpour M, El-Sheery NI, Brestic M (2019) Application of silicon nanoparticles in agriculture. *J Biotechnol* 9:1–11

Resham S, Khalid M, Kazi AG (2015) Nanobiotechnology in agricultural development. J Pharma Technol Res Manage:683–698

Riaz M, Zhao S, Kamran M, Ur Rehman N, Mora-Poblete F, Maldonado C, Hamzah Saleem M, Parveen A, Ahmed Al-Ghamdi A, Al-Hemaid FM (2022) Effect of nano-silicon on the regulation of ascorbate-glutathione contents, antioxidant defense system and growth of copper stressed wheat (Triticum aestivum L.) seedlings. Front Plant Sci 13:986991

Saed B, Ziaee M, Kiasat AR, Jafari Nasab M (2022) Preparation of nanosilica from sugarcane bagasse ash for enhanced insecticidal activity of diatomaceous earth against two stored-products insect pests. J Therm Anal Calorim 41(2):516–522

Safari S, Nazari F, Vafaee Y, Teixeira da Silva JA (2023) Impact of rice husk biochar on drought stress tolerance in perennial ryegrass (Lolium perenne L.). J Plant Growth Regul 42(2):810–826

Sandhya K, Prakash NB, Meunier JD (2018) Diatomaceous earth as source of silicon on the growth and yield of rice in contrasted soils of Southern India. J Soil Sci Plant Nutr 18(2):344–360

Sekifuji R, Tateda M (2019) Study of the feasibility of a rice husk recycling scheme in Japan to produce silica fertilizer for rice plants. J Sustain Energy Rep 29:1–9

Shaban M, Hasanzadeh M (2020) Biomedical applications of dendritic fibrous nanosilica (DFNS): recent progress and challenges. Recent Adv Nanotechnol 10(61):37116–37133

Shalata A, Neumann PM (2001) Exogenous ascorbic acid (vitamin C) increases resistance to salt stress and reduces lipid peroxidation. J Exp Bot 52(364):2207–2211

Shashi M, Mannan M, Islam M, Rahman M (2018) Impact of rice husk biochar on growth, water relations and yield of maize (Zea mays L.) under drought condition

Shi Y, Zhang Y, Han W, Feng R, Hu Y, Guo J, Gong H (2016) Silicon enhances water stress tolerance by improving root hydraulic conductance in Solanum lycopersicum L. Front Plant Sci 7:196

Silva LA, dos Santos IFS, de Oliveira MG, Tiago Filho GL, Barros RM (2021) Rice husk energy production in Brazil: an economic and energy extensive analysis. J Clean Prod 290:125188

Singh T, Singh P, Singh A (2021) Silicon significance in crop production: special consideration to rice: an overview. J Plant Interact 10:223–229

Snehal S, Lohani P (2018) Silica nanoparticles: its green synthesis and importance in agriculture. J Phytochem 7(5):3383–3393

Sutulienė R, Ragelienė L, Samuolienė G, Brazaitytė A, Urbutis M, Miliauskienė J (2021) The response of antioxidant system of drought-stressed green pea (Pisum sativum L.) affected by watering and foliar spray with silica nanoparticles. Hortic J 8(1):35

Takai-Yamashita C, Fuji M (2020) Hollow silica nanoparticles: a tiny pore with big dreams. Adv Polym Technol 31(2):804–807

Tateda M (2016) Production and effectiveness of amorphous silica fertilizer from rice husks using a sustainable local energy system. J Sustain Rep 9(3):1–12

Tayade R, Ghimire A, Khan W, Lay L, Attipoe JQ, Kim Y (2022) Silicon as a smart fertilizer for sustainability and crop improvement. Biosyst Eng 12(8):1027

Thabet AF, Boraei HA, Galal OA, El-Samahy MF, Mousa KM, Zhang YZ, Tuda M, Helmy EA, Wen J, Nozaki T (2021) Silica nanoparticles as pesticide against insects of different feeding types and their non-target attraction of predators. Sci Rep 11(1):14484

Tran TLC, Guirguis A, Jeyachandran T, Wang Y, Cahill DM (2023) Mesoporous silica nanoparticle-induced drought tolerance in Arabidopsis thaliana grown under in vitro conditions. Front Plant Biol 50(11):889–900

Vaculík M, Lukačová Z, Bokor B, Martinka M, Tripathi DK, Lux A (2020) Alleviation mechanisms of metal (loid) stress in plants by silicon: a review. J Environ Biol 71(21):6744–6757

Wang R, Fu W, Wang J, Zhu L, Wang L, Wang J, Ahmad Z (2019) Application of rice grain husk derived biochar in ameliorating toxicity impacts of Cu and Zn on growth, physiology and enzymatic functioning of wheat seedlings. Biosyst Eng 103:636–641

Wibowo D, Zhao C-X, Peters BC, Middelberg AP (2014) Sustained release of fipronil insecticide in vitro and in vivo from biocompatible silica nanocapsules. J Agric Food Chem 62(52):12504–12511

Wibowo EAP, Arzanto AW, Maulana KD, Hardyanti I, Septyaningsih H, Widiarti N (2017) Preparation and characterization of silica nanoparticles from rice straw ash and its application as fertilizer. J Chem Pharm Res 9:193–199

Williams R, Belo JB, Lidia J, Soares S, Ribeiro D, Moreira CL, Almeida L, Barton L, Erskine W (2023) Productivity gains in vegetables from rice husk biochar application in nutrient-poor soils in Timor-Leste. Sci Rep 13(1):10858

Yan G, Huang Q, Zhao S, Xu Y, He Y, Nikolic M, Nikolic N, Liang Y, Zhu Z (2024) Silicon nanoparticles in sustainable agriculture: synthesis, absorption, and plant stress alleviation. Front Plant Sci 15:1393458

Zargar SM, Mahajan R, Bhat JA, Nazir M, Deshmukh R (2019) Role of silicon in plant stress tolerance: opportunities to achieve a sustainable cropping system. J Bioremed Biodeg 9(3):73

Zhu Y, Gong H (2014) Beneficial effects of silicon on salt and drought tolerance in plants. Agric Food Sci Finland 34:455–472

Zou Y, Yang T (2019) Rice husk, rice husk ash and their applications. In: Rice bran and rice bran oil. Elsevier, pp 207–246

Zushi K, Matsuzoe N, Kitano M (2009) Developmental and tissue-specific changes in oxidative parameters and antioxidant systems in tomato fruits grown under salt stress. Sci Hortic 122(3):362–368

An Application of Biopesticides in Control of Pest and Crop Protection: An Eco-Friendly Management

6

Narendra Kumar and S. M. Paul Khurana

Abstract

Agriculture is significantly impacted by pests such as plant diseases, insects, nematodes, weeds, and other pests, causing reduced yields and quality. Biopesticides, derived from live organisms and their metabolites, offer a sustainable alternative to harmful chemical pesticides, which pose serious environmental and health risks. Biopesticides include fungi, bacteria, nematodes, and entomopathogenic viruses. The secondary metabolites of plants are becoming more important nowadays. It substitutes conventional pesticides, which are the backbone of many pest control programs. Biopesticides are effective, less toxic, biodegradable, and host specific, enabling them safer for humans, animals, and environment. Despite their potential, a global usage growth rate of 10% per annum, biopesticides currently represent about 2.8% of India's pesticide market. Greater awarness and adoption are also needed, supported through research into formultion,delivery and commercialization. With rising global concerns about environmental safety and health, promoting biopesticides is crucial. This chapter includes the significance of biopesticides, types of biopesticides, and future prospects, emphasizing the need for education and policy support to enhance their integration into sustainable agriculture practices which are essential for promoting their widespread adoption and addressing global concerns about environmental safety.

N. Kumar (✉)
Department of Botany, Guru Ghasidas Vishwavidyalaya (a Central University), Bilaspur, Chhattisgarh, India

S. M. P. Khurana (✉)
Amity Institute of Biotechnology, Amity University Haryana, Gurgaon, Manesar, Haryana, India

A. Kumar, M. K. Solanki (eds.), *Microbial Biocontrol Techniques*, Microorganisms for Sustainability 54,
https://doi.org/10.1007/978-981-97-8739-5_6

Keywords

Biopesticides · Microbial · Bioinsecticides · Biofungicides · Bioherbicides · Biochemical pesticides · Bionematicides · Mode of action · Biopesticide market

6.1 Introduction

A variety of pests, including weeds, insects, plant infections, and nematodes, have had a negative impact on agriculture, lowering the estimation about 45% of crop losses or over 290 billion annually. Conventional pesticides produce more food. However, they projected negative consequences on the ecosystem and non-target animals. Furthermore, volatile pesticide residues can occasionally cause domestic consumers to express concerns about the safety of their food and create trade barriers for export products. Biopesticides, which are formulations made from live microorganisms like fungi, bacteria, and viruses as well as natural resources including bacteria, animals, plants, and minerals, provide effective environmentally friendly tools for producing a new generation of agricultural goods that are sustainable. The global usage of biopesticides is steadily increasing over 10% annually (Aneja et al. 2000).

Global manufacturing of biopesticides exceeds 3000 tonnes annually (Gupta and Dixit 2010). India has enormous potential in both the production and use of biopesticides (Al-Zaidi et al. 2011). Microbial pesticides, botanical pesticides (also known as plant pesticides), and biochemical pesticides are the three main categories of biopesticides. Remarkably, *Bacillus thuringiensis* is the sole entomopathogenic endospore-producing bacterium on which 90% of microbial insecticides are based on it (Kumar and Singh 2015). Given the enormous diversity of microorganisms, there are many opportunities to find new or adapt existing biopesticides to protect the environment from the deadly consequences that agrochemicals have on non-target creatures, particularly people because weeds, fungus, and insects have caused devastating effects on agriculture, leading to a dramatic drop in productivity.

When chemical pesticides became available, this crisis was mostly remedied. However, the excessive reliance on chemical pesticides and their ultimate unrestrained usage has made the need for alternatives primarily related to environmental issues incredible. A nutritional imbalance and lack of productivity have been caused by degraded soils and groundwater pollution. In addition to causing trade barriers for export commodities, excessive pesticide residues can occasionally cause domestic consumers to express worries about food safety. As such, the urgent need is for an environmentally friendly substitute. A practical and environmentally friendly answer to pest issues are biological pesticides, often known as biopesticides, which are founded on harmful microbes specific to a given insect. Live organisms that are pathogenic and destructive to the target pest are frequently used biopesticides (Ali et al. 2024; Verma et al. 2022). They comprise *Trichoderma*, *Bacillus thuringiensis*, *Phytophthora* bioherbicides, as well as biofungicides. Using biopesticides recommends agriculture, public health plans, and several possible benefits.

Biopesticides are gaining popularity due to their associated benefits, which includes (1) lower environmental impact and inherent harm; (2) pest-specific targeting or a limited target organism; (3) high effectiveness in less amounts and rapid decomposition potential that may reduce exposure and prevent pollution issues; (4) biopesticides may make a substantial contribution to Integrated Pest Management (IPM) programs when they are included as a component.

6.2 Benefits of Biopesticides

See Table 6.1.

6.3 Biopesticides Categories

Biopesticides, also known as bio + pest + cide, are insecticide formulations that contain live things (natural enemies) and materials (metabolites) that come from bacteria, plants, animals, and minerals. They control pests in an environmentally benign and non-toxic way. Based on the type of active constituents used, three main categories of biopesticides are recognized by the EPA (US Environmental Protection Agency) microbiological pesticides, biochemical pesticides, and plant-incorporated protectants also known as plant-based pesticides (or botanical pesticides). The IOBC (International Organization for Biological Control) and the IBMA (International Biocontrol Manufacturers Association) 2008 advocate for the term biocontrol agents (BCAs) to be used in place of microbial pesticides. Biopesticides are categorized as microbiological, botanical, zooid, and genetically modified plant insecticides that are formulated using natural resources as their source. Plant pesticides are naturally arising plant matters. They control mechanisms of pest which are non-toxic. The PIPs (plant-incorporated protectants) are plant genetically modified in producing pesticidal substances, such as (Bt pesticidal protein), RNA

Table 6.1 Benefits of biopesticides

Features	Benefits of biopesticides
Effectiveness (cost)	Less costly
Persistence	Self-perpetuating, low residual effect, biodegradable
Knockdown effect	Mostly delayed
Bulkiness and handling	Carrier based and easy handling
Resurgence done to pest	It is less
Resistance	Usually lesser prone
Impact over beneficial flora	Usually lesser harmful over beneficial pests
Specificity on target	Host specific
Time of waiting	Almost nil
Control nature	More preventive
Shelf life	Usually less

Table 6.2 Some commercially accessible biochemical pesticides (Aneja et al. 2000)

Name of product	Types of pesticide	Name of active ingredients	Crops	Target organisms
AZATIN XL	Insecticide	Azadirachtin	Fruits, veggies, herbs, and ornamental crops	Weevils, whiteflies, leafhoppers, aphids, thrips
NEMA-Q	Nematicide	*Quillaja saponaria*	Ornamentals, orchards and vineyards field crops, tur	Plant parasitic nematodes
BARRIER H	Herbicide	Citronella oil	Grass land	Ragwort
BIOMITE	Attractant	Citronellol	Apples, grapes, ornamental and nursery plants, nuts, hops, pears, stone fruits, cucurbits	Tetranychid mites
EXOSEXCM	Attractant	(*E*,*E*)-8,10-dodecadien-I-ol	Pears, apple	Codling moth

interference pesticides are the four categories into which the EPA-(United States Environmental Protection Agency) has divided biopesticides. Microbial pesticides contain active constituents such as fungi, bacteria, or viruses. Biochemical pesticides, herbal pesticides, and plant pesticides (Table 6.2).

6.3.1 Microbial Pesticides

In all three types of biopesticides, more than 90% are based on microorganisms, including viruses, fungus, and bacteria. Microbial pesticide refers to the class of mass-produced microbiological formulations, regulated, sold, and administered like that of chemical pesticides. The four types of microbial pesticides are bioherbicides for weeds, biofungicides for crop fungal infections, bionematicides for plant parasitic nematodes, and bioinsecticides for arthropod pests (insects). Microorganisms that cause pests are naturally present, microbial insecticides are typically safe for domestic animals, plants, and people to use. They also have an elevated level of specificity that makes them non-pathogenic to non-target pests (Aneja 2000). According to Thakore (2006), approximately 74% of the global market for all crop types is occupied by bacterial biopesticides, followed by fungal biopesticides (10%), viral biopesticides (5%), predator biopesticides (8%), and other biopesticides (3%).

6.3.2 Bioinsecticides

Metchnikoff and Pasteur first proposed the concept of control of insect pests biologically in 1882. Investigations took place for the microbiological control of pest insects started to pick up speed in the 1940s, and in the 1950s, large-scale microbial preparations and formulations for use as microbial insecticides were started.

The following is an overview of various biocontrol agents, their countries of origin, products, and target pests (Mishra et al. 2014a, b).

In USA, *Bacillus popilliae* is used in the form of spore powder (Doom-Milky) to target Japanese beetle grubs. *B. sphaericus* H5ab strain 2363 (ATCC1170) is marketed as VectoLex for controlling blackflies and mosquitoes. Florbac also known as *B. thuringiensis* (subsp. *aizawai*) NB 200 is effective against larvae of moth. *B. thuringiensis* (subsp. *israelensis*) is the agent used by BMP to attack mosquitoes and blackflies. *B. thuringiensis* (subsp. *israelensis*) EG2215 is present in the products Aquabac and Gnatrol, which are used to repel flies and mosquitoes. The Colorado potato beetle is targeted by M-Trak, which uses the δ-endotoxin of *B. thuringiensis* (subsp. *aizawai*) in dead *P. fluorescens*. *Plutella xylostella*, the diamondback moth, is the target of Agree-WG, which contains *B. thuringiensis* (subsp. *aizawai*) GC-91.

Products from *Bacillus thuringiensis* (subsp. *Kurstaki*) such as Worm Whipper, Security Dipel Dust, Worm Spray, Ferti-Lome, Bonide, Britz BT, Foray, Javelin WG, Green Light, Hi-Yield Thuricide Forestry, Biobit HP, Wilbur-Ellis BT 320, Dust, Dipel, and Deliver are used against Lepidopteran larvae. BMP123 is a specific product for lepidopteran larvae. *B. thuringiensis* (subsp. *kurstaki*)-EG2348-containing condor also feeds on lepidopteran larvae. *B. thuringiensis* (subsp. *tenebrionis*) is what Novodor uses to treat Colorado potato beetles.

Beauveria bassiana HF23, under the brand bal. Ence, is used against houseflies. Tick-Ex, containing *Metarhizium anisopliae* F52, targets ticks and grubs. *Paecilomyces fumosoroseus* Apopka-97, branded as PFR-97, is effective against whiteflies and thrips. Nosema locustae products like Nolo-Bait and Semaspore Bait target grasshoppers and crickets. CLV-LC, containing *Anagrapha falcifera* NPV, targets lepidopteran larvae. CYD-X, using *Cydia pomonella* GV, is for codling moth control. Gypchek targets gypsy moths using Gypsy moth NPV.

Lepinox WDG, containing *B. thuringiensis* (subsp. *kurstaki*) EG7826, targets Lepidopteran larvae. Baits Motel Stay-awhile, using *Beauveria bassiana* 447, is effective against ants. Naturalis L., containing *Beauveria bassiana* ATCC 74040, targets various insects, mites, whiteflies, and thrips. Botanigard 22WP, Botanigard ES, Mycotrol ES, and Mycotrol O, using *Beauveria bassiana* GHA, are effective against various insects. ELCAR and GemStar, using *Heliothis zea* NPV, target cotton bollworm and tobacco budworm. FruitGuard, containing *Plodia interpunctella* GV, targets the Indian meal moth. Virosoft, using *Mamestra configurata* NPV (107308), targets the bertha armyworm. Virus Spod-X, containing *Spodoptera exigua* NPV, targets the beet armyworm. Bull Run, using Saccharomyces cerevisiae, acts as a fly attractant.

In Europe, Turex, containing *Bacillus thuringiensis* (subsp. *aizawai*) GC-91, targets Lepidopteran pests. VectoBac, containing *B. thuringiensis* (subsp. *israelensis*) AM65, targets Sciaridae. Gnatrol and Aquabac, using *B. thuringiensis* (subsp. *israelensis*) EG2215, target mosquitoes and flies. Dipel WP, containing *B. thuringiensis* (subsp. *kurstaki*) HD-1, targets Lepidopteran larvae. Batic and Delfin, using

B. thuringiensis (subsp. *kurstaki*) ABTS 351, PB54, SA11 & 12, and EG2348, target Lepidopteran larvae. BMP 123/Prolong, using *B. thuringiensis* (subsp. *kurstaki*) BMP123, targets Lepidopteran pests. Novodor, using *B. thuringiensis* (subsp. *tenebrionis*) NB176, targets Coleoptera pests.

Naturalis L., containing *Beauveria bassiana* ATCC 74040, targets mites, whiteflies, and thrips. BotaniGard, using *Beauveria bassiana* GHA fungus, targets whiteflies, aphids, and thrips. Mycotal and Vertalec, using *Lecanicillium muscarium* (Ve 6), target whiteflies, thrips, and aphids (excluding *Chrysanthemum aphid*). Preferal WG, containing *Paecilomyces fumosoroseus* Apopka-97, targets greenhouse whiteflies. Nofly, using *Paecilomyces fumosoroseus* Fe9901, targets whiteflies.

Capex, containing *Adoxophyes orana* BV-0001GV, targets the summer fruit tortrix. BioTepp, using *Cydia pomonella* GV, targets the codling moth. Spod-X GH, containing *Spodoptera exigua* NPV, targets *Spodoptera exigua*.

In Japan, products like Toarowaa, Esmark, Guartjet, Dipol, Tuneup, Fivestar, and Biomax, containing *Bacillus thuringiensis* subsp. kurstaki, target Lepidopteran larvae. Quark, Xen Tari, Florbac, and Sabrina, containing *Bacillus thuringiensis* (subsp. *aizawai*), also target Lepidopteran larvae. Bacilex, using *B. thuringiensis* (subsp. *aizawai* and *kurstaki*), targets Lepidopteran larvae. Hamaki-Tenteki, containing Adoxophyes orana GV and Homona magnanima GV, targets Odoxophyes honmai and Homona magnanima.

Vertalec, using *Lecanicillium longisporum*, targets aphids. Preferd, containing *Paecilomyces fumosoroseus*, targets whiteflies and aphids. Nofly, using *Paecilomyces fumosoroseus* Fe9901, targets whiteflies. Botanigard, containing *Beauveria bassiana*, targets whiteflies, diamondback moths, and thrips. Bio Safe, using *Steinernema carpocapsae*, targets weevils, back cut worms, and peach fruit moths.

In India, Tacibio and Technar, containing *B. thuringiensis* (subsp. israelensis), target Lepidopteran pests. Bio-Bart, Biolep, Halt, and Taciobio-Btk, using *B. thuringiensis* (subsp. *kurstaki*), target Lepidopteran pests. Biolarvex, Biogrubex, Biowonder, Vera, Phalada 101B, Bioguard, Bio-power, Myco-Jaal, Biosoft, ATEC Beauveria, Larvo-Guard, and Biorin, using *Beauveria bassiana*, target coffee berry borers, diamondback moths, thrips, grasshoppers, whiteflies, aphids, and codling moths.

Sun Agro Meta Bio-Magic, ABTEC, Meta-Guard, Biomet, and Biomagic Meta, containing *Metarhizium anisopliae*, target Coleoptera and Lepidoptera, termites, mosquitoes, leafhoppers, beetles, and grubs. Priority and Nemato-guard, containing *Paecilomyces fumosoroseus*, target whiteflies. Yorker, ABTEC, Paecil, Paceilomyces, Pacihit, ROM biomite, and Bio-Nematon, using *P. lilacinus*, target whiteflies.

Spodocide and Spodoterin Spoddi-cide, using *Spodoptera litura*, target the Oriental leaf worm. Biovirus-H, Helicop, Helicide, Virin-H, Helocide, and Heligard, containing *Helicoverpa armigera* NPV, target the cotton bollworm. Vert-Guard Bioline, Biosappex, Verisoft, ABTEC, Ecocil, *Verticillium*, *Versitile*, Phalada 107 V, Biovert Rich, ROM Verlac, ROM Gurbkill, Sun Agro Verti, and Biocatch, using Verticillium lecanii, target whiteflies, homopteran pests, and the coffee green bug.

6.3.3 Biofungicides

Biofungicides are mixtures of live organisms used to conquer growth of bacteria and fungi that cause damage to plants. Biofungicides are derived from remarks of natural methods in which advantageous microorganisms, typically extracted from soil, prevent plant diseases from doing their job (Solanki et al. 2024). Microorganisms classified as biocontrol are free-living plants, actinomycetes or fungi which are vigorous in soil, root environments, foliar (Kumar et al. 2024; Wang et al. 2023; Kashyap et al. 2023). These microbes form a wide range of antibiotic compounds, contest with supplementary fungi, and make limited or systemic resistance to some plant diseases (Kumari et al. 2022; Bastas and Kumar 2023). They also parasitize other fungus. Another example of this disease control strategy is the application of composts and suppressive growth media, containing living microorganisms, to treat diseases. The following criteria demonstrate the effectiveness of the biofungicidal agents:

1. **Parasitism**: Capacity of certain species to assault and ingest plant diseases has extensively researched. Growth, binding, and contact that is directed, hyphae coiling around the host fungus, penetration, and destruction are all examples of mycoparasitism in biocontrol microorganisms. Enzyme synthesis that breaks down cell walls is a crucial step in the biocontrol process.
2. **Antibiosis**: A single microbe produces chemicals that directly harm other species through growth inhibition or toxicity. Throughout their growth process, a variety of soil-dwelling bacteria frequently create these substances, which are known as antibiotics.
3. **Rhizosphere competence**: The most effective biocontrol microbial strains possess this feature, which allows them to colonize and thrive alongside plant roots (Solanki et al. 2023).
4. **Plant growth promotion**: Microorganisms that colonize beneficial roots encourage plant growth and productivity (Kumar et al. 2018). When disease-causing fungi are absent, a large number of bacteria and fungi that induce resistance encourage the growth of both roots and shoots.
5. **Inducing metabolic changes**: Plants can become more resistant to an array of harmful microorganisms (i.e., fungi and bacteria) that affect plants by means of metabolic changes induced by biocontrol microorganisms. By accelerating the metabolism of plant defense chemicals, systemic acquired resistance (SAR) enhances the reaction of plants to pathogenic invasion (Bhardwaj et al. 2024).

6.3.4 Bioherbicides (Mycoherbicides)

Since the early 1980s, the amount of research reports on the creation of bioherbicides has expanded dramatically. Equally, the targeted amount of weeds for management and the amount of pathogen research projects have enlarged. Globally, there has also been an upsurge in the practical listed or unlisted usage of bioherbicides. Parallel to this, there have been more US patents granted for applications of bioherbicidal for fungi and their technology, which may portend a future in which bioherbicide use will become more prevalent (El-Sayed 2005). Few commercial biological weed control solutions have shown long-term success; the majority of biological weed control products researched and registered worldwide are based on fungal species compositions. Three genera of fungi: *Fusarium* spp., *Chondrostereum* spp., and *Colletotrichum* spp. most of the research interest has been given to these genera as a potential herbicide (Aneja 2014). Numerous bacteria have been studied for possible biological agents to manage weeds. The two that garnered the greatest interest were *Pseudomonas fluorescens* and *Xanthomonas campestris*. According to some research, using bacteria for biological weed control has several benefits over using fungi. These benefits include faster bioherbicide agent growth, simpler propagation requirements, and increased suitability for gene transfer or mutagenesis-based genetic alteration (Li et al. 2003). Under the brand name CAMPERICO, in Japan, strain JT-P482 of *Xanthomonas campestris* takes a lot of interest as per a potential biocontrol agent for (*Poa annua*) annual bluegrass control (Imaizumi et al. 1997).

An additional strain of *Pseudomonas fluorescens* known as BRG100 has demon strated the ability to create extracellular compounds with phytotoxic effects. It has been demonstrated that BRG100 exhibits suppressive activity on green foxtail, a grassy plant. Above all, crop protection technology received a lot of attention during the annual Agrow Awards, which took place in London on September 17, 2016. This occasion honors global innovation in areas like crop protection, packaging, and formulation. The pesticidal treatment SolviNix, created by Gainesville, Florida-based BioProdex, is the year 2015 winner of Agrow's award for best innovative biopesticide. This year, in April, the FDA approved SolviNix, a bioherbicide that fights tropical soda, a weed that can stifle additional cattle feed and plants (Brachmann 2015). In certain instances, viruses that impact weeds have additionally suggested as potential sources of bioherbicides. According to Kazinczi et al. (2006), viruses' lack of host specificity and genetic heterogeneity make them unsuitable as candidates for inundative biological control. As of right now, roughly 24 bioherbicides have been registered worldwide, out of these 10 bioherbicides are listed in the United States, Canada, i.e. 8, from South Africa, i.e. 3, one from Japan, the Netherlands, India, and China each, respectively. Remarkably, the majority of commercially manufactured formulas exist in a liquid condition. Research reveals that the optimal preparation of a bioherbicide yields a product that is cost effective, having a long shelf life, effortlessness of application, and potency (Auld et al. 2003). Because fungi's reliance on dew to maintain their potency in dry environments, formulation continues to be a barrier to the commercial development of many promising bioherbicides. Several commercial bioherbicides have not caused issues since they are applied as wound inoculants or utilized in irrigated settings. Reducing reliance on dew is therefore a primary goal in the development of numerous possible bioherbicides. Before getting activated, solid formulations usually need to be able to endure in the field and wait for the right circumstances.

Generally, solid formulations have to be able to endure in the field and wait for the right circumstances in order to activate. As long as liquid formulations stay wet on the target plant surface, they may cause infections shortly after application. Numerous endeavors have been undertaken to enhance the ability of liquid compositions to retain water.

6.3.5 Plant Pesticides

Pesticides produced by plants using genetic material integrated into the plant are known as plant pesticides. For instance, scientists are able to ingest gene for Bt pesticidal protein on the genetic makeup of plants. Pest killing substances are then produced by plants, not the Bt bacteria. While it does not affect the plant itself, EPA controls the genetic makeup and protein.

6.3.6 Biochemical Pesticides

These are found naturally and manage pests in a non-toxic way. However, synthetic ingredients used in conventional insecticides typically result in the pest's death or inactivation. Plant growth regulators and pheromones are examples of biochemical pesticides that either attract or repel pests. They include compounds which interfere with mating or growth. The committee made by the Environmental Protection Agency (EPA) determine if a pesticide meets the criteria to be classified as a biochemical pesticide since regardless of whether a natural pesticide eliminates the pest in a not harmful manner, it can be difficult to determine. Demand and use for biopesticides are expanding along with the growth in the total amount produced worldwide. A gradual rise in the usage of biopesticides has been seen in India.

Biopesticide potential in India is enormous. To maximize benefits, farmers in India must be educated before adopting it. Though they only made up 2.89% of India's total pesticide market as of 2005, biopesticides are predicted to expand at a rate of roughly 2.3% annually over the next several years (Thakore 2006). The Insecticide Act of 1968 only permits the registration of 12 types of biopesticides in India to far.

6.3.7 Bionematicides

Entomopathogenic nematodes (EPN) are another form of biocontrol agent used to control pests including gnats, white grubs, weevils, and other bug species belonging to the *Sesiidae* family. In cryptic settings, these interesting organisms regulate the insects' population (including stem borers, soil-borne pests). *Heterorhabditis*, *Phasmarhabditis*, and *Steinernema* are the genera of nematodes and are frequently employed in management of pests which, as infectious juveniles, assault the hosts. Infective juveniles (IJS) are free-living creatures that enter insects by their mouths, spiracles, cuticles, or anus, according to Kaya and Gaugler (1993). The IJs depart whenever the nematodes have finished their three generations within the host, the cadaver searches for new hosts.

The following nematodes have been effectively employed as bionematicides: *Heterorhabditis bacteriophora*, *H. megidis*, *S. riobravis*, *S. glaseri*, and *S. scapterisci*.

Following are some commercially accessible bionematicides worldwide listed below:

1. *Steinernema carpocapsae* targets weevils, black cutworms, peach fruit moths, and common cutworms and is available under the product name BIO SAFE, provided by SDS Biotech Co. Ltd. *S. glaseri* is effective against bluegrass webworm, black cutworms, lawn grass cutworm, white grubs, and weevils and is marketed as BIO TOPIA by company SDS Biotech Co. Ltd.
2. *S. feltiae* is used against fungus gnats, sciarid flies, vine weevils, and other types of soil insects. This biocontrol agent is sold under multiple product names such as ENTON, NEMASYS, NEMA-PLUS, OTIENEM, NEEMASHIELD, GNAT NOT, and SCAN MASK. Companies dealing with these products include Novartis BCM, Becker Underwood, e-nema GmbH, Biobest, Biocontrol System, Biologic Co., Biocontrol Inc., Certis, and Koppert Integrated.
3. Lepidopteran larvae, soil insects, turf, and Japanese beetles are the targets of *Heterorhabditis bacteriophora*. Products like CRUISER HETEROMASK, NEMA-BIT, LARVANEM, TERRANEM and NEMA-TOP/-GREEN all include this agent. Companies include Integrated Biocontrol Systems, Koppert, BIT, Ecogen, BioLogic, e-nema, and Hydro-Gardens distribute these products.
4. *S. riobravis* is used to control citrus weevils and is available as BIOVECTOR 355 and DEVOUR, with Certis being the supplier.
5. *S. scapterisci* targets mole crickets and is sold as NEMATAC S by Becker Underwood Inc.
6. *H. megidis* is efficient against soil insects and black vine weevils, with products like LARVANEM and DICKMAULRUSSLER-NEMATODEN offered by Koppert Bio Syst. and Andermatt Biocontrol AG.
7. Lastly, *Phasmarhabditis hermaphrodita* is used to combat slugs and is available under the product name NEMASLUG from Becker Underwood Inc.

6.4 Main Biopesticides Formed and Its Application in India

1. **Neem**: Neem (*Azadirachta indica*) is the source of product, having various compounds in it, including "Azadirachtin," which has an impact on the reproductive and digestive systems of several significant pests. Effective neem formulations are being developed and commercially manufactured as a result of recent research conducted in India and beyond. Neem' s demand is expected to rise because it poses no threat to birds or mammals and does not cause cancer. Still, there isn't much demand at all right now. In India, over 100 companies are registered to manufacture insecticides derived from neem, although only a small

number actually do so. Additionally, the majority of the manufacturing is meant for export markets; relatively little is sold locally.

2. ***Bacillus thuringiensis* or Bt** is extensively applicable biopesticide in the world. Primarily, it is a disease that affects lepidopterous pests such as rice stem borers and cotton bollworms. Bt emits toxins that harm the insect's midgut and ultimately cause it to die when consumed by pest larvae. KT preparations are mostly made from strains of the (sub species *dendrolimus*, *kurstaki*, *galeriae*).
3. **Bacterioviruses**: These viruses are designed to specifically target and infect certain significant plant pests and especially effective against lepidopterous pests of rice, vegetables, and cotton. Due to the challenges associated with producing them on a wide scale, their application has been restricted to local regions. Since their large-scale manufacturing presents some problems, their application has been limited to specific sectors.They are formed on a modest scale from different IPM centers and state agriculture departments, but they are not commercially marketed in India.
4. ***Trichoderma***: A fungicide known as *Trichoderma* works well against diseases in the soil, like root rot (Guzmán-Guzmán et al. 2023). This is especially important for dryland crops that are prone to various diseases, like chickpea, black gram, green gram, and groundnut. *Trichoderma* biopesticide is inexpensive to prepare and just requires some microbiological understanding.
5. ***Trichogramma***: *Trichogramma* are tiny parasitic wasps that only feed on eggs. They deposit their eggs among the eggs of many lepidopteran pests. The host egg is consumed and destroyed by the *Trichogramma* larvae after they hatch. *Trichogramma* is especially efficient against lepidopteran pests such as stem borers in rice, cotton bollworms, pink bollworms, and sugarcane internode borers. They are also employed to combat pests of fruits and vegetables. In India, *Trichogramma* is widely utilized as biocontrol agent primarily because it eradicates the pest, while it is still in the egg stage, preventing crop harm from the parasite. *Trichogramma* eggs must be used quickly, ideally before they hatch.

   Trichogramma is not sold through dealers or store owners because of this, which also restricts its large-scale manufacture and marketing.

6.5 The Effective Application of Biopesticides and Biocontrol Entities in India: Success Stories

Following are a few instances of how biopesticides and biocontrol agents have been successfully used in India (Kalra and Khanuja 2007):

1. *B. thuringiensis* is responsible for controlling diamondback moths
2. Beauveria' s management of mealy bugs, mango hoppers, and coffee pod borer
3. *B. thuringiensis* in cotton, pigeonpea, and tomato to control Helicoverpa
4. Using neem treatments to control whiteflies on cotton
5. N.P.V. controls the gram of Helicoverpa
6. *Trichogramma*'s management of sugarcane borer
7. Use of *Trichoderma*-based treatments to prevent rots and wilts in a variety of crops

6.6 Current Status of Biopesticides

Globally biopesticides production exceeds over 3000 tonnes annually, and their application is growing at a pace of 10% annually (Kumar and Singh 2015). Globally, around 1400 products of biopesticides are manufactured and sold (Marrone 2007). According to EPA data, there are more over 200 products offered in the US market while there are just 60 comparable items available in the EU. The 30 OECD nations produce more than 225 different types of microbial biopesticides. About 45% of the biopesticides supplied worldwide are used in the NAFTA nations (the United States, Canada, and Mexico), while in Asia, i.e., 5% are only utilized (Bailey et al. 2010; Hubbard et al. 2014). Because of the advantages that come with these products, the biopesticide sector is expanding quickly. Some of these benefits include being naturally less harmful and burdening the environment; targeting a small number of target pests or a single pest; being effective in small amounts; and being non-toxic to humans. Additionally, when used in Integrated Pest Management (IPM) systems, biopesticides may drastically reduce the need for chemical pesticides while preserving crop yield levels.

6.7 Mode of Action Vis-Vis Insect Pests

6.7.1 Biopesticides

In nature, living creatures such as microorganisms (such as bacteria, fungus, and viruses), plants, and animals are basically used to make biopesticides. Due to their non-toxic, environmentally friendly mode of action, biopesticides are becoming increasingly important worldwide for controlling serious plant-damaging insect pests. The primary use of biopesticides and their by-products is the control of pests that damage the plants (Mazid et al. 2011).

Following three categories are used to classify biopesticides:

1. Protectants integrated to plants
2. Biochemical pesticides
3. Microbial pesticides

Regarding edible fruits and vegetables, in particular, residual problems are a big concern for customers, but they don't exist with them. These can be equally as successful as conventional pesticides after being applied in insect pest management, especially for vegetables and nuts including fruits and floras. These are a class of pesticides that combine the environmental safety and performance of synthetics to provide excellent resistance control potential and little usage limits (Senthil-Nathan 2013). According to researchers, working to create safe, environmentally friendly integrated crop management (ICM), compatible pest management strategies have been paying more attention to and interest in biopesticides (Copping and Menn 2000). Farmers' adoption of biopesticides, in particular, may be influenced because of the growing popularity of "food produced organically" and the successful launch of "products with a biological base," having an inclusive variety of biological actions directed in contrast to significant target species. Microorganism-based insecticides provide underdeveloped nations with an exceptional opportunity for research and the development of natural biopesticide resources for protection of crops. To stop target insect pests after becoming resistant to pesticides made of synthetic chemicals, the toxins generated by biopesticides, biopesticide programs must be used (Senthil-Nathan 2006).

Biopesticides do not have the same controlling issues as chemical pesticides do. Agronomic crops can be reintroduced shortly after subsequent treatment and are often target-specific biopesticides. Additionally, they are harmless to advantageous insects and have no adverse effects to the environment and water quality. There is no risk to human health when using naturally occurring microorganisms in organic farming. Furthermore, using biopesticides has some other benefits, such as the fact that various targeted insect pests are susceptible to their implications.

The biopesticides that are produced from bacteria, including *Bacillus thuringiensis* (Bt), as well as a range of beneficial nematodes, protozoa, fungi, and viruses, have been developed meant for use in greenhouses, fields, gardens, and orchards (Grewal et al. 1997). The EPA classifies some pesticides derived from living creatures as biopesticides, along with biocontrol microorganisms and their insecticidal metabolic products. According to there are hundreds of registered goods on their list (Fig. 6.1).

6.7.1.1 Microbial Pesticides

Bacteria

B. thuringiensis

Different molecular strategies have been used to increase commercial adoption of biopesticides starting in the 1980s and continuing to the current. Formerly, several initiatives were made for the development of microbial pesticides, such as Bt, in commercial use for more than 40 years (Gelernter and Schwab 1993). Subsequent research revealed that certain Bacillus species, including *Bacillus sphaericus*—2362 (Bs) and *Bacillus thuringiensis israelensis* (Bti) were more efficient opposition to other dipteran larvae and mosquitoes (Revathi et al. 2013; Devi et al. 2023). It was initially shown in 1975 that Bti exhibited heightened toxicity toward larvae of mosquito (Goldberg and Margalit 1977). Numerous species and subspecies of bacteria have been recognized as biopesticides, and they are mostly applied to manage infections in plants and insects. Notable examples of these include *Bacillus* and *Pseudomonas*. Between these, insecticides based on various *Bacillus thuringiensis* Berliner subspecies are the most notable. *B. thuringiensis* ssp. *kurstaki* and *aizawai*, which have extreme action in contrast to lepidopteran larval species, i.e., *B. thuringiensis israelensis*, which has action counter to black fly (Simuliidae) larvae, mosquito larvae, fungus gnats, i.e., *B. thuringiensis tenebrionis*, which has action counter to both adults and larvae of coleopterans, including the potato beetle

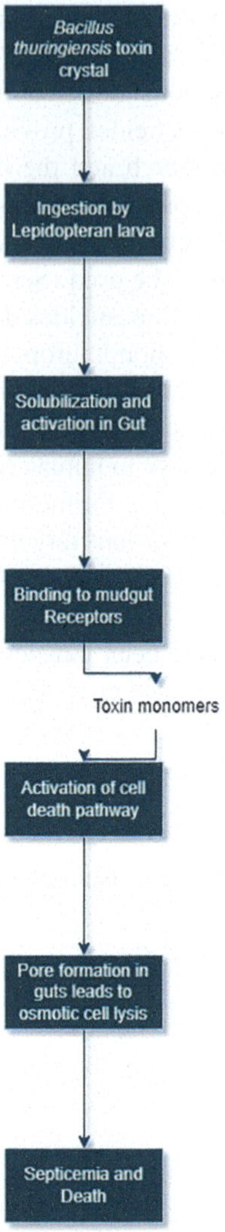

Fig. 6.1 Flow chart of how Bt toxin acts in opposition to lepidopteran insects (Senthil-Nathan 2014)

(Colorado) (*Leptinotarsa decemlineata*); and *B. thuringiensis japonensis* strain *Buibui*, which has activity counter to beetles (Carlton 1993; Copping and Menn 2000).

Bt forms crystalline proteins and terminates several pest species, including lepidopteran species, that are a nuisance. The attachment of Bt crystalline proteins near an insect gut receptor identifies targeted pest insects. The production of δ-endotoxins, or parasporal inclusion bodies, during sporulation it is widely given to the toxicity of Bti and certain additional dangerous strains. The larvae must absorb these endotoxins for toxicity to occur. Many insecticidal crystal proteins (δ-endotoxins) are produced by Bt and its subspecies, and their toxicity has been identified (Aronson and Shai 2001). When larvae absorb the mentioned poisons, gut tissue injury might lead to paralysis in gut. Subsequently, infected larvae cease nourishing and ultimately, an impaired midgut epithelium and malnutrition combine to cause their death. By outlasting organisms that are pest insects, certain other microbial insecticides work. It is imperative to provide ongoing supervision over microbial pesticides to prevent them from harming non-target organisms, such as people (Mazid et al. 2011) prior research indicates that the use of synhetic chemical insecticides has drastically decreased since the developement of microbial pesticides. The usage of synthetic chemical insecticides has significantly decreased (James 2009). It is documented that the production of Bt toxins by rhizobacteria that promote plant development and create chemicals called bacteriocins that have insecticidal properties. Bt is sold all over the world to control a variety of significant plant pests, primarily caterpillars, black flies, and mosquito larvae. The powders comprising a blend of dried spores and crystal poisons are among the commercial items derived from Bt. They are put on leaves or other surfaces that serve as food sources for insect larvae. Several crops have been genetically modified to include Bt toxin genes.

Fungi

Metarhizium anisopliae

M. anisopliae Sorokin var. *anisopliae* is an important entomopathogenic fungus. It spreads around the planet through soil, exhibiting a diverse array of insect hosts. This sub species was first identified as per a disease of the wheat cockchafer in year 1879 through scientist Metschnikoff under the name *Entomophthora anisopliae*. Sorokin later it was renamed as *M. anisopliae* in the year 1883 (Tulloch 1976).

As microbial insecticides, a number of entomopathogenic fungi and their compounds are employed. Worldwide, *M. anisopliae* is a hyphomycete entomopathogenic fungus, the most commonly utilized to manage insect pests (Roberts and St Leger 2004). This species has a vast array of distinct strains and isolates from a diversity of hosts and geographical places of origin.

Metarhizium is seen in natural habitats such as the soil, where moisture promotes filamentous growth, the formation of spores—infectious in nature, i.e., conidia that infect insects which are soil-dwelling on contact (Fig. 6.2). *M. anisopliae* is an appropriate species for further development and research. This may be applicable as a biocontrol agent, particularly for species that transmit the disease–malaria. Using sequencing data from the ITS and 28S rDNA D3 sections as well as in RAPD patterns, ten different clades were found and the genus *Metarhizium*'s taxonomy was revaluated. *M. anisopliae* var. *anisopliae* two entomopathogenic fungi are supposed to be safe and a good environmentally friendly substitute for pesticides made of synthetic chemicals. These entomopathogenic fungi were recognized as microbial agents and are developed commercially for controlling several pests and biological control.

According to several factors, together with highly varying toxicities strains of *M. anisopliae* are sourced from various geographic locations (Fegan et al. 1993). In nations such as Brazil, where about 100,000 ha of sugarcane are treated annually, *M. anisopliae* have been working extensively (Faria and Magalhães 2001). Once the strains have been thoroughly evaluated, they are released into the field. As per the findings of (Goettel et al. 2001), testing on mammalian models is necessary to assess potential dangers to human and animal health when cultivating and applying fungi as microbiological agents for biocontrol of insect pests.

Virus

Baculovirus

Baculoviruses are viruses with double strands of DNA that infect arthropods, primarily insects. Due to their great pathogenicity, baculoviruses are effectively applied as biocontrol agents in their natural state to manage a variety of dangerous insect pests (Moscardi 1999). Nevertheless, in horticulture and agriculture fields, wherever the thresholds for the damage of pest are typically lower, baculovirus treatment is still restricted. Only at the larval stage do baculoviruses induce mortality in Lepidoptera, the primary cluster from which they have been isolated (Cory 2000).

For infection to begin, the larvae must consume the baculoviruses. Following after ingestion, they go via the midgut of the insect and take hold throughout the entire body; in some insects, the infection may only impact the midgut. Granuloviruses (GVs) and nucleopolyhedroviruses (NPVs) are two different types of baculoviruses.

While occlusion bodies in GVs often contain only one virus particle, they typically include multiple virus particles in NPVs. Baculoviruses are occluded, meaning that the particles of virus are lodged in a protein matrix, which is a typical characteristic. Because occlusion bodies permit the virus to live exterior to host, they are crucial to the biology of baculoviruses (Cory 2000). Baculoviruses do not infect plants or animals, as has been clearly shown in recent times. They are restricted from which they were isolated as hosts within the insects as well. Currently, only Lepidoptera isolates have been used for genetic engineering of baculoviruses (Kolodny-Hirsch et al. 1997). Different bacteria viruses can infect different numbers of animals; nearly appear to be explicit to a single species and, as a

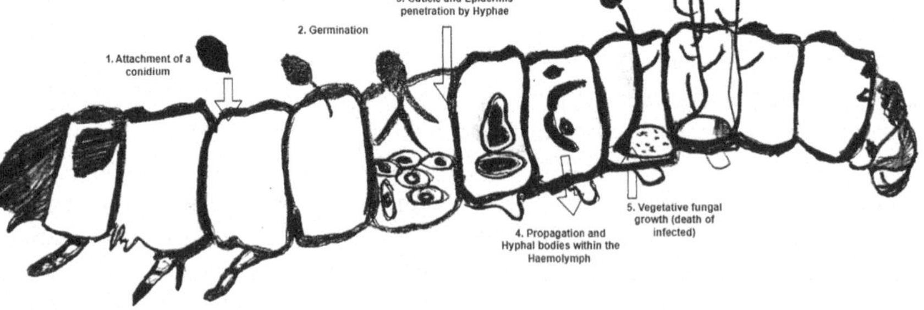

Fig. 6.2 MOA of entomopathogenic fungi in opposition to lepidopteran insects (Senthil-Nathan 2014)

result, are not likely to pose a risk in the event of genetic modification, while others can infect a diversity of hosts (Cory 2000). Toxins that are selective for specific insects are currently used in most new baculovirus recombinants. Nowadays, insect-selective poisons are used in most new baculovirus recombinants. The majority of genetic modification research has been done on *Autographa californica* NPV (AcNPV) also known as alfalfa looper virus. It has the most available molecular information, including its entire DNA sequence, that permits the exact insertion of foreign genes. Other strains of commercial or regional importance have recently been included in the genetically modified or GM-baculoviruses (Popham et al. 1997). Insect larvae infected with bacteria viruses have initially started a series of chemical and cellular appendages. Eventually, the larvae perish and develop enormous numbers of occlusion bodies of polyhedral shape made up virions of rod shaped (Miller et al. 1998). Recently, it has been suggested that *Agrotis ipsilon* multiple nucleopolyhedroviruses, or Agip MNPV, a naturally occurring baculovirus, may have the potential as a microbial pesticide for managing in turf of *A. ipsilon*. For the main purpose of controlling caterpillar pests, a variety of viral compositions are available. For example, moth granulosis virus (GV) Madex™, which marks the oriental fruit moth (OFM), was recently registered by Certis. Additionally, Certis sells Cyd-X™, which includes the codling moth GV and is a useful tool for managing codling moths. Certis offers Gemstar™, which comprises *Heliothis zea* NPV, and Spod-X™ that encompasses beet armyworm NPV, in addition to Madex and Cyd-X. According to Arthurs et al., Gemstar is approved for management of lepidopteran pests such as cotton bollworm and budworm, which are caterpillars that are mostly harmful insect-pests of Maize, soybeans, and many plants. A celery looper (*Syngrapha falcifera*) and an alfalfa looper (*Autographa californica*) have both been registered by Certis.

Nematodes

Steinernema (Rhabditida)

Nematodes are the newest and most popular biopesticide products. Rotation of crops, internalization of organic matter into the soil, and cover crops are effective ways to control pest nematodes. Early in the 1990s, many efficient entomopathogenic nematodes of the genera *Heterorhabditis* (Nematoda: Rhabditida) and *Steinernema* were identified, used as biocontrol agent against insects (Copping and Menn 2000). By pushing out harmful bacteria, insect-parasitic nematodes can invade stages of insects that are living in the soil and kill them in less than 48 h. Following the host' s death, the nematodes' infectious stages gets mature and leads to adults, giving rise to a next generation of infective juveniles (IJs) (Fig. 6.3). In addition to the widespread application of entomopathogenic nematodes to shield plants from harmful insect pests and illnesses, numerous initiatives are made to utilize IJs for spraying in the field to biocontrol insect pest populations. However, little is known about how native nematodes can affect populations of insect pests (Peters 1996). The third-stage IJs in nematodes start the parasitic cycle. These nonfeeding juveniles enter the body of a suitable insect host by natural openings like the mouth, anus, and spiracles. Later nematodes penetrate the host, discharge their symbiotic bacteria into the intestine, where they infest the hemocoel. The bacteria then cause septicemia, and it takes 24–48 h to kill the host (Fig. 6.3).

The bacteria quickly work the uptake of IJs, breaking down the tissues of host. The nematodes complete 2–3 generations in the host body. Earlier studies have demonstrated that entomopathogenic nematodes may serve as efficient biocontrol agents in contrast to several most important insect pest families originate in retail products, like the Curculionidae and Pyralidae (Shannag and Capinera 2000). Morris had previously shown that insect pests, like *Tenebrio molitor* L. and *Ephestia kuehniella* Zeller, that are found in stored goods are susceptible to an elevated nematode concentration. Georgian advised a field concentration of greater than 2.5 billion nematodes/ha to attack some of the most common insect pests of row crops;

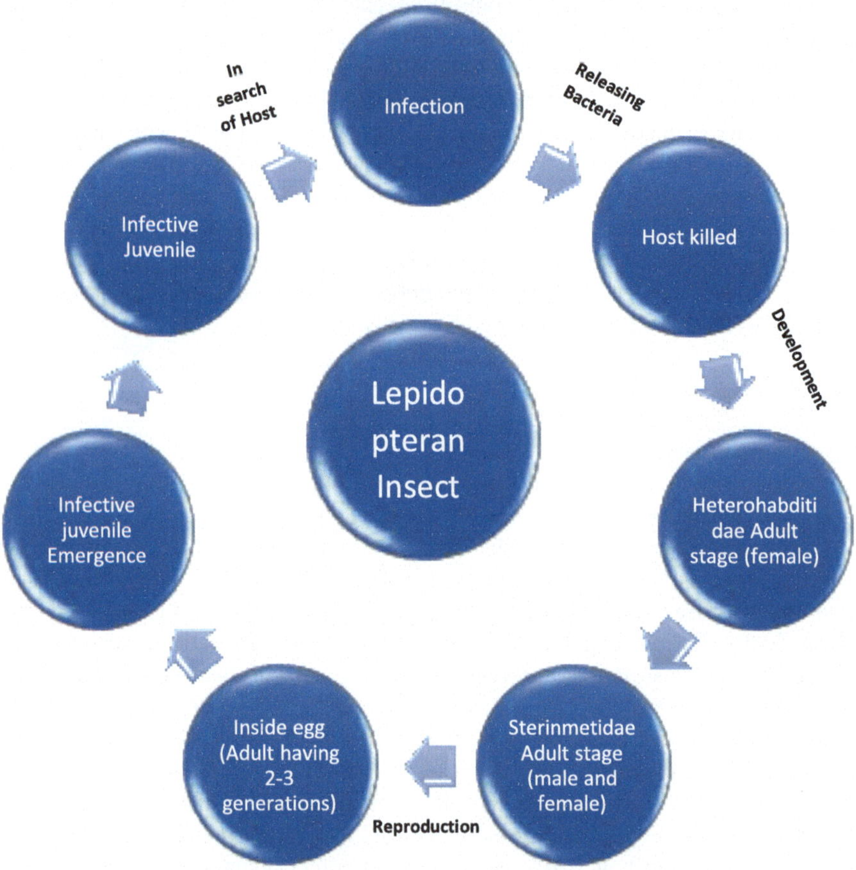

Fig. 6.3 MOA of entomopathogenic nematodes in opposition to lepidopteran insects (Senthil-Nathan 2014)

however, concentrations a few times higher—between 7 and 15 billion/ha—are necessary to attain to control the pest population (Loya and Hower 2002).

Protozoa

Nosema

Prior research has demonstrated that protozoan infections of insects are frequent in the natural world and show a crucial part in controlling populations of insect (Maddox 1987; Brooks 1988). Most frequently causing chronic infections, microsporidia, like *Nosema* spp., are slow-acting and host specific. For the most part, entomopathogenic protozoa have complex biological functions. Some species require an intermediary host, and they can only grow in living hosts. Among the most frequently seen species are microsporidia, which have the advantages of being persistent and recyclable in host populations moreover having a negative effect on target insects' ability to reproduce and be fit overall. Several species have shown some degree of efficacy when used as microbial control agents in flood situations (Solter and Becnel 2000).

Certain protozoan species, like the species which have been officially recognized and established for commercial use, Nosema locustae and grasshoppers are harmful (Henry and Oma 1981). The pathogen responsible for silkworm pébrine, *N. bombycis*, was the first microsporidium to be identified. It continued to exist in Europe, North America, and Asia in the middle of the 1800s (Becnel and Andreadis 2014). In nations like China that produce silk, the epidemic of phthalates continues to cause substantial economic losses (Cai et al. 2012). Numerous insect species, like heliothine moths and grasshoppers, are targeted by almost a thousand different types of protozoa, mostly microsporidia. The insect-pathogenic protozoan species *Vairimorpha necatrix* and *Nosema* spp. are virtually well known. Some insects that are susceptible to infection produce spores, which are made of protozoan spores that integrate with the host before developing in the midgut. After emerging from sporoplasm, germinating spores attack the cells of their host, causing widespread infection and the complete destruction of organs and tissues. A vulnerable host can produce an epizootic infection by ingesting the sporulation process, which recommences from the afflicted tissues. Naturally, insect predators and parasitoids frequently serve as disease vectors (Brooks 1988).

6.7.1.2 Botanical Insecticides

Plant-based natural compounds have been employed to provide protection against insect pests since ancient times. These compounds are generally effective. These substances were first identified scientifically in the nineteenth century and were then widely used in the early twentieth century (Morgan and Wilson 2001).

For about 3.5 million years, some insects and plants have cohabited on Earth, giving both sufficient amount of opportunity to grow offensive and defensive mechanisms. Plants have developed a range of defence strategies to help them survive against predator attacks.

Creating substances that are extremely harmful to insects is one example of a plant strategy.

Neem

Neem had a long past of usage in Asia, mostly in opposition to pests of the home and preservation, to a lesser amount, in contrast to agricultural insect-pests. However, Pradhan et al. (1962), made a breakthrough in the use of neem as an insecticide when they sprayed crops with a 0.1% neem seed kernel suspension at a low concentration during a locust infestation, effectively protecting the crops from insects. One of the most significant species in the Meliaceae family that produces limonoid compounds is *Azadirachta indica* also known as the Indian neem tree. Its products can be suggested for numerous crop pest management because of their relative selectivity and the significant insect control capability of some of their leaf and seed components. About 450–500 insect pest species have been tested for sensitivity to neem product activity in various nations worldwide; 413 of these species are apparently responsive at varied intensities (Schmutterer and Singh 1995).

Neem leaf activity against 12 nematodes, 103 insect pest species, and numerous harmful fungi has been evaluated in India alone. The potential of neem in pest management has been reviewed recently by Dhaliwal and Arora (2000). The majority of research has been on Azadirachtin which is produced in large quantities from seed extracts of neem and functions as a potent antifeedant and insect growth regulator.

As stated by Mordue and Blackwell (1993), Azadirachtin impacts the physiological functions of insects while having little effect on other biocontrol agents. Moreover, according to Senthil-Nathan (2013), neem products biodegrade naturally and pose no threat to non-target creatures. Neem activity against harmful insect pests in agricultural crop systems has been tested in several Asian nations through research conducted either unaided or in conjunction by existing insecticides and other biocontrol agents (Senthil-Nathan 2006). Field trials in India revealed that neem treatments were effective against various insects such as grasshoppers, green leafhoppers, yellow stem borer, rice leaf folder, and rice gall midge.

Melia azedarach

In light of alternative control strategies like helpful insects, which all need the integration of managed insect pest control, the promotion of botanicals as environmentally benign A primary concern has been insect growth regulators, microbiological treatments, and insecticides. Plant-based pesticides are made organically by extracting compounds from plants and using them for controlling harmful insect pests. Because of concerns about the ecological sustainability of synthetic pesticides and their potential toxicity to humans, non-target beneficial insects, and some domestic animals, interest in using alternative methods to manage insect pests has increased.

This finding suggests that developing biopesticides might be a sensible area for additional research. The family Meliaceae is known to have a wide variety of substances having antifeedant, insecticidal, growth-regulating, insect-deforming, and growth-modifying qualities.

Chinaberry, commonly known as *M. azedarach*, sometimes known as Persian lilac. It grows naturally in northwest India. It is a deciduous tree. It has been identified as having insecticidal qualities, albeit they have not yet been well studied. In addition to the tropical and subtropical regions of Asia, this tree is now grown in other warm regions of the world because of its exceptional climatic endurance. The insecticidal activity of plant leaves is used, despite the fact that *M. azedarach* fruit extracts have a variety of impacts on insects, such as growth retardation, decreased molting abnormalities, fecundity, and behavioral alterations (Ascher et al. 1995).

Additionally, many neem and other plant seed and leaf extracts show outstanding insecticidal qualities against vectors and are also highly environmental approachable. It is demonstrated that how effectively these neem chemicals keep mosquitoes away. Toxicants that are plant derived are unquestionably a useful origin of potential insecticides. According to Mordue (Luntz) and Blackwell (1993), plants and other natural pesticides may be extremely important in major insect control initiatives and programs aimed at reducing mosquito populations.

6.7.1.3 Biochemical Pesticides

Pheromones

Pheromones are substances formed by insects that cause other people to behave in a particular way. These pheromones are called alarm, aggregation, sex, and other names according on the evoked response they cause. They have a wide range of actions. Some pheromones act as sex attractants, helping entities to recognize and track down possible partners, whereas others promote the oviposition, trail-following, and accumulation of other congeners.

Over 1600 different sex attractants and pheromones have been identified, pheromones have emerged as indispensable tools to monitor and management of pest populations in the agriculture sector (Witzgall et al. 2004). Since there are more opportunities to look into the application of semiochemicals in controlling the problem of insect pests, a substantial quantity of research has been conducted on insect pheromones in recent decades. Over the last few years, a substantial amount of research has been conducted on insect pheromones, and there are now more opportunities to investigate the application of semiochemicals in pest management. Pheromone-based insect control has limitations when used alone, but it can be used in conjunction with other methods for integrated control).

A crucial constituent of the pheromone systems is known that many of the coleopteran species that have been studied up to this point are volatile plant species. Understanding how plant volatile mixtures and pheromones are incorporated into insect olfactory systems is becoming more and more crucial in order to help insects discern between odor plume strands and pheromone molecules alone and react behaviorally to these signals (Baker and Heath 2005).

There are now more opportunities to study the application of semiochemicals for pest management because considerable investigation on insect pheromones has been done in current years. If combined with other techniques, phenomena-based insect control can be utilized for integrated pest management, albeit it has drawbacks when applied alone.

The plant volatiles show a crucial role in the pheromone system of the many Coleopteran species been examined. An increasingly important question is how pheromone and plant volatile combinations are incorporated into insect olfactory systems, allowing them to discriminate, among pheromone molecules by themselves the issue regarding the way (pheromone + odor plume) strands react to signals through behavior is one that receives more and more attention (Baker and Heath 2005). Insecticide treatments that were calendar-based and broad-spectrum have recently been purposefully replaced in the pest management system with more comprehensive, integrated, and highly effective methods. Moreover, several important factors influencing the present pest management strategies in commercial agriculture include resistance management, food security, and conservation of environment (Witzgall et al. 2010). One of the most significant issues that modern agriculture is supplying the rapidly growing world population with food while combining sustainability and ecological stewardship. Broad-spectrum use of pesticides can be reduced in agricultural areas by using pheromones and/or allelochemicals as behavioral manipulation techniques in place of or in addition to current pest management strategies (Witzgall et al. 2008). Millions of hectares are being monitored for pest activity and managed through the use of pheromones and other semiochemicals. Pheromones provide several benefits when it comes to pest monitoring, such as reduced costs, specificity, simplicity, and extreme sensitivity (Witzgall et al. 2010). Pheromone lures for monitoring insect pests can yield management insights on when to use insecticides, for example.

Insects generate a pheromone that is highly species-specific. The pheromones that insects produce vary greatly by species. To prepare for mating, virgin female insects release sex pheromones, and males follow the concentration slope in search of the female producer. In order to signal to other insects that there is a good food source nearby, insects like wood-invading beetles produce aggregation pheromones.

Insects that are protected from predators release specific alarm pheromones, which aid in the pest bug's relocation and transform it into a dangerous hazard. All insects produce a pheromone in some approximate form, and companies exist that make a pheromone specifically for each and every customer, according to prior studies. The US Environmental Protection Agency has registered thirty products based on pheromones that interrupt mating, according to pest species' biocontrol agents which may threaten agriculture (Copping and Menn 2000).

6.8 Biopesticide Market

In 2008, the anticipated value of the worldwide pesticide market was 40 billion US dollars; by 2009, that amount had increased to roughly 43 billion US dollars in order to reach 51 billion dollars in the year 2014. This data is taken from the 2010 BBC report. There is a strong growth area in the biopesticide business within the worldwide market of pesticide, which was valued at 1.6 billion dollars in the year 2009 and is expected to increase to about 3.3 billion dollars at (15.6%) CAGR in 2014.

As per (Sinha and Biswas 2008) findings, North America accounts for 40% of the world's output of biopesticides, with Europe and Oceanic countries following closely behind with 20% each. India's trend of biopesticide consumption has demonstrated a sharp rise over time, with usage reaching 1920 metric tonnes in 2005–2006.

According to reports, just 2.89% of India' s total pesticide market is predicted to expand at a speed of 2.3% each year over the next several years (Thakore 2006). In India, the Insecticide Act of 1968 has authorized the use of 12 different forms of biopesticides (Table 6.3). The most often produced and used neem-based biopesticides in India are NPV, *Trichoderma*, *Bacillus thuringiensis* (Gupta and Dixit 2010).

6.9 Major Challenges and Constraints

A number of negative consequences were brought about by reliance on synthetic pesticides and their careless application such as pesticide residues in manufactured items, secondary pest outbreaks, soil fertility, pest resistance, and water quality (Al-Zaidi et al. 2011). With benefits including non-toxic mechanisms, eco-friendliness, efficacy, and suitability for integrated pest management programs, biopesticides unquestionably play a crucial role in managing desirable pests and

Table 6.3 List of biopesticides that India has listed in accordance with the Insecticides Act 1968 (Aneja et al. 2013)

S. no.	Biopesticide name
A.	*Bacillus thuringiensis* var. *galleriae*
B.	*B. thuringiensis* var. *israelensis*
C.	*B. thuringiensis* var. *kurstaki*
D.	*B. sphaericus*
E.	Neem-based pesticides
F.	*Spodoptera litura* NPV
G.	*Helicoverpa armigera* NPV
H.	Cymbopogon
I.	*Pseudomonas fluorescens*
J.	*Trichoderma viride*
K.	*Trichoderma harzianum*
L.	*Beauveria bassiana*

attracting public interest (Orozco-Mosqueda et al. 2023). Since biopesticides are typically made using living microbes, extreme caution and attention must be taken from the start of the developmental stage until the product is used. Globally, biopesticide production and application are expanding at the fastest possible rate.

The biggest obstacles to the development and marketing of any biopesticides appear to be the few drawbacks of utilizing them, such as their availability, sluggish effect, less persistence and broad-spectrum action, quick destruction by UV lamps, and poor water solubility. The main factors slowing down the development of biopesticides, however, are ignorance about the advantages of biopesticides, product knowledge, and farmer confidence. Due to inconsistent performance and unstable supplies, many farmers ceased using biopesticides (Alam 2000). One of the main reasons they were falling behind was a lack of confidence in the application and efficacy of biopesticides (Arora et al. 2011). About 45% of all biopesticides sold are used in Mexico, Canada, the United States, and Asia lags, using only 5% of all biopesticides sold globally (Bailey et al. 2010).

A significant challenge in implementing biotechnological applications for biopesticides is the lack of clarity in numerous nations' legislative frameworks concerning the creation and deployment of genetic engineering (GE) in biopesticide production. The producers of biopesticides frequently face unique challenges in the registration processes. Compared to conventional insecticides, such as *Bacillus thuringiensis Trichoderma*, viral pesticides have a few inherent advantages. These advantages include a small host range, the ability to infect only closely related species of the intended insect, and more friendliness toward the environment.

Despite this, there are still a number of issues with their marketing and development. Public acceptance and production circumstances are the main issues with viral biopesticide. These include the costs and duration associated with conducting testing in accordance with laws and regulations. Second, it can be grown on live hosts, tissues, or cell lines. The government agency must grant ethical approval before developing a tissue culture facility for the same (Lapointe et al. 2012). The process by which viral products are produced in insects is less accurate and more complex than that of chemical insecticides. The ultimate and most challenging phase of a microbial product's development, according to Ravensberg (2011), is commercialization. The two most important variables in product development are time to market and product cost.

A new business owner could expect to spend between 14 and 21 million US dollars, and it will take at least 5–7 years to go to market, including registration. Therefore, during the product development process, each of these essential components for the effective commercialization of microbial pest management products must be considered.

6.10 Discussion

Research has been done on the soil's physicochemical features both earlier and later chemical pesticide treatment (Goyal 2014). It is revealed that chemical pesticides have shown to have adverse impacts on agriculture and human well-being, according to reports by Goyal (2014). According to Sudha and Ambuja (2011), the chief source of water pollution is runoff from agriculture. Sharma (2011) reports of several insecticides to grasshoppers on the toxicity. Chaturvedi et al. (2012) described the benefits of environmentally friendly insecticides. According to Vasundriya et al. (2012), there is an effect of (dichloro-diphenyl-trichloroethane) commonly called as DDT and dimethonate on the proliferation of bacteria which are non-symbiotic in nature. Biopesticides provide an environmentally friendly and efficient substitute to chemical pesticides. Promoting biopesticides was highly advised by the National Farmer Policy of 2007 in order to maintain a safe environment, boost agricultural productivity, and protect farmer health. Biopesticides are highly specialized, less expensive, and do not damage non-targeted species or contaminate the environment. However, their effectiveness may change depending on several environmental conditions.

6.11 Conclusion

Biopesticides are of interest because of their high degree of specificity, which prevents them from destroying non-target species or leaving behind hazardous residues. While the chemical pesticides are not environmentally friendly and pose serious health risks due to residues found in food and fodder. Pesticides with chemicals have the potential to reduce beneficial species and genetically vary plants. To optimize profits, farmers must employ biopesticides, which in some cases are great substitutes for chemical pesticides. Organic food is in demand due to rising health consciousness. It shows that biopesticides have enormous potential for expansion. Considerable conventional wisdom could offer important hints for creating more advanced and potent biopesticides. Farmers will ultimately use more biopesticides if organic farming is prioritized. The investigation of biopesticides as an environmentally acceptable means of controlling pests is highly suggested by the National Farmer Policy NFP (2007). Research on biopesticides is still in its early stages, thus more funding and regularity are required. To produce biopesticides, extensive in the fields of formulation, distribution, marketing, and screening of potential control agent or agents, research is needed. Despite this, it is still necessary for individuals generally, and farmers specifically, should possess expertise in managing goods and implementing preventative measures.

The expanding discipline of biopesticides offers a safer, more ecologically friendly means of managing pest populations, such as weeds, insects, and plant diseases, which are less dangerous to humans, animals, and the environment. We need to remind producers, governments, legislators, farmers, and the general public about the necessity of switching to biopesticides for pest control needs because environmental safety is a global priority. The expanding discipline of biopesticides offers a safer, more ecologically friendly means of managing populations of pests like plant diseases, weeds, and insects, which are less detrimental for the environment, humans, and animals. We need to inform producers, governments, legislators, farmers, and the general public about the necessity of switching to biopesticides for pest control needs because environmental safety is a global priority. Farmers in developing countries should pay particular attention to this.

References

Alam G (2000) A study of biopesticides and biofertilizers in Haryana, India Gatekeeper Series No. 93. IIED, London

Ali A, Ölmez F, Zeshan MA, Mubeen M, Iftikhar Y, Sajid A, Abid M, Kumar A, Divvela PK, Solanki MK (2024) Yeast-based solutions in controlling plant pathogens. Biocatal Agric Biotechnol 58:103199. https://doi.org/10.1016/j.bcab.2024.103199

Al-Zaidi AA, Elhag EA, Al-Otaibi SH, Baig MB (2011) Negative effects of pesticides on the environment and the farmers awareness in Saudi Arabia: a case study. J Anim Plant Sci 21:605

Aneja KR (2000) Biopesticides (microbial pesticides). Pragati Prakashan, Meerut, pp 20–27

Aneja KR (2014) Exploitation of phytpathogenic fungal diversity for the development of bioherbicides. Kavaka 42(1):7–15

Aneja KR, Khan SA, Kaushal S (2000) Horse purslane (*Trianthema portulacastrum* L.) and its biocontrol with fungal pathogens: an overview. In: Biocontrol potential and its exploitation in sustainable agriculture: crop diseases, weeds, and nematodes. Springer, pp 187–198

Aneja KR, Kumar V, Jiloha P, Kaur M, Sharma C, Surain P et al (2013) Potential bioherbicides: Indian perspectives. In: Biotechnology: prospects and applications. Springer, pp 197–215

Aronson AI, Shai Y (2001) Why *Bacillus thuringiensis* insecticidal toxins are so effective: unique features of their mode of action. FEMS Microbiol Lett 195(1):1–8

Arora NK, Khare E, Maheshwari DK (2011) Plant growth promoting rhizobacteria: constraints in bioformulation, commercialization, and future strategies. In: Plant growth and health promoting bacteria. Springer, pp 97–116

Ascher KRS, Schmutterer H, Zebitz CPW, Naqvi SNH (1995) The Persian lilac or chinaberry tree: Melia azedarach L. In: The neem tree: source of unique natural products for integrated pest management, medicine, industry and other purposes. VCH, Weinheim, pp 605–642

Auld BA, Hetherington SD, Smith HE (2003) Advances in bioherbicide formulation. Weed Biol Manag 3(2):61–67

Bailey KL, Boyetchko SM, Längle TJBC (2010) Social and economic drivers shaping the future of biological control: a Canadian perspective on the factors affecting the development and use of microbial biopesticides. Biol Control 52(3):221–229

Baker TC, Heath JJ (2005) Pheromones: function and use in insect control. In: Comprehensive molecular insect science. Elsevier, pp 407–459

Bastas KK, Kumar A (2023) Omics technologies in the plant–microbe interactions. In: Bastas KK, Kumar A, Sivakumar U (eds) Microbial biocontrol: molecular perspective in plant disease management. Springer Nature, Singapore, pp 257–282. https://doi.org/10.1007/978-981-99-3947-3_13

Becnel JJ, Andreadis TG (2014) Microsporidia in insects. In: Microsporidia: pathogens of opportunity. Wiley, pp 521–570

Bhardwaj A, Kaur K, Kumar A, Singh SK, Kharshiing B (2024) Development of microbial biotechnology products for phytopathogen management. In: The potential of microbes for a circular economy. Academic Press, pp 105–130. https://doi.org/10.1016/B978-0-443-15924-4.00013-8

Brachmann S (2015) The future of agricultural pest control is biopesticides, IoT insect monitoring systems
Brooks FM (1988) Entomogenous protozoa. In: Ignoffo CM, Mandava MB (eds) Handbook of natural pesticides, Microbial insecticides, part A, entomogenous protozoa and fungi, vol V. CRC Press Inc., Boca Raton, pp 1–149
Cai SF, Lu XM, Qiu HH, Li MQ, Feng ZZ (2012) Phagocytic uptake of *Nosema bombycis* (Microsporidia) spores by insect cell lines. J Integr Agric 11(8):1321–1326
Carlton BC (1993) Genetics of Bt insecticidal crystal proteins and strategies for the construction of improved strains. In: Pest control with enhanced environmental safety. ACS symposium series, vol 524. American Chemical Society, pp 326–337
Chaturvedi DA, Jain P, Malik S (2012) Study of ecofriendly pesticides. Res Hunt VII(iii):48–50
Copping LG, Menn JJ (2000) Biopesticides: a review of their action, applications and efficacy. Pest Manag Sci (formerly Pestic Sci) 56(8):651–676
Cory JS (2000) Assessing the risks of releasing genetically modified virus insecticides: progress to date. Crop Prot 19(8–10):779–785
Devi S, Sharma S, Tiwari A, Bhatt AK, Singh NK, Singh M, Kaushalendra, Kumar A (2023) Screening for multifarious plant growth promoting and biocontrol attributes in Bacillus strains isolated from Indo-Gangetic soil for enhancing growth of rice crops. Microorganisms 11:1085. https://doi.org/10.3390/microorganisms11041085
Dhaliwal GS, Arora R (2000) Role of phytochemicals in integrated pest management. In: Phytochemical biopesticides. CRC Press, pp 92–109
El-Sayed W (2005) Biological control of weeds with pathogens: current status and future trends (Biologische Schadpflanzenbekämpfung mit Pathogenen: Aktueller Status und Trends von morgen). J Plant Dis Prot (Zeitschrift für Pflanzenkrankheiten und Pflanzenschutz) 112:209–221
Faria MD, Magalhães BP (2001) O uso de fungos entomopatogênicos no Brasil. Biotecnologia Ciência Desenvolvimento 22(1):18–21
Fegan M, Manners JM, Maclean DJ, Irwin JAG, Samuels KDZ, Holdom DG, Li DP (1993) Random amplified polymorphic DNA markers reveal a high degree of genetic diversity in the entomopathogenic fungus Metarhizium anisopliae var. anisopliae. Microbiology 139(9):2075–2081
Gelernter W, Schwab GE (1993) Transgenic bacteria, viruses, algae and other microorganisms as *Bacillus thuringiensis* toxin delivery systems
Goettel MS, Hajek AE, Siegel JP, Evans HC (2001) 13 Safety of fungal biocontrol agents. In: Fungi as biocontrol agents. CABI Publishing, p 347
Goldberg LJ, Margalit J (1977) A bacterial spore demonstrating rapid larvicidal activity against *Anopheles sergentii*, *Uranotaenia unguiculata*, *Culex univitattus*, *Aedes aegypti* and *Culex pipiens*
Goyal S (2014) Study of effects of pesticides on soil. Res Hunt IX(1):35–38
Grewal PS, Lewis EE, Gaugler R (1997) Response of infective stage parasites (Nematoda: Steinernematidae) to volatile cues from infected hosts. J Chem Ecol 23:503–515
Gupta S, Dixit AK (2010) Biopesticides: an ecofriendly approach for pest control. J Biopesticides (Special Issue) 1:186–188
Guzmán-Guzmán P, Kumar A, de Los Santos-Villalobos S, Parra-Cota FI, Orozco-Mosqueda MDC, Fadiji AE, Hyder S, Babalola OO, Santoyo G (2023) Trichoderma species: our best fungal allies in the biocontrol of plant diseases—a review. Plants 12:432. https://doi.org/10.3390/plants12030432
Henry JE, Oma EA (1981) Pest control by Nosema locustae, a pathogen of grasshoppers and crickets
Imaizumi S, Nishino T, Miyabe K, Fujimori T, Yamada M (1997) Biological control of annual bluegrass (*Poa annua* L.) with a Japanese isolate of *Xanthomonas campestris* pv. poae (JT-P482). Biol Control 8(1):7–14
James C (2009) Global Status of Commercialized Biotech/GM Crops

Kalra A, Khanuja SPS (2007) Research and development priorities for biopesticide and biofertiliser products for sustainable agriculture in India. In: Teng PS (ed) Business potential for agricultural biotechnology. Asian Productivity Organisation, pp 96–102

Kashyap N, Singh SK, Yadav N, Singh VK, Kumari M, Kumar D, Shukla L, Kaushalendra, Bhardwaj N, Kumar A (2023) Biocontrol screening of endophytes: applications and limitations. Plants 12:2480. https://doi.org/10.3390/plants12132480

Kaya HK, Gaugler R (1993) Entomopathogenic nematodes. Annu Rev Entomol 38(1):181–206

Kazinczi G, Lukacs D, Takacs A, Horvath J, Gaborjanyi R, Nadasy M, Nadasy E (2006) Biological decline of Solanum nigrum due to virus infections. Zeitschrift Fur Pflanzenkrankheiten Und Pflanzenschutz-Sonderheft 20:325

Kolodny-Hirsch DM, Sitchawat T, Jansiri T, Chenrchaivachirakul A, Ketunuti U (1997) Field evaluation of a commercial formulation of the Spodoptera exigua (Lepidoptera: Noctuidae) nuclear polyhedrosis virus for control of beet armyworm on vegetable crops in Thailand. Biocontrol Sci Technol 7(4):475–488

Kumar S, Singh A (2015) Biopesticides: present status and the future prospects. J Fertil Pestic 6(2):1–2

Kumar A, Singh VK, Tripathi V, Singh PP, Singh AK (2018) Plant growth-promoting rhizobacteria (PGPR): perspective in agriculture under biotic and abiotic stress. In: Crop improvement through microbial biotechnology. Elsevier, pp 333–342. https://doi.org/10.1016/B978-0-444-63987-5.00016-5

Kumar A, Solanki MK, Wang Z, Solanki AC, Singh VK, Divvela PK (2024) Revealing the seed microbiome: navigating sequencing tools, microbial assembly, and functions to amplify plant fitness. Microbiol Res 279:127549. https://doi.org/10.1016/j.micres.2023.127549

Kumari M, Qureshi KA, Jaremko M, White J, Singh SK, Sharma VK, Singh KK, Santoyo G, Puopolo G, Kumar A (2022) Deciphering the role of endophytic microbiome in postharvest diseases management of fruits: opportunity areas in commercial up-scale production. Front Plant Sci 13:1026575. https://doi.org/10.3389/fpls.2022.1026575

Lapointe R, Thumbi D, Lucarotti CJ (2012) Recent advances in our knowledge of baculovirus molecular biology and its relevance for the registration of baculovirus-based products for insect pest population control. In: Integrated pest management and pest control. IntechOpen, pp 481–522

Li Y, Sun Z, Zhuang X, Xu L, Chen S, Li M (2003) Research progress on microbial herbicides. Crop Prot 22(2):247–252

Loya LJ, Hower AA (2002) Population dynamics, persistence, and efficacy of the entomopathogenic nematode *Heterorhabditis bacteriophora* (Oswego strain) in association with the clover root curculio (Coleoptera: Curculionidae) in Pennsylvania. Environ Entomol 31(6):1240–1250

Maddox JV (1987) Protozoan diseases. In: Epizootiology of insect diseases, vol 1. Wiley, pp 417–452

Marrone PG (2007) Barriers to adoption of biological control agents and biological pesticides. CABI Rev:12. https://doi.org/10.1079/PAVSNNR20072051

Mazid S, Kalita JC, Rajkhowa RC (2011) A review on the use of biopesticides in insect pest management. Int J Sci Adv Technol 1(7):169–178

Miller LK, Kaiser WJ, Seshagiri S (1998) Baculovirus regulation of apoptosis. In: Seminars in virology, vol 8(6). Academic Press, pp 445–452

Mishra J, Tewari S, Singh S, Arora NK (2014a) Biopesticides: where we stand? In: Plant microbe's symbiosis: applied facets. New Delhi, Springer, pp 37–75

Mishra K, Kumar P, Basu S, Rai K, Aneja S (2014b) Risk factors for severe acute malnutrition in children below 5 y of age in India: a case-control study. Indian J Pediatr 81:762–765

Mordue AJ, Blackwell A (1993) Azadirachtin: an update. J Insect Physiol 39(11):903–924

Morgan ED, Wilson ID (2001) Insect hormones and insect chemical ecology. ChemInform 32(3):263–375. https://doi.org/10.1016/B978-0-08-091283-7.00053-9

Moscardi F (1999) Assessment of the application of baculoviruses for control of Lepidoptera. Annu Rev Entomol 44(1):257–289

Orozco-Mosqueda MDC, Kumar A, Fadiji AE, Babalola OO, Puopolo G, Santoyo G (2023) Agroecological management of the grey mould fungus Botrytis cinerea by plant growth-promoting bacteria. Plants 12:637. https://doi.org/10.3390/plants12030637

Peters A (1996) The natural host range of Steinernema and Heterorhabditis spp. and their impact on insect populations. Biocontrol Sci Technol 6(3):389–402

Popham HJ, Li Y, Miller LK (1997) Genetic improvement of Helicoverpa zeanuclear polyhedrosis virus as a biopesticide. Biol Control 10(2):83–91

Pradhan S, Jotwani MG, Rai BK (1962) The neem seed deterrent to locusts. Indian Farming 12(8):7–11

Ravensberg WJ (2011) A roadmap to the successful development and commercialization of microbial pest control products for control of arthropods, vol 10. Springer Science & Business Media

Revathi K, Chandrasekaran R, Thanigaivel A, Kirubakaran SA, Sathish-Narayanan S, Senthil-Nathan S (2013) Effects of Bacillus subtilis metabolites on larval Aedes aegypti L. Pestic Biochem Physiol 107(3):369–376

Roberts DW, St Leger RJ (2004) *Metarhizium* spp., cosmopolitan insect-pathogenic fungi: mycological aspects. Adv Appl Microbiol 54:1–70

Schmutterer H, Singh RP (1995) List of insect pests susceptible to neem products. In: Schmutterer H (ed) The neem tree: source of unique natural products for integrated pest management, medicine, industry and other purposes. VCH, Weinheim, pp 325–326

Senthil-Nathan S (2006) Effects of Melia azedarach on nutritional physiology and enzyme activities of the rice leaffolder Cnaphalocrocis medinalis (Guenée) (Lepidoptera: Pyralidae). Pestic Biochem Physiol 84(2):98–108

Senthil-Nathan S (2013) Physiological and biochemical effect of neem and other Meliaceae plants secondary metabolites against Lepidopteran insects. Front Physiol 4:359

Senthil-Nathan S (2014) A review of biopesticides and their mode of action against insect pests. In: Environmental sustainability: role of green technologies. Springer, pp 49–63

Shannag HK, Capinera JL (2000) Interference of *Steinernema carpocapsae* (Nematoda: Steinernematidae) with *Cardiochiles diaphaniae* (Hymenoptera: Braconidae), a parasitoid of melonworm and pickleworm (Lepidoptera: Pyralidae). Environ Entomol 29(3):612–617

Sharma T (2011) Toxicity of insecticides to the adults of choreodous illustris. J Environ Res Dev 6(2):286–290

Sinha B, Biswas I (2008) Potential of bio-pesticides in Indian agriculture vis-a-vis rural development. India Science and Technology

Solanki MK, Joshi NC, Singh PK, Singh SK, Santoyo G, de Azevedo LCB, Kumar A (2023) From concept to reality: transforming agriculture through innovative rhizosphere engineering for plant health and productivity. Microbiol Res 127553. https://doi.org/10.1016/j.micres.2023.127553

Solanki MK, Wang Z, Kaushik A, Singh VK, Roychowdhury R, Kumar M, Kumar D, Singh J, Singh SK, Dixit B, Kumar A (2024) From orchard to table: significance of fruit microbiota in postharvest diseases management of citrus fruits. Food Control 110698. https://doi.org/10.1016/j.foodcont.2024.110698

Solter LF, Becnel JJ (2000) Entomopathogenic microsporida. In: Lacey LA, Kaya HK (eds) Field manual of techniques in invertebrate pathology: application and evaluation of pathogens for control of insects and other invertebrate pests. Kluwer Academic, Dordrecht, pp 231–254

Sudha V, Ambuja NK (2011) Agriculture non point source pollution effects in fresh water reservoir, Krishnagiri, Tamilnadu, India. J Environ Res Dev 6(2):222–231

Thakore Y (2006) The biopesticide market for global agricultural use. Ind Biotechnol 2(3):194–208

Tulloch M (1976) The genus *Metarhizium*. Trans Br Mycol Soc 66:407–411

Vasundriya R, Malhotra M, Siddiqui SA, Pathrade M (2012) In vitro effect of DDT & dimetnonate on growth of non-symbiotic bacteria. J Environ Res Dev 7(1):66–69

Verma S, Azevedo LCB, Pandey J, Khusharia S, Kumari M, Kumar D, Kaushalendra, Bhardwaj N, Teotia P, Kumar A (2022) Microbial intervention: an approach to combat the postharvest pathogens of fruits. Plants 11:3452. https://doi.org/10.3390/plants11243452

Wang Z, Solanki MK, Kumar A, Solanki AC, Pang F, Ba ZX, Niu JQ, Ren ZX (2023) Promoting plant resilience against stress by engineering root microenvironment with Streptomyces inoculants. Microbiol Res 277:127509. https://doi.org/10.1016/j.micres.2023.127509
Witzgall P, Lindblom T, Bengtsson M, Toth M (2004) The Pherolist. http://www.pherolist.slu.se/pherolist.php. Accessed 23 Jul 2013
Witzgall P, Stelinski L, Gut L, Thomson D (2008) Codling moth management and chemical ecology. Annu Rev Entomol 53:503–522
Witzgall P, Kirsch P, Cork A (2010) Sex pheromones and their impact on pest management. J Chem Ecol 36:80–100

Beneficial Microbes in Agriculture: Recent Development and Outlook

7

Bushra Rasool, L. V. Ravishankar, Nisar Ahmad Dar, Amrish Vaid, Zaffar Bashir, Bisma Rashid, Sana Surma, Baby Summuna, M. D. Thabrez, and Sakshi Sharma

Abstract

Microbes are vital for establishment of agricultural approaches. To maintain crop growth, development, and production, their contribution to fertility and health of soil is essential. Microbes also have a negative impact on agriculture by causing diseases. Understanding crop-soil microbiome's enormous activities and anatomical complexity are required for effective deployment of these microbes in agriculture. Even though soil and crop microbiomes have been researched for many years, potential of inoculants, or beneficial microbes, to establish colonies in the soil and ensure ecosystem stability, is a major factor how effectively laboratory and greenhouse innovations can be used in the field. Furthermore, multitude and organization of crop and soil microbiomes are influenced by crop and surroundings. Biotechnologies of microbial applications in agriculture resist biotic and abiotic challenges, improve their health, and increase yield. The community description is critical for manipulating microbiome and identifying biofertilizers and bio control agents. Next-generation sequencing techniques determine cultural and non-cultural microorganisms connected with soil, and

B. Rasool (✉) · L. V. Ravishankar · A. Vaid · M. D. Thabrez · S. Sharma
Division of Plant Pathology, Faculty of Agriculture, SKUAST-Jammu, Jammu, Jammu and Kashmir, India

N. A. Dar
Maize Research Station, Udhampur, SKUAST-Jammu, Jammu, Jammu and Kashmir, India

Z. Bashir · B. Rashid
Division of Basic Sciences and Humanities, FoA, Wadura, SKUAST-Kashmir, Jammu, Jammu and Kashmir, India

S. Surma · B. Summuna
Division of Plant Pathology, FoH, SKUAST-Kashmir, Jammu, Jammu and Kashmir, India

A. Kumar, M. K. Solanki (eds.), *Microbial Biocontrol Techniques*, Microorganisms for Sustainability 54,
https://doi.org/10.1007/978-981-97-8739-5_7

crop microbiome has greatly advanced our understanding of this issue. Furthermore, genome-editing and multidimensional genomics approaches were given to scientists with a foundation for developing reliable and communities of microbes promoting crop production, resistance to disease, cycling of nutrient, and stress mitigation. This chapter provides a brief description of importance of benign microbes in agriculture, microbiome technology, adaptation of this technology in fields, and primary methodologies utilized by laboratories throughout the world for research plant–soil microbiome. These endeavors are critical for advancing smart agricultural technologies.

Keywords

Agriculture · Biotic · Biofertilizer · Biocontrol · Microbes

7.1 Introduction

Agriculture has traditionally proven an effective means to fulfill the worldwide food requirement. The populace continues to grow at a rapid pace; consequently, it is critical that agriculture fulfill this growing need. Thus, improving agricultural output and quality is critical (Sansinenea 2021). However, the plants are continually infected with various diseases caused by pathogens, resulting in crop damage and economic losses. The regular cultivation also causes the soil to lose a significant quantity of nutrients, which lowers crop quality and productivity. Pesticides and fertilizers with chemicals have been employed in this way to decrease these adverse effects due to severe health and environmental issues brought on by their prolonged use including harm to the ecosystem and increased insect resistance (Zambrano-Mendoza et al. 2021). Consequently, this is evident that innovative techniques need to be created and researched in order to optimize agricultural yield. Including beneficial microbes in agricultural techniques is one of these strategies. The relationship of microbiome and plant health still needs to be clarified, despite the fact that numerous research investigations have shown the function those microbes play in soil fertility, health, and plant fitness. These results will provide insight into the complete crop–microbe interaction process as well as the identification of microbes soon to be used to increase agricultural yield (Farrar et al. 2014; Bhattacharyya and Jha 2012). By exploring of plant microbiome, we may elaborate functional and structural array of microbial communities of linked to certain crops and environments. Researchers have provided ample documentation on the microbial variety found in the phyllosphere, rhizosphere, and endosphere across many areas and organs (Gupta et al. 2022; Fendrihan and Pop 2021). Structural barriers or existence of certain structures like trichomes to encourage and facilitate microbial colonization and biochemical responses development promote microbial associations. Research on a comprehensive strategy for understanding the population of microbes and its connection with crops is still being pursuing in multiple labs worldwide (Gupta et al. 2022; Sergaki et al. 2018). A host's species, selection pressure, habitat, stage of development, and agricultural techniques are some of elements that affect diversity of microbial community. Scientific advances have been made in technologies used to analyze variety of microbes and makeup of a particular plant microbiome. Recently, sophisticated technologies that provide higher-resolution images of the microbiome have supplanted culture-dependent identification, despite its limits in providing a complete view of the microbiome (Kaul et al. 2021).

Microbiome engineering serve as an alternative way for comprehension, altering and establishing combining technologies for growing communities of microbes, which are crucial for crop health and production in this scenario (Afridi et al. 2022). Novel and forthcoming advances are valuable in determining the extent of microbiome diversity in any particular environment (Cullen et al. 2020). In this chapter, we will discuss the microbiome's advantages; their interactions under different environment manipulate crop–microbe interactions.

7.1.1 Crop Microbiome Network

Naturally, soil and plant systems are invaded by microbes. Chemical signaling is connected to one another regulating these interactions between plants and microbes. This interaction betwixt plant roots and soil governs microbial diversity found in rhizosphere, one of these micro-communities. Microbial communities are recruited by root exudates. Additionally, the rhizosphere aids in formation of endophytic communities inside plants (Anand et al. 2023). Endophytes are unaffected by environment and have no detrimental impact on plant. Eventually, any portion of aboveground plant becomes invaded by phyllosphere (Brunel et al. 2020; Compant et al. 2019; Kashyap et al. 2023). The above and belowground ecosystem's corresponding microbial communities and diversities are linked by these micro-communities. Researchers have looked these micro-communities to learn more their potential for sustainable, expanding and developing agriculture (Santoyo et al. 2016; Vandenkoornhuyse et al. 2015; Wang et al. 2020). Chemical mediators that are generated in response to certain triggers, such as ethylene, flavonoids, and terpenoids, increase the advantageous connections between microorganisms and plants (Compant et al. 2019; Chiaranunt and White 2023).

7.2 Use of Microbes in Agriculture

7.2.1 Beneficial Microbes and Their Process of Action

Research in crop microbiomes reveals that microbes affect hosts directly or indirectly. Nitrogen fixation, iron and phosphate sequestration, hormonal balance (viz. gibberellic acid, GA, ABA, CK, and auxin, AUX), and enzyme synthesis (viz. 1-aminocyclopropane-1-carboxylic acid deaminase) are directly impacted on crop microbiome (Egamberdieva et al. 2017; Bashir et al. 2018a). In order to reduce harm that infection causes to host, strategies such as antibiotics (viz. bacteriocins, lipopeptides, proteases, and siderophores), volatile chemicals, competitive exclusion, predation, and microbe-mediated immunity are employed. These strategies either stimulate or repress the host's innate immune responses. This is indirect method by which microbes regulate host functions (Vannier et al. 2019).

7.2.2 Plant Growth Promoting Bacteria and Stressors (Biotic and Abiotic)

Excessive usage of agrochemicals in agriculture had a severe impact on the environment, harmed human health, and weakened soil fertility. Furthermore, hazardous microorganisms (*Ralstonia* sp., *Xanthomonas* sp., and *Rhizoctonia* sp.) initiate disease outbreak widespread when pesticides are misused (Ebrahimi-Zarandi et al. 2022). As a result, contemporary trend for sustainable agricultural methods, in which benign crop microbes are exploited, increases crop yield and minimizes diseases (Kumar et al. 2021). Generations of research have demonstrated that microorganisms may prevent disease and mitigate abiotic stressors on crops (Koza et al. 2022; Kumar et al. 2018). They are linked for managing soil health and cycles of nutrients within ecosystems and minimizing the detrimental effects of global warming. Blending these characteristics enables benign microbes a promising target for agricultural productivity enhancement (Munir et al. 2022). PGPB are microbes that suppress diseases and are classified as biocontrol agents (El-Saadony et al. 2022; Bastas et al. 2023; Bastas and Kumar 2023). They protect against disease by inhibiting microorganisms and reducing iron availability (Olanrewaju et al. 2017). Moreover, advances in plant engineering have made it possible to create resistant crops through transgenics or molecular breeding; these cultivars carry genes for disease resistance and enhance nutritional content and absorption and a host of other changes (Dong and Ronald 2019). Transgenics is an ecologically benign response to agriculture's biotic stresses (Shah et al. 2021). The post-genomic age provides new potential transgenic microbes of PGPB genes for abiotic and biotic stress tolerance (Singh et al. 2022a, b).

Abiotic stresses have an effect on crop physiology and biological process. Certain crops can adapt to stress, while others are overwhelmed by it. PGPB involved in disease regulation via hormone and enzyme production. PGPB release secondary metabolites which allow crops to adapt stress (Shah et al. 2021; Nadarajah and Rahman 2021). Certain species of bacteria (viz. *Pseudomonas, Rhizobium, Bacillus*, and *Enterobacter*) are recognized as promising stress (abiotic) manipulation options (Abdul Rahman et al. 2021). Microbes are swiftly growing creatures that acclimatize with their surroundings by producing biofilms and exopolysaccharides (EPS) and absorbing osmolytes or remain dormant (Nadarajah and Rahman 2021). When crops are exposed to various challenges, they generate hormones that aid in maintenance of growth and development (Verma et al. 2016). For example, *Azospirillum* species produce more ABA response to drought (Cohen et al. 2015). Under high salinity circumstances, PGPB take in and retain osmolytes in cell without harm, since EPS produced by these microorganisms binds to cations, resulting in turgidity in microbial cell (Kumawat et al. 2022).

7.2.3 Implementation of PGPB in Agriculture

Microbes like fungi and PGPB that have been noticed, identified, and gathered for utilization in field applications are vital for agriculture future. There are a variety of possible uses for PGPB, especially in enhancing nutrient absorption and sequestration (Kumawat et al. 2022). By expanding root development and enhancing surface area for water and nutrient absorption, these microbes form symbiotic relationships with plant root systems (Vacheron et al. 2013). Some of these microbes are beneficial for bioremediation, hazardous material degradation, growth enhancement, and biocontrols (Bala et al. 2022). Nitrogen-fixing microbes (viz. *Nitrosomonas* sp., *Nitrobacter* sp., and *Rhizobium* sp.) have proven boost nitrogen availability for plants, resulting in higher plant production and biomass (Gopalakrishnan et al. 2015). Fungi have been used to enhance yield, due to abundance availability of nutrients in the form of soluble phosphates, thereby limiting pathogen development (Tariq et al. 2022; Ali et al. 2024).

7.3 Microbiome Engineering

The crop diversification impacts the soil microbial population suggests that population of microbes in a niche ecosystem may be altered by genotypes and environment to improve host activities (Nadarajah and Rahman 2021). As a result, designing a rhizospheric system for plants can benefit those (Lumibao et al. 2020). The importance of microbiome engineering in sustainable agriculture will be described in more detail in the following sections.

7.3.1 Rhizosphere Microbiome

A diverse range of macro and microfauna is found in ecto and endorhizospheres). Numerous investigations have examined at microbial communities under these circumstances and determined that the microbes and the hosts' root systems interact (Dubey et al. 2019). Most of these crop relationships are beneficial for crop health. Numerous studies have been conducted on root endophytes (viz. *Rhizobium* spp. and *Brevibacillus* spp.) with nodulating and non-nodulating crops respectively, since these crops contribute significantly to nitrogen fixation (Mendes et al. 2013). Additionally, taxonomic diversity in distinct strains that colonize different plant tissues was addressed by researchers utilizing metagenome data. Numerous biological and physiological processes, such as stress tolerance, food uptake, and sequestration, are controlled by the microbial communities of the host. For this reason, understanding how plant–microbe interactions enhance plant function requires identifying and mapping the function of microbes in each kind of plant tissue (Lundberg et al. 2013).

7.3.2 Rhizosphere Engineering

The modern omics techniques have enabled us to make intentional alterations to the environment and plants. With the advances in science, we will be able to design the environment more successfully in the future. This technique, however, will stay the same: we need to understand positive mechanism influencing environment and construct equivalent conditions in field to mimic symbiotic interactions that occur in nature between soil, microorganisms, and plants (Pathak et al. 2022). Once this is done, bioinoculant will thrive in the seeded habitat (Moreira et al. 2023).

7.3.3 Remodeling of Microbiome

To modify the microbiome, efforts must be directed toward determining the identity of microbes and their functioning in the rhizosphere. Therefore, in order to make more informed decisions regarding consortia formation during design of agricultural applications, a database of microbes and their unique contributions to rhizosphere is necessary (Hayat et al. 2012). Crops and microbes have symbiotic connections should be highlighted in this database and used to classify species that carry out certain tasks. Furthermore, abundance of bacteria that produce phosphate and siderophores boosts plant's nutritional availability (Bashir et al. 2017, 2018a, b). Another group of species generates antimicrobials, which are useful in preventing soil infections. These organisms secrete antibacterial chemicals such as 2,4-diacetylphloroglucinol (DAPG) (Caulier et al. 2019).

Crop–microbe interactions and crop growth and development are further facilitated by phytohormones (Gupta et al. 2018). Numerous phytohormones, viz. auxins, cytokinins, and gibberellins (GA), are produced by microbial communities. Phytohormones are essential for plant growth, as transcriptome research has shown. In addition to creating systemic acquired resistance (SAR), these chemical interactions with signaling molecules like salicylic acid and jasmonate also cause systemic resistance (ISR) within crops. Injecting non-pathogenic bacteria into plants provides resistance against a broad spectrum of illnesses that affect them both above and below ground. The signaling pathways for ethylene and jasmonate are main foundation of ISR. Response to disease assaults, this enables plants to react more rapidly. MAMPs, or microbe-associated molecular patterns, are known to activate ISR in response to a range of microorganisms (Zamioudis and Pieterse 2012). These include flagella, cell envelope components, and siderophores. Noteworthy is fact that certain PGPB release volatile organic compounds (VOCs) that cause ISR reactions and encourage crop development (Singh et al. 2017, 2022a, b). Many well-known bacteria cause ISRs, viz. *Bacillus, Pseudomonas, Serratia, Azospirillum,* and *Trichoderma* species (Chowdhury et al. 2018; Devi et al. 2023; Guzmán-Guzmán et al. 2023).

7.3.4 Rules for Microbiome Engineering

Knowing ecological factors that influence the development, durability, and geographical adaptation of plant-associated microbiomes is critical for the effective practical implementation of microbiome therapeutics (Kumar et al. 2024). To create models that favor successful colonization, a thorough knowledge of the ecological mechanisms that build and sustain various plant-associated microbiomes is required. In accordance to the ecological (Fierer et al. 2012) and meta-community theories (Miller et al. 2018), four major co-evolutionary processes in microbiome–host interactions lead to multi-species assemblages. These processes are dispersal, diversification, selection, and drift. Drift and selection influence the relative abundance of microbial species, while these processes interact intricately with ecological features of microbial communities (e.g., resistance, resilience, and functional redundancy), all of which are essential for stability and development of the community. In contrast to dispersion and drift, which become significant during the stages of root and embryonic development, selection facilitates community assembling during latter stages of plant growth. The developmental stage affects relative significance of the processes. Priority effects refer to stability of arrival order to influence development and functioning of plant-associated microbiomes (Yuan et al. 2021). For example, rare taxa have a major impact on microbiome at different stages of crop growth by increasing prevalence of suppressive microbes or stimulating the plant defenses. The makeup of microbial communities may be examined using SynCom to examine influence of priority effects. The results of this research indicate that taxa of microbial communities have a persistent effect on their formation, rendering them immune to external perturbations.

Drawing from the above-mentioned findings, it can be deduced that early stages of development are favorable for microbiome inoculation. Community structure, internal ecological interactions, migration from various habitats, and abiotic stressors have an impact on community assembly (Miller et al. 2018).

7.3.5 Plant-Mediated Strategies for Microbiome Modulation

Genetic engineering and plant breeding are two distinct strategies used to manipulate plant characteristics. Trivedi et al. (2017) stated that methodologies for plant breeding evaluating microbial community are being researched as a way to boost agricultural productivity. Selection of certain taxa and activities that provide high yields is now a part of plant breeding strategies that involve microbiome selection. Yin et al. (2013) and Lebedev et al. (2022) found that transgenic plants exhibiting enhanced resistance to disease in phosphorus deficiency soils had superior plant development and disease suppression. Also, exudates from transgenic tobacco and

Arabidopsis plants altered the pH of the rhizosphere in an acidic state, increasing the plants' ability to mineralize minerals from the soil and develop more robustly (Gévaudant et al. 2007; Cameron et al. 2013).

7.4 Designing Plant Microbiomes for Agriculture

Microbial population diversity has reached a brink of changes in the soil ecosystem with rapid inoculation of domesticated microbial consortia rather than natural one, comprehending role of the native microbes acting as competitive suppressors to the domesticated ones. Recent meta-genomic analysis of soil samples under the artificial and natural ecosystem has distant variation with respect to the agronomic activities, and this situation can be best explained with microbial association of rhizosphere under different genotypes of canola (Taye et al. 2020; Wohor et al. 2022). Microbiome's application will become successful through compatible interaction of the beneficial microbiome-based breeding strategies in the concerned environmental and genotype interaction (Busby et al. 2017; Kumari et al. 2022). Identifying the genomic regions of plant with positive interaction on microbiome, and multiple field trails under different environments are critical to find relationship between microbes and crop (Wille et al. 2019).

Plant genomic regions are responsible for plant-associated microbiome, as demonstrated by genome-wide association studies (GWAS). These genomic regions, shared by a variety of plant kinds, are highly related to the abundance of certain rhizosphere microbiome subgroups (Wang et al. 2022). Designer plants can be created by changing these potential genes to have certain microbiomes and benefits. The overexpressing genotype causes alterations in the phyllosphere microbiota that are linked to structural changes and triggered immunological responses (Jochum et al. 2019). GWAS can help us understand the molecular foundation of relationships between the microbiome and the host, but these associations may be misleading and are not always able to establishing causal associations (Awany et al. 2019). The resulting assortment of soil microbiomes influences plant physiology by influencing the expression of several genes involved in plant hormone synthesis (Li et al. 2023). Chemical signaling between plants and nodule bacteria for symbiotic relation has transformed how plants preferentially recruit nitrogen-fixing *Rhizobia* (Geddes et al. 2019). Furthermore, plant defense hormones like salicylic acid, secondary metabolites coumarin of plants and their root microbiome under stress, exploited genetically for plant breeding (Lebeis et al. 2015).

7.5 Optimal Microbiome Management Practices

Although tiny quantities of favorable microbes under disease-suppressing soils blended to transmit disease-suppressive qualities, disease-suppressive soils can be achieved by particular management strategies, like tillage and crop rotation (Palojärvi et al. 2020; De Corato 2020). Adopting some management methods with beneficial microbial impacts can help agroecosystems in healthier directions. Intercropping leguminous and non-leguminous crops help in medium of exchange of nutrients due to microbe association (Chen et al. 2020). In general, intercropping positively increases soil microbes, leading to improved fitness and productivity. Organic farming and conservation tillage have a positive influence on the soil microbiome and, help in crop growth and production (Wei et al. 2019).

7.6 Genetically Engineered Microbiomes

Genome-based techniques have identified several genomic regions in microbes which govern interactions between plant and microbe (De Souza et al. 2019). Development of improved microbial inoculants has greater potential due to genetic manipulation of microbes with enhanced traits. *Snodgrassella alvi*, a symbiotic organism, was recently genetically engineered by RNA interference (RNAi) protecting honeybees from virus and parasitic mites. Genome-editing tools like CRISPR and CRISPR/Cas 9 have allowed for alteration of genes and genomes to attain beneficial plant- and microbe-associated characteristics (Leonard et al. 2020).

Banana Guard, genetically modified strains of *Pseudomonas putida*, was developed to tackle *Fusarium oxysporum*, responsible for Panama wilt. This altered bacterium recognizes the fusaric acid generated by *F. oxysporum* acting as fungal growth inhibitors. Bacterial genome has capability of inhibition effect for self-destruct once the fungus is no longer present (Kemal et al. 2015).

Nitrogen fixation has been made feasible by manipulating naturally occurring endophytic bacteria found in cereal crops (Ryu et al. 2019). N_2-fixing strains are genetically modified to express N_2-fixing genes. Nowadays, some research is underway to generate microbiomes with modified nitrogen-fixing capacities to successfully substitute artificial fertilizers in agriculture (Goold et al. 2018).

7.7 Methods for Microbiome Studies

7.7.1 CRISPR/Cas9

CRISPR/Cas9 gene editing tool is used for studying plant–microbe interactions (Prabhukarthikeyan et al. 2020). Furthermore, CRISPR/Cas9 edit microbes enhance their positive impacts on crops (Glandorf 2019). A few of phytopathogens investigated with genome-editing tool (viz. *Alternaria alternata, Colletotrichum sansevieriae, Fusarium proliferatum, Phytophthora* spp., and *Sclerotinia sclerotiorum*) (Prabhukarthikeyan et al. 2020). Fusarium is responsible for vascular wilts by secreting mycotoxins like fumonisins. Fusarium's polyketide synthase gene, FUM1, is responsible for fumonisin production. Gene silencing of FUM1 by CRISPR-Cas9, and mutants were unable to produce fumonisins (Ferrara et al. 2019). Yet another instance of editing can be seen in *Trichoderma atroviride*, ace1 gene increases polyketide biosynthesis increasing its influence over some soil pathogen diseases. As a consequence, activating gene clusters improved these microbes' potential for use as biocontrols. This research might reveal pathways in bacteria could be altered to enable better management of biotic and abiotic challenges (Fang and Chen 2018).

7.7.2 Genome-Wide Association Studies (GWAS)

Numerous researches on plant–microbe interactions have employed GWAS to study link between plant genotype and microbial recruitment (Kim and Reinke 2019). This GWAS technique was used to discover SNPs related with blast resistance in rice (Zhu et al. 2016). A single-nucleotide polymorphism (SNP) microarray of African rice varieties revealed genetic regions linked with blast resistance in rice (Mgonja et al. 2016). Furthermore, GWAS has been shown for useful in detecting disease resistance in maize (Bartoli and Roux 2017). Recent research has also demonstrated the power of GWAS to discover the link betwixt leguminous plants and nodulating bacteria (Oladzad et al. 2020).

7.7.3 Technologies for Sequencing Microbiome

Targeted and shotgun sequencing are two fundamental approaches in metagenomic research and are utilized to address wide spectrum of environmental questions. The Illumina platform is recommended approach for metagenomic sequencing as such technologies enable researchers to speed up the whole procedure at cheaper prices. It generates millions of short, very accurate random readings that may be aggregated or used as markers for certain metabolic processes and/or subsets of microbes. The Illumina platform persists as most efficient sequencing platform despite the development of other state-of-the-art technologies as PacBio and Ion Torrent (Bartram et al. 2011; Krishna et al. 2019). Genome sequencing used to characterize numerous PGPB with diverse phenotypic features for crop growth potential in recent years. For example, *Brevibacterium frigoritolerans* ZB201705, isolates of maize form salt and drought stress rhizospheres, produced significant number of proteins from a full genome (Zhang et al. 2019). DNA fingerprinting technique examined the endophytic population of plant tissues, amplified rDNA restriction analysis (ARDRA), denaturing gradient gel electrophoresis (DGGE), temperature gradient gel electrophoresis (TGGE), and terminal restriction fragment length polymorphism (T-RFLP). Alternative approaches for community fingerprinting include automated ribosomal intergenic spacer analysis (ARISA) analyses highly unstable region betwixt 16S and 23S rDNA (Johnston-Monje and Mejia 2020). However, metagenomics approaches based on NGS have far outpaced the advancement of all of these DNA fingerprinting methods (Afzal et al. 2019).

7.7.4 Exploring Microbiomes Through Metatranscriptomics, Metaproteomics, and Metabolomics

Plant systems such as cereal and legumes have been found to differ at the kingdom level by rhizosphere microbiome metatranscriptomics (Turner et al. 2013; Solanki et al. 2023). Enhanced transcriptional expression of genes linked to amino acid and glucose metabolism is observed in microbiome related to roots under stress, particularly during drought (Chen et al. 2022). Metagenomic and metatranscriptomic analyses have been used more recently to establish communities of microbes as biosensors (Xie et al. 2021). Because there is no connection between transcriptional and translational processes, proteomic and metabolomic techniques have been developed to augment transcriptome data. The genetic characteristics of microbial communities in the rhizospheres and phyllosphere of agricultural plants have been well understood by applying the technique of metaproteomic analysis (Srivastava et al. 2022). Metabolic pathways enable specialized colonization and adaptation in the rhizosphere sections which were also identified by metaproteomics-based inferences of the microbiomes' physiology (Mendes et al. 2013; Knief et al. 2011). The metabolome has been used extensively to detect plant diseases and their causal agents, even if microbiome science remains in its infancy (Zhimo et al. 2021). Root metabolome influences the plant's ability to function and engage with pests and diseases by influencing recruitment of certain population of microbes that colonize rhizobiomes and plant. Metabolome data are crucial for identification and quantification of small molecules that enable interactions and communication between the plant and microbiome. Agroecosystem characteristics, metabolites, microorganisms, and minerals are intricately correlated, as recent developments in multiomics and integrated informatics have revealed (Srivastava et al. 2022).

7.7.5 Culturomics to Assist Molecular Approaches

Cultured bacteria are used in order to employ microbiomes to create commercial inoculants (Martínez-Hidalgo et al. 2019). In order to advance into omic investigations and to supply bacterial reference genome sequences for functional studies, we also need microbial cultures (Land et al. 2015). Glick states that PGPB has the ability to directly and indirectly promoting plant development. Direct approaches refer to the utilization of bacterial characteristics such as organic nitrogen fixation, siderophore production, ACC deaminase activity, phytohormones, biofilm formation, and phosphate solubilization that directly aid in growth of plants. EPS exopolysaccharides that enhance soil aggregation protect against pathogens and promote water and nutrient retention (Bouremani et al. 2023). Moreover, synthesis of HCN by PGPB has antimicrobial qualities and stop spread of plant diseases (Rijavec and Lapanje 2016). Furthermore, a large number of PGPB possess secretion systems for production of antimicrobial agents, such as lytic enzymes, organic volatile chemicals, and antibiotics, which inhibit growth of potentially phytopathogenic microbes (Bhardwaj et al. 2024; Santoyo et al. 2021).

7.8 Conclusions

"Microbiome-driven cropping systems" might portend the coming revolution in agriculture and a more environmentally friendly way to grow plants. Moreover, the creation of genetically altered crops or creatures might result in the progress needed to eradicate hunger for the increasing number of people worldwide. The utilization of genome-editing methods like CRISPR in conjunction with multiomics methodologies will enable this improvement in agricultural output, disease resistance, and nutritional status. But more research is needed in the coming years to fully understand the genomics and designing of intricate metabolic and systems of ecology that control interactions between microbes and plants. For example, research should not only identify helpful or causal organisms, but also "how" this information may be used consistently to improve plant growth, lower the occurrence and spread of disease. Creating standardized methods and a hub for the collection and analysis of meta-data would also help us cut down on waste and duplication in research. These methods have to be designed such that any laboratory in the world may use them to translate goods that are raised in the lab into the field. In addition, we must concentrate on creating novel technologies that are faster, more precise, more economical, and easier to use in the field and in laboratories. The technologies that are created have to be affordable for small laboratories as well as well-funded research organizations. To boost the scope, speed, and volume of information gleaned from the data, bioinformaticians must constantly develop new tools or improve the platforms that already exist. Researcher up skilling must change in tandem with developments in the context of plant-associated microbiome research; improvements should not cease at equipment and methodology. There is much potential to advance knowledge and technology in this subject and the future seems bright.

References

Abdul Rahman NSN, Abdul Hamid NW, Nadarajah K (2021) Effects of abiotic stress on soil microbiome. Int J Mol Sci 22:9036

Afridi MS, Ali S, Salam A, Terra WC, Hafeez A, Sumaira Ali B, AlTami MS, Ameen F, Ercisli S (2022) Plant microbiome engineering: hopes or hypes. Biology 11:1782

Afzal I, Shinwari ZK, Sikandar S, Shahzad S (2019) Plant beneficial endophytic bacteria: mechanisms, diversity, host range and genetic determinants. Microbiol Res 221:36–49

Ali A, Ölmez F, Zeshan MA, Mubeen M, Iftikhar Y, Sajid A, Abid M, Kumar A, Divvela PK, Solanki MK (2024) Yeast-based solutions in controlling plant pathogens. Biocatal Agric Biotechnol 58:103199. https://doi.org/10.1016/j.bcab.2024.103199

Anand U, Pal T, Yadav N, Singh VK, Tripathi V, Choudhary KK, Shukla AK, Sunita K, Kumar A, Bontempi E, Ma Y (2023) Current scenario and future prospects of endophytic microbes: promising candidates for abiotic and biotic stress management for agricultural and environmental sustainability. Microb Ecol 86:1455–1486. https://doi.org/10.1007/s00248-023-02190-1

Awany D, Allali I, Dalvie S, Hemmings S, Mwaikono KS, Thomford NE, Gomez A, Mulder N, Chimusa ER (2019) Host and microbiome genome-wide association studies: current state and challenges. Front Genet 9:637

Bala S, Garg D, Thirumalesh BV, Sharma M, Sridhar K, Inbaraj BS, Tripathi M (2022) Recent strategies for bioremediation of emerging pollutants: a review for a green and sustainable environment. Toxics 10:484

Bartoli C, Roux F (2017) Genome-wide association studies in plant pathosystems: toward an ecological genomics approach. Front Plant Sci 8:763

Bartram AK, Lynch MDJ, Stearns JC, Moreno-Hagelsieb G, Neufeld JD (2011) Generation of multimillion-sequence 16s rRNA gene libraries from complex microbial communities by assembling paired-end illumina reads. Appl Environ Microbiol 77:3846–3852

Bashir ZMY, Zargar FA, Kousar MS, Husain M, Rasool F (2017) Phosphorus solubilizing microorganisms: mechanism and diversity. Int J Chem Stud 5(5):666–673

Bashir Z, Zargar MY, Baba ZA, Mohiddin FA, Peer FA, Mir SA, Najar GR (2018a) Isolation and characterization of phosphorus solubilizing bacteria (PSB) from rhizospheric soils of apple (*Malus Domestica*). J Pharmacogn Phytochem 7(2):2997–3000

Bashir Z, Zargar MY, Baba ZA, Mohiddin FA (2018b) Solubilization of potassium containing mineral by bacteria from Kashmir iris (*Iris kashmiriana*) rhizosphere. Pharma Innov J 17(4):747–750

Bastas KK, Kumar A (2023) Omics technologies in the plant–microbe interactions. In: Bastas KK, Kumar A, Sivakumar U (eds) Microbial biocontrol: molecular perspective in plant disease management. Springer Nature, Singapore, pp 257–282. https://doi.org/10.1007/978-981-99-3947-3_13

Bastas KK, Kumar A, Sivakumar U (eds) (2023) Microbial biocontrol: molecular perspective in plant disease management, vol 49. Springer Nature, Singapore. https://doi.org/10.1007/978-981-99-3947-3

Bhardwaj A, Kaur K, Kumar A, Singh SK, Kharshiing B (2024) Development of microbial biotechnology products for phytopathogen management. In: The potential of microbes for a circular economy. Academic Press, pp 105–130. https://doi.org/10.1016/B978-0-443-15924-4.00013-8

Bhattacharyya PN, Jha DK (2012) Plant growth-promoting rhizobacteria (PGPR): emergence in agriculture. World J Microbiol Biotechnol 28:1327–1350

Bouremani N, Cherif-Silini H, Silini A, Bouket AC, Luptakova L, Alenezi FN, Baranov O, Belbahri L (2023) Plant growth-promoting rhizobacteria (PGPR): a rampart against the adverse effects of drought stress. Water 15:418

Brunel C, Pouteau R, Dawson W, Pester M, Ramirez KS, Van Kleunen M (2020) Towards unraveling macroecological patterns in rhizosphere microbiomes. Trends Plant Sci 25:1017–1029

Busby PE, Soman C, Wagner M, Friesen M, Kremer J, Bennett A, Morsy M, Eisen JA, Leach JE, Dangl JL (2017) Research priorities for harnessing plant microbiomes in sustainable agriculture. PLoS Biol 15:e2001793

Cameron DD, Neal AL, Van Wees SC, Ton J (2013) Mycorrhiza-induced resistance: more than the sum of its parts? Trends Plant Sci 18:545

Caulier S, Nannan C, Gillis A, Licciardi F, Bragard C, Mahillon J (2019) Overview of the antimicrobial compounds produced by members of the *Bacillus subtilis* group. Front Microbiol 10:302

Chen Y, Bonkowski M, Shen Y, Griffiths B, Jiang Y, Wang X, Sun B (2020) Root ethylene mediates rhizosphere microbial community reconstruction when chemically detecting cyanide produced by neighbouring plants. Microbiome 8:4

Chen Y, Yao Z, Sun Y, Wang E, Tian C, Sun Y, Liu J, Sun C, Tian L (2022) Current studies of the effects of drought stress on root exudates and rhizosphere microbiomes of crop plant species. Int J Mol Sci 23:2374

Chiaranunt P, White JF (2023) Plant beneficial bacteria and their potential applications in vertical farming systems. Plants (Basel) 12:400

Chowdhury FT, Islam MR, Islam MR, Khan H (2018) Diversity of plant endophytic volatile organic compound (voc) and their potential applications. In: Endophytes and secondary metabolites. Springer International Publishing, Cham, pp 1–27

Cohen AC, Bottini R, Pontin M, Berli FJ, Moreno D, Boccanlandro H, Travaglia CN, Piccoli PN (2015) *Azospirillum brasilense* ameliorates the response of *Arabidopsis thaliana* to drought mainly via enhancement of ABA levels. Physiol Plant 153:79–90

Compant S, Samad A, Faist H, Sessitsch A (2019) A review on the plant microbiome: ecology, functions, and emerging trends in microbial application. J Adv Res 19:29–37
Cullen CM, Aneja KK, Beyhan S, Cho CE, Woloszynek S, Convertino M, McCoy SJ, Zhang Y, Anderson MZ, Alvarez-Ponce D (2020) Emerging priorities for microbiome research. Front Microbiol 11:136
De Corato U (2020) Soil microbiota manipulation and its role in suppressing soil-borne plant pathogens in organic farming systems under the light of microbiome-assisted strategies. Chem Biol Technol Agric 7:17
De Souza RSC, Armanhi JSL, Damasceno NDB, Imperial J, Arruda P (2019) Genome sequences of a plant beneficial synthetic bacterial community reveal genetic features for successful plant colonization. Front Microbiol 10:1779
Devi S, Sharma S, Tiwari A, Bhatt AK, Singh NK, Singh M, Kaushalendra KA (2023) Screening for multifarious plant growth promoting and biocontrol attributes in Bacillus strains isolated from Indo-Gangetic soil for enhancing growth of rice crops. Microorganisms 11:1085. https://doi.org/10.3390/microorganisms11041085
Dong OX, Ronald PC (2019) Genetic engineering for disease resistance in plants: recent progress and future perspectives. Plant Physiol 180:26
Dubey A, Malla MA, Khan F, Chowdhary K, Yadav S, Kumar A, Sharma S, Khare PK, Khan ML (2019) Soil microbiome: a key player for conservation of soil health under changing climate. Biodivers Conserv 28:2405–2429
Ebrahimi-Zarandi M, Riseh RS, Tarkka MT (2022) Actinobacteria as effective biocontrol agents against plant pathogens, an overview on their role in eliciting plant defense. Microorganisms 10:1739
Egamberdieva D, Wirth SJ, Alqarawi AA, Abd-Allah EF, Hashem A (2017) Phytohormones and beneficial microbes: essential components for plants to balance stress and fitness. Front Microbiol 8:2104
El-Saadony MT, Saad AM, Soliman SM, Salem HM, Ahmed AI, Mahmood M, El-Tahan AM, Ebrahim AAM, El-Mageed TAA, Negm SH (2022) Plant growth-promoting microorganisms as biocontrol agents of plant diseases: mechanisms, challenges and future perspectives. Front Plant Sci 13:923880
Fang C, Chen X (2018) Potential biocontrol efficacy of *Trichoderma atroviride* with cellulase expression regulator ace1 gene knock-out. 3 Biotech 8:302
Farrar K, Bryant D, Cope-Selby N (2014) Understanding and engineering beneficial plant–microbe interactions: plant growth promotion in energy crops. Plant Biotechnol J 12:1193–1206
Fendrihan S, Pop CE (2021) Biotechnological potential of plant associated microorganisms. Rom Biotechnol Lett 26:2700–2706
Ferrara M, Haidukowski M, Logrieco AF, Leslie JF, Mulè GA (2019) CRISPR-Cas9 system for genome editing of *Fusarium proliferatum*. Sci Rep 9:19836
Fierer N, Ferrenberg S, Flores GE, González A, Kueneman J, Legg T, Lynch RC, McDonald D, Mihaljevic JR, O'Neill SP (2012) From animalcules to an ecosystem: application of ecological concepts to the human microbiome. Annu Rev Ecol Evol Syst 43:137–155
Geddes BA, Paramasivan P, Joffrin A, Thompson AL, Christensen K, Jorrin B, Brett P, Conway SJ, Oldroyd GED, Poole PS (2019) Engineering transkingdom signalling in plants to control gene expression in rhizosphere bacteria. Nat Commun 10:3430
Gévaudant F, Duby G, von Stedingk E, Zhao R, Morsomme P, Boutry M (2007) Expression of a constitutively activated plasma membrane H^+-ATPase alters plant development and increases salt tolerance. Plant Physiol 144:1763–1776
Glandorf DCM (2019) Re-evaluation of biosafety questions on genetically modified biocontrol bacteria. Eur J Plant Pathol 154:43–51
Goold HD, Wright P, Hailstones D (2018) Emerging opportunities for synthetic biology in agriculture. Genes (Basel) 9:341
Gopalakrishnan S, Sathya A, Vijayabharathi R, Varshney RK, Gowda CL, Krishnamurthy L (2015) Plant growth promoting rhizobia: challenges and opportunities. 3 Biotech 5:355

Gupta A, Sharma A, Pathak R, Kumar A, Sharma S (2018) Solid state fermentation of non-edible oil seed cakes for production of proteases and cellulases and degradation of anti nutritional factors. J Food Biotechnol Res 2:1–6

Gupta A, Singh UB, Sahu PK, Paul S, Kumar A, Malviya D, Singh S, Kuppusamy P, Singh P, Paul D (2022) Linking soil microbial diversity to modern agriculture practices: a review. Int J Environ Res Public Health 19:3141

Guzmán-Guzmán P, Kumar A, de Los Santos-Villalobos S, Parra-Cota FI, Orozco-Mosqueda MDC, Fadiji AE, Hyder S, Babalola OO, Santoyo G (2023) Trichoderma species: our best fungal allies in the biocontrol of plant diseases—A review. Plants 12:432. https://doi.org/10.3390/plants12030432

Hayat R, Ahmed I, Sheirdil RA (2012) An overview of plant growth promoting rhizobacteria (PGPR) for sustainable agriculture. In: Crop production for agricultural improvement. Springer, Dordrecht, pp 557–579

Jochum MD, McWilliams KL, Pierson EA, Jo YK (2019) Host-mediated microbiome engineering (HMME) of drought tolerance in the wheat rhizosphere. PLoS One 14:e0225933

Johnston-Monje D, Mejia JL (2020) Botanical microbiomes on the cheap: inexpensive molecular fingerprinting methods to study plant-associated communities of bacteria and fungi. Appl Plant Sci 8:e11334

Kashyap N, Singh SK, Yadav N, Singh VK, Kumari M, Kumar D, Shukla L, Kaushalendra, Bhardwaj N, Kumar A (2023) Biocontrol screening of endophytes: applications and limitations. Plants (Basel) 12:2480. https://doi.org/10.3390/plants12132480

Kaul S, Choudhary M, Gupta S, Dhar MK (2021) Engineering host microbiome for crop improvement and sustainable agriculture. Front Microbiol 12:1125

Kemal R, Islamiah PWN, Lusiany T (2015). Synthetic biology for biocontrol: a mini-review. In: Proceedings of the 6th international conference on global resource conservation, Malang, Indonesia, 1–2 December, vol 6, pp 92–94

Kim SM, Reinke RF (2019) A novel resistance gene for bacterial blight in rice, Xa43(t) identified by GWAS, confirmed by QTL mapping using a bi-parental population. PLoS One 14:e0211775

Knief C, Delmotte N, Chaffron S, Stark M, Innerebner G, Wassmann R, von Mering C, Vorholt JA (2011) Metaproteogenomic analysis of microbial communities in the phyllosphere and rhizosphere of rice. ISME J 6:1378–1390

Koza NA, Adedayo AA, Babalola OO, Kappo AP (2022) Microorganisms in plant growth and development: roles in abiotic stress tolerance and secondary metabolites secretion. Microorganisms 10:1528

Krishna BM, Khan MA, Khan ST (2019) Next-generation sequencing (NGS) platforms: an exciting era of genome sequence analysis. In: Microbial genomics in sustainable agroecosystems. Springer, Singapore, pp 89–109

Kumar A, Singh VK, Tripathi V, Singh PP, Singh AK (2018) Plant growth-promoting rhizobacteria (PGPR): perspective in agriculture under biotic and abiotic stress. In: Crop improvement through microbial biotechnology. Elsevier, pp 333–342. https://doi.org/10.1016/B978-0-444-63987-5.00016-5

Kumar A, Singh SK, Kant C, Verma H, Kumar D, Singh PP, Modi A, Droby S, Kesawat MS, Alavilli H, Bhatia SK (2021) Microbial biosurfactant: a new frontier for sustainable agriculture and pharmaceutical industries. Antioxidants (Basel) 10:1472. https://doi.org/10.3390/antiox10091472

Kumar A, Solanki MK, Wang Z, Solanki AC, Singh VK, Divvela PK (2024) Revealing the seed microbiome: navigating sequencing tools, microbial assembly, and functions to amplify plant fitness. Microbiol Res 279:127549. https://doi.org/10.1016/j.micres.2023.127549

Kumari M, Qureshi KA, Jaremko M, White J, Singh SK, Sharma VK, Singh KK, Santoyo G, Puopolo G, Kumar A (2022) Deciphering the role of endophytic microbiome in postharvest diseases management of fruits: opportunity areas in commercial up-scale production. Front Plant Sci 13:1026575. https://doi.org/10.3389/fpls.2022.1026575

Kumawat KC, Sharma B, Nagpal S, Kumar A, Tiwari S, Nair RM (2022) Plant growth-promoting rhizobacteria: salt stress alleviators to improve crop productivity for sustainable agriculture development. Front Plant Sci 13:1101862

Land M, Hauser L, Jun S-R, Nookaew I, Leuze MR, Ahn T-H, Karpinets T, Lund O, Kora G, Wassenaar T (2015) Insights from 20 years of bacterial genome sequencing. Funct Integr Genomics 15:141

Lebedev V, Lebedeva T, Tikhonova E, Shestibratov K (2022) Assessing impacts of transgenic plants on soil using functional indicators: twenty years of research and perspectives. Plants (Basel) 11:2439

Lebeis SL, Paredes SH, Lundberg DS, Breakfield N, Gehring J, McDonald M, Malfatti S, Del Rio TG, Jones CD, Tringe SG (2015) Plant microbiome. Salicylic acid modulates colonization of the root microbiome by specific bacterial taxa. Science 349:860–864

Leonard SP, Powell JE, Perutka J, Geng P, Heckmann LC, Horak RD, Davies BW, Ellington AD, Barrick JE, Moran NA (2020) Engineered symbionts activate honey bee immunity and limit pathogens. Science 367:573

Li Y, Liang J, Deng B, Jiang Y, Zhu J, Chen L, Li M, Li J (2023) Applications and prospects of CRISPR/Cas9-mediated base editing in plant breeding. Curr Issues Mol Biol 45:918–935

Lumibao CY, Bernik BM, Formel SK, Kandalepas D, Mighell KL, Pardue J, Van Bael SA, Blum MJ (2020) Rhizosphere microbial communities reflect genotypic and trait variation in a salt marsh ecosystem engineer. Am J Bot 107:941–949

Lundberg DS, Yourstone S, Mieczkowski P, Jones CD, Dangl JL (2013) Practical innovations for high-throughput amplicon sequencing. Nat Methods 10:999–1002

Martínez-Hidalgo P, Maymon M, Pule-Meulenberg F, Hirsch AM (2019) Engineering root microbiomes for healthier crops and soils using beneficial, environmentally safe bacteria. Can J Microbiol 65:91–104

Mendes R, Garbeva P, Raaijmakers JM (2013) The rhizosphere microbiome: significance of plant beneficial, plant pathogenic, and human pathogenic microorganisms. FEMS Microbiol Rev 37:634–663

Mgonja EM, Balimponya EG, Kang H, Bellizzi M, Park CH, Li Y, Mabagala R, Sneller C, Correll J, Opiyo S (2016) Genome-wide association mapping of rice resistance genes against *Magnaporthe oryzae* isolates from four african countries. Phytopathology 106:1359–1365

Miller ET, Svanbäck R, Bohannan BJ (2018) Microbiomes as metacommunities: understanding host-associated microbes through metacommunity ecology. Trends Ecol Evol 33:926–935

Moreira ZPM, Chen MY, Ortuno DLY, Haney CH (2023) Engineering plant microbiomes by integrating eco-evolutionary principles into current strategies. Curr Opin Plant Biol 71:102316

Munir N, Hanif M, Abideen Z, Sohail M, El-Keblawy A, Radicetti E, Mancinelli R, Haider G (2022) Mechanisms and strategies of plant microbiome interactions to mitigate abiotic stresses. Agronomy 12:2069

Nadarajah K, Rahman NSNA (2021) Plant–microbe interaction: aboveground to belowground, from the good to the bad. Int J Mol Sci 22:10388

Oladzad A, González A, Macchiavelli R, De Jensen CE, Beaver J, Porch T, McClean P (2020) Genetic factors associated with nodulation and nitrogen derived from atmosphere in a middle American Common Bean Panel. Front Plant Sci 11:576078

Olanrewaju OS, Glick BR, Babalola OO (2017) Mechanisms of action of plant growth promoting bacteria. World J Microbiol Biotechnol 33:197

Palojärvi A, Kellock M, Parikka P, Jauhiainen L, Alakukku L (2020) Tillage system and crop sequence affect soil disease suppressiveness and carbon status in boreal climate. Front Microbiol 11:534786

Pathak P, Rai VK, Can H, Singh SK, Kumar D, Bhardwaj N, Roychowdhury R, de Azevedo LCB, Kaushalendra, Verma H, Kumar A (2022) Plant-endophyte interaction during biotic stress management. Plants (Basel) 11:2203. https://doi.org/10.3390/plants11172203

Prabhukarthikeyan SR, Parameswaran C, Keerthana U, Teli B, Jagannadham PTK, Cayalvizhi B, Panneerselvam P, Senapati A, Nagendran K, Kumari S (2020) Understanding the plant-microbe interactions in CRISPR/Cas9 era: indeed a sprinting start in marathon. Curr Genomics 21:429–443

Rijavec T, Lapanje A (2016) Hydrogen cyanide in the rhizosphere: not suppressing plant pathogens, but rather regulating availability of phosphate. Front Microbiol 7:1785

Ryu M-H, Zhang J, Toth T, Khokhani D, Geddes BA, Mus F, Garcia-Costas A, Peters JW, Poole PS, Ané J-M (2019) Control of nitrogen fixation in bacteria that associate with cereals. Nat Microbiol 5:314–330

Sansinenea E (2021) Application of biofertilizers: current worldwide status. In: Rakshit A, Meena VS, Parihar M, Singh HB, Singh AK (eds) Biofertilizers, Advances in bio-inoculants, vol 1. Elsevier, Amsterdam, pp 183–190

Santoyo G, Moreno-Hagelsieb G, del Carmen Orozco-Mosqueda M, Glick BR (2016) Plant growth-promoting bacterial endophytes. Microbiol Res 183:92–99

Santoyo G, Urtis-Flores CA, Loeza-Lara PD, Orozco-Mosqueda MDC, Glick BR (2021) Rhizosphere colonization determinants by plant growth-promoting rhizobacteria (PGPR). Biology 10:475

Sergaki C, Lagunas B, Lidbury I, Gifford ML, Schäfer P (2018) Challenges and approaches in microbiome research: from fundamental to applied. Front Plant Sci 9:1205

Shah A, Nazari M, Antar M, Msimbira LA, Naamala J, Lyu D, Rabileh M, Zajonc J, Smith DL (2021) PGPR in agriculture: a sustainable approach to increasing climate change resilience. Front Sustain Food Syst 5:667546

Singh M, Kumar A, Singh R, Pandey KD (2017) Endophytic bacteria: a new source of bioactive compounds. 3 Biotech 7:315. https://doi.org/10.1007/s13205-017-0942-z

Singh R, Pandey KD, Singh M, Singh SK, Hashem A, Al-Arjani ABF, Abd-Allah EF, Singh PK, Kumar A (2022a) Isolation and characterization of endophytes bacterial strains of Momordica charantia L. and their possible approach in stress management. Microorganisms 10:290. https://doi.org/10.3390/microorganisms10020290

Singh J, Saini DK, Kashyap R, Kumar S, Chopra Y, Sandhu KS, Goraya M, Aggarwal R (2022b) Omics technologies for agricultural microbiology research. In: Trends of applied microbiology for sustainable economy. Elsevier, Amsterdam, pp 343–394

Solanki MK, Joshi NC, Singh PK, Singh SK, Santoyo G, de Azevedo LCB, Kumar A (2023) From concept to reality: transforming agriculture through innovative rhizosphere engineering for plant health and productivity. Microbiol Res 279:127553. https://doi.org/10.1016/j.micres.2023.127553

Srivastava AK, Das AK, Jagannadham PTK, Bora P, Ansari FA, Bhate R (2022) Bioprospecting microbiome for soil and plant health management amidst huanglongbing threat in citrus: a review. Front Plant Sci 13:858842

Tariq MR, Shaheen F, Mustafa S, Ali S, Fatima A, Shafiq M, Safdar W, Sheas MN, Hameed A, Nasir MA (2022) Phosphate solubilizing microorganisms isolated from medicinal plants improve growth of mint. Peer J 10:e13782

Taye ZM, Helgason BL, Bell JK, Norris CE, Vail S, Robinson SJ, Parkin IAP, Arcand M, Mamet S, Links MG (2020) Core and differentially abundant bacterial taxa in the rhizosphere of field grown Brassica napus genotypes: implications for canola breeding. Front Microbiol 10:3007

Trivedi P, Schenk PM, Wallenstein MD, Singh BK (2017) Tiny microbes, big yields: enhancing food crop production with biological solutions. Microb Biotechnol 10:999

Turner TR, Ramakrishnan K, Walshaw J, Heavens D, Alston M, Swarbreck D, Osbourn A, Grant A, Poole PS (2013) Comparative metatranscriptomics reveals kingdom level changes in the rhizosphere microbiome of plants. ISME J 7:2248–2258

Vacheron J, Desbrosses G, Bouffaud ML, Touraine B, Moënne-Loccoz Y, Muller D, Legendre L, Wisniewski-Dyé F, Prigent-Combaret C (2013) Plant growth-promoting rhizobacteria and root system functioning. Front Plant Sci 4:356

Vandenkoornhuyse P, Quaiser A, Duhamel M, Le Van A, Dufresne A (2015) The importance of the microbiome of the plant holobiont. New Phytol 206:1196–1206

Vannier N, Agler M, Hacquard S (2019) Microbiota-mediated disease resistance in plants. PLoS Pathog 15:e1007740

Verma V, Ravindran P, Kumar PP (2016) Plant hormone-mediated regulation of stress responses. BMC Plant Biol 16:86

Wang X, Whalley WR, Miller AJ, White PJ, Zhang F, Shen J (2020) Sustainable cropping requires adaptation to a heterogeneous rhizosphere. Trends Plant Sci 25:1194–1202

Wang Y, Wang X, Sun S, Jin C, Su J, Wei J, Luo X, Wen J, Wei T, Sahu SK (2022) GWAS, MWAS and mGWAS provide insights into precision agriculture based on genotype-dependent microbial effects in foxtail millet. Nat Commun 13:5913

Wei Z, Gu Y, Friman VP, Kowalchuk GA, Xu Y, Shen Q, Jousset A (2019) Initial soil microbiome composition and functioning predetermine future plant health. Sci Adv 5:eaaw0759

Wille L, Messmer MM, Studer B, Hohmann P (2019) Insights to plant-microbe interactions provide opportunities to improve resistance breeding against root diseases in grain legumes. Plant Cell Environ 42:20–40

Wohor OZ, Rispail N, Ojiewo CO, Rubiales D (2022) Pea breeding for resistance to rhizospheric pathogens. Plants (Basel) 11:2664

Xie J, Dawwam GE, Sehim AE, Li X, Wu J, Chen S, Zhang D (2021) Drought stress triggers shifts in the root microbial community and alters functional categories in the microbial gene pool. Front Microbiol 12:744897

Yin C, Hulbert SH, Schroeder KL, Mavrodi O, Mavrodi D, Dhingra A, Schillinger WF, Paulitz TC (2013) Role of bacterial communities in the natural suppression of *Rhizoctonia solani* bare patch disease of wheat (*Triticum aestivum* L.). Appl Environ Microbiol 79:7428

Yuan H, Li T, Li H, Wang C, Li L, Lin X, Lin S (2021) Diversity distribution, driving factors and assembly mechanisms of free-living and particle-associated bacterial communities at a subtropical marginal sea. Microorganisms 9:2445

Zambrano-Mendoza JL, Sangoquiza-Caiza CA, Campaña-Cruz DF, Yánez-Guzmán CF (2021) Use of biofertilizers in agricultural production. In: Technology in agriculture. IntechOpen, London. https://www.intechopen.com/online-first/76918

Zamioudis C, Pieterse CMJ (2012) Modulation of host immunity by beneficial microbes. Mol Plant Microbe Interact 25:139–150

Zhang C, Li X, Yin L, Liu C, Zou H, Wu Z, Zhang Z (2019) Analysis of the complete genome sequence of *Brevibacterium frigoritolerans* ZB201705 isolated from drought- and salt-stressed rhizosphere soil of maize. Ann Microbiol 69:1489–1496

Zhimo VY, Kumar A, Biasi A, Salim S, Feygenberg O, Toamy MA, Abdelfattaah A, Medina S, Freilich S, Wisniewski M, Droby S (2021) Compositional shifts in the strawberry fruit microbiome in response to near-harvest application of Metschnikowia fructicola, a yeast biocontrol agent. Postharvest Biol Technol 175:111469. https://doi.org/10.1016/j.postharvbio.2021.111469

Zhu D, Kang H, Li Z, Liu M, Zhu X, Wang Y, Wang D, Wang Z, Liu W, Wang GL (2016) A genome-wide association study of field resistance to *magnaporthe oryzae* in rice. Rice 2016(9):44

CRISPR/Cas9-Based Genome Engineering in Plants for Enhancing Disease Resistance

8

Haris Butt and Kubilay Kurtulus Bastas

Abstract

Plant biotechnology, which includes a variety of scientific methods and techniques, is vital in identifying and modifying plant genes to create desired traits in plants or to achieve specific outcomes. The CRISPR (Clustered Regularly Interspaced Short Palindromic Repeats)- mediated genetic modification technique has arisen as a modern approach to increase plant resistance to various bacterial, viral, and fungal pathogens. Techniques to reduce off-target effects involve designing precise guide RNAs, utilizing high-accuracy Cas9 (CRISPR-associated protein 9) enzymes, and using the ribonucleoprotein transport approach. The use of CRISPR/Cas9 (CRISPR) technology in crop protection shows significant potential as a sustainable alternative to pesticides. Combining multiple resistance genes and adopting good farming practices are essential strategies for increasing durability. Disease resistance, in particular, stands out as a promising area for applying this technology in agriculture. CRISPR aids in the gene stacking process and provides opportunities for efficiently creating new resistances. Additionally, addressing technical challenges related to targeted mutagenesis and achieving true genetic modification in crops are critical. Improving the efficiency of homology-directed repair and developing multiplexing protocols are vital for broadening the range of CRISPR applications in plant breeding. This technique possesses the capability to transform crop protection, offering a way to produce superior, disease-resistant crops and reduce the environmental impact of pesticide use. This chapter examines diverse uses of the CRISPR technique in managing plant diseases.

H. Butt · K. K. Bastas (✉)
Department of Plant Protection, Faculty of Agriculture, Selcuk University, Konya, Turkey
e-mail: kbastas@selcuk.edu.tr

A. Kumar, M. K. Solanki (eds.), *Microbial Biocontrol Techniques*, Microorganisms for Sustainability 54,
https://doi.org/10.1007/978-981-97-8739-5_8

Keywords

Pathogen · CRISPR/Cas9 · Plant protection · Genome engineering

8.1 Introduction

The escalating global demand for food security, exacerbated by burgeoning population growth and the looming specter of climate change, underscores the critical imperative to create robust and effective techniques for ensuring a consistent and ample food supply worldwide (Haque et al. 2018; United Nations 2019). In this context, the role and application of plant biotechnology have become indispensable in addressing the challenges confronting global agriculture. Plant biotechnology, encompassing a spectrum of scientific methodologies and techniques, is instrumental in identifying and manipulating plant genes to develop preferred characteristics or produce targeted outputs in plants (Kalia 2018). Through the utilization of these methods, advantageous traits in crops can be developed and improved, and harmful elements like allergens in common foods such as rice and groundnuts are possible to be mitigated (Barh and Azevedo 2017).

The introduction of techniques like Zinc Finger Nucleases (ZFNs), Transcriptional Activator-like Effector Nucleases (TALENs), and the more recent CRISPR system has significantly advanced the precision and efficiency of plant genetic modification. CRISPR system has become the favored method as it is more accurate, efficient, easy and affordable, attracting considerable interest in the genetic engineering (Haque et al. 2018; Li et al. 2016; Sun et al. 2016a, b; Wang et al. 2018a, b). Initially discovered in bacteria, the CRISPR leverages the defense mechanism of *S. pyogenes* to make accurate alterations in the DNA of specific organisms (Ishino et al. 1987; Jansen et al. 2002). The central step in this process involves designing a single guide RNA (sgRNA) that complements a specific DNA sequence. This sgRNA then pairs with the Cas9 protein, forming a complex that facilitates the creation of accurate double-stranded breaks in the DNA (Costa et al. 2017). This breakthrough has undergone extensive research for genetic engineering applications across various organisms, paving the way for transformative applications in agriculture and beyond.

As global population projections soar to an estimated 9.8 billion by 2050, the imperative to bolster crop resistance to biotic stresses becomes paramount in meeting the burgeoning food demands of an expanding populace (Nejat et al. 2017; Dong and Ronald 2019). Plant diseases, caused by a variety of bacteria, viruses, and fungi pose significant threats to crop yields, necessitating the enhancement of plant resistance for sustainable agricultural production (Das et al. 2019; Langner et al. 2018; Kumar et al. 2021). Conventional breeding projects historically emphasized on incorporating natural alleles to enhance resistance. However, CRISPR method has transformed the landscape of plant pathology and crop improvement (Boyd and O'Toole 2012; Dracatos et al. 2018). By conferring precise genetic modifications, CRISPR facilitates the production of plants resistant to diseases with unparalleled efficacy, heralding a new phase of precision breeding strategies to combat plant diseases and mitigate yield losses.

Moreover, CRISPR-mediated gene editing holds tremendous potential in enhancing crop quality and nutritional value, thus aligning with the shifting focus of agricultural research from mere production increment to quality improvement (Wolter et al. 2019). The emergence of gene editing technologies, including CRISPR, ushers in a paradigm shift in crop breeding by enabling precise and heritable modifications in plant genomes with minimal off-target effects (Mohanta et al. 2017). This transformative approach not only expedites the development of desired agronomic traits but also circumvents the laborious and time-consuming nature of conventional mutagenesis-based breeding processes, thereby hastening the pace of crop enhancement efforts (Nelson et al. 2018).

In summary, the convergence of escalating food demand, environmental challenges, and technological advancements underscores the pivotal role of CRISPR system in revolutionizing agricultural practices and ensuring global food security. This review aims to elucidate the recent developments and applications of CRISPR system in crop enhancement, with a focus on improving disease resistance and addressing the challenges of sustainable agriculture amid mounting biotic stresses and shifting climatic conditions (Kumar et al. 2018; Pathak et al. 2022; Singh et al. 2022).

8.2 Advancements in Genome Engineering Using CRISPR/Cas9 System

The CRISPR/Cas system, stemming from the discovery of Cas proteins and CRISPR sequences, has evolved into a versatile tool for genome manipulation (Ishino et al. 1987; Richter et al. 2012). It is categorized into different types, with Type II CRISPR/SpCas9 from *S. pyogenes* being extensively used (Ding et al. 2018; Liu and Moschou 2018). The gRNA directs the protein to precise DNA sequences, which induces a break in the double strands of the DNA (DSBs) near the protospacer adjacent motif (PAM) (Brouns et al. 2008; Jinek et al. 2012). Gene knockout or precise genetic alterations are achieved through natural DNA repair mechanisms such as non-homologous end joining (NHEJ) for gene disruption or homology-directed repair (HDR) for precise genetic modifications (Cong et al. 2013). In plants, the CRISPR process includes target site selection, synthesis of sgRNA, and the subsequent screening of gene-edited plants, enabling various genome editing abilities like knock-in, knockout, and knockdown (Endo et al. 2016). The continuous exploration of CRISPR, along with efforts to broaden PAM sequences, promises further advancements in precise gene editing across different organisms and fields of research. Figure 8.1 illustrates the visual representation of the CRISPR system.

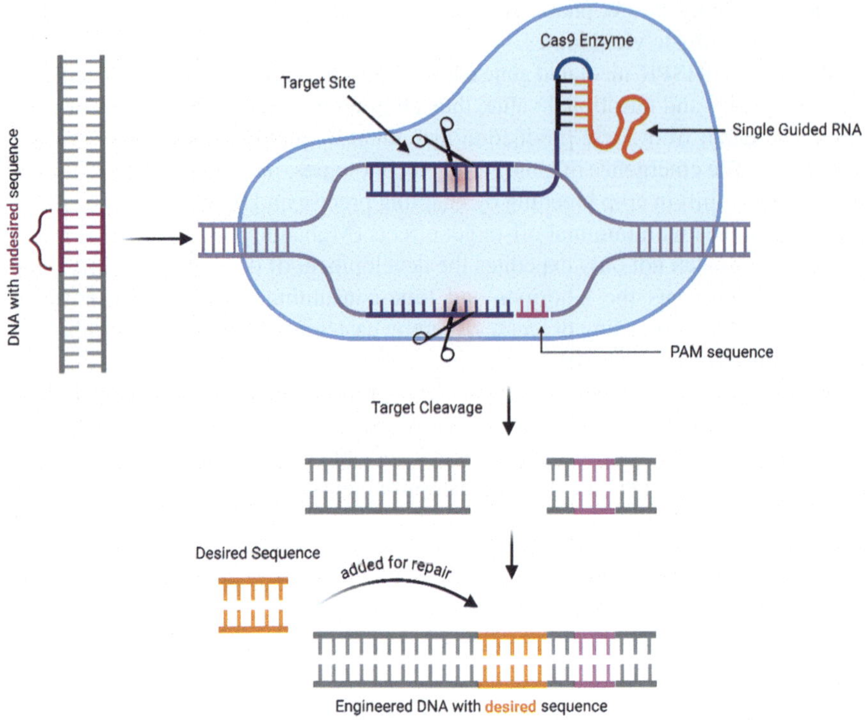

Fig. 8.1 Visual representation of CRISPR-Cas9 gene editing process. The diagram depicts the sequential steps involved in modifying a DNA sequence using the CRISPR technique. The process begins with the identification of a DNA sequence containing an undesired segment (highlighted in purple). A sgRNA guides the Cas9 enzyme which recognizes and attaches to the aimed site within the DNA. The sgRNA (depicted in orange and black) aligns with the target site via base pair sequence complementarity. The Cas9 enzyme then introduces a DSB at the target site, adjacent to a PAM sequence, leading to cleavage of the DNA. Containing the desired sequence, a repair template DNA (shown in orange) is introduced to guide the repair process. The desired sequence is incorporated into the DNA, creating an engineered DNA with the corrected sequence. This figure illustrates the precise mechanism of gene editing facilitated by the CRISPR technology, highlighting the roles of the sgRNA, Cas9 enzyme, and the DNA repair process in achieving targeted genetic modifications

8.3 CRISPR/Cas9 Delivery and Genome Editing Techniques for Plants

The method of delivering CRISPR machinery into plant cells encompasses diverse methodologies aimed at achieving precise genome editing. Atmospheric-pressure plasma strategy involves the direct introduction of CRISPR ribonucleoproteins, designed to target specific genes, into plant cells. This novel technique enables gene editing without the necessity of DNA introduction, thereby broadening its applicability across various plant species and paving the way for future advancements in plant breeding initiatives (Yanagawa et al. 2023). The Type IV Secretion System of

Agrobacterium also facilitates the transport of the Cas9 protein into plant cells. By leveraging this method, CRISPR/Cas9 mutagenesis can be achieved in plants without the requirement for DNA introduction, thus improving the effectiveness of genetic alterations in plant science (Schmitz et al. 2020). Moreover, recent breakthroughs in CRISPR technology have ushered in the era of genetic engineering in plants, made possible by using preassembled CRISPR ribonucleoprotein complexes. This cutting-edge approach eliminates the need for DNA delivery, thereby enabling precise genetic engineering in plants (Wada et al. 2020). In summary, the incorporation of CRISPR into plant cells involves the deployment of innovative techniques that facilitate targeted genome editing. These advancements signify a significant leap forward in agricultural biotechnology, offering a robust tool for enhancing crop traits and fortifying plants against diseases.

8.4 CRISPR/Cas9 for Plant Disease Management

The utilization of the CRISPR genome engineering system has surfaced as a contemporary strategy to bolster plant resilience against a spectrum of bacterial, viral, and fungal pathogens. This innovative technology not only facilitates the enhancement of plant immunity but also offers notable advantages such as ease of implementation and expedited results compared to traditional breeding methodologies. Moreover, CRISPR-mediated editing requires less time and resources, positioning it as a promising approach for accelerating crop improvement efforts. CRISPR/Cas tools have primarily been investigated for their effectiveness against viral infections, with subsequent endeavors focusing on enhancing resistance to fungal and bacterial diseases. Below, we will delve into recent research showcasing the effectiveness of the CRISPR method in conferring resistance to pathogens. Table 8.1 outlines various applications of the CRISPR technique in phytopathology.

CRISPR technology has revolutionized resistance to plant diseases caused by bacterial pathogens. For instance, Jia and Wang (2014) developed an innovative approach for introducing CRISPR reagents via *Xcc* infection, successfully knocking out the *CsWRKY22* gene and enhancing resistance to *Xcc* in sweet orange (Wang et al. 2019). Similarly, in apples, Pompili et al. (2020) employed CRISPR to knockout *MdDIPM4*, increasing resistance to *E. amylovora*. Additionally, CRISPR was used to target *CsLOB1* in citrus, enhancing resistance to *Xcc* infection (Jia et al. 2017; Peng et al. 2017). It is noteworthy that tomato equivalent of SlDMR6-1 shows increased expression when infected by *Pst, P. capsici*. By employing the CRISPR system, null mutants of SlDMR6-1 exhibited resistance to these pathogens, including *Xanthomonas* spp., while preserving normal tomato growth and development (de Toledo Thomazella et al. 2016). These results underscore the flexibility and effectiveness of CRISPR in engineering disease resistance across different plant species, offering promising solutions for sustainable agriculture.

Viruses pose a significant challenge in plant agriculture due to their rapid evolution and insect-mediated spread. Plant viruses are categorized into six primary groups according to their genome characteristics, including ssDNA, dsRNA,

Table 8.1 Applications of the CRISPR system in technique in phytopathology

Plants/crops	Pathogen	Target gene	Technique used	Outcome	References
Sweet orange	*Xcc*	*CsWRKY22*	CRISPR via *Xcc* infection	Enhanced resistance to citrus canker	Jia and Wang (2014), Wang et al. (2019)
Apple	*Erwinia amylovora*	*MdDIPM4*	CRISPR knockout	Enhanced resistance to the pathogen	Pompili et al. (2020)
Citrus	*Xcc*	*CsLOB1*	CRISPR knockout	Enhanced resistance to the pathogen	Jia et al. (2017), Peng et al. (2017)
Tomato	*Pst*, *P. capsici*, *Xanthomonas* spp.	*SlDMR6-1*	CRISPR knockout	Enhanced resistance	de Toledo Thomazella et al. (2016)
N. benthamiana, *Arabidopsis*	BSCTV, BeYDV	Viral genomes	CRISPR/Cas9 with sgRNAs targeting viral genomes	Reduction in viral load and symptom attenuation	Baltes et al. (2015), Ji et al. (2015), Ali et al. (2015, 2016)
Multiple plants	*Begomovirus*	Conserved viral sequences	CRISPR/Cas9 with sgRNAs targeting conserved sequences	Broad-spectrum immunity	Ali et al. (2015, 2016)
Cucumber, Tobacco, Turnip	RNA viruses (e.g., *CMV*, *TMV*, *TuMV*)	Viral RNA	RNA-targeting nucleases (*FnCas9*, *Cas13a*)	Potential resistance	Zhang et al. (2018), Aman et al. (2018)
Wheat	*Bgt*	*TaMLO*	CRISPR/Cas9 knock-out	Increased resistance to *Bgt*	Lyngkjær et al. (2000), Wang et al. (2014)
Grapevine	*Botrytis cinerea*	*VvWRKY52*	CRISPR/Cas9 knock-out	Improved resistance to the pathogen	Wang et al. (2018a, b)
Various plants	*Alternaria alternata*, *Leptosphaeria maculans*, *Fusarium oxysporum*, and *F. graminearum*	Various fungal genes	CRISPR/SpCas9-mediated genome editing	Enhanced resistance to multiple fungal pathogens	Wenderoth et al. (2017), Idnurm et al. (2017), Wang et al. (2018a, b), Gardiner and Kazan (2018)

ssRNA-, reverse-transcribing viruses, and ssRNA+ viruses (Roossinck et al. 2015). While conventional CRISPR systems mainly target double-stranded DNA, recent advancements have enabled the development of systems capable of targeting RNA viruses directly. For instance, FnCas9, originally targeting bacterial RNA, has been engineered to suppress human single-stranded RNA viruses through direct RNA-targeting (Price et al. 2015).

Geminiviruses, particularly those with single-stranded DNA genomes, have been a focal point for CRISPR-based plant virus resistance studies (Baltes et al. 2015). Among geminiviruses, the *Begomovirus* genus is economically significant, causing substantial crop losses worldwide (Gilbertson et al. 2015). CRISPR system has been employed to target specific genomic regions of geminiviruses, such as *BSCTV* and *BeYDV* by employing sgRNAs targeting specific motifs within the viral genomes, leading to a reduction in viral load and symptom attenuation in *N. benthamiana* and Arabidopsis plants (Baltes et al. 2015; Ji et al. 2015; Ali et al. 2015, 2016). Notably, the development of sgRNAs targeting conserved sequences in multiple viral strains has shown promise in conferring broad-spectrum immunity against geminiviruses (Ali et al. 2015, 2016). Unlike DNA viruses, targeting RNA viruses presents additional challenges due to the lack of recognition by conventional Cas9 enzymes. However, the discovery of RNA-targeting nucleases such as *FnCas9* and *Cas13a* has opened new avenues for combating *CMV* and *TMV* (Zhang et al. 2018; Aman et al. 2018). Cas13a, an RNA-directed ribonuclease, exhibits potential in combating *TuMV* (Aman et al. 2018). These enzymes have demonstrated efficacy in reducing RNA virus accumulation in plants by specifically targeting viral RNA sequences. Furthermore, utilizing catalytically inactive variants of these nucleases offers a strategy to mitigate the emergence of viral variants resistant to CRISPR-based interventions (Zhang et al. 2018). Another approach for achieving virus resistance involves editing host susceptibility genes critical for viral infection. By targeting genes encoding eukaryotic translation initiation factors like eIF4E and eIF(iso)4E, researchers have successfully generated plants resistant to various RNA viruses (Chandrasekaran et al. 2016; Pyott et al. 2016; Macovei et al. 2018). Mutations in these host factors disrupt the viral life cycle, rendering plants immune to infection. Importantly, segregating CRISPR-edited mutants from transgenic elements ensures regulatory compliance and facilitates the release of non-transgenic resistant plants into the environment (Chandrasekaran et al. 2016). In conclusion, the combination of transgenic approaches, CRISPR-mediated genome editing, and host factor manipulation offers versatile strategies for enhancing plant virus resistance. These advancements hold great promise for sustainable agriculture by decreasing dependence on synthetic pesticides and reducing crop losses caused by viral infections.

The understanding of molecular pathways involved in plant–fungus interactions has paved ways in developing methods aimed at enhancing fungal resistance in plant species (Bastas and Kumar 2023). Genes identified as candidates for plant resistance to fungi are now prime targets for CRISPR method (Borrelli et al. 2018). Fungi, as the most widespread phytopathogens, pose significant challenges in disease management due to their diverse lifestyles and high genetic adaptability (Ali et al. 2024; Bhardwaj et al. 2024). This flexibility allows them to readily infect new hosts and develop resistance to fungicides (Doehlemann et al. 2017). Manipulating host susceptibility genes through genetic engineering has shown promise as a potential technique to address this issue.

Powdery mildew affecting various plants has been a primary target for enhancing resistance. Notably, the discovery of barley *mlo* mutants marked a breakthrough in developing wheat cultivars with long-lasting resistance to powdery mildew (Lyngkjær et al. 2000; Wang et al. 2014). Enhanced resistance in wheat against powdery mildew was demonstrated by knocking out the *TaMLO* gene using CRISPR (Wang et al. 2014). Wang et al. (2014) reported that the manipulation of negative regulators and susceptibility genes involved in defense pathways has shown promise in conferring resistance to fungal pathogens such as *B. graminis* f. sp. *triti*ci. This method has also demonstrated success in conferring resistance to fungal pathogens in grapevine and apple. Targeting susceptibility genes like VvWRKY52 in grapevine has increased resistance against *B. cinerea* (Wang et al. 2018a, b). Likewise, transient leaf transformation combined with disease assays has enabled the validation of sgRNA design and observation of CRISPR-induced mutations in crops such as cacao (Fister et al. 2018).

Furthermore, there has been an effective use of CRISPR/SpCas9-mediated engineering in various fungal species, including *Alternaria alternata*, *Leptosphaeria maculans*, *Fusarium oxysporum*, and *F. graminearum* (Wenderoth et al. 2017; Idnurm et al. 2017; Wang et al. 2018a, b; Gardiner and Kazan 2018). The introduction of CRISPR/SpCas9 essentials in fungi can be achieved through stable or transient expression using various methods such as polyethylene glycol, *Agrobacterium*, electroporation, and biolistic transformation (Schuster and Kahmann 2019). Additionally, the SpCas9/sgRNA ribonucleoprotein (RNP) complex has shown effectiveness in editing fungal genomes, offering a rapid and efficient approach (Foster et al. 2018; Wang et al. 2018a, b). In conclusion, CRISPR-based genetic engineering holds immense potential for enhancing fungal resistance in plant species, offering a promising technique for improving crops and disease management in agriculture (Borrelli et al. 2018; Doehlemann et al. 2017).

8.5 Challenges and Future Prospects

The CRISPR system, while powerful, faces different challenges. Off-target effects remain a concern, as guide RNAs may match with unintended genomic sequences, necessitating optimization of Cas9 precision. New Cas proteins like Cas12a and Cas13a show promise in specific contexts, yet unintended changes such as translocations and inversions can still occur due to DNA repair pathways. Additionally, Cas9-induced double-strand breaks can trigger cell death pathways, hampering efficiency. Modifications like nuclease-deficient Cas9 proteins have been developed to mitigate these effects, allowing for base editing and site-specific recombination. Another challenge is the need for PAM sequence, limiting target options. Moreover, regulatory frameworks for gene-edited crops vary globally, hindering commercialization. Delivery methods for CRISPR cargoes, especially in monocots, are also problematic. Strategies to minimize off-target effects include designing specific guide RNAs, using high-fidelity Cas9 enzymes, and employing the ribonucleoprotein delivery method. Despite these hurdles, advances in genome-wide association studies and cross-population validation are driving progress in plant gene editing, offering insights into gene functions and interactions crucial for harnessing the complete capabilities of CRISPR technology in agriculture. The application of CRISPR in crop protection holds great potential as a long-term alternative to pesticide use. Disease resistance, in particular, emerges as a compelling avenue for leveraging this technology in agriculture. Several factors contribute to the appeal of CRISPR/Cas9 for disease resistance: firstly, our deep understanding of molecular mechanisms involved in pathogen interactions allows for targeted editing of genes implicated in resistance. Secondly, the modification of a single gene can often confer disease resistance, simplifying the technical challenges involved. Thirdly, targeted mutagenesis, the primary application of CRISPR/Cas9 in crops presently, lends itself well to disease resistance by inactivating susceptibility genes.

However, realizing the potential of CRISPR in crop protection faces significant challenges. First and foremost is the translation of promising results from controlled environments to field conditions. Field trials are essential to assess the efficacy and agronomic performance of edited crops under real-world conditions. Moreover, edited crops may exhibit unintended side effects on plant physiology, necessitating thorough evaluation under field conditions. Another challenge lies in ensuring the durability of disease resistances over time and managing them effectively. Stacking multiple resistance genes and implementing good agronomic practices are crucial strategies for enhancing durability. CRISPR facilitates the stacking process and offers opportunities for creating novel resistances efficiently. Furthermore, overcoming technical limitations related to targeted mutagenesis and implementing true genome editing in crops is imperative. Enhancing the efficiency of homology-directed repair (HDR) and developing multiplexing protocols are essential for expanding the scope of CRISPR applications in plant breeding. Moving forward, addressing these challenges will require interdisciplinary collaboration and continuous advancements in scientific knowledge. CRISPR technology holds immense potential to revolutionize crop protection, offering a pathway to surface elite, disease-resistant crops and mitigate the environmental impact of pesticide use.

References

Ali Z, Abulfaraj A, Idris A, Ali S, Tashkandi M, Mahfouz MM (2015) CRISPR/Cas9-mediated viral interference in plants. Genome Biol 16:1–11

Ali Z, Ali S, Tashkandi M, Zaidi SSEA, Mahfouz MM (2016) CRISPR/Cas9-mediated immunity to geminiviruses: differential interference and evasion. Sci Rep 6(1):26912

Ali A, Ölmez F, Zeshan MA, Mubeen M, Iftikhar Y, Sajid A, Abid M, Kumar A, Divvela PK, Solanki MK (2024) Yeast-based solutions in controlling plant pathogens. Biocatal Agric Biotechnol 58:103199. https://doi.org/10.1016/j.bcab.2024.103199

Aman R, Ali Z, Butt H, Mahas A, Aljedaani F, Khan MZ, Ding S, Mahfouz M (2018) RNA virus interference via CRISPR/Cas13a system in plants. Genome Biol 19:1–9

Baltes NJ, Hummel AW, Konecna E, Cegan R, Bruns AN, Bisaro DM, Voytas DF (2015) Conferring resistance to geminiviruses with the CRISPR–Cas prokaryotic immune system. Nat Plants 1(10):1–4

Barh D, Azevedo V (eds) (2017) Omics technologies and bio-engineering, vol 1: towards improving quality of life. Academic

Bastas KK, Kumar A (2023) Omics technologies in the plant–microbe interactions. In: Bastas KK, Kumar A, Sivakumar U (eds) Microbial biocontrol: molecular perspective in plant disease management. Springer Nature, Singapore, pp 257–282. https://doi.org/10.1007/978-981-99-3947-3_13

Bhardwaj A, Kaur K, Kumar A, Singh SK, Kharshiing B (2024) Development of microbial biotechnology products for phytopathogen management. In: The potential of microbes for a circular economy. Academic, pp 105–130. https://doi.org/10.1016/B978-0-443-15924-4.00013-8

Borrelli VM, Brambilla V, Rogowsky P, Marocco A, Lanubile A (2018) The enhancement of plant disease resistance using CRISPR/Cas9 technology. Front Plant Sci 9:407423

Boyd CD, O'Toole GA (2012) Second messenger regulation of biofilm formation: breakthroughs in understanding c-di-GMP effector systems. Annu Rev Cell Dev Biol 28:439–462

Brouns SJ, Jore MM, Lundgren M, Westra ER, Slijkhuis RJ, Snijders AP, Dickman MJ, Makarova KS, Koonin EV, Van Der Oost J (2008) Small CRISPR RNAs guide antiviral defense in prokaryotes. Science 321(5891):960–964

Chandrasekaran J, Brumin M, Wolf D, Leibman D, Klap C, Pearlsman M, Sherman A, Arazi T, Gal-On A (2016) Development of broad virus resistance in non-transgenic cucumber using CRISPR/Cas9 technology. Mol Plant Pathol 17(7):1140–1153

Cong L, Ran FA, Cox D, Lin S, Barretto R, Habib N, Hsu PD, Wu X, Jiang W, Marraffini LA, Zhang F (2013) Multiplex genome engineering using CRISPR/Cas systems. Science 339(6121):819–823

Costa JR, Bejcek BE, McGee JE, Fogel AI, Brimacombe KR, Ketteler R (2017) Genome editing using engineered nucleases and their use in genomic screening. Assay guidance manual [internet]

Das A, Sharma N, Prasad M (2019) CRISPR/Cas9: a novel weapon in the arsenal to combat plant diseases. Front Plant Sci 9:439983

de Toledo Thomazella DP, Brail Q, Dahlbeck D, Staskawicz B (2016) CRISPR-Cas9 mediated mutagenesis of a DMR6 ortholog in tomato confers broad-spectrum disease resistance. BioRxiv, p 064824

Ding D, Chen K, Chen Y, Li H, Xie K (2018) Engineering introns to express RNA guides for Cas9- and Cpf1-mediated multiplex genome editing. Mol Plant 11(4):542–552

Doehlemann G, Ökmen B, Zhu W, Sharon A (2017) Plant pathogenic fungi. In: The fungal kingdom, pp 701–726

Dong OX, Ronald PC (2019) Genetic engineering for disease resistance in plants: recent progress and future perspectives. Plant Physiol 180(1):26–38

Dracatos PM, Haghdoust R, Singh D, Park RF (2018) Exploring and exploiting the boundaries of host specificity using the cereal rust and mildew models. New Phytol 218(2):453–462

Endo A, Masafumi M, Kaya H, Toki S (2016) Efficient targeted mutagenesis of rice and tobacco genomes using Cpf1 from *Francisella novicida*. Sci Rep 6(1):38169

Fister AS, Landherr L, Maximova SN, Guiltinan MJ (2018) Transient expression of CRISPR/Cas9 machinery targeting TcNPR3 enhances defense response in Theobroma cacao. Front Plant Sci 9:329023

Foster AJ, Martin-Urdiroz M, Yan X, Wright HS, Soanes DM, Talbot NJ (2018) CRISPR-Cas9 ribonucleoprotein-mediated co-editing and counterselection in the rice blast fungus. Sci Rep 8(1):14355

Gardiner DM, Kazan K (2018) Selection is required for efficient Cas9-mediated genome editing in Fusarium graminearum. Fungal Biol 122(2–3):131–137

Gilbertson RL, Batuman O, Webster CG, Adkins S (2015) Role of the insect supervectors Bemisia tabaci and Frankliniella occidentalis in the emergence and global spread of plant viruses. Annu Rev Virol 2:67–93

Haque E, Taniguchi H, Hassan MM, Bhowmik P, Karim MR, Śmiech M, Zhao K, Rahman M, Islam T (2018) Application of CRISPR/Cas9 genome editing technology for the improvement of crops cultivated in tropical climates: recent progress, prospects, and challenges. Front Plant Sci 9:617

Idnurm A, Urquhart AS, Vummadi DR, Chang S, Van de Wouw AP, López-Ruiz FJ (2017) Spontaneous and CRISPR/Cas9-induced mutation of the osmosensor histidine kinase of the canola pathogen Leptosphaeria maculans. Fungal Biol Biotechnol 4:1–12

Ishino Y, Shinagawa H, Makino K, Amemura M, Nakata A (1987) Nucleotide sequence of the iap gene, responsible for alkaline phosphatase isozyme conversion in Escherichia coli, and identification of the gene product. J Bacteriol 169(12):5429–5433

Jansen R, Embden JDV, Gaastra W, Schouls LM (2002) Identification of genes that are associated with DNA repeats in prokaryotes. Mol Microbiol 43(6):1565–1575

Ji X, Zhang H, Zhang Y, Wang Y, Gao C (2015) Establishing a CRISPR–Cas-like immune system conferring DNA virus resistance in plants. Nat Plants 1(10):1–4

Jia H, Wang N (2014) Targeted genome editing of sweet orange using Cas9/sgRNA. PLoS One 9(4):e93806

Jia H, Zhang Y, Orbović V, Xu J, White FF, Jones JB, Wang N (2017) Genome editing of the disease susceptibility gene Cs LOB 1 in citrus confers resistance to citrus canker. Plant Biotechnol J 15(7):817–823

Jinek M, Chylinski K, Fonfara I, Hauer M, Doudna JA, Charpentier E (2012) A programmable dual-RNA–guided DNA endonuclease in adaptive bacterial immunity. Science 337(6096):816–821

Kalia A (2018) Nanotechnology in bioengineering: transmogrifying plant biotechnology. In Omics technologies and bio-engineering. Academic, pp 211–229

Kumar A, Singh VK, Tripathi V, Singh PP, Singh AK (2018) Plant growth-promoting rhizobacteria (PGPR): perspective in agriculture under biotic and abiotic stress. In: Crop improvement through microbial biotechnology. Elsevier, pp 333–342. https://doi.org/10.1016/B978-0-444-63987-5.00016-5

Kumar A, Singh SK, Kant C, Verma H, Kumar D, Singh PP, Modi A, Droby S, Kesawat MS, Alavilli H, Bhatia SK (2021) Microbial biosurfactant: a new frontier for sustainable agriculture and pharmaceutical industries. Antioxidants (Basel) 10:1472. https://doi.org/10.3390/antiox10091472

Langner T, Kamoun S, Belhaj K (2018) CRISPR crops: plant genome editing toward disease resistance. Annu Rev Phytopathol 56:479–512

Li T, Liu B, Chen CY, Yang B (2016) TALEN-mediated homologous recombination produces site-directed DNA base change and herbicide-resistant rice. J Genet Genomics 43(5):297–305

Liu C, Moschou PN (2018) Phenotypic novelty by CRISPR in plants. Dev Biol 435(2):170–175

Lyngkjær M, Newton A, Atzema J, Baker S (2000) The barley mlo-gene: an important powdery mildew resistance source. Agronomie 20(7):745–756

Macovei A, Sevilla NR, Cantos C, Jonson GB, Slamet-Loedin I, Čermák T, Voytas DF, Choi IR, Chadha-Mohanty P (2018) Novel alleles of rice eIF4G generated by CRISPR/Cas9-targeted mutagenesis confer resistance to Rice tungro spherical virus. Plant Biotechnol J 16(11):1918–1927

Mohanta TK, Bashir T, Hashem A, Abd-Allah EF, Bae H (2017) Genome editing tools in plants. Genes (Basel) 8(12):399

Nejat N, Rookes J, Mantri NL, Cahill DM (2017) Plant–pathogen interactions: toward development of next-generation disease-resistant plants. Crit Rev Biotechnol 37(2):229–237

Nelson R, Wiesner-Hanks T, Wisser R, Balint-Kurti P (2018) Navigating complexity to breed disease-resistant crops. Nat Rev Genet 19(1):21–33

Pathak P, Rai VK, Can H, Singh SK, Kumar D, Bhardwaj N, Roychowdhury R, de Azevedo LCB, Kaushalendra, Verma H, Kumar A (2022) Plant-endophyte interaction during biotic stress management. Plants (Basel) 11:2203. https://doi.org/10.3390/plants11172203

Peng A, Chen S, Lei T, Xu L, He Y, Wu L, Yao L, Zou X (2017) Engineering canker-resistant plants through CRISPR/Cas9-targeted editing of the susceptibility gene Cs LOB 1 promoter in citrus. Plant Biotechnol J 15(12):1509–1519
Pompili V, Dalla Costa L, Piazza S, Pindo M, Malnoy M (2020) Reduced fire blight susceptibility in apple cultivars using a high-efficiency CRISPR/Cas9-FLP/FRT-based gene editing system. Plant Biotechnol J 18(3):845–858
Price AA, Sampson TR, Ratner HK, Grakoui A, Weiss DS (2015) Cas9-mediated targeting of viral RNA in eukaryotic cells. Proc Natl Acad Sci U S A 112(19):6164–6169
Pyott DE, Sheehan E, Molnar A (2016) Engineering of CRISPR/Cas9-mediated potyvirus resistance in transgene-free Arabidopsis plants. Mol Plant Pathol 17(8):1276–1288
Richter C, Chang JT, Fineran PC (2012) Function and regulation of clustered regularly interspaced short palindromic repeats (CRISPR)/CRISPR associated (Cas) systems. Viruses 4(10):2291–2311
Roossinck MJ, Martin DP, Roumagnac P (2015) Plant virus metagenomics: advances in virus discovery. Phytopathology 105(6):716–727
Schmitz DJ, Ali Z, Wang C, Aljedaani F, Hooykaas PJ, Mahfouz M, de Pater S (2020) CRISPR/Cas9 mutagenesis by translocation of Cas9 protein into plant cells via the Agrobacterium type IV secretion system. Front Genome Ed 2:6
Schuster M, Kahmann R (2019) CRISPR-Cas9 genome editing approaches in filamentous fungi and oomycetes. Fungal Genet Biol 130:43–53
Singh R, Pandey KD, Singh M, Singh SK, Hashem A, Al-Arjani ABF, Abd-Allah EF, Singh PK, Kumar A (2022) Isolation and characterization of endophytes bacterial strains of Momordica charantia L. and their possible approach in stress management. Microorganisms 10:290. https://doi.org/10.3390/microorganisms10020290
Sun Y, Li J, Xia L (2016a) Precise genome modification via sequence-specific nucleases-mediated gene targeting for crop improvement. Front Plant Sci 7:233621
Sun Y, Zhang X, Wu C, He Y, Ma Y, Hou H, Guo X, Du W, Zhao Y, Xia L (2016b) Engineering herbicide-resistant rice plants through CRISPR/Cas9-mediated homologous recombination of acetolactate synthase. Mol Plant 9(4):628–631
United Nations (2019) World population prospects 2019. https://population.un.org/wpp/Download/Standard/Population/. Accessed 7 Sept 2023
Wada N, Ueta R, Osakabe Y, Osakabe K (2020) Precision genome editing in plants: state-of-the-art in CRISPR/Cas9-based genome engineering. BMC Plant Biol 20:1–12
Wang Y, Cheng X, Shan Q, Zhang Y, Liu J, Gao C, Qiu JL (2014) Simultaneous editing of three homoeoalleles in hexaploid bread wheat confers heritable resistance to powdery mildew. Nat Biotechnol 32(9):947–951
Wang Q, Cobine PA, Coleman JJ (2018a) Efficient genome editing in Fusarium oxysporum based on CRISPR/Cas9 ribonucleoprotein complexes. Fungal Genet Biol 117:21–29
Wang M, Wang S, Liang Z, Shi W, Gao C, Xia G (2018b) From genetic stock to genome editing: gene exploitation in wheat. Trends Biotechnol 36(2):160–172
Wang L, Chen S, Peng A, Xie Z, He Y, Zou X (2019) CRISPR/Cas9-mediated editing of CsWRKY22 reduces susceptibility to Xanthomonas citri subsp. citri in Wanjincheng orange (Citrus sinensis (L.) Osbeck). Plant Biotechnol Rep 13:501–510
Wenderoth M, Pinecker C, Voß B, Fischer R (2017) Establishment of CRISPR/Cas9 in Alternaria alternata. Fungal Genet Biol 101:55–60
Wolter F, Schindele P, Puchta H (2019) Plant breeding at the speed of light: the power of CRISPR/Cas to generate directed genetic diversity at multiple sites. BMC Plant Biol 19(1):176
Yanagawa Y, Suenaga Y, Iijima Y, Endo M, Sanada N, Katoh E, Toki S, Okino A, Mitsuhara I (2023) Genome editing by introduction of Cas9/sgRNA into plant cells using temperature-controlled atmospheric pressure plasma. PLoS One 18(2):e0281767
Zhang YZ, Shi M, Holmes EC (2018) Using metagenomics to characterize an expanding virosphere. Cell 172(6):1168–1172

Biocontrol Mechanisms by Root-Associated *Bacillus* Species

9

Muhammad Hamza Ashraf, Nazim Hussain,
Muhammad Zafar Saleem, and Rimsha Haqqi

Abstract

The production of multiple metabolites by *bacillus* species makes them potential candidate in sustainable agriculture, as they minimize the growth of pathogens by interrupting their cell membranes. The rapid colonization of the rhizosphere, the formation of biofilm, and the production of siderophores as strategies to store essential nutrients allows plant to be displaced from the competition, while genomic research showed that these antimicrobial substances are produced by large biosynthetic gene clusters (BGCs) regulated by complex networks that employ quorum sensing and sigma factors for adaptation to various contexts and are specifically in response to specific chemicals. The acquisition of the plasmid and horizontal gene transfer can improve the adaptability and biocontrol spectrum of *bacillus*. Although the variation in the performance of biopesticides, among other reasons due to environmental factors, and application methods, field applications in crops have demonstrated substantial disease reduction and higher yields. *Bacillus* has to be formulated with the best formulation techniques like encapsulation and sustained release to maintain the viability and effectiveness of *bacillus*. Continual research and advancements in synthetic biology and metabolic engineering play a crucial role in improving *bacillus* biocontrol capabilities. The incorporation of *bacillus* species into sustainable agriculture offers a promising approach to decrease reliance on chemical pesticides, enhance soil health, and boost crop productivity, thereby contributing to the development of resilient agricultural ecosystems.

M. H. Ashraf · N. Hussain (✉) · M. Z. Saleem · R. Haqqi
Centre for Applied Molecular Biology (CAMB), University of the Punjab, Lahore, Pakistan
e-mail: nazim.camb@pu.edu.pk; zafar.camb@pu.edu.pk

A. Kumar, M. K. Solanki (eds.), *Microbial Biocontrol Techniques*, Microorganisms for Sustainability 54,
https://doi.org/10.1007/978-981-97-8739-5_9

Keywords
Plant diseases · Integrated disease management · *Bacillus* species · Environmental factors · Plant–microbe interaction

9.1 Introduction

9.1.1 Overview of Plant–Microbe Interactions

Plant–microbe interactions embody a complex and dynamic relationship that is vital for plant health, and influence various soil ecosystem functions (Bastas and Kumar 2023; Pathak et al. 2022). These interactions are generally characterized as mutualistic, communalistic, or pathogenic. Plant nutrient acquisition and stress tolerance are facilitated by mutualistic interactions in which plants associate with *mycorrhizal* fungi or nitrogen-fixing bacteria. Most often these are relationships between microbes that reap benefits from the plant but do not have substantial impacts on the plant directly, and are more communalistic in nature. In contrast, pathogenic interactions can cause disease and are a major public health issue, with serious consequences for plant yield and productivity (S. S. Bhattacharyya and Furtak 2022).

Since the rhizosphere is a hotspot for interactions mediated primarily by exudates, (i.e., secreted factors produced by the root and its associated microbial community), most of these interactions are noticed in the rhizosphere. Exudates in SSD comprise those organic compounds which irrespective of their route of origin are present in high amounts in soil and plant, exogenously delivered to root systems and serve as substrates for the growth of other microbes including sugars, amino acids, and secondary metabolites (Li et al. 2024). Microbes can then provide benefit to plants via nutrient solubilization, phytohormone production, and by inducing plant pathogen suppression, among other mechanisms these microbes use to influence growth and development in their host plant. The balance between these interactions is imperative for plant health and sustainable agricultural productivity (Nadarajah and Abdul Rahman 2021; Raheem et al. 2019; Zeng et al. 2022).

9.1.2 Importance of Root-Associated Bacteria in Plant Health

The root-associated bacteria, specifically the rhizosphere, are primarily involved in the promotion of plant growth. Most of these bacteria, they help in cycling the nutrients, by mineralizing the organic matter to solubilize phosphorus and other essential nutrients to be more readily available to the plants. In addition, they are able to secrete phytohormones like indole-3-acetic acid (IAA), which increase the growth and development of the root and maximize the absorption of water and nutrients by the plant. Suppression of soil-borne pathogens by root-associated bacteria is one widely recognized important function (Huang et al. 2019). This conclusion is consistent with various mechanisms of biocontrol through production of antibiotics, competition for space and nutrients, and induction of systemic resistance in the host plant (Devi et al. 2023; Guzmán-Guzmán et al. 2023). These bacteria could be highly effective for controlling plant diseases, because they inhibit pathogen growth and infection (Bonaterra et al. 2022).

Moreover, *rhizospheric* bacteria could help plants in the abiotic stress management including drought, salinity, and toxicity of heavy metals. It does this by generating stress-metabolites, strengthening the plant antioxidant activities, and modifying root architecture (Dargiri and Naeimi 2024; Kumar et al. 2018; Singh et al. 2022; Wang et al. 2023a, b).

9.2 *Bacillus* Species as Biocontrol Agents

Bacillus species are spore forming bacteria that have been commonly utilized as biocontrol agents due to their capacity to enhance plant growth and suppress plant pathogens. The soil bacterium employs these mechanisms to improve its chances of survival, persistence, and growth in the ever-changing conditions of the rhizosphere (Miljaković et al. 2020). This includes a broad variety of antifungal agents, such as bacillomycins, iturins, and fengycins, which can disrupt the cell membranes of harmful fungus. To further enhance their persistence in the rhizosphere, Bacillus spp. produce lipopeptides like surfactin, which serve as biosurfactants and have biocontrol functions. Additionally, these lipopeptides have a direct favorable influence on root colonization, such as biofilm development (Zhang et al. 2023).

The interplay between bacillus species' competitive mechanisms is another critical factor. When pathogens have trouble growing iron-dependent organisms because of a lack of iron in their environment, siderophores step in to mediate competitive exclusion by storing iron. When certain species of Bacillus colonize plants, they can trigger defense mechanisms that increase the plant's resistance to various pathogens. This process involves the production of proteins and secondary metabolites that are related to pathogenesis (Ilham et al. 2019). Furthermore, they are a promising option for environmentally friendly farming because they work well with IPM techniques (Karačić et al. 2024). Table 9.1 summarizes the multiple ways in which the plant microbiome, which includes *Bacillus goodina* root-associated bacteria, interacts to plant health and yield. They are very suitable for sustainable agriculture due to their multi-partite biocontrol mechanisms, which encompass antibiotic synthesis, competition, and ISR induction (Lee et al. 2022).

9.2.1 Biology and Ecology of Root-Associated Bacillus Species

Bacillus species are Gram positive, rod-shaped bacteria that exhibit impressive physiological and metabolic versatility. Facultative aerobes can live with or without oxygen societies. The main feature of the *Bacillus* species is to be able to produce endospores which are also usually heat, desiccation, radiation, and chemical disinfectant (resulting in the endopores) tolerant and other stresses resistant. Due to the fact that they are able to form spores, they are able to survive in many different

Table 9.1 Bacillus species utilized in crop protection

Bacillus species	Mechanism of action	Benefits	Current challenges
Bacillus pumilus	Antibiosis, ISR	Reduces Alternaria infections, enhances growth	Variable effectiveness, environmental conditions
Bacillus licheniformis	Antibiosis, competition	Suppresses bacterial pathogens, improves crop yield	Formulation challenges, environmental variability
Bacillus cereus	Antibiosis, parasitism	Targets fusarium, improves soil health	Non-target effects, persistence in soil
Bacillus vallismortis	Antibiosis, ISR	Effective against various pathogens, promotes resistance	Environmental adaptation, consistency in effectiveness
	Antibiosis, ISR	Induces systemic resistance, broad pathogen control	Consistency in field conditions, regulatory hurdles
Bacillus polymyxa	Antibiosis, nitrogen fixation	Fixes nitrogen, improves soil fertility	Complex soil interactions, formulation stability
	Antibiosis, nutrient solubilization	Solubilizes phosphate, enhances nutrient uptake	Complex interactions in soil, application methods
	Antibiosis, competition	Suppresses bacterial pathogens, improves crop yield	Formulation challenges, environmental variability
	Controls lepidopteran pests, reduces pesticide use	Controls lepidopteran pests, reduces pesticide use	Resistance development, regulatory compliance
	Antibiosis, competition, ISR	Effective against Botrytis, improves plant health	Formulation stability, application timing
	Antibiosis, ISR, competition	Broad-spectrum antifungal activity, plant growth promotion	Variable field performance, environmental sensitivity

environments and are often found in great numbers associated with the rhizosphere (Biermann and Beutel 2023).

Bacillus is a diverse genus consisting of a large number of species, each with specific genetic and phenotypic features. These bacteria are characterized by a high genomic plasticity with genomes usually between 3 and 6 megabases. Bacillus carries a tremendous pool of genes associated with biocontrol activities required for the secondary metabolites, lytic enzymes, and antibiotics. The cell wall of Bacillus spp. consists of thick peptidoglycan layer, which gives structural integrity and resistance to osmotic stress (de Jesus Silva et al. 2019).

In terms of metabolism, they are likewise very versatile. Organic acids, simple sugars, and complex polysaccharides are some of the carbon sources that they employ. Their metabolic flexibility gives them an edge when colonizing the rhizosphere, as these exudates are continuously fed and contain a wealth of organic chemicals that promote development (Chen et al. 2021; Likhitha et al. 2023).

9.2.2 Habitat and Colonization of Plant Roots

Root exudation of carbohydrates and rhizosphere microorganisms are essential for plant nutrition uptake and biogeochemical process control. As a result of the sloughing off of basic carbon sources from plant roots, it becomes a microbe hotspot. The rhizosphere can be colonized by other species of Bacillus by the use of root exudates and competitive success with other soil microorganisms. Initially, Bacillus cells move by swarming motility toward root exudate. Through chemotaxis, Bacillus species use the signal of some chemicals to bind to specific receptors on the cell membrane and move toward suitable areas (Zhou et al. 2021). The spores of Bacillus adhere to root epidermal cells by synthesizing extracellular polymeric substances (EPS) to form biofilm upon root surface. Biofilms offer a protective microenvironment in which the Bacillus can exist and therefore improve their rhizosphere persistence (Bhattacharyya et al. 2023).

In interaction of Bacillus spp with plant root, a mechanism of signaling molecules took place, as every playing their own role. As Bacillus are able to produce phytohormones such as indole-3-acetic acid (IAA) that stimulate root growth and development. It not only helps to bail out the plant by helping in nutrient and water use but also to increase the surface area for the colonization of bacillus. Bacillus strains also can trigger systemic resistance (ISR) in plants, which activates the plant immune system in such a way that it becomes more difficult for any plant pathogen to take hold and colonize (Tsotetsi et al. 2022).

9.2.3 Diversity and Distribution in the Rhizosphere

The rhizosphere provides colonization for myriad of bacillus species with each bacillus associated contributing to the complex regulation that underlies the health and balance of the soil-plant system. The compositions of Bacillus communities in the rhizosphere are closely related to soil type, plant species, root exudate, and environmental conditions. Many *Bacillus* spp., including *Bacillus subtilis, Bacillus amyloliquefaciens, Bacillus pumilus, Bacillus thuringiensis* have been identified by molecular techniques using the 16S rRNA gene sequencing (Navarro-Noya et al. 2022). The distribution of Bacillus species in the rhizosphere is frequently spotty, with individuals of some species preferentially located in specific root zones. For instance, *B. subtilis* colonizes the root surface and rhizoplane, whereas Bacillus megaterium is found in the root interior or endosphere. This spatial distribution is driven by the specialized properties and metabolic capacities of each Bacillus species, as well as its associations with the host plant and more general rhizosphere microorganisms (Du et al. 2022).

Bacillus species in the rhizosphere illustrate not only their ecological adaptability but also their functional versatility. Bacillus species also produce a multitude of antimicrobial compounds, enzymes, and other metabolites that are pivotal in promoting plant health and enriching the rhizosphere ecosystem. This functional diversity appears to enable Bacillus to perform a variety of plant-benefitting functions such as pathogen suppression, nutrient solubilization, and growth promotion. Adopting a more classical notion, we also have to consider that Bacillus species in the rhizosphere are interconnected, not only because of their interactions with plants there but also due to their numerous interactions with other soil microorganisms (Wang et al. 2022a). There, they can interact positively, such as Bacillus between mycorrhizal fungi, or negatively, such as competition against pathogenic bacteria and fungi. The diversity seems low, but the effects of these interactions can greatly shape the nature and variation of the rhizosphere microbial community (Sun et al. 2022; Solanki et al. 2023).

9.3 Mechanisms of Biocontrol by Root-Associated Bacillus Species

9.3.1 Antibiosis

Antibiosis is the major mechanism of biocontrol by *Bacillus* strains owing to antibiotics and other secondary metabolites that inhibit the growth of pathogens (Boulahouat et al. 2023). Due to the diverse metabolic capabilities, several species of Bacillus synthesize a large variety of antimicrobial and antifungal metabolites such as polyketides, no ribosomal peptides, lipopeptides, and bacteriocins between 52 and 200 amino acids. Such compounds are potentially able to inhibit and even kill pathogenic microbes by interfering with their cellular processes. These antibiotics are synthesized by biosynthetic gene clusters that are often located within operons and are subject to complex regulation by transcriptional regulators that sense environmental signals or respond to microbial population dynamics (Caulier et al. 2019). The direct clash with pathogens is not the only method that results in the vast antibiotic production of Bacillus species, but also affects the process of modulating the microbial community within the rhizosphere. Bacillus species can change the competitive landscape favoring other bacteria that beneficially support plant health by selectively inhibiting some microorganisms. Moreover, some Bacillus secondary metabolites have signaling activities associated with promoting plant growth and defense (Andrić et al. 2020).

9.3.1.1 Production of Antibiotics and Secondary Metabolites

Bacillomycin belongs to the iturin family of lipopeptides produced by several Bacillus species such as *B. subtilis* and *B. amyloliquefaciens* (Lu et al. 2021b, 2022). The iturin compounds exert strong antifungal activities acting by altering the membrane integrity of fungal cells, through pore formation, which results in leakage and subsequent fungal death. Its specificity factor to fungal plant pathogens makes this a valuable biocontrol agent against a broad diversity of pathogenic fungi in genus Fusarium including *Fusarium* spp. and *Rhizoctonia solani* (Wang et al. 2022b).

For resistance against fungal pathogens, iturin is a well-studied lipopeptide with broad antifungal spectrum. It lost its activity against *F. graminearum,* whereas iturin has an action mode where it is inserted into the fungal cell membrane abrogating ion channels that cause disruption in the membrane potential resulting in cell osmolysis. Iturin-producing Bacillus strains have been demonstrated to have disease control efficacy against a number of plant-associated diseases including powdery mildew and gray mold in multiple crops (Dang et al. 2019).

Surfactin is a polypeptide consisting of an N-terminal cycle with a lipophilic fatty acid side chain responsible for its strong surfactant properties. It disrupts cell membranes, which provides it with antimicrobial activity against many species of bacteria and fungi. In addition to changing surface tension of the cell wall, surfactin can facilitate biofilm formation in the rhizosphere, contributing to colonization and persistence of Bacillus. Its dual action renders surfactin, an indispensable part of Bacillus-mediated biocontrol mechanisms (Stoll et al. 2021).

9.3.2 Competition

The nutrient competition is the basic biocontrol mechanism used by Bacillus strains which are able to multiply faster than the pathogenic microorganisms on release environment and therefore to outcompete them for the nutrients and for the specific ecological niches. Root exudates, such as sugars, amino acids, organic acids, and vitamins, render the rhizosphere nutrient rich. These exudates are efficiently taken up and utilized by Bacillus species, thereby denying the pathogens the necessary nutrients to grow and establish themselves in the host. Bacillus species are equipped with a number of nutrient acquisition strategies, including the production of siderophores, such as high-affinity iron-chelating compounds (Zhu et al. 2023). By sequestering iron away from the environment, siderophores make it difficult for iron-dependent pathogens to get at their preferred element. These ACE interactions could result in the cellular uptake of iron-siderophore complexes by Bacillus cells through specific receptors, which give them a competitive advantage. This system is particularly efficacious toward iron-dependent fungal pathogens (Rizzi et al. 2019).

Apart from nutrient competition, Bacillus species also compete for root colonization sites and space within the rhizosphere. The capacity of this genus to colonize biofilms on roots is a most important and determinant for the competition where various Bacillus species predominate (Hu et al. 2023). Exopolysaccharide (EPSs) are responsible for biofilm formation, produce by a bacterial layer known as EPS matrix surrounding Bacillus cells on the root surface, and provide protection from environmental stresses and antimicrobial compounds. Biofilms also allow the initial colonization of beneficial microorganisms, providing a stable local population able to outcompete pathogenic microorganisms (Netrusov et al. 2023). Bacillus biofilms that occupy space in vitro physically reduce the number of sites available to binding of pathogens, presumably reducing pathogen infection of the host plant. Moreover, the biofilm matrix reduces pathogen colony development and virulence by sequestering and degrading pathogen-derived signals (Hu et al. 2023).

9.4 Induced Systemic Resistance (ISR)

Some rhizosphere microorganisms have the capability to activate a plant defense mechanism known as induced systemic resistance (ISR), induced by Bacillus species. Upon the colonization of root systems, Bacillus species triggers a systemic signaling cascade that can modulate systemic resistance and enhance plant defense mechanisms against a variety of pathogens. The whole mechanism of ISR is shown in Fig. 9.1. This is in contrast to more localized resistance responses, such as ISR, which provides broad-spectrum protection that has the ability to move systemically throughout the plant (Kloepper et al. 2004).

Bacillus-induced ISR is mediated by several signaling molecules and pathways. MAMPs produced by Bacillus species including flagellin, lipopolysaccharides, and peptidoglycan fragments are recognized by plant pattern recognition receptors (PRRs). Recognition leads to the ignition of a signaling cascade and the activation of other downstream responses to fight off an invader. In addition, volatile organic compounds (VOC) and other metabolites secreted by Bacillus spp. have been reported to indirectly manipulate plant signal transduction mechanisms and strengthen immunity (Rumyantsev et al. 2023).

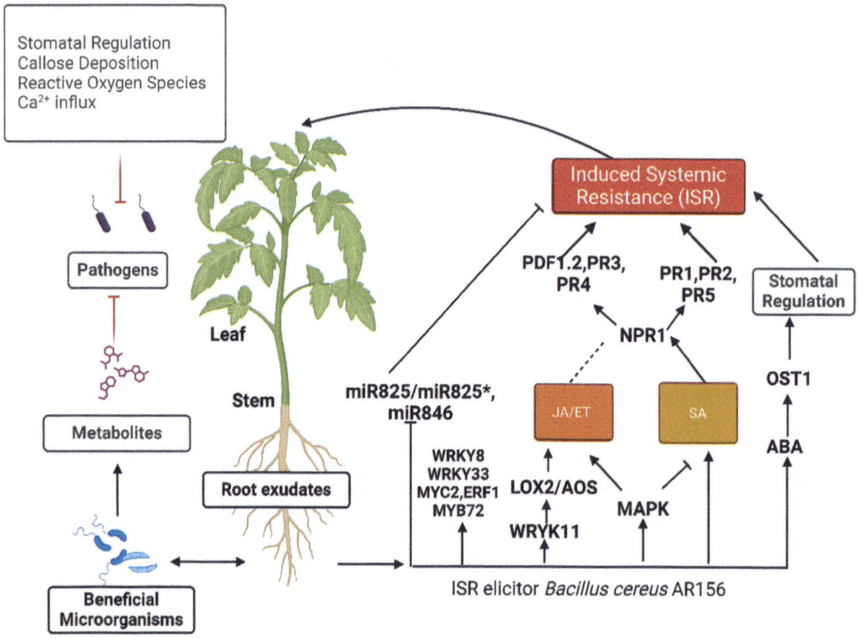

Fig. 9.1 A complicated network is formed by the transcriptional factors and genes partaking in the defense response by signaling crosstalk. Beneficial microorganisms activate JA and SA in an NRP1-dependent pathway. MAPK, SA, ET, and JA coordinate in the downstream defense response. Additionally, beneficial microorganisms induce non-coding RNAs that play a critical role in the regulation of the host development and resistance to the pathogen

9.4.1 Parasitism and Predation

9.4.1.1 Direct Parasitism of Plant Pathogens

Direct parasitism is one of important mechanisms by which Bacillus species control plant pathogens. It involves a direct and targeted attack on the pathogen species, leading to their breakdown and eventual suppression or death. This is due to the production of hydrolytic enzymes and catabolic compounds by Bacillus species which degrade the structural features of pathogenic organism which is the major mode of action of the isolate (Yang et al. 2023).

9.4.1.2 Hydrolytic Enzymes

1. Chitinases: Chitinases breakdown the chitin which is the main component of the fungal cell walls. Chitinases hydrolyze chitin to low molecular weight oligomers and monomers, which result in damage to the fungal cell wall and cytolysis of the fungal cell mediated by homologous chitinolytic enzymes produced by Bacillus sp. One possible targeted response mechanism is the upregulation of chitinases following exposure to fungal pathogens since chitin is a structural component of fungi. However, *Bacillus cereus* and *B. thuringiensis* have been reported to exhibit strong chitinase activity, a requirement for their antifungal activity against important fungal pathogens such as *Fusarium* spp. and *Sclerotinia sclerotiorum* (Al-Rajhi et al. 2023).
2. Proteases: Another group of enzymes produced by Bacillus sp. which helps in breaking down of proteins including those present in the cell walls of pathogens and other structural proteins important for survival of the pathogen. The extracellular proteases of *B. subtilis* and *Bacillus licheniformis* include serine proteases, metalloproteases, and cysteine proteases. The enzymes work by targeting proteins in the cell walls of the pathogens, which weakens those walls and sensitizes the pathogens to environmental stresses and host plant defenses (Keshapaga et al. 2023; Nogueira et al. 2022).
3. Glucanases: Glucans make up an important polysaccharide structural component in fungal cell walls that are hydrolyzed by glucanase. Bacillus species use glucans to break down the structural integrity of fungal pathogens, causing division of the cell wall and death of the pathogen. One of them is the β-1,3-glucanase produced by *B. amyloliquefaciens* known to be active against Phytophthora infestans, and *Rhizoctonia solani* among others (Garcia-Rubio et al. 2019).

9.4.1.3 Antimicrobial Compounds

1. Lipopeptides: Lipopeptides produced by Bacillus species include, in particular, iturins, fengycins, and surfactins. The lipopeptides are potent antifungal agents that become integrated into the surface of the pathogen cell membranes producing pores and ultimately, loss of ions and cell death. In particular, iturins and fengycins have demonstrated activity against a spectrum of fungal pathogens, disrupting membrane integrity with an in vitro mechanism that has consisted of challenging the Candida with the elicitor. Prominent producers of these lipopep-

tides are *B. subtilis* and *B. amyloliquefaciens,* and this contributes to their biocontrol efficacy (Wang et al. 2023a, b).

2. Antibiotics: Besides lipopeptides, Bacillus spp. also synthesize other antibiotics including bacilysin and difficidin, which have antagonistic activities against various bacterial and fungal pathogens. All these antibiotics disrupt processes normally occurring in the cell: protein synthesis and formation of cell walls, essential for pathogens. The production of these compounds is tightly regulated by quorum sensing systems so that cells only produce antibiotics when the population reaches a certain density, maximizing the use of resources as well as biocontrol potential (Bilal et al. 2017; Guo et al. 2020; Khatoon et al. 2022; Liu et al. 2018; Peng et al. 2018).

The primary method in which Bacillus species are said to exert control over pathogens is through direct parasitism. The substrates that increase the release of hydrolytic enzymes and compounds exert antimicrobial activity, causing degradation of the cell wall and membrane, this results in an interference in the structural and metabolic functions of the pathogenic agent. This multi-action approach kills the pathogen and prevents growth and dissemination. Direct parasitism by Bacillus species is dependent on environmental factors, such as the general plant health of the Bacillus cells and the presence of antagonistic microorganisms in the rhizosphere or other in vitro systems (Bahaddad et al. 2023). Bacillus activity is best in fields with adequate moisture, moderate temperatures, and a pH range that favors Bacillus. Expression of hydrolytic enzymes and antimicrobial compounds is favored under such conditions.

9.5 Molecular Basis of Biocontrol by Root-Associated Bacillus Species

9.5.1 Genetic Regulation of Biocontrol Traits

Early research suggested that the genetic control of biocontrol traits in Bacillus species is a complex system of genes and regulatory pathways controlling the synthesis of antibiotics, lytic enzymes, and other biocontrol-related molecules as illustrated in Fig. 9.2. At the center of this regulatory network are the operons (Collections of genes transcribed into single messenger RNA and regulatable by a single promoter). An example of which is the *B. subtilis* operon is responsible for the production of the iturins, a family of highly active lipopeptide antibiotics. Regulation of these operons is tightly controlled in response to environmental signals and intracellular conditions (Rocha et al. 2023).

The regulation of biocontrol traits involves the cell-to-cell communication mechanism quorum sensing. Bacillus species generate hydrophobic peptides known as autoinducers that can be secreted and accumulate in the environment based on bacterial growth. When a critical threshold of quorum sensing molecules is reached, it initiates an orchestrated response within the bacterial population, turning on or off

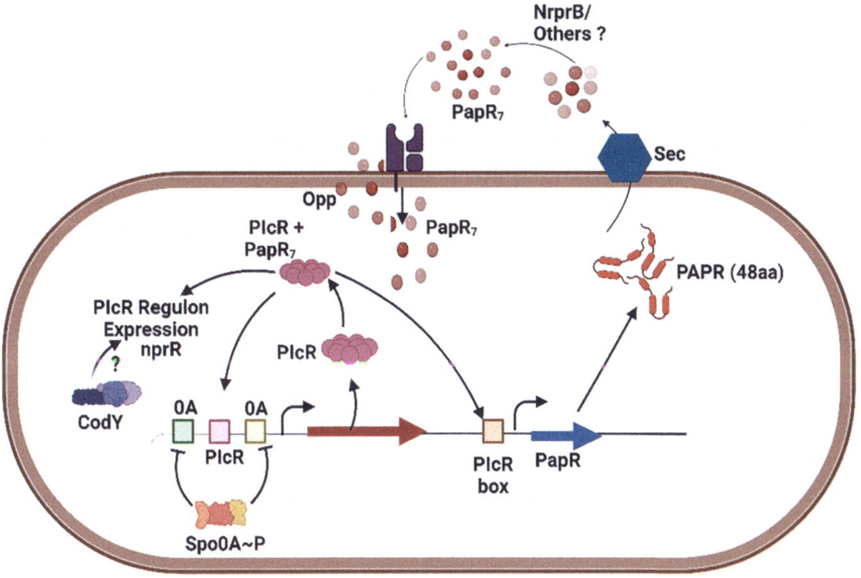

Fig. 9.2 Visual representation of PlcR-PapR QS system, its activity and regulation. The product of papR gene, 48 amino acids, belongs to PlcR regulon. Sec machinery secretes the PapR and NprB protease as well as other peptidases to process it. PapR is present in the extracellular component and bacterial cytoplasm. Oligopeptide permease Opp imports PapR in the cell. PapR binds to PlcR permitting the activation of PicR regulon expression. plcR is autoregulated and under the Spo0A-P negative control. CodY controls the expression of PlcR-dependent genes positively via unknown mechanism

a specific set of genes. This mechanism will ensure that the production of antibiotics, and other types of secondary metabolites, is coordinated with the population density so that the resource is used as efficiently as possible, and biocontrol activity is improved as much as it can (Shah et al. 2024).

Moreover, sigma factors are significant regulators of biocontrol traits and are the proteins that interact with RNA polymerase and guide it to specific promoters. The Bacillus species contain numerous sigma factors that direct the transcription of different groups of genes in response to various environmental signals. For instance, the alternative sigma factor can become mobilized in the response to a diversity of environmental challenges and mediate the activity of several antibiotics and stress resistance genes.

9.5.2 Genomic Insights into Bacillus Biocontrol Capabilities

The rapid and significant developments made in genomic sequencing have contributed strikingly to the knowledge of biocontrol mechanisms in Bacillus spp. Genome synteny has uncovered countless gene clusters involved in producing antibiotics and includes both no ribosomal peptide synthetases (NRPS) and polyketide synthetases

(PKS) being the most prevalent. Here, we describe two self-sufficient *B. subtilis* enzymes responsible for the biosynthesis of complex antibiotics and lipopeptides that are essential for the biocontrol activity of this bacterium. Large numbers of biosynthetic gene clusters (BGCs) for secondary metabolites are harbored by diverse Bacillus species identified through whole-genome sequencing (WGS) efforts, including certain with antimicrobial properties. *B. amyloliquefaciens* FZB42 genome alone is reported to possess more than 10 BGCs responsible for the biosynthesis of antibiotic compounds, like bacillomycin, fengycin, and surfactin (e.g., regulatory genes that allow these genes to be expressed in response to environmental cues often cluster around these sets of genes (Hu et al. 2023).

9.5.3 Role of Plasmids and Horizontal Gene Transfer

Plasmids, extrachromosomal DNA molecules, are a very important part of the genetic diversity and adaptability of Bacillus species. They frequently encode genes that allow the host to obtain selective advantage under specific environmental conditions (e.g., antibiotic/antimicrobial resistance, heavy metal resistance, and capability to degrade a wide range of complex organic compounds). Plasmids may contain genes encoding antimicrobial compounds and cell wall-degrading enzymes in the biocontrol context (Kim et al. 2022).

Bacillus species can play many roles in various environments, including extreme environments, in which they may be harsh environments or hostile environments. Horizontal gene transfer (HGT) is a principle mean by which new genes are introduced into Bacillus species as shown in Fig. 9.3. Bacillus can evolve rapidly in response to environmental change; a process that can take place through conjugation, transformation, or transduction, enabling bacillus rapidly adapt to changing environmental conditions so that they can enhance their biocontrol potential. Whereas conjugation is the direct transfer of plasmids from one bacterial cell to another bacterial cell via a conjugative pilus, transformation is the uptake of plasmid or free DNA from the environment. Transduction, transfer of genetic material between bacteria is mediated by bacteriophages. Thus, HGT of biosynthetic gene clusters can increase the biocontrol abilities in bacillus species. These, the genes involved in non-ribosomal peptide synthetases (NRPS) and polyketide synthases (PKS) including bacillibactin and difficidin, respectively, could be transferred to Bacillus species enhancing the development of a wide spectrum of antimicrobial compounds, improving their ability to inhibit the growth of a variety of plant pathogens, as tipiliae, gliocladium and phytophthora. Moreover, HGT allows certain traits to move from one bacillus species to another, which could broaden the range of bacillus species that might be used as more efficient biocontrol agents (Scott et al. 2023).

Plasmids also carry the regulatory elements that either act in a positive (pTi elements) or negative (pFi elements) manner mediating the differential gene expression of biocontrol-related genes. For example, plasmid-encoded transcription factors have been found to inhibit chromosomally encoded promoters, providing

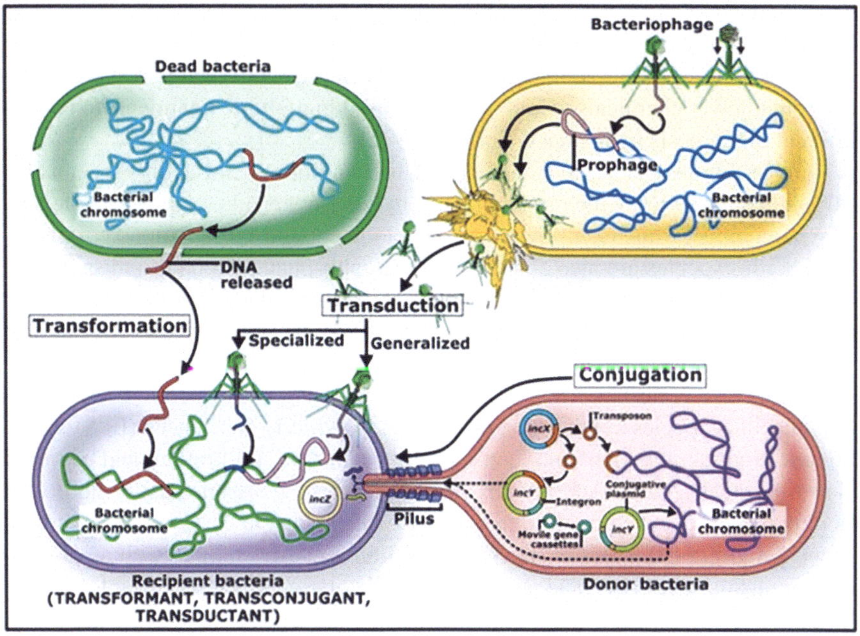

Fig. 9.3 Mechanisms involved in the horizontal gene transfer. The main mechanisms are transduction, conjugation, and transformation can mobilize and share genetic material with both related and non-related species. Bacteria acquire virulent traits, antibiotic resistance mechanisms, and others to guarantee its survival. (Adopted from (Bello-López et al. 2019) with permission; an open access article distributed under the Creative Commons Attribution (CC BY) license (http://creativecommons.org/licenses/by/4.0/))

regulation of potential antibiotic and lytic enzyme production. The contrasting patterns of mobilization and the interaction of plasmid and chromosomal genes that plasmids often seem to play a critical role in the molecular mechanisms of bacillus-mediated biocontrol further complicate the regulation of biocontrol traits (Thompson et al. 2023).

9.6 Role of Bacillus Species in Sustainable Agriculture

9.6.1 Utilization of Bacillus Species in Crop Protection

Bacillus species are important for plant pathogen biocontrol, which plays a vital role in agriculture and economy sustainability. Utilization of bacillus species for crop protection has a lot of benefits over the application of commercial chemical pesticides; many of them include improvement in the environmental and agricultural performance (compared to the conventional chemical pesticides) (Miljaković et al. 2020; Rocha et al. 2023). Bacillus species have a plenty of broad-spectrum antimicrobial activities that are quite same unlike other potential one, making one of the major benefits of using bacillus species. Bacillus species produce several desirable antibiotics, lipopeptides, and enzymes that can inhibit the growth of a broad spectrum of plant pathogens including bacteria, fungi, and nematodes. This means that fewer chemicals would have to be used in multiple treatments to control different insect pests in an area (Zhou et al. 2021).

Another thing is plant yield, and health has been improved. Bacillus sp. can also activate plant growth by the production of phytohormones like indole-3-acetic acid (IAA) and gibberellins resulting in root development and nutrient uptake. In addition, bacillus species are also able to solubilize phosphate and other indispensable nutrients, thus accessible to plants. Bacillus species are, therefore, of great interest in the context of integrated pest management (IPM) because they can offer double characteristics as antagonists and plant growth promoters (Castaldi et al. 2021).

Bacillus also contributes to soil health by improving soil structure and fertility. Some bacillus species synthesize extracellular polymeric substances (EPS) which promote soil aggregation and consequently moisture infiltration and aeration (Netrusov et al. 2023). Additionally, the decomposing organic matter ability of bacillus species also help in nutrient cycling, which helps in enriching the soil with essential nutrients. Bacillus sp. in crop protection is eco-friendly as well as sustainable. Compared with chemical pesticides, bacillus-based biocontrol agents are biodegradable and have low impact on non-target organisms (including beneficial insects, wildlife, and human health), in turn, reduces the environmental impact of agriculture and contributes to the protection of biodiversity (Karačić et al. 2024).

Different bacillus species have been widely used as biocontrol agents in diverse agricultural systems, with success in managing plant disease and promoting crop health. Some of them are enlisted below:

1. *B. subtilis*: *B. subtilis* is a widely orchestrated and used bacillus species in biocontrol. It is also effective against a wide spectrum of soil-borne and foliar pathogens. For example, *B. subtilis* strains may repress the growth of the pathogenic *fungus Fusarium* spp., which cause wilt or root rot diseases on a wide range of plants. The lipopeptides of *B. subtilis* such as surfactin and iturin are responsible for its antifungal activity. Organic and conventional farming has likewise adopted the commercialization of *B. subtilis*-based products, like Serenade®, to protect tomato, pepper, and strawberry crops from devastating fungal diseases (Karačić et al. 2024).
2. *B. amyloliquefaciens*: This biocontrol agent is known for producing many antibiotics and secondary metabolites. It has also been used to fight fungal pathogens, including *Botrytis cinerea*, which causes gray mold in grapes and other fruits. Antifungal compounds that disrupt the cell membranes of pathogens: bacillomycin, fengycin, and surfactin are also all produced by *B. amyloliquefaciens*. This species was also effective for the control of bacterial pathogens mentioned above, which cause leaf spots and blights, by *Xanthomonas and Pseudomonas* (Ling et al. 2023).
3. *B. thuringiensis*: *B. thuringiensis* (Bt) is famous for its efficacy as a biocontrol agent against insect pests due to its insecticidal nature crystal proteins (Cry pro-

teins) produced by Bt but toxic to certain species of insect larvae. When the larvae consume these proteins, they attach to receptors located in the insect gut, triggering cell lysis and eventually death. There are many products based on Bt (for example, dipel and thuricide) which are heavily used to control lepidopteran pests in crops such as corns, cotton, and vegetables. Bt has also had a gene inserted into crops with it that makes the crop resistant to pests, and this in turn reduces the need for chemical insecticides (Ramírez-Pool et al. 2024).

4. *B. pumilus: B. pumilus* has shown biocontrol activity against different fungal pathogens. It is typically utilized for the control diseases of vegetable crops caused by *Alternaria* spp. *B. pumilus* produces a series of antifungal compounds, pumilacidins and *bacilysins*, which exhibit inhibitory effect on hyphae and spores of fungi. This species is also able to stimulate systemic resistance in plants, making them more resistant to these diseases (Soliman et al. 2022).

9.6.2 Integration into Existing Agricultural Practices

Utilization of bacillus species in agriculture should also be well planned and an accounting to maximize efficacy and sustainability. Selective strains of bacillus species, method of application, and compatibility with different combination of other agricultural inputs and practices determine the successful integration.

The selection of bacillus strains is a key to the success of any biocontrol project. Strains should be selected depending on their biocontrol efficacy for a given crop and environmental conditions. Further, a discussion regarding the stability, viability of the bacillus formulations is equally important as they are important to their performance in the field. There are many ways to apply bacillus species to crops like seed treatments, soil inoculation, foliar sprays, and drench applications. The second is a series of seed treatments to address the soil-borne disease sooner through early season protection using bacillus treatments applied to the seed before planting. Soil inoculation—bacillus is applied directly to the soil where it colonizes the rhizosphere and gives root protection. A type of foliar spray controls foliar pathogens and improves immunity of plant. Bacillus-based biocontrol agents ought to be compatible with other agricultural inputs such as fertilizers, pesticides, and mode of irrigation. This would need to be ensured without negative impact of chemical pesticides on the survival as well as the activity of the bacillus strains. Mixing bacillus with organic fertilizers and other organic amendments enhances its activity by adding necessary nutrients and organic matter which are crucial for microbial proliferation.

Continuous monitoring and evaluation are very important to determine the efficiency of bacillus-based biocontrol agents with regular adjustments when required. This includes monitoring pathogen populations, disease incidence, and also the health of the crop. Economic benefits of applying bacillus such as yield increase and reduced cost in pesticide application would not only include in evaluation.

9.7 Challenges and Future Directions

9.7.1 Limitations and Challenges in Field Application

Though the bacillus species used for biocontrol have some advantages, there are certain limitations and challenges that have to be overcome to optimize their performance in the field. A key limitation is the inconsistency of biocontrol performance depending on field-specific environmental conditions. Because bacillus species are living organisms, they are affected by surrounding conditions such as soil composition, temperature, humidity, and pH. These types of environmental conditions can heavily regulate the survival, colonization, and activity of a given bacillus isolate in the rhizosphere. The inconsistent field performances of bacillus-based biocontrol agents across different conditions result in very limited commercial potential (Ayaz et al. 2023).

However, the delivery of the formulation for the bacillus-based products may have important limitations as well. It is important to keep bacillus strains stable and viable during storage, transport, and application so as to maintain their functionality. How to formulate bacillus spores for long-term stability (in terms of spore viability) against environmental stress, allowing for their sustained activity in the field. Further, application methods such as seed treatments, soil inoculum, and foliar sprays have to be adapted according to the crop and the pathogen in order to achieve an efficacy for biocontrol.

A further issue is related to the activity of bacillus species in interaction with the indigenous rhizosphere bacteria. The rhizosphere, a densely populated soil-plant interface, hosts a complex microbial community that forms specific and dynamic associations with plant roots (Solanki et al. 2024). Since the exogenous application of Bacillus spps. may interfere with such microbial interactions one way or the other, it is suggested that their usage may not be beneficial due to resulting side effects via either exploitation of opportunistic pathogens or blocking the exertion of beneficial organisms (Comeau et al. 2021). To maintain this successful performance of bacillus-based biocontrol agents, it is essential to have a full understanding of bacterial interactions and their control.

In addition to these aspects, the economic factors play a crucial part in directing the adoption of bacillus biocontrol solutions. The cost of production, formulation, and application of the bacillus-based product must be competitive with the chemical pesticides currently in use and financially attractive to the farmers. For farmers to adopt Bacillus-based biocontrol products, they must be cost-effective, competitive with chemical pesticides, and perceived as reliable and effective. Thus, establishing cost-benefit ratios and long-term positive effects of Bacillus-based biocontrol using field trials and extension services would be a key to push their use for an environmental-friendly agriculture.

9.7.2 Potential for Resistance Development in Pathogens

Resistance development in the plant pathogen to Bacillus-based biocontrol agents is a major constraint for potential use of these organisms in long term. As with chemical pesticides, antimicrobial compounds produced by Bacillus can select for pathogen populations capable of withstanding exposure to the compounds through the development of resistance. This resistance can be driven by mutations in target sites, activation of efflux pump mechanisms, or horizontal gene transfer of resistance genes (Zhang et al. 2023).

9.7.3 Future Research Needs and Technological Advancements

Further studies and technological innovation are essential to enhance the application of Bacillus species in sustainable agriculture, future research and development for improved efficacy, reliability and adoption of bacillus-based biocontrol agents.

1. Genomic and metagenomic studies: Genome sequencing of various Bacillus strains could uncover additional biosynthetic gene clusters and regulatory pathways vital for biocontrol. However, metagenomic studies may reveal effects of Bacillus introduction to the native soil microbiome and in positive facilitation of biocontrol efficacy by specific microbial interactions (Lu et al. 2021a; Kumar et al. 2024; Kumari et al. 2022; Zhimo et al. 2021).
2. Functional genomics and proteomics: These results may provide the means to elucidate better the key genes and proteins that underpin the biocontrol mechanisms. Transcriptomics, proteomics, and metabolomics are valuable tools that be applied for investigating the expression and regulation of biocontrol-associated genes and pathways in terms of environmental factors. Such insights could guide the engineering of Bacillus to create strains with improved biocontrol and ecological fitness (Shahid et al. 2021).
3. Synthetic biology and metabolic engineering: Synthetic biology and metabolic engineering have provided a wealth of tools with which to potentially augment the biocontrol capacity of Bacillus species. The same holds true for the production of antimicrobial compounds, phytohormones, or stress tolerance factors by Bacillus strains and can be enhanced to a higher level that enables Bacillus strains to act as more efficient and robust biocontrol agents. Biocontrol molecules production might be positively controlled by building biosensors and regulatory circuits that respond to specific environmental cues which can be developed using synthetic biology tools (Lee et al. 2023).
4. Advanced formulation technologies: Development of advanced formulation technologies is a key to enhancing the stability, viability, and delivery of Bacillus-based products. Said encapsulation techniques like microencapsulation and nanoencapsulation could safeguard Bacillus spores from environmental stress and provide sustained release into the field. There is also potential for these genetically improved strains to be formulated together with organic carriers or

soil amendments that would combine to be part of strategies that may be associated with better traction with existing agriculture practices.

5. Further field trials and demonstration projects: Large scale field trials and demonstration projects are necessary to prove the efficiency of Bacillus-based biocontrol agents for diverse agricultural situation. This allows the trials to provide data on the performance, the cost-effectiveness, and the long-term benefits of Bacillus formulations. Bacillus-based biocontrol and sustainable agriculture practices can be enhanced through farmers, extension, and agricultural organizations collaboration efforts.

9.8 Conclusion

Several Bacillus species have been identified as important allies for sustainable agriculture and provide numerous benefits including pathogen suppression and plant growth promotion. The great diversity of biocontrol mechanisms employed by these bacteria antibiosis, competition, induced systemic resistance (ISR), and parasitism highlights the efficacy and adaptability that had driven the farmers' success. Bacillus species have been shown to effectively control diseases and pests in crops such as tomatoes, grapes, and corn. These examples demonstrate a need for appropriate management strategies, supported by early and strategic application and integration in combination with other IPMs. This does not mean that Bacillus-based biocontrol agents do not face any down side in the field. These properties will address key performance hurdles such as non-uniformity in performance caused by environmental conditions, stability of the formulation, and resistance of pathogens. Future research with respect to genomic information, technological advances in formulation, and experimentation on a bigger field scale will be beneficial in overcoming such adverse circumstances. This includes the careful planning, monitoring, and farmer education required to effectively and sustainably incorporate Bacillus species into existing agricultural practices. Due to the economic and environmental importance of these species as biocontrol agents, it allows for reduced use of chemical pesticides, better soil health, and increased growth of crops they are used in combined with other components of IPM (Integrated Pest Management). The increasing adoption of Bacillus-based products is expected to lead to more Bacillus-based solutions included as sustainable and resilient solutions in agricultural systems as regulatory frameworks and policies are evolving to enable biocontrol solutions. Thus, an exploitation of the biocontrol capacity of Bacillus species for sustainable agriculture can be concluded in short. Conducting further research and practicality by overcoming the current limitations and incorporating Bacillus biocontrol agent with other sustainable practices can strike at the long run toward viable agricultural productivity and eco-viability.

References

Al-Rajhi AM, Alawlaqi MM, Ghany TMA, Moawad H (2023) Amanita sp. from subtropical region of Saudi Arabia as a source of chitinase enzyme and its antifungal activity. Bioresources 18(2):2928

Andrić S, Meyer T, Ongena M (2020) Bacillus responses to plant-associated fungal and bacterial communities. Front Microbiol 11:1350

Ayaz M, Li C-H, Ali Q, Zhao W, Chi Y-K, Shafiq M et al (2023) Bacterial and fungal biocontrol agents for plant disease protection: Journey from lab to field, current status, challenges, and global perspectives. Molecules 28(18):6735

Bahaddad SA, Almalki MHK, Alghamdi OA, Sohrab SS, Yasir M, Azhar EI, Chouayekh H (2023) Bacillus species as direct-fed microbial antibiotic alternatives for Monogastric production. Probiotics Antimicrob Proteins 15(1):1–16. https://doi.org/10.1007/s12602-022-09909-5

Bastas KK, Kumar A (2023) Omics technologies in the plant–microbe interactions. In: Bastas KK, Kumar A, Sivakumar U (eds) Microbial biocontrol: molecular perspective in plant disease management. Springer Nature, Singapore, pp 257–282. https://doi.org/10.1007/978-981-99-3947-3_13

Bello-López JM, Cabrero-Martínez OA, Ibáñez-Cervantes G, Hernández-Cortez C, Pelcastre-Rodríguez LI, Gonzalez-Avila LU, Castro-Escarpulli G (2019) Horizontal gene transfer and its association with antibiotic resistance in the genus Aeromonas spp. Microorganisms 7(9):363

Bhattacharyya SS, Furtak K (2022) Soil–plant–microbe interactions determine soil biological fertility by altering rhizospheric nutrient cycling and biocrust formation. Sustainability 15(1):625

Bhattacharyya A, Mavrodi O, Bhowmik N, Weller D, Thomashow L, Mavrodi D (2023) Bacterial biofilms as an essential component of rhizosphere plant-microbe interactions. Methods Microbiol 53:3–48. https://doi.org/10.1016/bs.mim.2023.05.006

Biermann R, Beutel S (2023) Endospore production of Bacillus spp. for industrial use. Eng Life Sci 23(11):e2300013. https://doi.org/10.1002/elsc.202300013

Bilal M, Guo S, Iqbal HM, Hu H, Wang W, Zhang X (2017) Engineering Pseudomonas for phenazine biosynthesis, regulation, and biotechnological applications: a review. World J Microbiol Biotechnol 33:1–11

Bonaterra A, Badosa E, Daranas N, Francés J, Roselló G, Montesinos E (2022) Bacteria as biological control agents of plant diseases. Microorganisms 10(9):1759. https://doi.org/10.3390/microorganisms10091759

Boulahouat S, Cherif-Silini H, Silini A, Bouket AC, Luptakova L, Alenezi FN, Belbahri L (2023) Biocontrol efficiency of Rhizospheric bacillus against the plant pathogen Fusarium oxysporum: a promising approach for sustainable agriculture. Microbiol Res 14(3):892–908

Castaldi S, Petrillo C, Donadio G, Piaz FD, Cimmino A, Masi M et al (2021) Plant growth promotion function of Bacillus sp. strains isolated from salt-pan rhizosphere and their biocontrol potential against Macrophomina phaseolina. Int J Mol Sci 22(7):3324. https://doi.org/10.3390/ijms22073324

Caulier S, Nannan C, Gillis A, Licciardi F, Bragard C, Mahillon J (2019) Overview of the antimicrobial compounds produced by members of the Bacillus subtilis group. Front Microbiol 10:435128

Chen Y-J, Leung PM, Wood JL, Bay SK, Hugenholtz P, Kessler AJ et al (2021) Metabolic flexibility allows bacterial habitat generalists to become dominant in a frequently disturbed ecosystem. ISME J 15(10):2986–3004

Comeau D, Balthazar C, Novinscak A, Bouhamdani N, Joly DL, Filion M (2021) Interactions between Bacillus spp., Pseudomonas spp. and Cannabis sativa promote plant growth. Front Microbiol 12:715758

Dang Y, Zhao F, Liu X, Fan X, Huang R, Gao W et al (2019) Enhanced production of antifungal lipopeptide iturin A by Bacillus amyloliquefaciens LL3 through metabolic engineering and culture conditions optimization. Microb Cell Factories 18:1–14

Dargiri SA, Naeimi S (2024) Plant growth-promoting Rhizobacteria in management of biotic and abiotic stresses

de Jesus Silva F, Ferreira LC, Campos VP, Cruz-Magalhães V, Barros AF, Andrade JP et al (2019) Complete genome sequence of the biocontrol agent Bacillus velezensis UFLA258 and its comparison with related species: diversity within the commons. Genome Biol Evol 11(10):2818–2823

Devi S, Sharma S, Tiwari A, Bhatt AK, Singh NK, Singh M, Kaushalendra, Kumar A (2023) Screening for multifarious plant growth promoting and biocontrol attributes in Bacillus strains isolated from Indo-Gangetic soil for enhancing growth of rice crops. Microorganisms 11:1085. https://doi.org/10.3390/microorganisms11041085

Du E, Chen Y, Li Y, Sun Z, Gui F (2022) Rhizospheric Bacillus-facilitated effects on the growth and competitive ability of the invasive plant Ageratina adenophora. Front Plant Sci 13:882255

Garcia-Rubio R, de Oliveira HC, Rivera J, Trevijano-Contador N (2019) The fungal cell wall: Candida, Cryptococcus, and Aspergillus species. Front Microbiol 10:2993. https://doi.org/10.3389/fmicb.2019.02993

Guo S, Wang Y, Bilal M, Hu H, Wang W, Zhang X (2020) Microbial synthesis of antibacterial phenazine-1, 6-dicarboxylic acid and the role of PhzG in Pseudomonas chlororaphis GP72AN. J Agric Food Chem 68(8):2373–2380

Guzmán-Guzmán P, Kumar A, de Los Santos-Villalobos S, Parra-Cota FI, Orozco-Mosqueda MDC, Fadiji AE, Hyder S, Babalola OO, Santoyo G (2023) Trichoderma species: our best fungal allies in the biocontrol of plant diseases—A review. Plants 12:432. https://doi.org/10.3390/plants12030432

Hu G, Wang Y, Blake C, Nordgaard M, Liu X, Wang B, Kovács ÁT (2023) Parallel genetic adaptation of Bacillus subtilis to different plant species. Microbial Genomics 9(7):001064

Huang CM, Chen WC, Lin SH, Wang YN, Shen FT (2019) Exploration of root-associated bacteria from the medicinal plant Platycodon grandiflorum. Microbes Environ 34(4):413–420. https://doi.org/10.1264/jsme2.ME19030

Ilham B, Noureddine C, Philippe G, Mohammed EG, Brahim E, Sophie A, Muriel M (2019) Induced systemic resistance (ISR) in Arabidopsis thaliana by Bacillus amyloliquefaciens and Trichoderma harzianum used as seed treatments. Agriculture 9(8):166

Karačić V, Miljaković D, Marinković J, Ignjatov M, Milošević D, Tamindžić G, Ivanović M (2024) Bacillus species: excellent biocontrol agents against tomato diseases. Microorganisms 12(3):457. https://doi.org/10.3390/microorganisms12030457

Keshapaga UR, Jathoth K, Singh SS, Gogada R, Burgula S (2023) Characterization of high-yield Bacillus subtilis cysteine protease for diverse industrial applications. Braz J Microbiol 54(2):739–752

Khatoon Z, del Carmen Orozco-Mosqueda M, Huang S, Nascimento FX, Santoyo G (2022) Peptide antibiotics produced by bacillus species: first line of attack in the biocontrol of plant diseases. In: Islam MT, Rahman M, Pandey P (eds) Bacilli in agrobiotechnology: plant stress tolerance, bioremediation, and bioprospecting. Springer International Publishing, Cham, pp 31–46

Kim MS, Jeong D-E, Choi S-K (2022) Bacillus integrative plasmid system combining a synthetic gene circuit for efficient genetic modifications of undomesticated Bacillus strains. Microb Cell Fact 21(1):259. https://doi.org/10.1186/s12934-022-01989-w

Kloepper JW, Ryu CM, Zhang S (2004) Induced systemic resistance and promotion of plant growth by Bacillus spp. Phytopathology 94(11):1259–1266. https://doi.org/10.1094/phyto.2004.94.11.1259

Kumar A, Singh VK, Tripathi V, Singh PP, Singh AK (2018) Plant growth-promoting rhizobacteria (PGPR): perspective in agriculture under biotic and abiotic stress. In: Crop improvement through microbial biotechnology. Elsevier, pp 333–342. https://doi.org/10.1016/B978-0-444-63987-5.00016-5

Kumar A, Solanki MK, Wang Z, Solanki AC, Singh VK, Divvela PK (2024) Revealing the seed microbiome: navigating sequencing tools, microbial assembly, and functions to amplify plant fitness. Microbiol Res 279:127549. https://doi.org/10.1016/j.micres.2023.127549

Kumari M, Qureshi KA, Jaremko M, White J, Singh SK, Sharma VK, Singh KK, Santoyo G, Puopolo G, Kumar A (2022) Deciphering the role of endophytic microbiome in postharvest diseases management of fruits: opportunity areas in commercial up-scale production. Front Plant Sci 13:1026575. https://doi.org/10.3389/fpls.2022.1026575

Lee JH, Anderson AJ, Kim YC (2022) Root-associated bacteria are biocontrol agents for multiple plant pests. Microorganisms 10(5):1053

Lee J, Kim S, Jung H, Koo B-K, Han JA, Lee H-S (2023) Exploiting bacterial genera as biocontrol agents: mechanisms, interactions and applications in sustainable agriculture. J Plant Biol 66(6):485–498. https://doi.org/10.1007/s12374-023-09404-6

Li A, Jin K, Zhang Y, Deng X, Chen Y, Wei X et al (2024) Root exudates and rhizosphere microbiota in responding to long-term continuous cropping of tobacco. Sci Rep 14(1):11274

Likhitha P, Undirwade D, Kulkarni U, Kolhe A, Moharil M (2023) Cry toxin expression in different plant parts of Bt cotton at different phenological stages. Egypt J Biol Pest Control 33(1):98

Ling L, Li Y, Jiang K, Wang Y, Luo H, Cheng W et al (2023) Volatile organic compounds of Bacillus spp. as an emerging antifungal resource play a significant role in fruit postharvest disease control. Food Biosci 56:103201

Liu Y, Wang Z, Bilal M, Hu H, Wang W, Huang X et al (2018) Enhanced fluorescent siderophore biosynthesis and loss of phenazine-1-carboxamide in phenotypic variant of Pseudomonas chlororaphis HT66. Front Microbiol 9:759

Lu P, Jiang K, Hao YQ, Chu WY, Xu YD, Yang JY et al (2021a) Profiles of Bacillus spp. isolated from the rhizosphere of Suaeda glauca and their potential to promote plant growth and suppress fungal phytopathogens. J Microbiol Biotechnol 31(9):1231–1240. https://doi.org/10.4014/jmb.2105.05010

Lu H, Li R, Yang P, Luo W, Chen S, Bilal M et al (2021b) iTRAQ-BASED proteomic analysis of the mechanism of fructose on improving fengycin biosynthesis in Bacillus amyloliquefaciens. Molecules 26(20):6309

Lu H, Xu H, Yang P, Bilal M, Zhu S, Zhong M et al (2022) Transcriptome analysis of Bacillus amyloliquefaciens reveals fructose addition effects on fengycin synthesis. Genes (Basel) 13(6):984

Miljaković D, Marinković J, Balešević-Tubić S (2020) The significance of Bacillus spp. in disease suppression and growth promotion of field and vegetable crops. Microorganisms 8(7):1037. https://doi.org/10.3390/microorganisms8071037

Nadarajah K, Abdul Rahman NSN (2021) Plant-microbe interaction: aboveground to belowground, from the good to the bad. Int J Mol Sci 22(19):10388. https://doi.org/10.3390/ijms221910388

Navarro-Noya YE, Chávez-Romero Y, Hereira-Pacheco S, de León Lorenzana AS, Govaerts B, Verhulst N, Dendooven L (2022) Bacterial communities in the rhizosphere at different growth stages of maize cultivated in soil under conventional and conservation agricultural practices. Microbiol Spectr 10(2):e0183421. https://doi.org/10.1128/spectrum.01834-21

Netrusov AI, Liyaskina EV, Kurgaeva IV, Liyaskina AU, Yang G, Revin VV (2023) Exopolysaccharides producing bacteria: a review. Microorganisms 11(6):1541. https://doi.org/10.3390/microorganisms11061541

Nogueira LS, de Carvalho Tavares IM, Santana NB, Ferrão SPB, Teixeira JM, Costa FS et al (2022) Thermostable trypsin-like protease by Penicillium roqueforti secreted in cocoa shell fermentation: production optimization, characterization, and application in milk clotting. Biotechnol Appl Biochem 69(5):2069–2080

Pathak P, Rai VK, Can H, Singh SK, Kumar D, Bhardwaj N, Roychowdhury R, de Azevedo LCB, Kaushalendra, Verma H, Kumar A (2022) Plant-endophyte interaction during biotic stress management. Plants (Basel) 11:2203. https://doi.org/10.3390/plants11172203

Peng H, Zhang P, Bilal M, Wang W, Hu H, Zhang X (2018) Enhanced biosynthesis of phenazine-1-carboxamide by engineered Pseudomonas chlororaphis HT66. Microb Cell Fact 17:1–12

Raheem A, Sajid M, Iqbal MS, Aslam H, Bilal M, Rafiq F (2019) Microbial inhabitants of agricultural land have potential to promote plant growth but they are liable to traditional practice of wheat (T. aestivum L) straw burning. Biocatal Agric Biotechnol 18:101060

Ramírez-Pool JA, Calderón-Pérez B, Ruiz-Medrano R, Ortiz-Castro R, Xoconostle-Cazares B (2024) Bacillus strains as effective biocontrol agents against phytopathogenic bacteria and promoters of plant growth. Microb Ecol 87(1):76

Rizzi A, Roy S, Bellenger JP, Beauregard PB (2019) Iron homeostasis in Bacillus subtilis requires siderophore production and biofilm formation. Appl Environ Microbiol 85(3):e02439–e02418. https://doi.org/10.1128/aem.02439-18

Rocha GT, Queiroz PRM, Grynberg P, Togawa RC, de Lima Ferreira ADC, do Nascimento IN et al (2023) Biocontrol potential of bacteria belonging to the Bacillus subtilis group against pests and diseases of agricultural interest through genome exploration. Antonie Van Leeuwenhoek 116(7):599–614. https://doi.org/10.1007/s10482-023-01822-3

Rumyantsev SD, Veselova SV, Burkhanova GF, Alekseev VY, Maksimov IV (2023) Bacillus subtilis 26D triggers induced systemic resistance against Rhopalosiphum padi L. by regulating the expression of genes AGO, DCL and microRNA in bread spring wheat. Microorganisms 11(12):2983

Scott TW, West SA, Dewar AE, Wild G (2023) Is cooperation favored by horizontal gene transfer? Evol Lett 7(3):113–120

Shah SD, Patel H, Saiyad SM, Bajpai B (2024) Effect of a phthalate derivative purified from Bacillus zhangzhouensis SK4 on quorum sensing regulated virulence factors of Pseudomonas aeruginosa. Microb Pathog 191:106664

Shahid I, Han J, Hanooq S, Malik KA, Borchers CH, Mehnaz S (2021) Profiling of metabolites of Bacillus spp. and their application in sustainable plant growth promotion and biocontrol. Front Sustain Food Syst 5:605195

Singh R, Pandey KD, Singh M, Singh SK, Hashem A, Al-Arjani ABF, Abd-Allah EF, Singh PK, Kumar A (2022) Isolation and characterization of endophytes bacterial strains of Momordica charantia L. and their possible approach in stress management. Microorganisms 10:290. https://doi.org/10.3390/microorganisms10020290

Solanki MK, Joshi NC, Singh PK, Singh SK, Santoyo G, de Azevedo LCB, Kumar A (2023) From concept to reality: transforming agriculture through innovative rhizosphere engineering for plant health and productivity. Microbiol Res 279:127553. https://doi.org/10.1016/j.micres.2023.127553

Solanki MK, Wang Z, Kaushik A, Singh VK, Roychowdhury R, Kumar M, Kumar D, Singh J, Singh SK, Dixit B, Kumar A (2024) From orchard to table: significance of fruit microbiota in postharvest diseases management of citrus fruits. Food Control 165:110698. https://doi.org/10.1016/j.foodcont.2024.110698

Soliman SA, Khaleil MM, Metwally RA (2022) Evaluation of the antifungal activity of Bacillus amyloliquefaciens and B. velezensis and characterization of the bioactive secondary metabolites produced against plant pathogenic fungi. Biology (Basel) 11(10):1390. https://doi.org/10.3390/biology11101390

Stoll A, Salvatierra-Martínez R, González M, Araya M (2021) The role of Surfactin production by Bacillus velezensis on colonization, biofilm formation on tomato root and leaf surfaces and subsequent protection (ISR) against Botrytis cinerea. Microorganisms 9(11):2251. https://doi.org/10.3390/microorganisms9112251

Sun X, Xu Z, Xie J, Hesselberg-Thomsen V, Tan T, Zheng D et al (2022) Bacillus velezensis stimulates resident rhizosphere Pseudomonas stutzeri for plant health through metabolic interactions. ISME J 16(3):774–787

Thompson CMA, Hall JPJ, Chandra G, Martins C, Saalbach G, Panturat S et al (2023) Plasmids manipulate bacterial behaviour through translational regulatory crosstalk. PLoS Biol 21(2):e3001988. https://doi.org/10.1371/journal.pbio.3001988

Tsotetsi T, Nephali L, Malebe M, Tugizimana F (2022) Bacillus for plant growth promotion and stress resilience: what have we learned? Plants (Basel) 11(19):2482. https://doi.org/10.3390/plants11192482

Wang N, Li H, Wang B, Ding J, Liu Y, Wei Y et al (2022a) Taxonomic and functional diversity of rhizosphere microbiome recruited from compost synergistically determined by plant species and compost. Front Microbiol 12:798476

Wang J, Qiu J, Yang X, Yang J, Zhao S, Zhou Q, Chen L (2022b) Identification of lipopeptide iturin A produced by Bacillus amyloliquefaciens NCPSJ7 and its antifungal activities against Fusarium oxysporum f. sp. niveum. Foods 11(19):2996. https://doi.org/10.3390/foods11192996

Wang Z, Liu C, Shi Y, Huang M, Song Z, Simal-Gandara J et al (2023a) Classification, application, multifarious activities and production improvement of lipopeptides produced by Bacillus. Crit Rev Food Sci Nutr 64:7451–7464

Wang Z, Solanki MK, Kumar A, Solanki AC, Pang F, Ba ZX, Niu JQ, Ren ZX (2023b) Promoting plant resilience against stress by engineering root microenvironment with Streptomyces inoculants. Microbiol Res 277:127509. https://doi.org/10.1016/j.micres.2023.127509

Yang P, Zhao Z, Fan J, Liang Y, Bernier MC, Gao Y et al (2023) Bacillus proteolyticus OSUB18 triggers induced systemic resistance against bacterial and fungal pathogens in Arabidopsis. Front Plant Sci 14:1078100

Zeng Q, Ding X, Wang J, Han X, Iqbal HM, Bilal M (2022) Insight into soil nitrogen and phosphorus availability and agricultural sustainability by plant growth-promoting rhizobacteria. Environ Sci Pollut Res 29(30):45089–45106

Zhang N, Wang Z, Shao J, Xu Z, Liu Y, Xun W et al (2023) Biocontrol mechanisms of Bacillus: improving the efficiency of green agriculture. Microb Biotechnol 16(12):2250–2263. https://doi.org/10.1111/1751-7915.14348

Zhimo VY, Kumar A, Biasi A, Salim S, Feygenberg O, Toamy MA, Abdelfattaah A, Medina S, Freilich S, Wisniewski M, Droby S (2021) Compositional shifts in the strawberry fruit microbiome in response to near-harvest application of Metschnikowia fructicola, a yeast biocontrol agent. Postharvest Biol Technol 175:111469. https://doi.org/10.1016/j.postharvbio.2021.111469

Zhou L, Song C, Li Z, Kuipers OP (2021) Antimicrobial activity screening of rhizosphere soil bacteria from tomato and genome-based analysis of their antimicrobial biosynthetic potential. BMC Genomics 22(1):29. https://doi.org/10.1186/s12864-020-07346-8

Zhu J, Chen Y, Imre K, Arslan-Acaroz D, Istanbullugil FR, Fang Y et al (2023) Mechanisms of probiotic Bacillus against enteric bacterial infections. One Health Adv 1(1):21

Cyanobacteria as Mediators of Systematic Resistance for Sustainable Disease Control in Agriculture

10

Muhammad Hamza Ashraf, Nazim Hussain, Rimsha Haqqi, Aramish Naz, and Syed Wajahat Ullah

Abstract

Plant diseases pose a serious threat to global food security, and the search for alternative and sustainable approaches for managing plant diseases in agriculture is imperative. Cyanobacteria have been considered as an alternative candidate due to its broad range of metabolic abilities and their mutualistic role for plants as possible biocontrol agents inducing systemic resistance (ISR) against multiple pathogens. This chapter mainly focuses on a comprehensive explanation of the part of cyanobacteria in eliciting systemic resistance in crops and its application in controlling diseases. Bioactive compounds that can either directly inhibit herbivory or trigger responses in other signaling molecules as a method of defense against herbivory are discussed in detail. Further it elaborates various applications of cyanobacteria for the management of diseases (foliar spray, soil amendment, seed treatment) and highlights the advantages and constraints associated with its utilization. By investigating complex molecular mechanisms going on between cyanobacteria and plants and turning into optimization of new strategies to combat diseases, researchers can alleviate the harmful impact of plant diseases on crop yield and address food security at a global scale.

Keywords

Plant diseases · Food security · Cyanobacteria · Signaling pathways · Disease management

M. H. Ashraf · N. Hussain (✉) · R. Haqqi · A. Naz · S. W. Ullah
Centre for Applied Molecular Biology (CAMB), University of the Punjab, Lahore, Pakistan
e-mail: nazim.camb@pu.edu.pk

A. Kumar, M. K. Solanki (eds.), *Microbial Biocontrol Techniques*, Microorganisms for Sustainability 54,
https://doi.org/10.1007/978-981-97-8739-5_10

10.1 Introduction

10.1.1 Plant Diseases and Their Impact on Agriculture: An Overview

Plant diseases pose a significant threat to agriculture on a global scale impacting crop productivity, quality, and economic viability. These diseases are caused by several pathogens like fungi, bacteria, viruses, nematodes, and parasitic plants (Raheem et al. 2019; Singh et al. 2023). All pathogens have their own infection strategies and ways of life which can pose tremendous difficulty to treat and control them. There are multiple kinds of fungal infections, and they can do a lot of harm to plants. Rusts, smuts, mildews, and blights are among the plant maladies for which they are responsible. Fungal spores instinctively have a virulence capacity to spread due to their consummative modes of transitivity through the mediums of air, water, and soil, meaning that they are able to journey long distances (Rabaan et al. 2023).

Bacterial infections also pose a particular problem because once they establish themselves, they are usually quite adept at moving around and hard to eradicate. For example, bacterial wilt caused by *Ralstonia solanacearum* (Wang et al. 2023) and fire blight caused by *Erwinia amylovora* are major diseases affecting a wide range of crops. They are responsible for infections in *solanaceous* plants and fruit trees, respectively. Bacterial pathogens often penetrate plants through wounds or natural openings and proceed with systemic infections that are difficult to eradicate. However, viral diseases are more difficult problem because they are transmitted by aphids and whiteflies, which act like a bridge, and also can be established in seeds and plant material (Wu et al. 2022). The *Tobacco Mosaic Virus (TMV)* and *the Tomato Yellow Leaf Curl Virus*, TYLCV are quite problematic in that they can result in significant stunting, chlorosis, and yield losses. Invasive vector management and development of virus-resistant crop varieties are widespread strategies to tackle viral diseases (Augustine et al. 2022).

10.1.2 Plant Diseases Linkage with Food Security

Besides the immediate crop yield losses, plant diseases have a broader impact within agriculture. Resistant pests and diseases mean the farmer may need additional inputs, e.g., pesticides and fungicides and other monitoring and management services, which will be reflected in production cost. This could also affect the quality of crops, thereby jeopardizing their marketability as well and may further lead to financial losses for farmers. It also points to major outbreaks potentially leading to widespread food chain disruptions, therefore increasing food insecurity and higher market prices both leading to poor health (Ristaino et al. 2021). However, plant diseases may also have serious impacts on the environment as well as the economy. Dependence on chemical control measures increases the likelihood of the development of pesticide resistance, damage to beneficial organisms, and environmental pollution, this because the impacts of diseases need to tackled through sustainable disease management strategies. These practices are integrated pest management (IPM), crop rotation, and resistant varieties (Scortichini 2022).

The issue of plant diseases and their linkage to food shortages can jeopardize agricultural sustainability and global food security. As they reduce yields, hamper crop quality, and restrict access to healthy foods by what they target, plant diseases that emerge due to pathogens such as nematodes, bacteria, viruses, and fungi top the list of the most serious threats to agricultural productivity. Chronic food shortages and the persistent threat of hunger, a concern of epic proportions that continues to be exacerbated and sustained as numerous plant diseases, especially in emerging countries where agriculture is the primary source of livelihood, worsen poverty and socioeconomic disparities (Ristaino et al. 2021). Food shortages, and consequential food insecurity, come in various forms occurring simultaneously or at various global locations but every manifestation entails a decline in yield, nutrient quality, marketability, and accessibility to the world's most vulnerable populations (Vaughan 2020).

Plant diseases spread beyond colonial boundaries, and massive crop failures leading to food crises are more likely to occur because it has been able to cross continents as a result of the globalization of trade and farming. This potential risk factor and the phenomenon of climatic change are modifying environmental variables and contributing to worsening conditions, making it easier for existing diseases to propagate and resurrecting old diseases. There are several systems for addressing the relationship between plant diseases and food shortage. First is integrated disease management, responsible for coordinated action across sectors to halt the spread of plant diseases. Second is improved early detection and surveillance systems that allow for the identification and treatment of diseases at all levels of development (Spence et al. 2020; Ali et al. 2024; Bastas and Kumar 2023). The third component is plant-centric breeding, which considers crop traits that have been linked to successful plagues. Finally, building farmers' capacity and the existing information to respond to diseases and more broadly control risk. Resilient food systems that may counteract the threat of plant plagues and ensure food security for current and subsequent generations require investment in study and innovation, encouragement of sustainable farming manners, initiative framework investment, and mobilization of international collaboration and collaboration. Furthermore, understanding the complicated relationship between plant health, agricultural output, and food security will ensure that all citizens have fair access to safe foods (Vilvert et al. 2022).

10.1.3 Conventional Approaches to Disease Control

In traditional plant disease control practices, such as those handed down through the generations, the main objective is to keep the influence of pathogens away from crop production is minimized. These practices include a wide variety of methods, among them cultural measures and pesticide application as well as physical barriers.

Plant diseases are eliminated or reduced in intensity merely by changing the growing environment. Crop rotation is a widely used practice and one that serves to prevent the accumulation of soil-borne pathogens. Crop rotation became a valuable technique for preventing crops from pathogens. These not only serve to help reduce disease spread through harmful parasites but also play an important role in preventing the seeds from cultivated land from contamination. Reasonable methods of irrigation must be chosen, but not exceeded, since that will increase the chances for disease (Café-Filho et al. 2019).

In controlling plant diseases, chemical fungicides, bactericides, and nematicides have long been the most important means of prevention. These substances are spread onto plants to prevent, kill, or retard the growth these pathogens. Systemic fungicides are absorbed and distributed around the plant to protect it from the inside out. But some fungicides just remain outside the plant and constitute a barrier. However, the general application of chemical pesticides has given rise to problems such as resistance to pathogens burgeoning, environmental pollution, and deleterious effects on non-target organisms in addition to those which they target including beneficial insects and soil microorganisms (Fenta and Mekonnen 2024).

10.1.4 Biological Control

The biological control method is a more sustainable, less polluting means of managing diseases. Organisms are used for combating plant pathogens like those which cause diseases in carriageways, thus eliminating or lessening the need to use chemical pesticides (Heydari et al. 2021; Vu et al. 2022). In this way, it is possible when utilizing this method to mix natural enemies of pathogens-predaceous insects, parasitic fungi, and beneficial bacteria against each other for the purpose of ultimately maintaining pathogen populations at levels which are tolerable or even low enough so that they never become serious problems indeed. Biological control is important because it offers sustainable disease management solutions while at the same time, it also maintains the ecological balance by desirable means (He et al. 2021). Some biological control agents (BCAs) can be specific. Further, BCAs are epibiotic in nature and are also phenotypically plastic, i.e., with their release, they can be used to respond to evolving disease environments and varying pathogen populations resulting in a built-in temporal feedback for the rise of disease as and when it occurs. Biological control is part of an IPM approach where a range of strategies is employed to manage pests and diseases (Guzmán-Guzmán et al. 2023; Kumari et al. 2022; Solanki et al. 2024). Additionally, the integration of BCAs into IPM programs could decrease farmers' dependency on chemical pesticides and reduce production costs and the environmental footprint of agriculture (Tariq et al. 2020).

10.2 Advantages and Drawbacks of Biological Control Techniques

10.2.1 Advantages

Biological control methods can be less dangerous to human health as they reduce the risk of pollution and are unlikely to result in the selection of resistant strains of pathogens. BCAs have a high degree of selection power to specifically target the pathogen and hence act with minimal harm to non-targets including pollinators and indigenous suppression organisms (Becker et al. 2024). In addition, this strategy can be used to develop sustainable biology-based methods for disease control by enhancing agro-ecosystem resilience through biological control. They may be established and will reproduce, spreading naturally to form extended protection with no requirement for reapplication every few weeks (He et al. 2021). Usage of lesser chemical inputs can subsequently lead to cheaper production cost, helping to make farmers lives slightly more economically sound. Also, when biological control measures are employed, the crops are healthier and the yields are more substantial (Timsina 2018).

10.2.2 Drawbacks

The effectiveness of BCAs may depend on certain environmental parameters, such as temperature, humidity, and soil pH as well as chemical controls. The effectiveness of BCAs may depend on certain environmental parameters, such as temperature, humidity, and soil pH as well as chemical controls. While biological control mechanisms are far less destructive to wildlife, they are generally slower to act than chemical pesticides. And during fast-spreading epidemics, this delay can be substantive (Daraban et al. 2023). The successful use of biological control requires to study relatively complex interaction between the biological control agents (BCAs), pathogens, and the hosts exist in the environment at the same time. The complexity of this complicates widespread adoption (Bielza et al. 2020). This might be in part because controlling the production, formulation, and regulation of BCAs could be more complex, and even more strict when compared to chemical pesticides. There is a large amount of research and development needed to assure the quality and consistency of BCAs (Boro et al. 2022).

10.3 Cyanobacteria

The photosynthesis abilities of cyanobacteria allow them to generate oxygen from sunlight have been explored in a broad range of aquatic and terrestrial environments including some where they occur together with some green algae. These prokaryotes are mainly known for their unique ability to perform oxygenic photosynthesis, with a significant impact on the global carbon and nitrogen cycles. The ancestors of cyanobacteria have existed for more than 2.5 billion years as they played a key role in oxygenating earth's then-lethal atmosphere (Aiyer 2022). Halophytes like Cyanobacteria are morphologically diverse; they are solitary celled or chain-like filamentous than colonies. They characterize a diverse group of organisms, existing in chauvinistic habitats from freshwater and marine ecosystems to extreme environments like hot springs and deserts. These capacities of physiology and metabolism that allow these organisms to survive in low nutrient environments is due to their adaptability in nitrogen fixation (Fujita and Uesaka 2022).

In recent years, there has been a renewed interest in exploring the applications in biopharmaceutical, bioenergy, biofuels, biochemicals (Khan et al. 2019; Saeed et al. 2022), and agriculture and has been found suitable for agro-technological point of views, mostly for plant growth, crop productivity improvement, and disease control. Animal diversities have a large collection of bioactive compounds like polysaccharides, proteins, vitamins, and secondary metabolites. They are known to promote plant growth and resistance to some pathogens (Devi et al. 2023; Kumar et al. 2018). This concept of ISR of cyanobacteria has been now been universally well established, and an innovative way for enhancing soil fertility and thereby pest control in agriculture using live cyanobacteria is emerging (Abo-Shady et al. 2023; Elagamey et al. 2023).

10.4 Induced Systemic Resistance (ISR)

Induced systemic resistance (ISR) is the state by which plants exhibit resistance to infection by a broad spectrum of pathogens to which they have been previously exposed. Although intrinsic immunity is, well, intrinsic, the ISR is an adopted response that only fires after a priming event. This typically includes utilizing non-pathogenic bacteria or other microorganisms that interact with the plant to initiate defense pathways. Since the early reports of the ISR phenomenon in the twentieth century, it is known that a single (or multiple) earlier exposure of plants to non-lethal stresses may confer the ability of plants to train resistance against a number of stresses (Kamle et al. 2020; Olowe et al. 2020). Research during the 1970s and 1980s also established that the presence profile of *rhizobacteria* around plant roots could elicit systemic resistance (Zeng et al. 2022). Such kind of resistance was represented by Systemic Acquired Resistance (SAR) which was induced by pathogenic infections and related to localized cell death. Research using *Pseudomonas* and *Bacillus species* is among the evidence that demonstrated the existence of a mechanism behind until most recently, the concept of ISR was not acknowledged very widely. These studies showed that these bacteria could confer resistance to plants without being pathogenic. The early studies laid the ground work for the understanding of these molecular and biochemical processes involved in ISR. These are complex signaling network and hormonal control processes (Messa 2021).

10.4.1 Mechanisms of Induced Systemic Resistance (ISR)

Induced Systemic Resistance (ISR) mechanisms involves a series of subsequent steps to unravel the manner in which plants perceive, trigger, and boost immune defense responses when colonized by beneficial microorganisms or challenged with abiotic elicitors of selected chemical origin as shown in Fig. 10.1.

The whole process can be divided into the following steps:

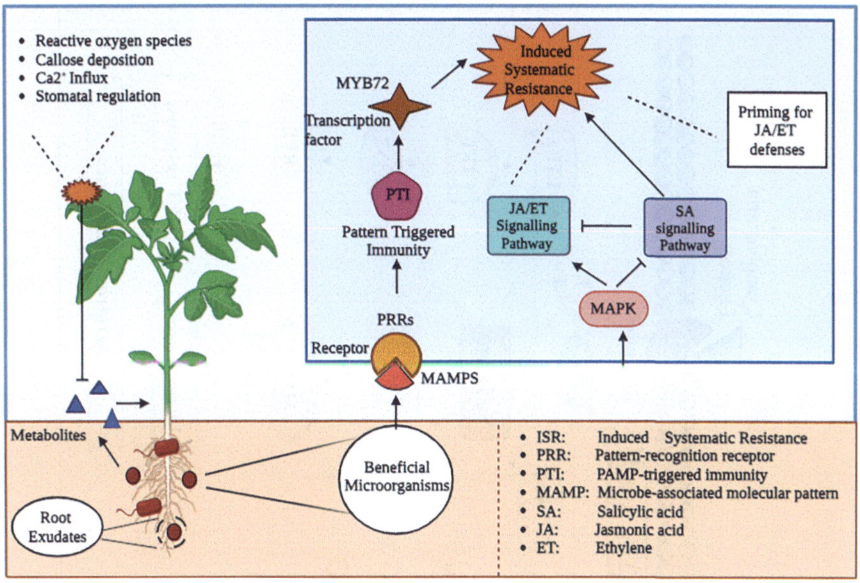

Fig. 10.1 Mechanisms of Induced Systemic Resistance (ISR): Induced systemic resistance (ISR) by beneficial microorganisms. JA and ET are central regulators phytohormones of ISR, and transcription factors (e.g., MYC2) mediate the increased responsiveness of this pathway to stimulation, known as priming. Transcription factor MYB72, as a root-specific transcription factor and early signaling factor, functions as a node of convergence in ISR elicited by beneficial microbes. *ET* ethylene, *JA* jasmonic acid, *NPR1*, NONEXPRESSOR OF PR GENES1, *MAMPs* microbe-associated molecular patterns, *PRRs* plant recognition receptors, *PTI* PAMP-triggered immunity, *TFs* transcription factors

10.4.2 Identifying Beneficial Microorganisms

1. **Microbe-Associated Molecular Patterns (MAMPs)**: The process is initiated by recognizing MAMPs (Microbe-Associated Molecular Patterns) that are exclusive to beneficial bacteria, fungi, and very importantly cyanobacteria. Plant receptors called Pattern Recognition Receptors (PRRs) detect different MAMPs (such as proteins, lipopolysaccharides, and other microbial exudates), it can also be a best strategy of work (Nishad et al. 2020).
2. **Root Colonization**: Cyanobacteria, among other beneficial microorganisms, often colonize on the root surface or in the rhizosphere to form a mutualistic relationship with plants. The idea of colonization drives a continuously replenishing reservoir of MAMPS to feed the defense arsenal of the plant immune system (Álvarez et al. 2023). The whole process can be divided into following steps:

10.4.3 Signal Transduction Pathways (STPs)

STPs are central for plant defense, and whole molecular mechanism is summarized in Fig. 10.2.

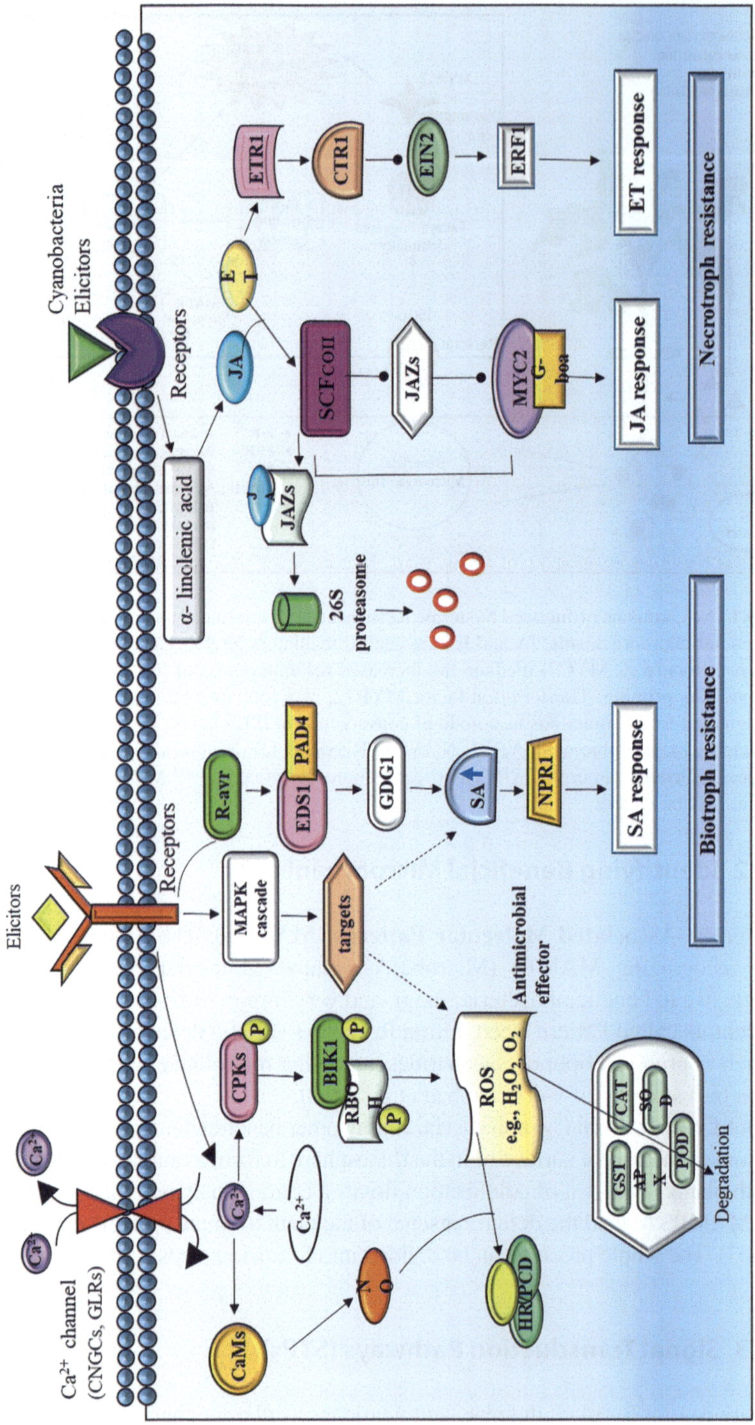

Cyanobacteria
Elicitors
Receptors
Elicitors
Receptors
Ca2+ channel
(CNGCs, GLRs)
Ca2+
CaMs
CPKs
P
BIK1
RBOH
MAPK cascade
targets
R-avr
EDS1
PAD4
GDG1
SA
NPR1
SA response
Biotroph resistance
α- linolenic acid
JA
ET
ETR1
CTR1
EIN2
ERF1
ET response
SCFCOI1
JAZs
MYC2
G-box
JA response
Necrotroph resistance
26S proteasome
ROS
e.g., H2O2, O2
Antimicrobial effector
NO
HR/PCD
GST
APX
CAT
SOD
POD
Degradation

Fig. 10.2 Schematic representation of the plant defense signal transduction network: Plants establish their defenses in response to pathogens by activating a cascade of signals. In plants, signal transduction networks are formed by the interplay of several, intricate signal channels. An MAPK cascade reaction, the activation of R genes, and the release of Ca^{2+} are all results of the signal transduction system being activated after the resistant host identifies pathogen-produced elicitors. Initiation of a primary immunological response, which includes HR, can be triggered by Ca^{2+} moving into the cytoplasm, which can activate CaMs and CMLs and induce downstream NO production. A synergistic action between NO and ROS may also regulate an HR/PCD. In addition to signaling SAR activation, reactive oxygen species (Café-Filho et al. 2019) have direct antibacterial effector functions and can boost host structural resistance. A robust resistance to some biotrophic infections can be achieved through the induction of a cascade of defense responses by SA, including the HR/PCD and SAR, through the interaction of the R gene with the avirulent gene (Avr)xx. Pathogen-induced SA biosynthesis requires GDG1, and EDS1 and PAD4 control its expression. Furthermore, SA has the ability to regulate GDG1, EDS1, and PAD4 expression via a positive feedback loop. High levels of SA accumulation are responsible for GDG1's positive feedback control. Plant resistance to necrotic diseases and wounds is primarily regulated by JA/ET signaling pathways. The resistance that root microbes cause, known as induced systemic resistance (ISR), is likewise mediated by JA and ET. One positive regulatory gene of the SA system, NPRl, is activated when SA blocks the JA/ET pathway. In wound reactions and defense against pathogen infections, ERF1 is situated at the intersection of the JA and ET signaling pathways. The JA signal can trigger the SCFCOI1 ubiquitin ligase to bind with JAZ, leading to the ubiquitination and destruction of the JAZ protein by the 26S proteasome. This, in turn, activates transcription factors such as MYC2, which initiate JA responses. In order to combat pathogen infection, a cascade of downstream responses is set in motion, which includes strengthening physical defensive structures, producing secondary metabolites, inhibiting growth pathogens through the induction of defensive proteins, activating the ROS scavenging system, and other disease resistance factors. Circuits that are gated by cyclic nucleotides; calcium-dependent protein kinases (CPKs), also known as calmodulins (CaMs), GLR stands for glutamate receptor-like genes. The acronyms NO, POD, SOD, CAT, APX, GST, and ROS stand for nitric oxide, peroxidase, superoxide dismutase, catalase, ascorbate peroxidase, glutathione S-transferase, and reactive oxygen species, respectively. The acronyms RBOHD, H_2O_2, O_2^-, OH, BOTRYTIS-INDUCED KINASE1, and MAPK stand for respiratory burst oxidase homolog D, hydrogen peroxide, hydroxyl radical, and mitogen-activated protein kinase, respectively. The R-Avr relationship encompasses the following: the pathogen's virulence (Avr) gene and the host's resistance (R) gene; EDS stands for enhanced disease susceptibility, PAD for phytoalexin-deficient 4, NPR for non-expresser of PR genes 1, GDG for GH3-like defense gene 1, EIN for ethylene-insensitive 2, ISR for induced systemic resistance, SCF for Skp/Cullin/F-box, COI for coronatine-insensitive 1, LOX for lipoxygenase 2, VSP for vegetative storage protein 2, JAZ for jasmonate ZIM-domain, CTR for copper transport protein 1, ETR for ethylene receptor gene 1, ERF for ethylene response factor, PR for pathogenesis-related protein, PGIP for polygalacturonase-inhibitory protein, and OXO for oxalate oxidase. Positive regulation is indicated by the arrows, and negative regulation is represented by the open blocks. Interactions that are indirect or perhaps probable are indicated by dashed lines

- **Activation of Signaling Molecules**: When MAMPs are perceived, multiple plant defense-related signaling pathways are activated. Jasmonic Acid (JA) and Ethylene (ET) are key signaling molecules in this path; however, other signal metabolites such as Salicylic Acid (SA) and Nitric Oxide (NO) are also believed to participate (Pérez-Llorca et al. 2023).
- **The pathway of Jasmonic Acid (JA)**: The jasmonic acid (JA) pathway in cyanobacteria is an interesting illustration of cross-talk between plant signaling chemicals and microbial metabolism. JA has been connected to higher plants for a long time, but now it is emerging within some cyanobacterial species, as well. The discovery suggests that a common set of biochemical reactions was present before plants and photosynthetic bacteria evolved along their independent paths (Ghorbel et al. 2021; Gomi 2020; Liu and Timko 2021). Here we discuss an in-depth study of the JA pathway in cyanobacteria to shed light on its evolutionary history, as well as possible ecological functions of this signaling molecule. α-linolenic acid (ALA) an abundant precursor of membrane lipids, important for JA biosynthesis, plays α-linolenic acid (ALA) in the JA biosynthesis pathway in cyanobacteria. Lipoxygenase (LOX) enzymes mediate the oxygenation of ALA to 13-hydroxylinolenic acid (13-HPOT), a critical step in the biosynthesis of JA. These enzymes function to metabolize ALA. OPDA is converted enzymatically via successive enzymatic steps, including those catalyzed by allene oxide synthase (AOS) and allene oxide cyclase (AOC), to JA. Cyanobacteria house homologs of key JA biosynthetic enzymes such as LOX, AOS, and AOC, indicating that JA biosynthesis pathway is widely conserved across organisms (Gomi 2021; Wang et al. 2021). Moreover, a previous study discovered that cyanobacterial genomes harbor genes encoding JA biosynthetic enzymes, a molecular evidence that cyanobacteria are capable of synthesizing JA too (Griffiths 2020).

Further investigation of the environmental cues and physiological conditions that modulate cyanobacterial development of JA and its concomitant activity is required, as to date few details are available of how JA production is regulated and the role of JA production. Together with the role of a plant signal molecule, JA and its analogues seem to exhibit a variety of physiological effects in cyanobacteria such as cellular processes like growth, defense, multi-species interactions, etc. (Griffiths 2020). There are a number of potential commonalities in the roles of JA in cyanobacteria and higher plants; for instance, JA has been associated with photosynthesis, nitrogen metabolism, and as defenses against oxidative stress, some features of which can be stimulated by JA alone in plants. For the interaction with microbial consortia, cyanobacteria exert the functions to shape the dynamics of microbial communities and ecosystem functioning through the cross-talk between cyanobacteria and other microbes, which may be mediated by JA as a signaling molecule.

The production of JA by cyanobacteria poses some interesting questions regarding the ecological and adaptive role of this compound in nature. Understanding JA ecological functions in cyanobacterial physiology and interspecies interactions would potentially inform ecosystem dynamics and biogeochemical cycling. That is because cyanobacteria exist in diverse ecosystems including freshwater lakes and oceans, as well as in terrestrial ecosystems (Mutalipassi et al. 2021). The new insights into jasmonic acid (JA) biosynthesis and signaling in cyanobacteria could lead to biotechnological advancements in agriculture, bioremediation, and pharmaceuticals, highlighting the importance of further exploring the regulatory mechanisms involved. Among other such mechanisms were stress tolerance, or the production of bioactive compounds. Deciphering the complex JA biosynthesis, metabolism, and signaling pathways of cyanobacteria should provide insight into how the physiology of diverse microbes as well as interspecies interactions within the context of natural environments are governed at the molecular level. Additionally, more long-term biotechnological solutions are potentially feasible by harnessing metabolic potential of cyanobacteria to make JA, and could be applied in a range of fields such as human health and environmental management.

Defense response collaboration between ET and JA upon the perception of PAMPs, ET biosynthesis is induced, and the ET signal is sensed by receptors that include ETR1. The trigger to initiate such processes is the perception of this phenomenon, which then leads to a cascade of signaling networks that include EIN2 as well as transcriptional regulators like EIN3/EIL1. This in the end results in defense gene activation (Binder 2020; Zhao et al. 2021).

Plants use calcium ions (Ca^{2+}) as signals that regulate many functions, including the immune response against invasion by a pathogen. Cyanobacterial elicitors are also able to trigger rapid Ca^{2+} ion influxes in plant cells, which activate downstream plant defense signaling pathways. An increase in calcium levels in the cell nucleus triggers a specific cascade of events (Negi et al. 2023). Food conditioned responses include processes related to food-gene expression, ion flux, and reactive oxygen species (Café-Filho et al. 2019) production, which are proteins phosphorylated by CDPKs and calmodulin-dependent protein kinases. The Ca^{2+} signal path is integrated with the multiple layers of immunological control in plants and interacts with the other defense signaling networks SA and JA/ET signaling. Plants use LNQRs to "prime" responses downstream rather than outright triggering them, thereby allowing the activation of Ca^{2+} signaling by cyanobacteria to be part of pre-defense preparedness against pathogen and environmental challenges as previously suggested (Edel et al. 2017).

10.4.4 Defense Responses

1. ISR often involves elevating the cell walls of the plant to create a physical defense, by enhancing the cell wall to prevent pathogens. This involves deposition of callose and lignin in the cell walls which strengthen the cell walls and hinder the process of pathogen penetration.
2. Upon ISR induction by *rhizobacteria*, diverse antimicrobial compounds like phytoalexins, defensins, and other pathogenesis-related (PR) proteins are synthesized. This mix contains some chemicals that exert a direct antibiotic effect, thereby restricting pathogens.

3. ISR favors the appearance of many genes that activate different defense mechanisms. These transcription factors bind to specific promoter regions of defense genes and are activated by JA/ET signaling pathways that operon this gene. These genes release enzymes important for producing secondary metabolism proteins intended to strengthen in cell walls and peptides endowed with antimicrobial properties.

10.4.5 Cyanobacteria-Specific Mechanisms

Several classes of secondary metabolites have been identified as being produced by cyanobacteria including cyclic peptides, alkaloids, and polyketides. These may act as or be converted to MAMPs (microbe-associated molecular patterns) or signaling molecules. These items will hit receptors on plants assisting in signaling and the start of ISR (Kumar et al. 2019). Generally, cyanobacteria produce polysaccharides (glycogen, starch, and mucilage) and exopolysaccharides to assist with root colonization and function as a signaling molecule. These molecules actually bind to receptors in plant cells that trigger ISR.

Cyanobacteria fix the atmospheric nitrogen and solubilize phosphate by mobilizing ore form of reserves making the nutrients available to the plant. Greater nutrient levels supply stronger plant health which results in robust plant defense mechanisms. The gene expression of the same level for the plant hormones responsible for stress response can be affected by cyanobacteria. By modulating different hormones, cyanobacteria were able to precisely tune plants' immune responses. These alterations are made through the action of plant hormones called cytokinins, which have well-known roles in stress response. By mediating a number of hormones, a fine flywheel of plant immune responses is properly set by cyanobacteria (Abdelaziz et al. 2022).

10.5 Cyanobacteria-Mediated ISR in Plants

The successful utilization of cyanobacteria for ISR begins with the rigorous evaluation and selection of strain isolation for bioactivity identification. To this end, numerous methodologies have been developed, ranging from classical culture-based methods to state-of-the-art molecular methods. Isolation of cyanobacterial strains from a variety of natural habitats, i.e., soil, waterbodies, and plant surface are the conventional methods. Strains are tested using extensive screening assays to evaluate their potential for triggering ISR in model or economically important host plants. Moreover, advances in molecular biology have helped to discover cyanobacteria-specific ISR elicitors through prescribed ways. Researchers performed culture-independent metagenomic analyses of bacterial communities associated with plants and identified particularly interesting novel cyanobacterial candidates for ISR. Furthermore, the functional genomics tools enable us to identify the key genes and pathways associated with cyanobacteria-mediated ISR. Then this information can be utilized to select strains that work better.

10.5.1 Cyanobacteria Capable of Inducing ISR

Various serotypes of cyanobacteria have been thoroughly explored for their capacity to elicit ISR in plants and therefore can be a repertoire of prospective candidates for agricultural application. Among these strains that has acquired high attentiveness are such involved in plant growth promoting display combined with their ISR prompting in different plant species and environmental situations.

1. *Nostoc* spp. have proven extremely potent inducers of ISR in a range of crop plants, due to their remarkable nitrogen-fixing abilities. Studies have shown that the use of the *nostoc* strains may effectively up-regulate plant defense responses against a broad spectrum of pathogens including bacteria, fungi, and viruses. Finally, the mechanisms of specific plant–symbiont interactions involving cyanobacteria provide additional benefits, demonstrating their multifaceted roles in plant–microbe interactions (Belton et al. 2020).
2. These cyanobacteria, *Anabaena* and *Gloeocapsa,* demonstrated remarkable ISR-inducing activities. It is suggested that this is influenced by the bioactive metabolites produced and the plant defense pathways affected by these fungi (Kamle et al. 2020; Messa 2021).

10.6 Mechanisms of Action

10.6.1 Production of Bioactive Compounds

The bioactive compounds are important in causing the plant's own defense mechanism to be turned on thereby more effectively resisting pathogen attack. For instance, bioactive compounds of several classes synthesized by cyanobacteria have stimulated and are active as ISR promoters. For instance, peptides derived from cyanobacteria, including microcystins and nostopeptins, have been demonstrated to modulate plant defense signaling pathways. These peptides can act as elicitors triggering the activation of defense-related genes and synthesis of antimicrobial compounds in plants. Cyanobacterial LPS are also proved to initiate SAR in plants secondarily, by inducing the synthesis of salicylic acid (SA) which acts as key signaling molecule in defense strategies. Furthermore, cyanobacteria can synthesize various secondary metabolites including cyanotoxins, alkaloids, and polyketides. These compounds possess an effect on pathogens on varying scales of bioactivity. These metabolites possess direct anti-pathogenic growth (examples include camalexin and scopolin) or indirect plant physiological effects that promote resistance (e.g., OPDA or 4R-ticopurine) (Messa 2021). For example, biofilms are strengthened by cyanobacterial exopolysaccharides (EPS) that are established as construction elements that prevent pathogens from entering plants. The volatile organic compounds (VOCs) emitted by cyanobacteria can prime neighboring plants to elicit better defense-related processes (Morcillo and Manzanera 2021).

10.6.2 Improvement of Plant Immune Responses

Cyanobacteria produce a large number of secondary metabolites, plant defensive hormones, growth regulators, and defective enzymes. Via elaborate molecular communication, cyanobacteria control the expression of defense genes in plants, activate their signaling pathways, and promote the accumulation of defense metabolites. Cyanobacteria activate plant PTI and ETI pathways leading to immunity and play as an immunogenic biomolecule. The cell wall components as well as secreted metabolites of cyanobacteria are recognized by plant cells via Pattern Recognition Receptors (PRRs). This leads to a cascade of signaling events that induces defense responses. These processes include the rapid elevation of calcium ions, the activation of mitogen-activated protein kinases (MAPKs) by phosphorylation, and alterations in the expression of defense-associated genes (Yu et al. 2024). In addition, the induction of ISR by cyanobacteria is associated with the possible activation of systemic defense responses. This primes plants to defend themselves more quickly against future pathogen invasions as well, in a more efficient manner. This response includes pile up of defense hormones, i.e., Salicylic acid (SA), Jasmonic acid (JA), and Ethylene (ET) against infection effect. At the transcriptional level, these hormones regulate the expression of signaling and defense genes and activate the synthesis of antimicrobial compounds. Additionally, they stimulate the production of ROS and phytoalexins responsible for the resistance of plants against invading pathogens (Liu and Timko 2021).

10.7 Application of Cyanobacteria in Disease Management

Cyanobacteria are widely known as biocontrol agents to suppress plant diseases in agro-ecosystems because of their various metabolic pathways and symbiotic relationships with plants. It can be used successfully, for example, using foliar sprays, incorporate them in the soil and treat seeds with them. Each method had its advantages and constraints, and depending on the crop being grown, environmental conditions and disease management objectives one or more would be suitable for most growers. In the following, we revisit these methods and their implications for disease management in agriculture and crop health improvement.

1. **Foliar Sprays**: Foliar application is directly spraying of cyanobacterial suspensions on leaves of the plants. This method has many advantages due to its ease of use, plant tissues suffering a high absorption rate for metabolites from cyanobacteria and also for delivering metabolites to the infection site directly. This can apply by prophylactic foliar sprays to protect the plants from potential patho-

genic attacks or a curative foliar in case if the infection has already occurred during the cropping period. It is the easy way of delivering both nutrients and bioactive compounds directly and faster to the inside plants which in turn boost up the plant health (Chabili et al. 2024).

2. **Soil Amendments**: Soil amendment with cyanobacteria is another strategy for managing diseases effectively. Cyanobacteria can be directly applied to the soil either in the form of liquid suspension or dried formulations (inoculants). When applied, the cyanobacteria form symbiotic partnerships with the plant roots; they colonize the rhizosphere and rhizoplane. This colonization process increases nutrient uptake, promotes root growth, and stimulates a plant defense system against certain damaging soil pathogens. In addition, the soil fertility and structure of soil are to a great extent enhanced by various cyanobacteria. This he does through a nitrogen-fixing process and the production of extracellular polymeric

Table 10.1 Advantages and limitations of various techniques of cyanobacteria-based disease management techniques

Technique	Advantages	Limitations	References
Foliar sprays	Rapid and accurate spray-application of cyanobacterial metabolites to plant foliage	Vulnerability to natural elements such as rain, wind, and temperature changes	Chabili et al. (2024)
	Can apply at any stage of plant growth	Under severe disease pressure, need to reapply the product more often to keep it working	
	Suitable for existing spraying techniques, compatible with Agri-practices	At higher concentrations or with sensitive plants, plant damage may occur with this technique	
Soil amendments	Maintaining durable mutualistic interactions between cyanobacteria and plant roots	The treatment may work better under different soil quality, microorganism interactions, or management practices	Yadav et al. (2023)
	Improving the fertility, structure, and water holding capacity of soil	This dissemination of cyanobacterial inoculants in to the soil profile is a slow and relaxed process	
	High resistance that confers upon the plant, a large texture of protection	Dried formulations are short lived and easily gas degraded	
Seed treatments	Quickly establish beneficial rhizosphere microbes	Gear requirements to aid seed coat and inoculant placement	Elnahal et al. (2022)
	Improved seedling vigor, uniformity, and tolerance to harsher growing conditions	It works only against the foliar pathogens and pests that are sensitive to systemic resistance	
	Reduced the risk of seedling diseases and damping-off, particularly in the high-density planting	High concentrations of inoculant can lead to seed phytotoxicity or diminished seed viability	

substances (EPS), that are beneficial invigorate the general well-being of the soil (Yadav et al. 2023).

3. **Seed Treatments**: The use of inoculants to promote plant growth, before planting of the seeds, is a prophylactic alternative in the management of diseases. This continues to be the classic example of seed treatment practice, where cyanobacterial suspension or formulations is basically applied to the outer layer of seeds, thus being able to buoy the formation of desirable microbial populations since the onset of the plant life cycle. As cyanobacteria establish, they basically seed the building root system, initiate the plants immune responses, and protect the plants from pathogens in the soil. Services such as seed treatments help increase seed germination rates and facilitate more consistent seedling emergence, reducing dampening-off and seedling diseases (Elnahal et al. 2022).

10.7.1 Advantages and Limitations of Various Techniques

The advantages and challenges of each method should be weighed up against each other in order to realistically apply cyanobacteria for disease management as described in Table 10.1. By using these behaviors in farming, farmers can harness the power of cyanobacteria to improve the health, yield, and sustainability of their crops (Pathak et al. 2018). Furthermore, it is important to focus on finding new biocontrol strategies with cyanobacteria for agro-ecosystems that should be based on continuous research to improve the efficacy, reliability, and utility of this ecologically friendly tactic in different agricultural settings and environmental conditions (Poveda 2021).

10.8 Conclusion

This chapter discusses the burgeoning field of application of cyanobacteria-derived biocontrol agents for use in inducing systemic resistance (ISR) for the control of plant disease in agriculture. The chapter first summarizes the immune responses (ISR) induced by cyanobacteria mediated by the production and detection of bioactive molecules and signaling molecules (e.g., salicylic acid, jasmonic acid, ethylene, and calcium ions). Cyanobacteria, which play a large number of metabolic roles, induced complex molecular interactions in plants to prepare them to defend against infections. However, the advantages and disadvantages of the two approaches determine which one is most suitable for which crops and climate. Soil additives enhance nutrient absorption and resistance in plants, while cyanobacteria efficiently deliver bioactive compounds to plant shoots through foliar spray. In contrast, seed treatments lead to the development of an early microbiome (a beneficial microbial community) within the soil, both helping the vitality of the seedling and reducing soil-borne diseases. The chapter also explores the possibility of using cyanobacteria-mediated ISR for enhancement of agricultural ecosystem sustainability as well as for improving crop health and production along with the limitations of bio-based control approaches.

References

Abdelaziz AM, Attia MS, Salem MS, Refaay DA, Alhoqail WA, Senousy HH (2022) Cyanobacteria-mediated immune responses in pepper plants against *Fusarium* Wilt. Plants (Basel) 11(15):2049

Abo-Shady AM, Osman ME-AH, Gaafar RM, Ismail GA, El-Nagar MM (2023) Cyanobacteria as a valuable natural resource for improved agriculture, environment, and plant protection. Water Air Soil Pollut 234(5):313

Aiyor K (2022) The great oxidation event: how cyanobacteria changed life. American Society for Microbiology

Ali A, Ölmez F, Zeshan MA, Mubeen M, Iftikhar Y, Sajid A, Abid M, Kumar A, Divvela PK, Solanki MK (2024) Yeast-based solutions in controlling plant pathogens. Biocatal Agric Biotechnol 58:103199. https://doi.org/10.1016/j.bcab.2024.103199

Álvarez C, Jiménez-Ríos L, Iniesta-Pallarés M, Jurado-Flores A, Molina-Heredia FP, Ng CK, Mariscal V (2023) Symbiosis between cyanobacteria and plants: from molecular studies to agronomic applications. J Exp Bot 74(19):6145–6157

Augustine SM, Tzigos S, Snowdon R (2022) Heat-killed tobacco mosaic virus mitigates plant abiotic stress symptoms. Microorganisms 11(1):87. https://doi.org/10.3390/microorganisms11010087

Bastas KK, Kumar A (2023) Omics technologies in the plant–microbe interactions. In: Bastas KK, Kumar A, Sivakumar U (eds) Microbial biocontrol: molecular perspective in plant disease management. Springer Nature, Singapore, pp 257–282. https://doi.org/10.1007/978-981-99-3947-3_13

Becker JS, Ruegger PM, Borneman J, Becker JO (2024) Indigenous populations of a biological control agent in agricultural field soils predicted suppression of a plant pathogen. Phytopathology 114(2):334–339. https://doi.org/10.1094/PHYTO-07-23-0221-R

Belton SP, McCabe PF, Ng CK (2020) The nitrogen-fixing symbiotic cyanobacterium, Nostoc punctiforme can regulate plant programmed cell death. bioRxiv 2020.2008. 2013.249318

Bielza P, Balanza V, Cifuentes D, Mendoza JE (2020) Challenges facing arthropod biological control: identifying traits for genetic improvement of predators in protected crops. Pest Manag Sci 76(11):3517–3526. https://doi.org/10.1002/ps.5857

Binder BM (2020) Ethylene signaling in plants. J Biol Chem 295(22):7710–7725

Boro M, Sannyasi S, Chettri D, Verma AK (2022) Microorganisms in biological control strategies to manage microbial plant pathogens: a review. Arch Microbiol 204(11):666

Café-Filho AC, Lopes CA, Rossato M (2019) Management of plant disease epidemics with irrigation practices. In: Irrigation in agroecosystems, vol 123

Chabili A, Minaoui F, Hakkoum Z, Douma M, Meddich A, Loudiki M (2024) A comprehensive review of microalgae and cyanobacteria-based biostimulants for agriculture uses. Plants (Basel) 13(2):159. https://doi.org/10.3390/plants13020159

Daraban GM, Hlihor RM, Suteu D (2023) Pesticides vs. biopesticides: from pest management to toxicity and impacts on the environment and human health. Toxics 11(12):983. https://doi.org/10.3390/toxics11120983

Devi S, Sharma S, Tiwari A, Bhatt AK, Singh NK, Singh M, Kaushalendra, Kumar A (2023) Screening for multifarious plant growth promoting and biocontrol attributes in Bacillus strains isolated from Indo-Gangetic soil for enhancing growth of rice crops. Microorganisms 11:1085. https://doi.org/10.3390/microorganisms11041085

Edel KH, Marchadier E, Brownlee C, Kudla J, Hetherington AM (2017) The evolution of calcium-based signalling in plants. Curr Biol 27(13):R667–R679

Elagamey E, Abdellatef MA, Flefel HE (2023) Perspective chapter: cyanobacteria—A futuristic effective tool in sustainable agriculture. In: Cyanobacteria-recent advances and new perspectives. IntechOpen

Elnahal AS, El-Saadony MT, Saad AM, Desoky E-SM, El-Tahan AM, Rady MM et al (2022) The use of microbial inoculants for biological control, plant growth promotion, and sustainable agriculture: a review. Eur J Plant Pathol 162(4):759–792
Fenta L, Mekonnen H (2024) Microbial biofungicides as a substitute for chemical fungicides in the control of phytopathogens: current perspectives and research directions. Scientifica (Cairo) 2024:5322696. https://doi.org/10.1155/2024/5322696
Fujita Y, Uesaka K (2022) Nitrogen fixation in cyanobacteria. In: Cyanobacterial physiology, pp 29–45
Ghorbel M, Brini F, Sharma A, Landi M (2021) Role of jasmonic acid in plants: the molecular point of view. Plant Cell Rep 40:1471–1494
Gomi K (2020) Jasmonic acid: an essential plant hormone, vol 21. MDPI, pp 1261
Gomi K (2021) Jasmonic acid pathway in plants 2.0, vol 22. MDPI, pp 3506
Griffiths G (2020) Jasmonates: biosynthesis, perception and signal transduction. Essays Biochem 64(3):501–512
Guzmán-Guzmán P, Kumar A, de Los Santos-Villalobos S, Parra-Cota FI, Orozco-Mosqueda MDC, Fadiji AE, Hyder S, Babalola OO, Santoyo G (2023) Trichoderma species: our best fungal allies in the biocontrol of plant diseases—A review. Plants 12:432. https://doi.org/10.3390/plants12030432
He DC, He MH, Amalin DM, Liu W, Alvindia DG, Zhan J (2021) Biological control of plant diseases: an evolutionary and eco-economic consideration. Pathogens 10(10):1311. https://doi.org/10.3390/pathogens10101311
Heydari M, Yousefi AR, Nikfarjam N, Rahdar A, Kyzas GZ, Bilal M (2021) Plant-based nanoparticles prepared from protein containing tribenuron-methyl: fabrication, characterization, and application. Chem Biol Technol Agric 8:1–11
Kamle M, Borah R, Bora H, Jaiswal AK, Singh RK, Kumar P (2020) Systemic acquired resistance (SAR) and induced systemic resistance (ISR): role and mechanism of action against phytopathogens. In: Fungal biotechnology and bioengineering, pp 457–470
Khan AZ, Bilal M, Mehmood S, Sharma A, Iqbal HM (2019) State-of-the-art genetic modalities to engineer cyanobacteria for sustainable biosynthesis of biofuel and fine-chemicals to meet bio-economy challenges. Life (Basel) 9(3):54
Kumar A, Singh VK, Tripathi V, Singh PP, Singh AK (2018) Plant growth-promoting rhizobacteria (PGPR): perspective in agriculture under biotic and abiotic stress. In: Crop improvement through microbial biotechnology. Elsevier, pp 333–342. https://doi.org/10.1016/B978-0-444-63987-5.00016-5
Kumar G, Teli B, Mukherjee A, Bajpai R, Sarma B (2019) Secondary metabolites from cyanobacteria: a potential source for plant growth promotion and disease management. In: Secondary metabolites of plant growth promoting rhizomicroorganisms: discovery and applications, pp 239–252
Kumari M, Qureshi KA, Jaremko M, White J, Singh SK, Sharma VK, Singh KK, Santoyo G, Puopolo G, Kumar A (2022) Deciphering the role of endophytic microbiome in postharvest diseases management of fruits: opportunity areas in commercial up-scale production. Front Plant Sci 13:1026575. https://doi.org/10.3389/fpls.2022.1026575
Liu H, Timko MP (2021) Jasmonic acid signaling and molecular crosstalk with other phytohormones. Int J Mol Sci 22(6):2914
Messa VR (2021) Biocontrol by induced systemic resistance using plant growth promoting rhizobacteria. Rhizosphere 17:100323
Morcillo RJL, Manzanera M (2021) The effects of plant-associated bacterial exopolysaccharides on plant abiotic stress tolerance. Metabolites 11(6):337. https://doi.org/10.3390/metabo11060337
Mutalipassi M, Riccio G, Mazzella V, Galasso C, Somma E, Chiarore A et al (2021) Symbioses of cyanobacteria in marine environments: ecological insights and biotechnological perspectives. Mar Drugs 19(4):227. https://doi.org/10.3390/md19040227
Negi NP, Prakash G, Narwal P, Panwar R, Kumar D, Chaudhry B, Rustagi A (2023) The calcium connection: exploring the intricacies of calcium signaling in plant-microbe interactions. Front Plant Sci 14:1248648

Nishad R, Ahmed T, Rahman VJ, Kareem A (2020) Modulation of plant defense system in response to microbial interactions. Front Microbiol 11:514909

Olowe OM, Akanmu AO, Asemoloye MD (2020) Exploration of microbial stimulants for induction of systemic resistance in plant disease management. Ann Appl Biol 177(3):282–293

Pathak J, Maurya PK, Singh SP, Häder D-P, Sinha RP (2018) Cyanobacterial farming for environment friendly sustainable agriculture practices: innovations and perspectives. Front Environ Sci 6:311475

Pérez-Llorca M, Pollmann S, Müller M (2023) Ethylene and jasmonates signaling network mediating secondary metabolites under abiotic stress. Int J Mol Sci 24(6):5990

Poveda J (2021) Cyanobacteria in plant health: biological strategy against abiotic and biotic stresses. Crop Prot 141:105450

Rabaan AA, Alfaraj AH, Alshengeti A, Alawfi A, Alwarthan S, Alhajri M et al (2023) Antibodies to combat fungal infections: development strategies and progress. Microorganisms 11(3):671

Raheem A, Sajid M, Iqbal MS, Aslam H, Bilal M, Rafiq F (2019) Microbial inhabitants of agricultural land have potential to promote plant growth but they are liable to traditional practice of wheat (T. aestivum L) straw burning. Biocatal Agric Biotechnol 18:101060

Ristaino JB, Anderson PK, Bebber DP, Brauman KA, Cunniffe NJ, Fedoroff NV et al (2021) The persistent threat of emerging plant disease pandemics to global food security. Proc Natl Acad Sci U S A 118(23):e2022239118. https://doi.org/10.1073/pnas.2022239118

Saeed MU, Hussain N, Shahbaz A, Hameed T, Iqbal HM, Bilal M (2022) Bioprospecting microalgae and cyanobacteria for biopharmaceutical applications. J Basic Microbiol 62(9):1110–1124

Scortichini M (2022) Sustainable management of diseases in horticulture: conventional and new options. Horticulturae 8(6):517

Singh BK, Delgado-Baquerizo M, Egidi E, Guirado E, Leach JE, Liu H, Trivedi P (2023) Climate change impacts on plant pathogens, food security and paths forward. Nat Rev Microbiol 21(10):640–656

Solanki MK, Wang Z, Kaushik A, Singh VK, Roychowdhury R, Kumar M, Kumar D, Singh J, Singh SK, Dixit B, Kumar A (2024) From orchard to table: significance of fruit microbiota in postharvest diseases management of citrus fruits. Food Control 165:110698. https://doi.org/10.1016/j.foodcont.2024.110698

Spence N, Hill L, Morris J (2020) How the global threat of pests and diseases impacts plants, people, and the planet. Plants, People, Planet 2(1):5–13

Tariq M, Khan A, Asif M, Khan F, Ansari T, Shariq M, Siddiqui MA (2020) Biological control: a sustainable and practical approach for plant disease management. Acta Agric Scand Sect B Soil Plant Sci 70(6):507–524

Timsina J (2018) Can organic sources of nutrients increase crop yields to meet global food demand? Agronomy 8(10):214

Vaughan A (2020) Global food crisis looms, vol 246. Elsevier, pp 7

Vilvert E, Stridh L, Andersson B, Olson Å, Aldén L, Berlin A (2022) Evidence based disease control methods in potato production: a systematic map protocol. Environ Evid 11(1):1–8

Vu CM, Ghotekar S, Pham TD, Viet NM, Oza R, Bilal M, Roy A (2022) Plant-mediated eco-friendly synthesis of platinum nanoparticles and their applications. In: Plant and nanoparticles. Springer, pp 155–170

Wang Y, Mostafa S, Zeng W, Jin B (2021) Function and mechanism of jasmonic acid in plant responses to abiotic and biotic stresses. Int J Mol Sci 22(16):8568

Wang Z, Luo W, Cheng S, Zhang H, Zong J, Zhang Z (2023) Ralstonia solanacearum—A soil borne hidden enemy of plants: research development in management strategies, their action mechanism and challenges. Front Plant Sci 14:1141902. https://doi.org/10.3389/fpls.2023.1141902

Wu W, Shan HW, Li JM, Zhang CX, Chen JP, Mao Q (2022) Roles of bacterial symbionts in transmission of plant virus by hemipteran vectors. Front Microbiol 13:805352. https://doi.org/10.3389/fmicb.2022.805352

Yadav P, Singh RP, Hashem A, Abd Allah EF, Santoyo G, Kumar A, Gupta RK (2023) Enhancing biocrust development and plant growth through inoculation of desiccation-tolerant cyano-

bacteria in different textured soils. Microorganisms 11(10):2507. https://doi.org/10.3390/microorganisms11102507
Yu XQ, Niu HQ, Liu C, Wang HL, Yin W, Xia X (2024) PTI-ETI synergistic signal mechanisms in plant immunity. Plant Biotechnol J 22:2113
Zeng Q, Ding X, Wang J, Han X, Iqbal HM, Bilal M (2022) Insight into soil nitrogen and phosphorus availability and agricultural sustainability by plant growth-promoting rhizobacteria. Environ Sci Pollut Res 29(30):45089–45106
Zhao H, Yin CC, Ma B, Chen SY, Zhang JS (2021) Ethylene signaling in rice and Arabidopsis: new regulators and mechanisms. J Integr Plant Biol 63(1):102–125

11 Microbial Inoculants for the Management of Pesticide Toxicity in Plants

Dharmendra Kumar, Som Dutt, Dinesh Kumar, Kailash Chand Naga, Vikas Mangal, and Brajesh Singh

Abstract

The widespread use of pesticides in contemporary agriculture has sparked worries about their negative impact on plant health and the environment. A thorough analysis of the function of microbial inoculants in reducing pesticide toxicity in plants is provided in this book chapter. Microbial inoculants, including bacteria and fungi, have shown promising potential in enhancing plant resilience to pesticide stress and promoting sustainable agricultural practices. The chapter explores the mechanisms by which these microorganisms influence the fate and impact of pesticides in the soil-plant system and the importance of understanding the mechanisms underlying the interaction between microbial inoculants and pesticides. Through various studies and experiments, scientists have elucidated the intricate processes involved in microbial degradation, enzymatic activities, and alleviation of oxidative stress in plants. This knowledge serves as a foundation for the development of targeted strategies for the application of microbial inoculants in different agricultural systems.

Keywords

Microbial inoculants · Pesticides · Detoxification · Sustainable agriculture

D. Kumar (✉) · S. Dutt · D. Kumar · K. C. Naga · V. Mangal · B. Singh
Division of Crop Physiology, Biochemistry and Postharvest Technology, ICAR-Central Potato Research Institute, Shimla, Himachal Pradesh, India
e-mail: dharmendra.kumar1@icar.gov.in; som.dutt1@icar.gov.in; Dinesh.Kumar3@icar.gov.in; Kailash.Naga@icar.gov.in; Vikas.Mangal@icar.gov.in; director.Cpri@icar.gov.in

A. Kumar, M. K. Solanki (eds.), *Microbial Biocontrol Techniques*, Microorganisms for Sustainability 54,
https://doi.org/10.1007/978-981-97-8739-5_11

11.1 Introduction

Chemicals known as pesticides are used to kill, repel, or control pests that pose a hazard to agricultural crops. These pests, which can harm plants and reduce their yield, include weeds, fungus, insects, and pathogens (Santos et al. 2021). Modern agriculture now cannot function without the use of pesticides, which help farmers protect their crops and increase yields. Pesticides play a crucial role in modern agriculture by helping to control pests and diseases that can negatively impact crops (Rajmohan et al. 2020). These chemical substances are designed to repel, kill, or otherwise manage pests that can harm or destroy crops. The use of pesticides has become widespread in agricultural practices, contributing significantly to increased food production and protection of crops (Koza et al. 2022). Furthermore, the practical implications of incorporating microbial inoculants into agricultural practices are significant. Farmers and agricultural practitioners stand to benefit from reduced reliance on chemical pesticides, leading to lower environmental impact, improved soil health, and sustainable crop production (Bélanger et al. 2011). The potential economic advantages of adopting microbial inoculants, coupled with their positive impact on crop yield and quality, make them a viable option for integrated pest management strategies (Tudi et al. 2021). However, it is crucial to acknowledge that challenges and gaps in knowledge still exist. The effectiveness of microbial inoculants can be influenced by various factors, including soil type, climate, and specific pesticide used. Further research is needed to optimize microbial formulations, application methods, and dosage for different agricultural scenarios (Kumar et al. 2021a).

This book chapter underscores the need for a holistic and integrated approach to agricultural management. By embracing microbial inoculants as a tool for mitigating pesticide toxicity, we can move towards more sustainable and resilient agricultural systems, promoting both environmental health and food security in the face of evolving challenges.

The history of pesticides can be traced back thousands of years, with the use of various substances to control pests and protect crops. Here is a brief overview of the pesticide history (Jafir et al. 2023):

Ancient Civilizations:

Sulfur: Ancient Sumerians used sulfur compounds to control insects as early as 4500 B.C.

Arsenic: Chinese civilizations used arsenic compounds around 1000 B.C. to control pests (Oberemok et al. 2015).

Middle Ages:

Nicotine: Europeans began using nicotine sulfate as an insecticide in the eighteenth century.

Nineteenth Century:

Paris Green: This copper acetoarsenite compound was widely used in the nineteenth century as an insecticide (Oberemok et al. 2015; Abubakar et al. 2020).

Lead Arsenate: Lead arsenate was introduced in the early twentieth century and became a popular insecticide for fruit crops.

Post-World War II: The discovery of dichloro-diphenyl-trichloroethane (DDT) in the 1930s by Paul Müller marked a turning point. DDT was efficient against a wide range of pests and played an important role in preventing insect-borne diseases such as malaria (Giri et al. 2021).

Organophosphates and Carbamates: These synthetic chemicals emerged as alternatives to organochlorines like DDT in the post-World War II era.

Silent Spring (1962): The "Silent Spring" book by Rachel Carson increased public awareness of the dangers to the environment and human health posed by the careless use of pesticides, especially DDT (Hayes and Hansen 2017).

Environmental Movement and Regulation: In response to concerns raised by the environmental movement, governments around the world started implementing regulations to control and monitor the use of pesticides. Several harmful pesticides, including DDT, were banned or restricted.

The Emergence of Integrated Pest Management (IPM): IPM became a popular approach in the late twentieth century, emphasizing the use of multiple strategies, including biological control and cultural practices, to manage pests in an environmentally sustainable way (Robles-González et al. 2008).

Modern Pesticides: The development of newer classes of pesticides, such as neonicotinoids, pyrethroids, and biopesticides, has continued into the twenty-first century. Biopesticides are derived from natural materials like plants, bacteria, and certain minerals.

Concerns about the environmental impact, resistance development, and health risks associated with certain pesticides have led to ongoing research and discussions about sustainable and safer alternatives.

Organic Farming: The organic farming movement promotes the use of natural and non-synthetic methods for pest control, minimizing reliance on conventional pesticides (Berruti et al. 2016).

The history of pesticides reflects a continuous evolution in response to the need for pest control, along with growing awareness of the environmental and health consequences associated with their use. Today, there is a greater emphasis on developing sustainable and environmentally friendly pest management strategies.

11.2 Types of Pesticides and Their Impact on Plants

In today's agriculture, pesticides are essential since they protect crops from diseases and pests that may significantly reduce yields. Use of pesticides comes with its own set of challenges, particularly regarding their impact on nontarget organisms, including plants. In this discussion, we will explore the various types of pesticides and their effects on plants.

Insecticides: Insecticides are pesticides used to control insect infestations. They can be divided into several categories based on their chemical composition (Jafir et al. 2023). Pyrethroids, organophosphates, carbamates, and neonicotinoids are common classes of insecticides. While these chemicals are effective in controlling pests, they can also impact plants directly. Some insecticides, particularly those with systemic properties, are absorbed by plants and distributed throughout their tissues. This can affect the plant's metabolism and growth. Moreover, the broad-spectrum nature of many insecticides means they may target not only harmful pests but also beneficial insects like pollinators. Bees, for example, can be negatively impacted by certain insecticides, affecting pollination processes essential for plant reproduction (Oberemok et al. 2015; Santos et al. 2021).

Fungicides: To manage fungal diseases in crops, fungicides are applied. They can be applied as sprays or systemic treatments. While their primary function is to combat fungi, the chemicals in fungicides can also affect plant health. Some fungicides may lead to phytotoxicity, causing damage to plant tissues. Additionally, the repeated use of fungicides can result in the development of resistant strains of fungi, posing long-term challenges for plant disease management (Odukkathil and Vasudevan 2013; Santos et al. 2021).

Herbicides: Herbicides are designed to control weeds, which compete with crops for nutrients, water, and sunlight. The impact of herbicides on plants is significant, as they are specifically formulated to target plant species. Nonselective herbicides can affect a wide range of plants, including crops, leading to unintended damage. Glyphosate, a widely used herbicide, has faced scrutiny due to its potential impact on nontarget plants. It inhibits an enzyme essential for the synthesis of amino acids in plants, but it can also affect nontarget plants, leading to issues like reduced crop yield and altered nutrient uptake (Tudi et al. 2021; Tarfeen et al. 2022).

Rodenticides: Rodenticides are used to control rodent populations in agricultural settings. While they are not directly applied to plants, the consequences of rodenticide use can indirectly affect vegetation. For instance, secondary poisoning may occur when predators consume rodents that have ingested rodenticides. This can lead to a cascade effect in ecosystems, potentially impacting plant-pollinator interactions and overall biodiversity (Boopathy 2000).

Impact on soil health: Pesticides, when applied to crops, can leach into the soil. This can affect the microbial communities in the soil that play a crucial role in nutrient cycling and overall soil health. Some pesticides can persist in the soil for extended periods, leading to long-term impacts on plant growth. Soil degradation can occur, impacting the fertility and structure of the soil.

Residue in edible parts: Pesticide residues can accumulate in the edible parts of plants, posing potential risks to human health. The residues may exceed acceptable levels, leading to concerns about the safety of consuming certain crops. Regulatory bodies set limits on pesticide residues in food to protect consumers, but the risk persists, particularly with the prolonged use of certain pesticides (Yadav et al. 2015).

Development of pesticide resistance: Target pests may become resistant to insecticides if they are used often and consistently. This phenomenon occurs when a population of pests evolves to survive exposure to a particular pesticide. In response, farmers may increase pesticide application rates or resort to using more potent pesticides. This cycle can have unintended consequences on nontarget plants and organisms in the ecosystem (Ann et al. 2013; Hayes and Hansen 2017).

While pesticides are crucial for ensuring global food security by controlling pests and diseases, their use necessitates a careful balance. Sustainable agriculture practices, integrated pest management, and development of environmentally friendly alternatives are essential to mitigate the negative impacts of pesticides on plants and ecosystems. For the long-term health of agricultural systems and the environment, it is imperative to find a balance between effective pest management and minimizing harm to nontarget organisms, including plants.

11.3 Classification of Pesticides

Pesticides are chemical or biological substances used to manage, repel, or eliminate pests that can harm crops, animals, or humans. They are classified based on different criteria, including their chemical composition, target pests, and mode of action. Here is a general classification of pesticides (Abubakar et al. 2020; Grella et al. 2020):

Based on Target Pests

- Insecticides: Target and control insects
- Herbicides: Control or kill unwanted plants (weeds)
- Fungicides: Combat fungal diseases in plants
- Rodenticides: Designed to control rodents
- Bactericides: Target bacteria causing plant diseases

Based on Chemical Composition

- Organochlorines: Contain carbon and chlorine atoms (e.g., DDT)
- Organophosphates: Contain carbon and phosphorus atoms (e.g., malathion)
- Carbamates: Contain carbamate functional groups (e.g., carbaryl)
- Pyrethroids: Synthetic chemicals modeled after pyrethrins found in chrysanthemum flowers
- Neonicotinoids: A class of neuroactive insecticides modeled after nicotine
- Biopesticides: Animals, plants, microorganisms, and some minerals that are all natural sources

Based on Mode of Action

- Contact pesticides: Kill or repel pests upon direct contact
- Systemic pesticides: Absorbed by plants and transported throughout the plant, making it toxic to pests that feed on it
- Stomach or ingestion pesticides: Ingested by pests and disrupt their physiological processes
- Fumigants: Gaseous pesticides that permeate the target area

Persistence and Residual Activity

- Residual pesticides: Remain effective for an extended period after application
- Non-residual pesticides: Lose effectiveness relatively quickly

Chemical Nature

- Synthetic pesticides: Chemically manufactured
- Natural pesticides: Plants, microorganisms, and some minerals that are all natural sources

Broad-Spectrum vs. *Narrow-Spectrum*

- Broad-spectrum pesticides: Effective against a wide range of pests
- Narrow-spectrum pesticides: Target specific pests

Preventive vs. *Curative Pesticides*

- Preventive pesticides: Applied before the onset of pest infestation
- Curative pesticides: Applied after the pest problem is evident

It is important to note that the use of pesticides should be carefully managed to minimize environmental impact, potential harm to nontarget organisms, and development of pesticide resistance in target pests. Additionally, integrated pest management (IPM) practices, combining biological, cultural, and mechanical control methods, are encouraged to reduce reliance on chemical pesticides.

Overview of different types of pesticides (insecticides, herbicides, fungicides) and their chemical nature:

Insecticides:

Chemical nature: The purpose of insecticides is to manage and control insect infestations. They can be classified based on their chemical composition (Racke et al. 1997; Vischetti et al. 2004).

Organochlorines: e.g., DDT, Lindane (historically used but phased out due to environmental concerns).

- Organophosphates: e.g., malathion, parathion
- Carbamates: e.g., carbaryl, aldicarb
- Pyrethroids: e.g., permethrin, cypermethrin (synthetic versions of natural pyrethrins)
- Neonicotinoids: e.g., imidacloprid, clothianidin
- Biopesticides: Derived from natural materials, such as *Bacillus thuringiensis* (Bt) and neem extracts

Herbicides:

Chemical nature: Herbicides are used for removing undesirable plants or weeds. They can be categorized based on their chemical composition:

- Glyphosate: A broad-spectrum herbicide that inhibits the synthesis of amino acids
- 2,4-Dichlorophenoxyacetic acid (2,4-D): A synthetic auxin that disrupts plant growth
- Paraquat: A nonselective herbicide that interferes with photosynthesis
- Imidazolinones: e.g., imazethapyr, imazamox
- Triazines: e.g., atrazine
- Bioherbicides: Plant-derived substances or microbes that control weeds

Fungicides:

Chemical nature: Fungicides are used to combat fungal infections in crops, plants, and other materials. They can be grouped based on their chemical composition:

- Benzimidazoles: e.g., thiabendazole
- Triazoles: e.g., tebuconazole, propiconazole
- Strobilurins: e.g., azoxystrobin, trifloxystrobin
- Copper-based compounds: e.g., copper sulfate
- Dithiocarbamates: e.g., mancozeb
- Biofungicides: utilize living organisms, such as bacteria (e.g., *Bacillus subtilis*) or fungi (e.g., *Trichoderma*), to control fungal diseases

It is important to note that the use of pesticides raises environmental and health concerns. The development and use of more environmentally friendly and targeted alternatives, as well as integrated pest management (IPM) practices, are areas of ongoing research and development in agriculture. Additionally, the regulatory landscape for pesticides varies across regions, with different countries having their own set of regulations and guidelines to ensure the safe use of these substances.

11.4 Mechanisms of Pesticide Toxicity in Plants

Pesticides are substances specifically designed for preventing, destroying, repelling, or mitigating any pest. While pesticides are intended to target pests, they can also have unintended effects on nontarget organisms, including plants. The mechanisms of pesticide toxicity in plants can vary depending on the specific type of pesticide and the plant species. Here are some general mechanisms (Kumar et al. 2015, 2016):

Disruption of cell membranes: Pesticides may interfere with the structure and function of cell membranes. This disruption can affect the permeability of the membranes, leading to leakage of cellular contents and ultimately cell death.

Inhibition of enzymes: Many pesticides act by inhibiting essential enzymes involved in various metabolic processes. Enzymes are crucial for the proper functioning of cells, and their inhibition can disrupt cellular activities, including energy production and synthesis of essential biomolecules.

Interference with photosynthesis: Some pesticides can interfere with the process of photosynthesis, which is vital for plant growth. Herbicides, in particular, may target enzymes involved in photosynthetic pathways, leading to a reduction in the production of carbohydrates and energy.

Oxidative stress: Pesticides can cause oxidative stress by releasing reactive oxygen species (ROS) into plant cells. ROS may damage biological components such as proteins, lipids, and DNA, causing cell death (Shelton et al. 2012).

DNA damage: Certain pesticides can cause direct damage to plant DNA. This can lead to mutations, disruptions in cell division, and, ultimately, cell death. DNA damage can have long-term effects on the growth and development of plants (Mnif et al. 2011).

Disruption of hormonal regulation: Pesticides may interfere with the hormonal regulation of plant growth and development. Hormones play a crucial role in processes such as seed germination, root development, and flowering. Pesticides that disrupt hormonal balance can have profound effects on plant physiology (Tudi et al. 2021).

Impairment of nutrient uptake: Pesticides can interfere with the uptake of essential nutrients by plants. This can lead to nutrient deficiencies, affecting various metabolic processes and overall plant health (Giri et al. 2021).

Altered microbial communities: Some pesticides can affect the soil microbiome, disrupting the beneficial interactions between plants and soil microorganisms. This disruption can impact nutrient cycling, plant-microbe symbiosis, and overall soil health.

Accumulation in plant tissues: Pesticides may accumulate in different plant tissues, posing a risk of toxicity to consumers of these plants, including humans and animals (Sharma et al. 2017).

Understanding the mechanisms of pesticide toxicity in plants is essential for developing sustainable agricultural practices and minimizing the environmental impact of pesticide use. Additionally, researchers aim to design pesticides that are more selective and have fewer adverse effects on nontarget organisms.

11.5 Pesticides Affect Plant Physiology and Metabolism

Chemicals known as pesticides are made to prevent or get rid of pests that might damage crops. They can have various effects on plant physiology and metabolism, and their impact depends on the type of pesticide and its mode of action. Here are some general ways in which pesticides can affect plants (Helbling 2015; Radhakrishnan et al. 2017):

Interference with photosynthesis: Herbicides are pesticides designed to control weeds. Some herbicides interfere with the photosynthetic process in plants. For example, they may inhibit enzymes involved in photosynthesis, disrupt electron transport, or interfere with the synthesis of chlorophyll. This can lead to a reduction in the plant's ability to produce energy through photosynthesis.

Disruption of cell division: Herbicides and insecticides: Certain pesticides can disrupt cell division in plants. This interference can affect the growth and development of plant tissues. For example, mitosis, the process by which cells divide, may be disrupted, leading to stunted growth or abnormal development (Gianfreda and Rao 2008).

Inhibition of enzyme activity: Pesticides may inhibit specific enzymes involved in plant metabolism. Insects and fungi have different physiological processes than plants, but some pesticides can affect enzymes common to both pests and plants. This interference disrupts vital metabolic pathways in the plant, affecting its overall health (Yadav et al. 2015).

Alteration of hormonal balance: Herbicides may interfere with plant hormones, particularly auxins, gibberellins, and cytokinins. This disruption can lead to abnormal growth patterns, inhibition of root development, or changes in leaf structure. Growth regulators, which are sometimes used as pesticides, can also influence hormonal balance to control plant growth (Kumar et al. 2018; Zhang et al. 2018).

Disruption of membrane integrity: Some pesticides can affect the integrity of cell membranes in plants. This disruption can lead to leakage of cellular contents and compromise the overall health of the plant. Insecticides and fungicides may have specific mechanisms for disrupting the membranes of insects or fungi, but these can also affect plant cells (Kumar et al. 2019; Kumar Arora et al. 2020).

Impact on nutrient uptake: Certain herbicides can affect the uptake of essential nutrients by plants. They may interfere with nutrient transport mechanisms or affect the availability of nutrients in the soil, leading to nutrient deficiencies in plants.

Induction of stress responses: Pesticides can induce stress responses in plants. This can trigger the production of reactive oxygen species (ROS) and activate defense mechanisms. Prolonged exposure to stress can, however, lead to negative impacts on plant health and productivity (Kumar et al. 2015).

It is important to note that while pesticides are designed to target pests, unintended effects on nontarget organisms, including plants, can occur. Additionally, the long-term use of pesticides can contribute to the development of resistance in pest populations and have broader environmental implications. Sustainable and integrated pest management practices aim to minimize these negative impacts while effectively controlling pests.

Long-term effects: Long-term exposure to pesticides can have profound and complex effects on plant physiology and metabolism. These effects can manifest over extended periods, impacting the overall health, growth, and sustainability of plant populations. Here are some long-term effects of pesticide exposure on plants.

Development of resistance: Prolonged use of pesticides can exert selective pressure on pest populations, leading to the development of resistance. Over time, pests can evolve mechanisms to tolerate or detoxify the chemicals used against them. As a result, the effectiveness of certain pesticides diminishes, requiring the use of higher concentrations or different chemicals (Moisan et al. 2019).

Shifts in plant communities: Persistent pesticide use can alter the composition of plant communities. Some species may be more resilient or resistant to certain pesticides, leading to changes in the abundance and diversity of plant populations. This can have cascading effects on ecosystems, affecting the organisms that depend on these plants (Lal et al. 2006).

Disruption of soil microbial communities: Pesticides can influence the composition and activity of soil microbial communities. Long-term exposure may lead to changes in the abundance of beneficial soil microbes, affecting nutrient cycling, soil structure, and overall soil health. This, in turn, can impact plant growth and productivity (Cartwright et al. 2000).

Accumulation in plant tissues: Residual pesticide residues may accumulate in plant tissues over time. This can occur in both treated and untreated plants through processes like bioaccumulation and biomagnification. The accumulation of pesticides in plant tissues can have implications for food safety and may affect organisms that consume these plants, including herbivores and humans.

Impacts on nontarget organisms: Long-term pesticide use can harm nontarget organisms, such as pollinators and natural enemies of pests. For example, certain pesticides may reduce populations of beneficial insects, leading to imbalances in ecosystems and making plants more susceptible to pests (Schuster and Schröder 1990).

Disruption of ecosystem services: Pesticides can interfere with ecosystem services provided by plants, such as carbon sequestration, water purification, and soil stabilization. Changes in plant physiology and population dynamics can have cascading effects on these ecosystem services, influencing the overall functioning of ecosystems (Lesk et al. 2017).

Soil and water contamination: The accumulation of pesticides in soil can lead to long-term soil contamination. Pesticides may persist in the environment, affecting subsequent plant generations and potentially leaching into water sources. This contamination can have widespread ecological impacts beyond the immediate area of pesticide application.

Altered plant physiology and genetics: Continuous exposure to pesticides may induce changes in the physiology and genetics of plants. This can include alterations in gene expression, stress responses, and activation of defense mechanisms. Over time, these changes may become heritable, potentially affecting the evolutionary trajectory of plant populations (Schroll et al. 2006).

To mitigate these long-term effects, sustainable and integrated pest management practices are essential. These approaches prioritize the use of a combination of strategies, such as crop rotation, biological control, and judicious use of pesticides, to minimize negative impacts on plant health and the broader environment.

Prolonged pesticide use can have cumulative and multifaceted impacts on both soil and plant health. The effects are often interconnected and can lead to a range of environmental and ecological challenges.

To mitigate these cumulative impacts, sustainable agricultural practices, integrated pest management strategies, and judicious use of pesticides are crucial. These approaches aim to minimize the reliance on chemical inputs, promote ecological balance, and maintain the long-term health and productivity of both soil and plant ecosystems.

The challenge lies in finding a balance between effectively managing pests to ensure food security and minimizing the adverse effects of pesticides on plant health and the environment. Sustainable agricultural practices, integrated pest management (IPM), and exploration of alternative methods, such as microbial inoculants, are crucial for mitigating the impact of pesticides on plants and promoting long-term agricultural sustainability.

11.6 Importance of Pesticide Management

Pesticide management is crucial for several reasons, as pesticides play a significant role in modern agriculture and public health. Here are some key aspects highlighting the importance of pesticide management (Kookana et al. 1998; Gałuszka et al. 2011; Tudi et al. 2021):

Crop protection: Pesticides are essential for protecting crops from pests, diseases, and weeds. They help ensure a reliable and sufficient food supply by preventing yield losses caused by various agricultural pests.

Increased agricultural productivity: Effective pest management through pesticides leads to increased agricultural productivity. Farmers can produce more food per unit of land, contributing to food security for growing populations.

Disease control: Pesticides are used to control vectors of diseases that affect both plants and animals. For instance, controlling mosquito populations helps prevent the spread of diseases like malaria, dengue, and Zika virus.

Economic impact: Pesticides contribute to the economic well-being of farmers and the agricultural sector. By protecting crops from pests, farmers can generate higher yields, which, in turn, can boost their income and support rural economies (Webster et al. 1999).

Human health: Proper pesticide management is critical to safeguarding human health. Unregulated or improper use of pesticides can lead to health issues, including acute poisoning, chronic health problems, and environmental contamination.

Biodiversity conservation: The indiscriminate use of pesticides can harm nontarget organisms, including beneficial insects, birds, and other wildlife. Effective pesticide management involves minimizing the impact on nontarget species, thus contributing to biodiversity conservation.

Environmental protection: Pesticides, if not managed properly, can contaminate soil, water, and air. Sustainable pesticide management practices aim to reduce environmental pollution, protect ecosystems, and ensure long-term environmental health (Malik and Grohmann 2012).

Residue management: Pesticide residues on food products can pose risks to human health. Proper pesticide management includes setting and enforcing maximum residue limits (MRLs) to ensure that the food supply is safe for consumption (Gałuszka et al. 2011).

Resistance management: Prolonged and indiscriminate use of pesticides can lead to the development of resistance in pest populations. Effective pesticide management involves strategies to minimize resistance, such as rotating pesticides, using integrated pest management (IPM) approaches, and promoting the development of new pest control technologies.

Regulatory compliance: Pesticide management involves the development and enforcement of regulations to ensure that pesticides are used safely and responsibly. Regulatory frameworks help set standards, monitor pesticide use, and penalize those who violate guidelines (Gani and Kazmi 2020).

In summary, pesticide management is essential for balancing the benefits of pest control with the need to protect human health, the environment, and biodiversity. It involves a holistic approach that considers the economic, social, and ecological implications of pesticide use. Integrated pest management (IPM) is a sustainable approach that combines biological, cultural, physical, and chemical control methods to minimize the impact of pests while reducing reliance on pesticides.

11.7 Effective Management Strategies to Mitigate Pesticide Toxicity

Effective management strategies to mitigate pesticide toxicity are crucial for several reasons, encompassing environmental conservation, human health, and sustainable agriculture. Pesticides serve an important role in contemporary agriculture, protecting crops from diseases and pests. However, their use can lead to unintended consequences if not managed properly (Karpouzas et al. 2005; Delcour et al. 2015; Zhang et al. 2018).

Mitigating pesticide toxicity is crucial to protect the environment, human health, and nontarget organisms. Implementing effective management strategies can help reduce the negative impacts of pesticides. Here are some strategies:

1. Integrated Pest Management (IPM):
   - Biological control: Encourage the use of natural predators, parasites, and pathogens to control pest populations.
   - Crop rotation: Rotate crops to disrupt the life cycle of pests and reduce the need for constant pesticide use.
   - Companion planting: Planting crops that naturally deter pests or attract beneficial insects can reduce the need for pesticides.
2. Selective Pesticide Use:
   - Use pesticides that target specific pests rather than broad-spectrum chemicals.
   - Choose pesticides with shorter residual effects to minimize their presence in the environment.
   - Biodiversity: Nontarget creatures, such as beneficial insects, birds, and aquatic life, can be harmed by the use of pesticides. Managing pesticide use helps minimize the impact on biodiversity.
   - Soil and water contamination: Pesticides can leach into soil and contaminate water bodies, affecting ecosystems. Effective management strategies aim to reduce runoff and contamination.
3. Risk Assessment:
   - Conduct thorough risk assessments before using any pesticide to understand its potential impact on human health and the environment.
   - Consider the toxicity, persistence, and mobility of the pesticide.
   - Residue on food: Improper pesticide use can result in residues on crops, leading to potential health risks for consumers. Effective management ensures adherence to safety guidelines and acceptable residue levels.
   - Occupational exposure: Farmworkers and individuals involved in pesticide application are at risk of exposure. Proper training, protective gear, and adherence to safety protocols are essential to mitigate these risks (Zhang et al. 2017; Wang et al. 2024).
4. Alternative Pest Control Methods:
   - Explore and promote alternative pest control methods, such as pheromones, traps, and cultural practices.
   - Use physical methods like mulching, weeding, and proper irrigation to reduce the need for chemical control.
5. Education and Training:
   - Provide farmers, pesticide applicators, and the public with education on the proper use and handling of pesticides.
   - Raise public knowledge of the potential hazards of using pesticides and the significance of adhering to safety regulations.
6. Precision Agriculture:
   - Use modern technologies like GPS-guided equipment and drones to apply pesticides more precisely, reducing overuse and minimizing off-target effects.

- Reducing dependency: To minimize dependency on chemical pesticides, effective management requires the promotion of alternative and integrated pest management approaches, such as biological control and crop rotation.
- Long-term viability: Sustainable agricultural practices contribute to the long-term viability of farming systems, preserving soil health and reducing the need for chemical interventions (Zhang et al. 2020; Jafir et al. 2023).

7. Buffer Zones:
   - Establish buffer zones between treated areas and water bodies, sensitive ecosystems, or residential areas to reduce the risk of pesticide runoff.
8. Monitoring and Surveillance:
   - Regularly monitor pesticide residues in soil, water, and food to assess the effectiveness of mitigation measures and detect potential issues promptly.
   - Pesticide resistance: Overuse of some pesticides can result in the emergence of resistant pest populations. Implementing integrated pest management (IPM) strategies, which include careful application of pesticides, helps avoid the onset of resistance.
9. Legislation and Regulation
   - Enforce and update regulations related to pesticide use, ensuring that only approved and safe pesticides are available in the market.
   - Implement restrictions on high-risk pesticides, and promote the use of alternatives.
   - Legal and ethical considerations: Adherence to regulations and ethical standards is essential to ensure that pesticides are used responsibly. Effective management includes monitoring and enforcing compliance with these regulations (Jess et al. 2014; Jiang 2018).
10. Research and Innovation:
    - Invest in research to develop and promote safer and more environmentally friendly pesticides.
    - Encourage the development of genetically modified crops that are resistant to pests, reducing the need for external chemical interventions.
    - Safer alternatives: Ongoing research is essential to develop and promote safer alternatives to conventional pesticides. This includes the development of biopesticides and precision agriculture technologies.
11. Emergency Response Planning:
    - Develop and communicate emergency response plans in case of pesticide spills or accidents to minimize the impact on the environment and public health.
    - Cross-border impact: Pesticide use and its consequences often extend beyond national borders. International collaboration is crucial to address global challenges related to pesticide toxicity and environmental impact.
    - Remember, a combination of these strategies, tailored to specific agricultural and environmental contexts, will contribute to a more sustainable and less toxic approach to pest management. Collaboration among farmers, researchers, policymakers, and the public is essential to successfully implement these strategies.

In conclusion, the need for effective management strategies to mitigate pesticide toxicity is paramount for the sustainability of agriculture, the protection of ecosystems, and the well-being of both the environment and human populations. These strategies should encompass a holistic approach, integrating scientific knowledge, regulatory frameworks, and community involvement to strike a balance between pest control and environmental preservation.

11.8 Role of Microbial Inoculants

Microbial inoculants play a crucial role in agriculture and environmental management by harnessing the beneficial activities of microorganisms. These inoculants typically consist of bacteria, fungi, or other microorganisms that are introduced into the soil, plant, or other environments to enhance specific biological processes. Here are some key roles of microbial inoculants:

Nitrogen fixation: Some microbial inoculants contain nitrogen-fixing bacteria (e.g., *Rhizobium* species for legumes) that can convert atmospheric nitrogen into a form that plants can use, thereby improving soil fertility (Yilihamu et al. 2020).

Phosphorus solubilization: Certain bacteria and fungi can solubilize phosphorus in the soil, making it more available to plants. This enhances phosphorus uptake and promotes plant growth (Kour et al. 2020).

Biocontrol of pathogens: Microbial inoculants can act as biopesticides, helping to control the growth of harmful plant pathogens. This is often achieved through the production of antimicrobial compounds or by outcompeting pathogenic organisms for nutrients and space.

Improving soil structure: Some microorganisms contribute to the formation and stabilization of soil aggregates, leading to improved soil structure. This helps in water retention, aeration, and root penetration.

Enhancing nutrient availability: Microbial inoculants can release nutrients from organic matter and make them available to plants. This includes the breakdown of complex organic compounds into simpler forms that plants can absorb.

Plant growth promotion: Some microbes may create plant growth hormones such as auxins, gibberellins, and cytokinins, which drive plant growth and development. Bioremediation: Microbial inoculants are used in bioremediation processes to degrade or detoxify pollutants in soil or water. Certain bacteria and fungi can break down contaminants into harmless substances (Egamberdieva et al. 2015).

Symbiotic relationships: Some microbial inoculants form symbiotic relationships with plants, providing them with nutrients and receiving sugars in return. Mycorrhizal fungi, for example, form symbiotic associations with plant roots and enhance nutrient uptake.

Stress tolerance: Microbial inoculants enhance a plant's capacity to absorb water and nutrients, which helps the plant survive a variety of environmental challenges like salinity and drought.

Reducing dependency on chemical inputs: By enhancing nutrient availability and providing protection against pathogens, microbial inoculants can contribute to reducing the reliance on synthetic fertilizers and pesticides, promoting more sustainable agricultural practices.

In summary, microbial inoculants play diverse and essential roles in agriculture, contributing to soil health, plant growth, and environmental sustainability. Their application can lead to increased crop yields, improved resource-use efficiency, and reduced environmental impact.

11.9 Microbial Inoculants as a Potential Solution for Pesticide Toxicity

Microbial inoculants represent a promising and eco friendly solution to address concerns related to pesticide toxicity in agricultural practices. Pesticides, while effective in controlling pests and diseases, often come with detrimental effects on the environment, nontarget organisms, and even human health. Microbial inoculants offer an alternative approach that harnesses the power of beneficial microorganisms to enhance plant growth, suppress diseases, and reduce the reliance on chemical pesticides (Kumar et al. 2015).

Microbial inoculants are compositions containing beneficial microorganisms including bacteria, fungus, and other microbes that, when applied to seeds, soil, or plants, improve their health and growth. These microbes may form symbiotic interactions with plants, increasing nutrition availability, reducing diseases, and promoting overall plant resilience (Kumar et al. 2016).

Types of Microbial Inoculants:

Rhizobacteria:

Function: Plant development is promoted by fixing atmospheric nitrogen, solubilizing phosphorus, and generating growth-promoting substances.

Examples: *Rhizobium, Azotobacter, Pseudomonas*

Mycorrhizal Fungi:

Function: Form symbiotic relationships with plant roots, increasing nutrient absorption (especially phosphorus and micronutrients).

Example: *Glomus, Rhizophagus*

Biopesticides:

Function: Consists of microorganisms that directly or indirectly suppress plant pests and diseases (Egbuna et al. 2019; Riaz et al. 2021).

Example: *Bacillus thuringiensis* (Bt), *Trichoderma*

Advantages of Microbial Inoculants:

Environmentally friendly: Microbial inoculants offer a sustainable and environmentally friendly alternative to chemical pesticides, reducing the overall ecological impact.

Reduced pesticide dependency: By enhancing plant resistance and promoting natural defense mechanisms, microbial inoculants can help reduce the need for chemical pesticides.

Improved

Soil health: Beneficial microorganisms contribute to soil health by improving nutrient cycling, organic matter decomposition, and overall soil structure.

Enhanced plant growth: Microbial inoculants can stimulate plant growth through increased nutrient availability, leading to improved yield and crop quality.

Biotic stress management: Some microbial inoculants induce systemic resistance in plants, making them more resilient to various environmental stresses, including pests and diseases (Kumar et al. 2020, 2021b).

Implementation and challenges:

Application methods: Microbial inoculants can be applied through seed treatment, soil application, or foliar spray. Proper application methods and timing are crucial for their effectiveness.

Research and education: Continued research and education are essential to understanding the specific interactions between microorganisms, plants, and pests. Farmers need access to information and training to effectively integrate microbial inoculants into their practices.

Commercial availability: Availability and accessibility of high-quality microbial inoculant products can be a challenge. Increased commercial production and distribution can address this issue.

Microbial inoculants are formulations containing beneficial microorganisms that are applied to seeds, soil, or plant surfaces to enhance plant growth, health, and productivity. These microorganisms, which can include bacteria, fungi, and other microbes, form symbiotic relationships with plants or contribute to the improvement of soil conditions. Microbial inoculants refer to formulations containing beneficial microorganisms, such as bacteria, fungi, or other microbes, that are applied to seeds, soil, or plant surfaces to enhance plant growth, health, and overall productivity. These microorganisms establish symbiotic or mutually beneficial relationships with plants, promoting nutrient uptake, protecting against pathogens, and contributing to overall soil fertility. The goal of using microbial inoculants is to support sustainable and environmentally friendly agricultural practices by reducing the reliance on synthetic fertilizers and pesticides.

In conclusion, microbial inoculants represent a promising avenue for sustainable agriculture, offering a viable solution to mitigate pesticide toxicity while promoting plant health and productivity. Embracing these natural alternatives aligns with the growing global emphasis on environmentally conscious farming practices.

11.10 Microbial Inoculants as a Tool for Pesticide Management

Microbial inoculants play a crucial role in sustainable agriculture, and they can be used as a tool for pesticide management. Here are several ways in which microbial inoculants contribute to reducing the reliance on chemical pesticides:

Biological control of pests: Some microbial inoculants act as biopesticides, providing a natural and environmentally friendly alternative to chemical pesticides.

These can be bacteria, fungi, or viruses that specifically target and control pest populations. They may infect, parasitize, or produce toxins harmful to pests.

Induction of systemic resistance (ISR): Certain beneficial microbes can induce systemic resistance in plants. This means that the plants become more resistant to pathogens, pests, and diseases without the microbe directly attacking the pests. This systemic resistance response can reduce the need for chemical pesticides (Choudhary and Johri 2009).

Reduced environmental impact: Unlike chemical pesticides, microbial inoculants generally have minimal environmental impact. They are often highly specific in their action, targeting particular pests without affecting nontarget organisms. This reduces the risk of ecological imbalances associated with broad-spectrum chemical pesticides.

Integration with integrated pest management (IPM): Microbial inoculants can be integrated into broader IPM strategies. IPM combines multiple pest management techniques, including biological control, cultural practices, and chemical control where required. Microbial inoculants contribute to the biological control aspect of IPM (Baker et al. 2002).

Sustainable agriculture practices: The use of microbial inoculants aligns with the principles of sustainable agriculture by promoting a more balanced and resilient agroecosystem. This approach reduces the environmental impact associated with the use of synthetic pesticides.

Promotion of biodiversity: Microbial inoculants contribute to soil health and biodiversity by fostering a diverse and balanced microbial community. A diverse soil microbiome can enhance ecosystem services, including natural pest control.

In summary, microbial inoculants offer a range of mechanisms to manage pests in an environmentally sustainable manner. When integrated into holistic pest management strategies, they contribute to reducing the reliance on chemical pesticides, promoting long-term agricultural sustainability.

11.11 Microbial Inoculants to the Detoxification of Pesticides in Soil and Plant Systems

Microbial inoculants can contribute to the detoxification of pesticides in soil and plant systems through several mechanisms, primarily involving the activities of certain microorganisms. Here are key ways in which microbial inoculants aid in pesticide detoxification:

Biodegradation: Microbial inoculants, such as certain bacteria and fungi, possess the ability to break down pesticides into less harmful or nontoxic substances through a process called biodegradation. Microorganisms can produce enzymes that cleave pesticide molecules, making them more amenable to degradation.

Metabolism and transformation: Microbes can metabolize pesticides, transforming them into metabolites that may be less toxic or more easily degraded. The microbial metabolic processes can involve the modification of pesticide structures, rendering them less harmful to the environment (Das and Mukherjee 2000; Ferguson and Pietari 2000).

Enzymatic activities: Microbial enzymes, such as hydrolases, oxidases, and dehydrogenases, play a crucial role in the breakdown of pesticides. These enzymes facilitate the cleavage of pesticide molecules, breaking them down into simpler and less toxic compounds.

Rhizosphere microbial communities: The area of soil where plant roots are present, known as the rhizosphere, is abundant in microbial diversity. Certain plant-associated microbes in the rhizosphere contribute to pesticide detoxification by actively metabolizing or degrading pesticides. The root exudates released by plants can stimulate the growth and activity of these beneficial microbes.

Conjugation reactions: Microorganisms can facilitate conjugation reactions with pesticides, leading to the formation of conjugated compounds. Conjugation reactions often result in the detoxification of pesticides, making them less harmful to plants and soil organisms (Meliani 2015).

Adaptation and evolution: Microbial populations can adapt and evolve in response to the presence of pesticides. Over time, microbial communities may develop enhanced capabilities for degrading specific pesticides through natural selection and evolutionary processes.

Phytoremediation: Certain plant-microbe partnerships, such as mycorrhizal associations, can contribute to the detoxification of pesticides. Mycorrhizal fungi can enhance the ability of plants to take up and accumulate pesticides, reducing their concentration in the soil (Alori and Fawole 2017).

It is important to note that the effectiveness of microbial inoculants can vary depending on factors such as soil conditions, pesticide type, and specific microorganisms used. Integrated pest management (IPM) strategies that combine microbial inoculants with other sustainable practices can offer comprehensive solutions for managing pesticide toxicity while promoting environmental and agricultural sustainability.

11.12 Mechanisms of Action

Microbial processes play a crucial role in breaking down or neutralizing pesticides in the environment. The mechanisms of action can vary depending on the type of pesticide and the microbial species involved (Gao and Wen 2016). Here are some general microbial processes involved in pesticide degradation:

Hydrolysis: Hydrolysis involves the breakdown of chemical compounds in the presence of water. Many pesticides are susceptible to hydrolysis, and microbes can produce enzymes that catalyze hydrolytic reactions (Thangadurai and Suresh 2014).

Example: Organophosphates, a class of pesticides, can be hydrolyzed by microbial enzymes.

Oxidation-reduction (redox) reactions: Microorganisms can facilitate redox reactions, where electrons are transferred between molecules. Redox reactions are common in the breakdown of pesticides, especially those containing sulfur or nitrogen (Van Schie and Young 2000).

Example: Some bacteria can perform denitrification or sulfide oxidation, contributing to the breakdown of pesticides containing nitrogen or sulfur.

Biodegradation: Microorganisms can utilize pesticides as a carbon and energy source through biodegradation. They produce enzymes that break down complex pesticide molecules into simpler compounds that can be further metabolized (Muhamad et al. 2017).

Example: Certain bacteria and fungi can biodegrade pesticides such as atrazine and glyphosate.

Conjugation reactions: Microorganisms can facilitate the conjugation of pesticide molecules with endogenous compounds, making them less toxic or more easily excreted.

Example: Glucosylation and glutathione conjugation are common conjugation reactions involved in the microbial detoxification of pesticides.

Enzymatic transformation: Microbes can produce a variety of enzymes that facilitate the transformation of pesticide molecules. These enzymes may add functional groups or modify the pesticide structure.

Example: Cytochrome P450 enzymes in bacteria and fungi can catalyze the oxidative transformation of pesticides (Luo et al. 2018).

Co-metabolism: Some microorganisms can co-metabolize pesticides, meaning that the pesticides are transformed as by-products during the metabolism of other organic compounds.

Example: *Methylobacterium* spp. can co-metabolize a variety of pesticides, contributing to their degradation (Jones et al. 2001).

Sorption and sequestration: Microorganisms can interact with pesticides through sorption and sequestration, effectively reducing their bioavailability.

Example: Microbial exopolysaccharides can sequester pesticides in soil, limiting their movement and uptake by plants.

It is important to note that the effectiveness of microbial processes in breaking down or neutralizing pesticides depends on factors such as microbial diversity, environmental conditions, and specific characteristics of the pesticides involved. Additionally, the combination of multiple microbial processes often contributes to the overall degradation of pesticides in the environment.

11.13 Factors Influencing the Efficacy of Microbial Inoculants

Microbial inoculants are formulations containing beneficial microorganisms, such as bacteria, fungi, or other microbes, that are applied to seeds, soil, or plants to enhance plant growth and health. The efficacy of microbial inoculants can be influenced by various factors, and understanding these factors is crucial for maximizing their effectiveness. Microbial inoculants are mixtures that are added to soils to improve plant development and general soil health. These mixtures can include fungus or bacteria that are beneficial to the soil. Several factors, including soil pH, temperature, and moisture, play crucial roles in determining the effectiveness of microbial inoculants (Berg 2009; Etesami and Beattie 2018).

Here are some key factors that can influence the efficacy of microbial inoculants:

Strain selection: The selection of appropriate microbial strains is critical. Different strains may have varying abilities to promote plant growth or provide protection against diseases.

Compatibility with host plants: Microbial inoculants should be compatible with the specific plant species they are intended for. Some microbes may form symbiotic relationships with certain plants but not others.

Environmental conditions: Factors such as temperature, humidity, and pH can affect the survival and activity of microbial inoculants. It is essential to consider the environmental conditions of the target site.

Soil conditions: Soil type and composition play a significant role. Microbial inoculants may perform differently in different soil types, and some may be more effective in specific soil conditions.

Application methods: The method of application, whether seed coating, soil drench, or foliar spray, can impact the distribution and survival of microbial inoculants. The application should be done in a manner that ensures contact with the target plant roots or foliage (Uqab et al. 2016).

Storage and shelf life: Proper storage conditions are essential to maintain the viability of microbial inoculants. The shelf life of the product can be influenced by factors such as temperature, moisture, and exposure to light.

Competitive microflora: The presence of other microorganisms in the soil can compete with the applied inoculants for resources. Assessing the microbial community in the target environment is crucial for understanding potential competition.

Disease and pest interactions: Microbial inoculants may have interactions with plant diseases and pests. Some strains may have biocontrol properties against certain pathogens, while others may enhance the plant's resistance.

Genetic variability: The genetic diversity within microbial populations can impact their ability to adapt to changing environmental conditions. In some cases, selecting strains with higher genetic diversity may enhance efficacy (Gyaneshwar et al. 2002).

Field history: The history of previous agricultural practices in a field, including the use of fertilizers and pesticides, can influence the microbial community and, consequently, the efficacy of inoculants.

Soil pH:

Effect on microbial activity: Soil pH is a measure of the acidity or alkalinity of the soil. Different microorganisms thrive in specific pH ranges. Some microbial inoculants are tailored for acidic soils, while others are more effective in neutral to alkaline soils (Anjum et al. 2012).

Survival and viability: Extreme pH conditions can impact the survival and viability of microorganisms. Acidic soils may limit the survival of alkaline-loving bacteria and vice versa. Microbial inoculants are often formulated to withstand a certain range of pH conditions.

Temperature:

Microbial growth: Temperature has a significant impact on the proliferation and activity of microbes. The majority of microorganisms have a range of temperatures in which they grow efficiently; going outside of this range might cause them to become less active or slow down (Giri et al. 2021).

Seasonal variation: Microbial inoculants may be more effective during specific seasons when temperatures are within the optimal range. For instance, some bacteria are more active in warmer soils, while certain fungi thrive in cooler conditions.

Moisture:

Water availability: Microorganisms require water for survival and activity. Adequate moisture levels are crucial for the establishment and functioning of microbial inoculants. Insufficient water can lead to desiccation of the microorganisms, reducing their effectiveness (Giri et al. 2021).

Waterlogging: Conversely, excessive moisture can lead to waterlogged conditions, reducing oxygen availability in the soil. This can negatively impact aerobic microorganisms and may favor anaerobic species, altering the microbial community dynamics (Grobelak et al. 2015).

Nutrient availability:

Nutrient interactions: Soil nutrients can interact with microorganisms in complex ways. Some microbial inoculants are designed to enhance nutrient availability to plants by fixing atmospheric nitrogen or solubilizing phosphorus. The availability of these nutrients in the soil can influence the success of microbial inoculants.

Soil composition:

Texture and structure: Soil texture (sand, silt, clay) and structure influence the distribution of microorganisms in the soil. Microbial inoculants may interact differently with soils of varying textures, affecting their colonization and activity (Purnomo et al. 2011).

Organic matter content: A microbial population that is active and diversified is often supported by a high amount of organic materials. Microbial inoculants may perform better in soils with adequate organic matter, as they can utilize organic compounds for energy and growth.

Pesticides and fertilizers: Some chemical inputs, such as certain pesticides and high doses of synthetic fertilizers, can have adverse effects on microbial populations. The compatibility of microbial inoculants with these inputs is essential for their effectiveness (Singh et al. 2017).

Understanding the interactions between microbial inoculants and these environmental factors is crucial for optimizing their performance in different agricultural or ecological contexts. It emphasizes the importance of selecting appropriate microbial strains and formulations that align with specific soil conditions and management practices.

Regulatory approvals and compliance: Compliance with regulatory standards and approvals is crucial for ensuring the safety and efficacy of microbial inoculants. Meeting regulatory requirements is essential for their successful commercialization and use.

11.14 Challenges and Future Prospects

Highlight existing challenges in the widespread adoption of microbial inoculants for pesticide management. The use of microbial inoculants for pesticide management has garnered attention as a sustainable and environmentally friendly alternative to traditional chemical pesticides. However, several challenges hinder the widespread adoption of microbial inoculants. Here are some key challenges:

Variable efficacy: Microbial inoculants may show variable efficacy under different environmental conditions, limiting their reliability.

Consistency: Maintaining consistent performance across diverse agricultural settings and against various pests and diseases is a challenge.

Regulatory standards: There is a lack of standardized testing and regulatory frameworks for microbial inoculants, making it difficult to ensure product quality and efficacy.

Identification and characterization: Identifying and characterizing the most effective strains for specific pests and diseases is a complex task, and standardization in this area is lacking.

Stability: Microbial inoculants may have limited shelf life and stability, especially under different storage conditions, which poses challenges for distribution and commercial viability.

Logistical issues: Transportation and storage logistics may affect the quality and viability of the inoculants, leading to potential issues in their effectiveness.

Farmer education: Lack of awareness and education among farmers about the benefits and proper application of microbial inoculants can hinder their adoption.

Market perception: Some farmers may be skeptical about the effectiveness of microbial inoculants compared to traditional pesticides, impacting market acceptance.

Compatibility: Integrating microbial inoculants into existing agricultural practices and systems can be challenging due to differences in application methods, timing, and compatibility with other inputs.

Distribution networks: Developing efficient distribution networks for microbial inoculants, especially in remote or resource-constrained areas, can be a challenge.

Access to technology: Some farmers may lack access to the necessary equipment and technology for the proper application of microbial inoculants.

Addressing these challenges will require collaborative efforts from researchers, policymakers, industry stakeholders, and farmers to promote the development, adoption, and sustainable use of microbial inoculants in pesticide management. Research and development in the field of microbes for the management of pesticide toxicity is an important area that can have significant implications for agriculture, environmental sustainability, and human health. Here are some suggested areas for further exploration:

Microbial degradation mechanisms: Investigate and understand the microbial degradation pathways of different pesticides. This includes identifying key enzymes and microbial species involved in the breakdown of specific pesticides.

Microbial consortia: Explore the synergistic effects of microbial consortia in pesticide degradation. Understanding how different microbial species work together can enhance the efficiency and effectiveness of pesticide remediation.

Genetic engineering of microbes: Utilize genetic engineering techniques to enhance the pesticide-degrading capabilities of microbes. This could involve modifying microbial genomes to produce enzymes that degrade a broader range of pesticides or improve degradation rates.

Bioremediation strategies: Develop innovative bioremediation strategies using microbes to remediate pesticide-contaminated environments. This could include the design of bioaugmentation and biostimulation approaches to enhance microbial activity in contaminated soils or water bodies.

Microbial biotechnology for pesticide production: Explore the use of microbes in the sustainable production of biopesticides. This could involve engineering microbes to produce natural compounds with pesticidal properties.

Interdisciplinary collaboration among microbiologists, geneticists, ecologists, environmental scientists, and agricultural experts is crucial for addressing the complexities of pesticide toxicity management using microbial approaches. Additionally, integrating emerging technologies such as artificial intelligence and machine learning for data analysis and prediction could further enhance the efficiency of research in this field.

11.15 Conclusion

In conclusion, the utilization of microbial inoculants presents a promising and sustainable approach for mitigating pesticide toxicity in plants. This book chapter has explored the intricate relationship between pesticides, soil microorganisms, and plant health, shedding light on the potential of microbial inoculants as an eco-friendly solution to counteract the adverse effects of chemical pesticides. The research findings discussed in this chapter emphasize the multifaceted role of microbial communities in the soil ecosystem. Microbial inoculants, such as beneficial bacteria and fungi, have been demonstrated to enhance the detoxification of pesticides, promoting the degradation of harmful chemical compounds. Moreover, these inoculants contribute to the restoration of soil microbial diversity and activity, fostering a balanced and resilient agroecosystem.

Acknowledgements The authors would like to acknowledge the Directors of ICAR-Central Potato Research Institute, Shimla, for their constant support and guidance.

References

Abubakar Y, Tijjani H, Egbuna C et al (2020) Pesticides, history, and classification. In: Natural remedies for pest, disease and weed control, pp 29–42. https://doi.org/10.1016/B978-0-12-819304-4.00003-8

Alori ET, Fawole OB (2017) Microbial inoculants-assisted phytoremediation for sustainable soil management. In: Phytoremediation: management of environmental contaminants, vol 5. Springer International Publishing, pp 3–17

Anjum R, Rahman M, Masood F, Malik A (2012) Bioremediation of pesticides from soil and wastewater. In: Environmental protection strategies for sustainable development, pp 295–328. https://doi.org/10.1007/978-94-007-1591-2_9

Ann MN, Cho YE, Ryu HJ et al (2013) Growth promotion of tobacco plant by 3-hydroxy-2-butanone from Bacillus vallismortis EXTN-1. Korean J Pestic Sci 17:388–393. https://doi.org/10.7585/kjps.2013.17.4.388

Bélanger PA, Beaudin J, Roy S (2011) High-throughput screening of microbial adaptation to environmental stress. J Microbiol Methods 85:92–97. https://doi.org/10.1016/j.mimet.2011.01.028

Berg G (2009) Plant-microbe interactions promoting plant growth and health: perspectives for controlled use of microorganisms in agriculture. Appl Microbiol Biotechnol 84:11–18

Berruti A, Lumini E, Balestrini R, Bianciotto V (2016) Arbuscular mycorrhizal fungi as natural biofertilizers: let's benefit from past successes. Front Microbiol 6:1559

Boopathy R (2000) Factors limiting bioremediation technologies. Bioresour Technol 74:63–67. https://doi.org/10.1016/s0960-8524(99)00144-3

Cartwright CD, Thompson IP, Burns RG (2000) Degradation and impact of phthalate plasticizers on soil microbial communities. Environ Toxicol Chem 19:1253–1261. https://doi.org/10.1002/etc.5620190506

Choudhary DK, Johri BN (2009) Interactions of Bacillus spp. and plants—With special reference to induced systemic resistance (ISR). Microbiol Res 164:493–513

Das AC, Mukherjee D (2000) Influence of insecticides on microbial transformation of nitrogen and phosphorus in typic Orchragualf soil. J Agric Food Chem 48:3728–3732. https://doi.org/10.1021/JF990811Q

Delcour I, Spanoghe P, Uyttendaele M (2015) Literature review: impact of climate change on pesticide use. Food Res Int 68:7–15. https://doi.org/10.1016/J.FOODRES.2014.09.030

Egbuna C, Sawicka B, Tijjani H, et al (2019) Biopesticides, safety issues and market trends. In: Natural remedies for pest, disease and weed control, pp 43–53. https://doi.org/10.1016/B978-0-12-819304-4.00004-X

Etesami H, Beattie GA (2018) Mining halophytes for plant growth-promoting halotolerant bacteria to enhance the salinity tolerance of non-halophytic crops. Front Microbiol 9:148

Ferguson JF, Pietari JMH (2000) Anaerobic transformations and bioremediation of chlorinated solvents. Environ Pollut 107:209–215. https://doi.org/10.1016/S0269-7491(99)00139-6

Gałuszka A, Migaszewski ZM, Manecki P (2011) Pesticide burial grounds in Poland: a review. Environ Int 37:1265–1272. https://doi.org/10.1016/j.envint.2011.04.009

Gani KM, Kazmi AA (2020) Ecotoxicological risk evaluation and regulatory compliance of endocrine disruptor phthalates in a sustainable wastewater treatment scheme. Environ Sci Pollut Res 27:7785–7794. https://doi.org/10.1007/s11356-019-07418-7

Gao DW, Wen ZD (2016) Phthalate esters in the environment: a critical review of their occurrence, biodegradation, and removal during wastewater treatment processes. Sci Total Environ 541:986–1001

Gianfreda L, Rao MA (2008) Interactions between xenobiotics and microbial and enzymatic soil activity. Crit Rev Environ Sci Technol 38:269–310. https://doi.org/10.1080/10643380701413526

Giri BS, Geed S, Vikrant K et al (2021) Progress in bioremediation of pesticide residues in the environment. Environ Eng Res 26:200446. https://doi.org/10.4491/EER.2020.446

Grella M, Miranda-Fuentes A, Marucco P et al (2020) Development of drift-reducing spouts for vineyard pneumatic sprayers: measurement of droplet size spectra generated and their classification. Appl Sci 10:1–21. https://doi.org/10.3390/APP10217826

Grobelak A, Napora A, Kacprzak M (2015) Using plant growth-promoting rhizobacteria (PGPR) to improve plant growth. Ecol Eng 84:22–28. https://doi.org/10.1016/j.ecoleng.2015.07.019

Gyaneshwar P, Naresh Kumar G, Parekh LJ, Poole PS (2002) Role of soil microorganisms in improving P nutrition of plants. Plant Soil 245:83–93

Hayes TB, Hansen M (2017) From silent spring to silent night: agrochemicals and the anthropocene. Elementa 5:57. https://doi.org/10.1525/ELEMENTA.246/GALLEY/999/DOWNLOAD/

Helbling DE (2015) Bioremediation of pesticide-contaminated water resources: the challenge of low concentrations. Curr Opin Biotechnol 33:142–148. https://doi.org/10.1016/J.COPBIO.2015.02.012

Jafir M, Irfan M, Zia-ur-Rehman M et al (2023) The global trend of nanomaterial usage to control the important agricultural arthropod pests: a comprehensive review. Plant Stress 10:100208. https://doi.org/10.1016/j.stress.2023.100208

Jess S, Kildea S, Moody A et al (2014) European Union policy on pesticides: implications for agriculture in Ireland. Pest Manag Sci 70:1646–1654. https://doi.org/10.1002/PS.3801

Jiang JQ (2018) Occurrence of microplastics and its pollution in the environment: a review. Sustain Prod Consump 13:16–23

Karpouzas DG, Fotopoulou A, Menkissoglu-Spiroudi U, Singh BK (2005) Non-specific biodegradation of the organophosphorus pesticides, cadusafos and ethoprophos, by two bacterial isolates. FEMS Microbiol Ecol 53:369–378. https://doi.org/10.1016/J.FEMSEC.2005.01.012

Kookana RS, Baskaran S, Naidu R (1998) Pesticide fate and behavior in Australian soils in relation to contamination and management of soil and water: a review. Aust J Soil Res 36:715–764. https://doi.org/10.1071/s97109

Koza NA, Adedayo AA, Babalola OO, Kappo AP (2022) Microorganisms in plant growth and development: roles in abiotic stress tolerance and secondary metabolites secretion. Microorganisms 10:1528. https://doi.org/10.3390/MICROORGANISMS10081528

Kumar Arora N, Fatima T, Mishra J et al (2020) Halo-tolerant plant growth promoting rhizobacteria for improving productivity and remediation of saline soils. J Adv Res 26:69–82. https://doi.org/10.1016/j.jare.2020.07.003

Kumar D, Kaushik R, Lata, Saxena AK (2015) Biotransformation of pentachlorophenol to intermediary metabolites by bacteria isolated from pulp and paper mill effluent irrigated cultivable land. J Pure Appl Microbiol 9:201–208

Kumar D, Kaushik R, Singh S (2016) Molecular and biochemical profiling of pentachlorophenol utilizing bacteria from pulp and paper mill effluent irrigated soil in northern India. J Pure Appl Microbiol 10:2045–2054

Kumar D, Shukla L, Singh S et al (2018) Assessment of diverse phthalate esters (PAEs) from irrigated agriculture soil under protected cultivation in IARI, New Delhi. Int J Chem Stud 6:3432–3435

Kumar A, Patel JS, Meena VS, Srivastava R (2019) Recent advances of PGPR based approaches for stress tolerance in plants for sustainable agriculture. Biocatal Agric Biotechnol 20:101271. https://doi.org/10.1016/J.BCAB.2019.101271

Kumar D, Dutt S, Raigond P, et al (2020) Potato probiotics for human health. In: Potato. Springer Singapore, pp 271–287

Kumar D, Dutt S, Raigond P, et al (2021a) Beneficial microorganisms in crop growth, soil health, and sustainable environmental management: current status and future perspectives. In: Plant-microbial interactions and smart agricultural biotechnology, pp 11–32. https://doi.org/10.1201/9781003213864-2

Kumar D, Dutt S, Raigond P, et al (2021b) Different biofertilizers and their application for sustainable development. In: Microbial technology for sustainable environment, pp 31–48. https://doi.org/10.1007/978-981-16-3840-4_3

Lal R, Dogra C, Malhotra S et al (2006) Diversity, distribution and divergence of lin genes in hexachlorocyclohexane degrading sphingomonads. Trends Biotechnol 24:121–129. https://doi.org/10.1016/j.tibtech.2006.01.005

Lesk C, Coffel E, D'Amato AW et al (2017) Threats to north American forests from southern pine beetle with warming winters. Nat Clim Chang 7:713–717. https://doi.org/10.1038/NCLIMATE3375

Malik A, Grohmann E (2012) Environmental protection strategies for sustainable development. In: Environmental protection strategies for sustainable development, pp 1–605. https://doi.org/10.1007/978-94-007-1591-2

Meliani A (2015) Bioremediation strategies employed by pseudomonas species, pp 351–383

Mnif W, Hassine AIH, Bouaziz A et al (2011) Effect of endocrine disruptor pesticides: a review. Int J Environ Res Public Health 8:2265–2303. https://doi.org/10.3390/IJERPH8062265

Moisan K, Cordovez V, van de Zande EM et al (2019) Volatiles of pathogenic and non-pathogenic soil-borne fungi affect plant development and resistance to insects. Oecologia 190:589–604. https://doi.org/10.1007/s00442-019-04433-w

Muhamad SAS, Jamilah B, Russly AR, Faridah A (2017) The antibacterial activities and chemical composition of extracts from Carica papaya cv. Sekaki/Hong Kong seed

Oberemok VV, Volodymyrivna Laikova K, Gninenko YI et al (2015) A short history of insecticides. J Plant Protect Res 55(3):221–226

Odukkathil G, Vasudevan N (2013) Toxicity and bioremediation of pesticides in agricultural soil. Rev Environ Sci Biotechnol 12:421–444. https://doi.org/10.1007/s11157-013-9320-4

Purnomo AS, Mori T, Takagi K, Kondo R (2011) Bioremediation of DDT contaminated soil using brown-rot fungi. Int Biodeterior Biodegrad 65:691–695. https://doi.org/10.1016/J.IBIOD.2011.04.004

Racke KD, Skidmore MW, Hamilton DJ et al (1997) Pesticide fate in tropical soils. Pure Appl Chem 69:1349–1371. https://doi.org/10.1351/PAC199769061349

Radhakrishnan R, Hashem A, Abd Allah EF (2017) Bacillus: a biological tool for crop improvement through bio-molecular changes in adverse environments. Front Physiol 8:667. https://doi.org/10.3389/fphys.2017.00667

Rajmohan KS, Chandrasekaran R, Varjani S (2020) A review on occurrence of pesticides in environment and current technologies for their remediation and management. Indian J Microbiol 60:125–138. https://doi.org/10.1007/s12088-019-00841-x

Riaz U, Murtaza G, Anum W, et al (2021) Plant growth-promoting rhizobacteria (PGPR) as biofertilizers and biopesticides. In: Microbiota and biofertilizers. Springer International Publishing, pp 181–196

Robles-González IV, Fava F, Poggi-Varaldo HM (2008) A review on slurry bioreactors for bioremediation of soils and sediments. Microb Cell Fact 7:5. https://doi.org/10.1186/1475-2859-7-5

Santos MS, Rodrigues TF, Nogueira MA, Hungria M (2021) The challenge of combining high yields with environmentally friendly bioproducts: a review on the compatibility of pesticides with microbial inoculants. Agronomy 11:870. https://doi.org/10.3390/agronomy11050870

Schroll R, Becher HH, Dörfler U et al (2006) Quantifying the effect of soil moisture on the aerobic microbial mineralization of selected pesticides in different soils. Environ Sci Technol 40:3305–3312. https://doi.org/10.1021/ES052205J

Schuster E, Schröder D (1990) Side-effects of sequentially-applied pesticides on non-target soil microorganisms: field experiments. Soil Biol Biochem 22:367–373. https://doi.org/10.1016/0038-0717(90)90115-G

Shelton JF, Hertz-Picciotto I, Pessah IN (2012) Neurodevelopmental disorders and prenatal residential proximity to agricultural pesticides: the CHARGE study. Environ Health Perspect 122(10):1103–1109

Singh R, Kumar S, Kumar H et al (2017) Effect of irrigation and integrated nutrient management on growth and yield of chickpea (Cicer arietinum L.). Plant Arch 17:1319–1323

Tarfeen N, Nisa KU, Hamid B et al (2022) Microbial remediation: a promising tool for reclamation of contaminated sites with special emphasis on heavy metal and pesticide pollution: a review. Process 10:1358. https://doi.org/10.3390/PR10071358

Thangadurai P, Suresh S (2014) Biodegradation of endosulfan by soil bacterial cultures. Int Biodeterior Biodegrad 94:38–47. https://doi.org/10.1016/J.IBIOD.2014.06.017

Tudi M, Ruan HD, Wang L et al (2021) Agriculture development, pesticide application and its impact on the environment. Int J Environ Res Public Health 18:1–24. https://doi.org/10.3390/IJERPH18031112

Uqab B, Mudasir S, Nazir R (2016) Review on bioremediation of pesticides. J Bioremed Biodegr 7:1–5. https://doi.org/10.4172/2155-6199.1000343

van Schie PM, Young LY (2000) Biodegradation of phenol: mechanisms and applications. Bioremediat J 4:1–18. https://doi.org/10.1080/10588330008951128

Vischetti C, Capri E, Trevisan M et al (2004) Biomassbed: a biological system to reduce pesticide point contamination at farm level. Chemosphere 55:823–828. https://doi.org/10.1016/J.CHEMOSPHERE.2003.11.042

Wang C, Jiang Y, Shao Y et al (2024) The influence and risk assessment of multiple pollutants on the bacterial and archaeal communities in agricultural lands with different climates and soil properties. Appl Soil Ecol 193:105130. https://doi.org/10.1016/j.apsoil.2023.105130

Webster JPG, Bowles RG, Williams NT (1999) Estimating the economic benefits of alternative pesticide usage scenarios: wheat production in the United Kingdom. Crop Prot 18:83–89. https://doi.org/10.1016/S0261-2194(98)00085-4

Yadav IC, Devi NL, Syed JH et al (2015) Current status of persistent organic pesticides residues in air, water, and soil, and their possible effect on neighboring countries: a comprehensive review of India. Sci Total Environ 511:123–137. https://doi.org/10.1016/J.SCITOTENV.2014.12.041

Zhang Q, Xia Z, Wu M et al (2017) Human health risk assessment of DDTs and HCHs through dietary exposure in Nanjing, China. Chemosphere 177:211–216. https://doi.org/10.1016/J.CHEMOSPHERE.2017.03.003

Zhang C, Sun Y, Hu R et al (2018) A comparison of the effects of agricultural pesticide uses on peripheral nerve conduction in China. Sci Rep 8:9621. https://doi.org/10.1038/S41598-018-27713-6

Zhang H, Qi L, Wu Y et al (2020) Numerical simulation of airflow field from a six–rotor plant protection drone using lattice Boltzmann method. Biosyst Eng 197:336–351. https://doi.org/10.1016/J.BIOSYSTEMSENG.2020.07.018

Screening of Microbial Biocontrol Agents Using Traditional and *In Silico* Approaches

12

Vishal, Navaneet Chaturvedi, Nischal Pradhan, Aamna Hassan, Kshitij Parmar, Monika Singh, Rachana Singh, and Ajay Kumar

Abstract

In the recent past, achieving food security for the rising global population has been one of the severe challenges due to changing climatic conditions or the food loss due to pest or pathogen attack during harvesting or postharvest storage conditions. However to control the losses caused by plant diseases or postharvest storage loss, farmers or the industries generally rely upon the chemical pesticides. But the uncontrolled and regular application of chemicals emerges as a serious concern for the soil texture, crop quality, human health, or the surrounding environments. In addition, the emergence of chemical resistance pest and alteration in the natural microflora of soil or plants are some other major adverse issues with the utilization of chemical pesticides in the agricultural field. Microbial biocontrol agents (BCAs) offer sustainable alternatives to chemical pesticides for managing plant diseases. This study evaluates BCAs using in vivo, in vitro, and in silico analyses to identify effective strains for agricultural application. In vivo screening in natural and greenhouse conditions measured lesion diameter, disease severity, and plant physiological status. In vitro assays assessed the antagonistic activity of BCAs against pathogens, focusing on inhibition

Vishal · N. Chaturvedi · N. Pradhan · R. Singh · A. Kumar (✉)
Amity Institute of Biotechnology, Amity University, Noida, Uttar Pradesh, India
e-mail: nischal.pradhan@s.amity.edu; rsingh2@amity.edu

A. Hassan
Jamia Hamdard University, New Delhi, India

K. Parmar
Amity Institute of Organic Agriculture, Amity University, Noida, Uttar Pradesh, India

M. Singh
Department of Biotechnology, School of Applied and Life Sciences, Uttaranchal University, Dehradun, Uttarakhand, India

A. Kumar, M. K. Solanki (eds.), *Microbial Biocontrol Techniques*, Microorganisms for Sustainability 54,
https://doi.org/10.1007/978-981-97-8739-5_12

zones and pathogen growth suppression. In silico analysis involved genomic and metabolomic data to predict BCA efficacy and interaction mechanisms with host plants.

The results demonstrate that integrating these methodologies provides a comprehensive understanding of BCA potential, highlighting the influence of plant genotype, endophyte colonization, and environmental factors on biocontrol efficiency. This multifaceted approach enhances the selection process for robust BCAs, paving the way for their practical use in sustainable agriculture.

Keywords

Biocontrol · Sustainable plant health · In vivo · In vitro · In silico

12.1 Introduction

As per the published report, global population will cross nine billion by 2050, and this rising population needs extra food from the current level even in the limited or shrinking natural resources to achieve food security for all (Bahar et al. 2020). However, each year, plant diseases contribute to significant losses, amounting to 15–30% of agricultural production Kumari et al. 2022). These losses are influenced by factors such as severity, weather conditions, pests, and crop compensation ability. Moreover, postharvest diseases also contribute significantly to quality losses, throughout the world (Sharma et al. 2017; Kumari et al. 2022; Verma et al. 2022). Among these, pests pose a significant threat to sustainability, as they are recognized for their detrimental impact on productivity.

However, to meet the needs of food security, farmers extensively utilize chemical fertilizers or chemical pesticides to meet the requirements of excessive foods. However, their utilization adversely affects the soil texture, plant productivity, pesticide residue in crops, human health, and environment. In addition, it also impacts the natural microflora of soil or plants that is essentially required for the optimum and sustainable crop production (Singh et al. 2017a, b; Kashyap et al. 2023; Bastas et al. 2023).

Presently, plant disease management relies predominantly on nonchemical measures, particularly cultural practices aimed at preventing and controlling diseases (Kumar 2022a, b). These practices include crop rotation, utilization of agricultural residues as bio-manure, and utilization of microbial based products such as biofertilizers or biopesticides. Utilization of these biological means is not only eco-friendly and economic but also favorable for the human health and gut microbiome. Adding to these complexities, the extensive utilization of pesticides has led to the emergence of resistant pests and pathogens, and management of the pest-resistant pathogen is again a severe challenge (Singh et al. 2017a, b; Solanki et al. 2023, 2024).

However, in the last two decades, to mitigate the challenges of biotic stresses or phytopathogen management, farmers or the researchers' community has been utilizing microbial biocontrol agents to combat the phytopathogen growth or for plant disease management (Kumar 2022a, b; Bastas et al. 2023). These microbial biocontrol agents include the strains of bacteria, fungi, yeast, actinomycetes, or cyanobacteria. Although the effectiveness of the biocontrol agents is similar with the chemical pesticides, it is slightly slower. However, the eco-friendly and economic nature and no presence of toxic residue have made biocontrol agents popular in phytopathogen management since the last few years. The utilization of microbial biocontrol agents in agriculture provides some extra benefit to the plants by enhancing the plant growth and mitigation against the different abiotic stresses (Wang et al. 2023; Kumar et al. 2024; Verma et al. 2022).

The microbial biocontrol agents including the bacteria, fungi, and cyanobacteria follow various direct or indirect mechanisms such as competition for nutrients and space, hyperparasitism, antibiosis, synthesis of antibiotics or bioactive compounds, induced systemic resistance to manage the phytopathogen growth, or plant disease management (Kumari et al. 2022; Köhl et al. 2019; Lahlali et al. 2022; Bonaterra et al. 2022). However, these microbial strains also possess the growth-promoting attributes such as phosphate solubilization, indole acetic acid production, and phytohormone modulation. Thus, the utilization of microbial biocontrol agents not only protects the plant from pathogen invasion but also plays a crucial role in enhancing plant growth (Kumar 2022a, b).

This chapter has been compiled to provide an overview of biocontrol agent, action mechanism of the biocontrol agents, and different screening methods, viz. in vitro, in vivo, or in silico, of the microbial biocontrol agents.

12.2 Action Mechanism of Microbial Biocontrol Agents

The effective management of phytopathogens or plant diseases depends upon several factors, but it primarily depends upon the host, nature of the pathogens, or environment. These three factors work interactionally as a triangle, which includes the host, pathogen, and environment, with the biological control agents (BCAs) (Kumari et al. 2022; Köhl et al. 2019). And being an organism, the biocontrol agent interacts with these components to manage the phytopathogen of plant diseases effectively. Many studies emphasize the importance of understanding the mechanisms between biological control agents (BCAs) and pathogens in order to select and develop effective biocontrol strategies (Kashyap et al. 2023). However, understanding of these interactions helps in selecting and developing effective BCAs.

A range of mechanisms involved in the biological control of plant diseases include hyperparasitism, competition for nutrients and space, antibiotic production, volatile production, and induction of plant systematic resistance.

12.3 Hyperparasitism

Hyperparasitism, a form of direct competitive interaction, is reported widespread among the flowering plants, in which the haustoria plays a crucial role in nutrient absorption (Sharma 2021). In general, mycoparasitism is of two types: the first one is biotrophic mycoparasitism, and the second one is necrotrophic hyperparasites. In biotrophic mycoparasitism, hyperparasite engages with a living host fungus and obtained nutrients from the host cells through haustoria without causing the host's death (Kiss 2001). This process involves the formation of infection structures and specialized hyphae by the hyperparasite (Moricca et al. 2005). The haustorium is essential for establishing compatibility with the host and facilitating prolonged nutrient absorption (Mendgen et al. 2006).

However, necrotrophic hyperparasites derive nutrients from dead host cells offering more commercial potential as microbial biocontrol agents (MBCAs). Necrotrophic hyperparasites invade and destroy host spores or hyphal cells mainly by secreting cell wall-degrading enzymes such as chitinases and glucanase and occasionally secondary metabolites. These enzymes cause breaches in the cell wall and disrupt the cytoplasmic organization (Köhl et al. 2019; Abdel Rahim and Abo-Elyousr 2018).

However, among the various biocontrol agents, the most studied mycoparasites are *Trichoderma* and *Clonostachys* species and inhibit the pathogen growth via secreting and antimicrobial secondary metabolites or cell wall-degrading enzymes (Lopes 2012; Rajani et al. 2023). *Trichoderma* strains, in particular, exhibit significant antagonistic potential against various plant pathogens, including *Sclerotinia sclerotiorum* and *Erwinia mallotivora* (Savchuk and Dilantha Fernando 2004; Tamizi et al. 2022). These strains also produce secondary metabolites that enhance plant growth and disease resistance (Vinale et al. 2014). Abdel-Rahim and Abo-Elyousr (2018) also reported about the biocontrol agent *Talaromyces acidophilus* which inhibits the growth of *Botrytis cinerea* via secreting enzymes chitinase, lipase, etc. Similarly, Cao et al. (2009) reported about the strain *Choiromyces aboriginum,* which inhibit the growth of the *pathogen Pythium* sp. via secreting the enzymes β-1,3-glucanase.

12.4 Competition for Nutrient and Spaces

Competition for nutrients and space is one of the common mechanisms followed by most of the biocontrol agents during phytopathogen management as during the infection or plant diseases, both the pathogens and the biocontrol agents need food or energy source for the growth and survival. The application of the biocontrol agents limits the pathogens from getting nutrients and occupying the space due to fast growth rate and deprives the pathogens from nutrient availability, resulting in the death of pathogenic microorganisms (Kumari et al. 2022). In the previous studies, authors reported about the biocontrol agents which follow the mechanisms of competition of nutrients and pace during pathogen elimination. For example,

Sarangi and Ramakrishnan (2023) reported about the *Bacillus* sp. which inhibits the pathogenic growth by following the same mechanisms. Similarly, Di Francesco et al. (2017) reported about the biocontrol yeast *Aureobasidium pullulans*, which inhibits the growth of *Monilinia laxa* by following the mechanism of nutrients and space competition. Liu et al. (2013) reported competition for nutrients during the control of pathogen *Penicillium italicum* by the biocontrol strain *Kloeckera apiculata*.

12.5 Antibiosis and Secretion of Bioactive Compounds

Antibiosis is a pivotal mechanism in biocontrol, which involves microorganisms, producing various specific or nonspecific metabolites, lytic agents, enzymes, and other toxic substances to inhibit pathogen growth (Fravel 1988; Loper and Lindow 1993; Saraf et al. 2014). In general, the biocontrol strains of either bacteria or fungi generate diverse types of antimicrobial compounds, such as lipopeptides and antibiotics, which effectively control the growth of a wide range of pathogens (Raaijmakers et al. 2010). Antibiotics are crucial in antibiosis-mediated biocontrol or suppressing soilborne plant pathogens (Saraf et al. 2014; Nehra et al. 2022). However, their production has been influenced by the plant genotypes, host plant, presence of pathogens, and density of cells in the producing strain and other biotic and abiotic stresses (Haas and Keel 2003).

In the previous studies, various authors reported the role of biocontrol agents in suppressing phytopathogen growth by secreting antibiotics or bioactive compounds or by the mechanism of antibiosis (Haas and Keel 2003; Arseneault and Filion 2017; Nehra et al. 2022). For example, *Bacillus*-secreted antimicrobial lipopeptides are recognized as natural defenses used in biocontrol against plant pathogens (Dimkić et al. 2022). Mageshwaran et al. (2012) reported the antibiosis mechanism in suppressing the blight disease in French bean by the biocontrol agent *Paenibacillus polymyxa*. Similarly, Romero et al. (2019) reported about the endophytic strain *Pseudomonas viridiflava* which inhibits the growth of various pathogens such as *Xanthomonas campestris* and *Sclerotinia sclerotiorum* by the mechanism of antibiosis. In the previous studies, various authors reported about the alkaloids or the bioactive compounds produced by the microbial biocontrol agents (Afzal et al. 2019; Singh et al. 2018; Kumari et al. 2022). For example, Zhang et al. (2022) reported about the bioactive compounds peptaibols produced by the *Trichoderma* sp. which play a pivotal role in controlling the bacterial leaf blight disease in rice.

12.5.1 Induction of Systemic Resistance

Induction of systemic resistance is one of the mechanisms induced by the host plant to protects itself from the pathogen invasion. In general, the systemic resistance is categorized into two types: the first one is induced systemic resistance (ISR) which is induced by nonpathogenic or symbiotic microbes, and the second one is systemic acquired resistance (SAR) induced by pathogenic microbes. The induction of ISR offers the host plant more resistance against the pathogenic attack (Alstrom 1991; Yu et al. 2022). However, the pathogenic microbes after attacking the plant host induce the pathogenic proteins or salicylic acid (SA)-dependent plant defense. In the previous studies, various authors reported the significance of ISR and SAR in producing immunity to the host plants against the phytopathogen attack (Knoester et al. 1999; Kumari et al. 2022; Yu et al. 2022).

12.5.2 Screening of Biocontrol Agents

The screening of biocontrol agents is crucial for developing sustainable pest management strategies. This procedure comprises identifying and carefully assessing microorganisms that can control pest or pathogens such as fungi, bacteria, and insects (Kurhekar et al. 2019; Kumari et al. 2022). The primary advantage of this screening is that it offers safe, sustainable alternatives for chemical pesticides, which are frequently hazardous to both human health and ecosystems (Pathak et al. 2022). In biocontrol research, in vitro, in vivo, and nowadays in silico approaches have been commonly practiced to evaluate the biocontrol efficacy of the microbial strains (Mischke 2004; Kurhekar et al. 2019).

12.5.3 In Vitro Screening

For the selection of biocontrol agents, the screening of microorganism to evaluate or screen their pathogen growth inhibition potential and disease management ability is the primary step. In general, the screening of microbial biocontrol potential has been characterized through various ways such as in vitro test, in vivo test analysis, or in silico analysis.

In vivo test analysis generally includes the glass test which has been carried out in the laboratory. During the screening, the pathogens or the biocontrol agents are grown either together or nearby to evaluate their pathogen inhibition potential (Mukherjee et al. 2012; Kashyap et al. 2023). The in vivo test analysis generally incudes the dual culture test in which direct screening of biocontrol agent has been carried out by growing the biocontrol agent and pathogen together in the semisolid media in the same petri plate at a particular distance. The inhibition of mycelium growth, inhibition of fungal spore growth considered as biocontrol action, and this inhibition potential have been evaluated by measuring the lesion diameter (Kashyap et al. 2023; Kumari et al. 2022).

However, in another in vitro approach, the pathogen has been uniformly spread over the petri plate by applying any fixed concentration of pathogen. For example, for the *Penicillium* pathogen species, 10^5 to 10^6 is usually used and the biocontrol agent is applied in the optimum known concentration in the center or spotted over the center. The clear halo zone present around the biocontrol agents shows their inhibitory potential, which is evaluated by measuring the zone diameter (Shehata et al. 2018).

In the other methods, the screening of biocontrol agents has been performed on the basis of volatile production. In these methods, the biocontrol agents and the pathogen are grown or placed in two different plates, and the plates are sealed with the tape of parafilm to ensure no leakage of the volatiles. The growth inhibition of the pathogen by the secreted volatiles confirms biocontrol potential of microbial strains (Fialho et al. 2011).

12.5.4 In Vivo Screening

The typical approach for assessing potential biological control agents (BCAs) in natural or greenhouse settings involves various parameters, including measuring lesion diameter, disease severity, or a specific disease index (Lecomte et al. 2016). In vivo screening not only evaluates the antagonistic activity of BCAs but also considers the plant's physiological condition. This assessment includes water status (e.g., transpiration, stomatal conductance), variations in antioxidant activity (e.g., enzymatic activity levels), production of plant defense compounds (e.g., phytoalexins), and growth parameters such as plant height, dry or fresh weight of certain plant parts, or flowering date (Kumari et al. 2022).

The antagonistic potential of BCAs varies with plant genotype or species, as different host genotypes can influence physiological functions in ways that affect infection rates and the host's immune response. Similarly, the colonization ability of endophytes, influenced by the physicochemical properties of plant exudates, affects their effectiveness in biocontrol against pathogens (Kashyap et al. 2023; Arnold and Lutzoni 2007).

12.6 In Silico Approaches

In the realm of biocontrol agents, in silico approaches have emerged as valuable tools for identifying and characterizing genes responsible for biocontrol properties (Raguchander et al. 2011). These methodologies have also found application in the design and synthesis of compounds with antifungal activity, exemplified by nitrodihydroquinolinone pesticides (Vaz et al. 2021; Amber-Vitos et al. 2016; Chaturvedi et al. 2017). Modeling pest-enemy interactions represents another in silico approach, where virtual simulations are created to forecast how various biocontrol agents might interact with a pest species. This aids in identifying potential candidates for field trials and assessing their effectiveness before real-world implementation. In the previous study, Plouvier and Wajnberg (2018) underscore the importance of modeling pest-enemy interactions to optimize biocontrol strategies.

12.7 In Silico: Docking Analysis

The unique and divergent microbial anti-pathogenic chemicals have led the researchers' interest in developing the biopesticides from these microbial sources to replace the hazardous synthetic agrochemicals (Shafi et al. 2017). The development of biopesticides from microbial sources is a promising alternative to synthetic agrochemicals, offering effective, environmentally safe, and sustainable pest management solutions (Nawaz et al. 2016; Manda et al. 2020). These biopesticides, derived from a variety of microorganisms including fungi, bacteria, and viruses, have shown potential in controlling plant pathogens and insect pests (Ruiu 2018). In particular, the chemical diversity of antifungal secondary metabolites produced by fungal endophytes presents a valuable resource for the development of biopesticides. The continued research and development in this field are crucial for the successful replacement of hazardous synthetic agrochemicals with safer and more sustainable alternatives.

On this account, the computational broad-scale approach has achieved significant efficacy in identifying novel antimicrobial compounds by determining their targets against pathogens in silico. McPhillie et al. (2015) and Volker and Brown (2002) both further emphasize the crucial role of in silico tools in the process of target identification, with the former highlighting the use of cheminformatic approaches and the latter describing a two-level strategy for target identification and validation. Chavali et al. (2012) further underscore the significance of systems biology tools, particularly constraint-based modeling of metabolic networks, in the identification of effective drug targets. Freiberg (2001) provides a comprehensive review of novel computational methods, including genome comparison methods and cross-genome analyses, that aid in target identification and prioritization. Collectively, these studies demonstrate the effectiveness of a computational broad-scale approach in identifying novel antimicrobial compounds and their corresponding targets.

As it is well established, the target proteins, BcPMT2 and PdSlt2, are critical for the survival and virulence of two major postharvest pathogens, *Botrytis cinerea* and *Penicillium digitatum*, respectively, (González-Sánchez et al. 2013; de Ramón-Carbonell and Sánchez-Torres 2017). The protein O-mannosyltransferase BcPMT2 plays a crucial role in both cell morphology and virulence maintenance in *Cryptococcus neoformans* (Willger et al. 2009; Shimizu et al. 2014). Similarly, the mitogen-activated protein kinase PdSlt2 regulates sporulation and virulence during citrus fruit infection caused by *Penicillium digitatum* (de Ramón-Carbonell and Sánchez-Torres 2017). Considering their significance, these proteins serve as potential targets for antivirulence strategies that aim to decrease the selective pressure for resistance (Scheuplein et al. 2020). However, selected antagonists were 2-propylacrylic acid (PPA) from *Aureobasidium pullulans* (Zain et al. 2009), fengycin from *Bacillus subtilis* (Sur et al. 2018), and phenazine-1-carboxylic acid (PCA) from *Pseudomonas aeruginosa* (Simionato et al. 2017), based on the literature survey of their antagonistic bioactivities and acceptable usage.

PPA has demonstrated its ability to inhibit the growth of diverse bacteria and fungi (Zhang et al. 2017). Fengycin, a lipopeptide, exhibits both antifungal and antibacterial properties (Xiong et al. 2017). PCA, a phenazine derivative, possesses broad-spectrum antifungal and antibacterial activities (Puopolo et al. 2013). Additionally, it has been shown to hinder the growth of various plant pathogens (Rane et al. 2007). Widely used systemic fungicides imazalil (IMZ) and thiabendazole (TBZ) were used as the reference compounds to infer the inhibitory potential of natural antagonists against them. These compounds hold significant potential for applications in agriculture, aquaculture, and other industries due to their remarkable bioactivities.

The literature and pathway analysis database, Kyoto Encyclopedia of Genes and Genomes (KEGG, http://www.genome.jp/pathways.html), have been searched for selected target protein annotations, their significant role in sustaining the pathogens' survival, and their full virulence in different hosts. *Botrytis cinerea* protein O-mannosyltransferases (BcPMT) play a significant role in fungal growth and its virulence. BcPMTs, specially BcPMT2, are vitally significant for the stable fungal cell wall, sporulation, and extracellular matrix (ECM) formation. Gene knockout study conducted by González et al. (2013) showed the distorted morphology and severe defects in fungal growth and virulence in all the Δbcpmt2 mutants of *B. cinerea*.

Mitogen-activated protein kinases (MAPKs) have a significant role in signal transduction to regulate developmental and differentiation processes in eukaryotes, governing various developmental and differentiation processes (Zhao et al. 2007). Their activation occurs through dual phosphorylation, and they are involved in regulating gene expression and cytoplasmic activities (Schaeffer and Weber 1999). Additionally, MAPKs are key players in cell cycle progression, meiosis control (Pelech and Sanghera 1992), and exerting physiological functions in gene expression, cell proliferation, and programmed cell death (Chang and Karin 2001). de Ramón-Carbonell and Sánchez-Torres (2017) reported the Slt2 MPAK homologue of *Penicillium digitatum* (PdSlt2) and elucidated its significance in fungal reproduction and citrus infection. The absence of PdSlt2 was found to have a significant impact on virulence during citrus infection, leading to reduced virulence and impaired sporulation. These findings were supported by Vilanova et al. (2016), who observed similar outcomes with the PdSte12 transcription factor, which also plays a critical role in conidiation and virulence during citrus fruit infection. Additionally, Ramón-Carbonell (2017) emphasized the involvement of PdSUT1, a putative sucrose transporter, in fungicide sensitivity and virulence during citrus fruit infection. Furthermore, they used the gene elimination approach to characterize the function of the PdSlt2 in *P. digitatum*.

However, for the docking analysis, three-dimensional structures of target proteins and compound molecules were required. Therefore, due to the lack of crystal structures of *BcPMT2* and *PdSlt2* in PDB databases, the SWISS-MODEL workspace (Waterhouse et al. 2018) was utilized for the homology modeling of the same using their amino acid sequences retrieved from the NCBI protein database (accession numbers XP_024553369.1 and XP_014534667, respectively). *Saccharomyces cerevisiae* protein O-mannosyltransferase (PDB ID: 6P25.B) and *Magnaporthe oryzae* mitogen-activated protein kinase (PDB ID: 5Z33.A) have been selected as the template structures for the modeling of *BcPMT2* and *PdSlt2*, respectively. The generated models of *BcPMT2* and *PdSlt2* had been validated using the PROCHECK module of SAVES 6.0 web facility (Laskowski et al. 1993) and found acceptable. However, the 3D structure of selected compounds (ligands) had been extracted from the PubChem database (https://pubchem.ncbi.nlm.nih.gov) with the CID numbers 443591 (fengycin), 5165043 (2-propylacrylic acid), 95069 (phenazine-1-carboxylic acid), 37175 (imazalil), and 5430 (thiabendazole).

Docking simulation can be performed using the AutoDock Vina (Trott and Olson 2010) to predict the binding potential of the selected compounds in the vicinity of target proteins. We docked ligands into the prepared docking environment around the target proteins and scored their binding energy (kcal/mol) by the Vina program. The observed binding energies and docking pose of selected compounds against target proteins have been shown in Fig. 12.1a and b, respectively. As per the comparative analysis, all the selected compounds showed remarkable inhibition potential against the target proteins compared to the reference compounds (IMZ and TBZ). PCA showed the highest binding affinity against both the target proteins, followed by fengycin and PPA (Fig. 12.1a). Also, PCA and fengycin exhibited better binding efficacy than both IMZ and TBZ. Visualization of nonbonded interactions of receptor-ligand complexes has been carried out using the Discovery Studio software.

Further, binding patterns of the ligands were investigated in the context of the functional motifs of the target proteins. As per the González et al. (2013) report, *BcPMT2* has mainly three motifs, PMT, MIR, and PMT_4TMC (Fig. 12.1c). Fencing and PPA had been bounded to the PMT motif of the *BcPMT2*, while IMZ

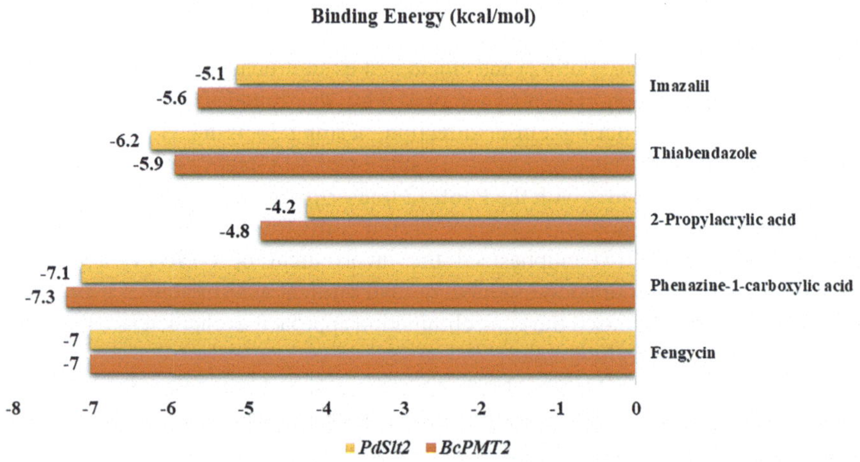

Fig. 12.1 (**a**) Binding energy of all the docked complexes. (**b**) Docking poses of all the docked complexes. (**c**) Interactive pattern of selected compounds against *BcPMT2*. (**d**) Interactive pattern of selected compounds against *BcPMT2*

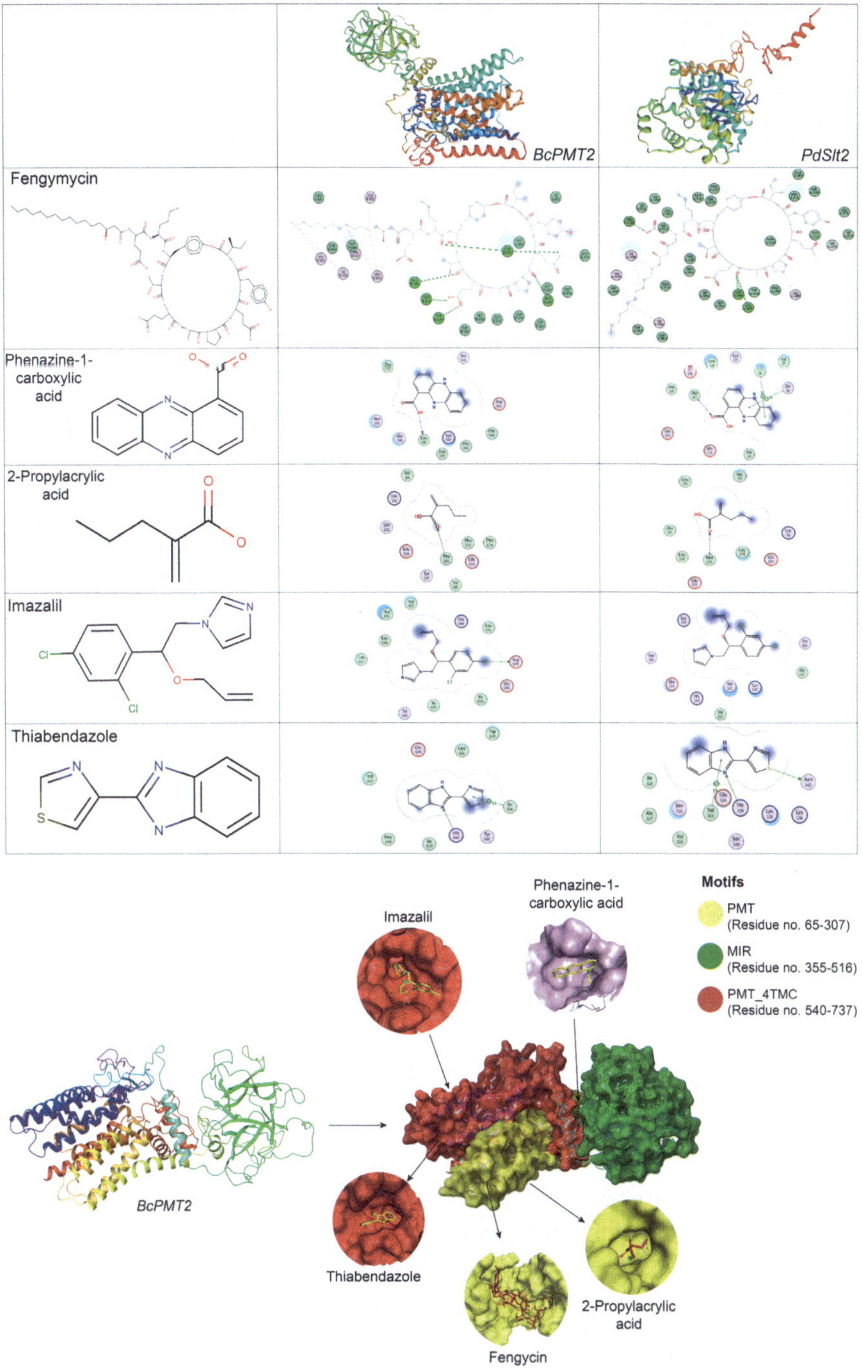

Fig. 12.1 (continued)

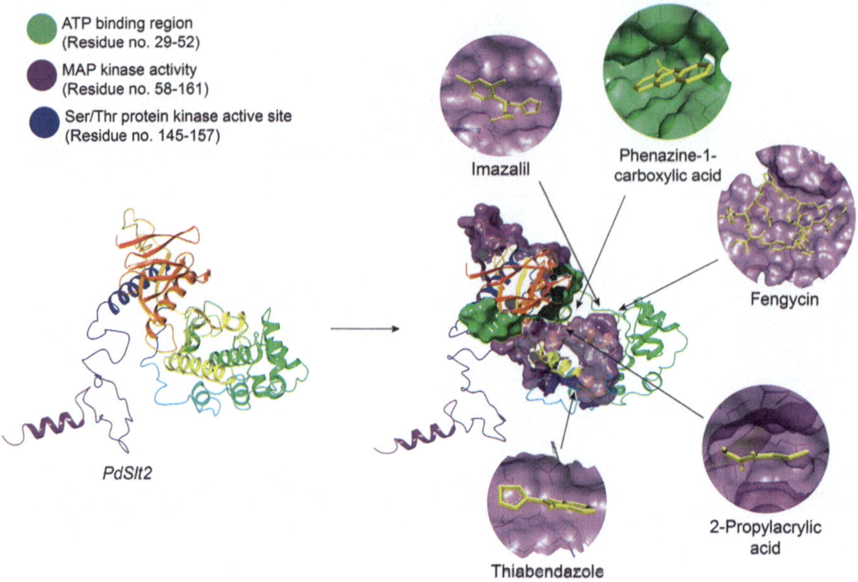

Fig. 12.1 (continued)

and TBZ were bounded to its transmembrane region, thought to disrupt the mannosyltransferases activity of the same (Fig. 12.1c). PCA was bounded in the interlinking void of the PMT and MIR domain, which might be inhibitory for both the motifs.

On the other hand, de Ramón-Carbonell and Sánchez-Torres (2017) exhibited a specific MAP kinase domain, an ATP-binding region, and a Ser/Thr protein kinase active site in the *PdSlt2*. With the pattern shown in Fig. 12.1d, all compounds except PCA were bounded in the MAP kinase domain of *PdSlt2*, suggesting it as their primary site of action, while PCA bounded with the amino acids of both MAP kinase domain and ATP-binding region. Based on the inhibitory potential of selected compounds compared to reference compounds in silico, they could be used against pathogens.

The screening of microbial biocontrol agents (BCAs) involves evaluating their potential to control plant pathogens and pests. However, this screening can be done using different approaches, in vivo, in vitro, and in silico methods, and each technique has its own significance and applications. However, combining in vivo, in vitro, and in silico approaches provides a comprehensive evaluation of BCAs. In vitro and in silico screenings can be used for initial high-throughput screening and mechanistic studies, while in vivo testing can validate the findings under realistic conditions. Moreover, the integrated approaches enhance the efficiency and accuracy of screening programs, enabling the identification and development of effective and reliable BCAs for sustainable agriculture. In addition, by leveraging the strengths of each method, researchers can develop a robust pipeline for the discovery and implementation of microbial biocontrol agents.

12.8 Conclusion

The screening of microbial biocontrol agents is evolving rapidly, integrating both traditional methods and cutting-edge in silico approaches. This synergy promises to enhance the efficiency, accuracy, and scope of discovering new MBAs for sustainable agriculture. However, the future of screening microbial biocontrol agents lies in the seamless integration of traditional and in silico approaches. This hybrid strategy will not only accelerate the discovery and development of new MBAs but also improve their precision, efficacy, and acceptance. As these technologies continue to advance, they will play a pivotal role in shaping sustainable and eco-friendly agricultural practices worldwide.

References

Abdel-Rahim IR, Abo-Elyousr KA (2018) Talaromyces pinophilus strain AUN-1 as a novel mycoparasite of Botrytis cinerea, the pathogen of onion scape and umbel blights. Microbiol Res 212:1–9

Afzal I, Shinwari ZK, Sikandar S, Shahzad S (2019) Plant beneficial endophytic bacteria: mechanisms, diversity, host range and genetic determinants. Microbiol Res 221:36–49

Alstrom S (1991) Induction of disease resistance in common bean susceptible to halo blight bacterial pathogen after seed bacterization with rhizosphere *Pseudomonads*. J Gen Appl Microbiol 37:495–501

Amber-Vitos O, Chaturvedi N, Nachliel E, Gutman M, Tsfadia Y (2016) The effect of regulating molecules on the structure of the PPAR-RXR complex. Biochim Biophys Acta 1861(11):1852–1863

Arnold AE, Lutzoni F (2007) Diversity and host range of foliar fungal endophytes: are tropical leaves biodiversity hotspots? Ecology 88(3):541–549

Arseneault T, Filion M (2017) Biocontrol through antibiosis: exploring the role played by subinhibitory concentrations of antibiotics in soil and their impact on plant pathogens. Can J Plant Pathol 39(3):267–274

Bahar NH, Lo M, Sanjaya M, Van Vianen J, Alexander P, Ickowitz A, Sunderland T (2020) Meeting the food security challenge for nine billion people in 2050: what impact on forests? Glob Environ Chang 62:102056

Bastas KK, Kumar A, Sivakumar U (eds) (2023) Microbial biocontrol: molecular perspective in plant disease management, vol 49. Springer Nature, Singapore. https://doi.org/10.1007/978-981-99-3947-3

Bonaterra A, Badosa E, Daranas N, Francés J, Roselló G, Montesinos E (2022) Bacteria as biological control agents of plant diseases. Microorganisms 10(9):1759

Cao R, Liu X, Gao K, Mendgen K, Kang Z, Gao J, Dai Y, Wang X (2009) Mycoparasitism of endophytic fungi isolated from reed on soilborne phytopathogenic fungi and production of cell wall-degrading enzymes in vitro. Curr Microbiol 59:584–592

Chang L, Karin M (2001) Mammalian MAP kinase signalling cascades. Nature 410:37–40

Chaturvedi N, Yadav BS, Pandey PN, Tripathi V (2017) The effect of β-glucan and its potential analog on the structure of Dectin-1 receptor. J Mol Graph Model 74:315–325

Chavali AK, D'Auria KM, Hewlett EL, Pearson RD, Papin JA (2012) A metabolic network approach for the identification and prioritization of antimicrobial drug targets. Trends Microbiol 20(3):113–123

de Ramón-Carbonell M, Sánchez-Torres P (2017) PdSlt2 Penicillium digitatum mitogen-activated-protein kinase controls sporulation and virulence during citrus fruit infection. Fungal Biol 121(12):1063–1074

Di Francesco A, Ugolini L, D'Aquino S, Pagnotta E, Mari M (2017) Biocontrol of Monilinia laxa by Aureobasidium pullulans strains: insights on competition for nutrients and space. Int J Food Microbiol 248:32–38

Dimkić I, Janakiev T, Petrović M, Degrassi G, Fira D (2022) Plant-associated Bacillus and Pseudomonas antimicrobial activities in plant disease suppression via biological control mechanisms—A review. Physiol Mol Plant Pathol 117:101754

Fialho MB, Moraes MH, Tremocoldi AR, Pascholati SF (2011) Potential of antimicrobial volatile organic compounds to control Sclerotinia sclerotiorum in bean seeds. Pesq Agrop Brasileira 46:137–142

Fravel DR (1988) Role of antibiosis in the biocontrol of plant diseases. Annu Rev Phytopathol 26(1):75–91

Freiberg C (2001) Novel computational methods in anti-microbial target identification. Drug Discov Today 6:72–80

González-Sánchez MÁ, de Vicente A, Pérez-García A, Pérez-Jiménez R, Romero D, Cazorla FM (2013) Evaluation of the effectiveness of biocontrol bacteria against avocado white root rot occurring under commercial greenhouse plant production conditions. Biol Control 67(2):94–100

Haas D, Keel C (2003) Regulation of antibiotic production in root-colonizing Pseudomonas spp. and relevance for biological control of plant disease. Annu Rev Phytopathol 41(1):117–153

Kashyap N, Singh SK, Yadav N, Singh VK, Kumari M, Kumar D, Shukla L, Kaushalendra, Bhardwaj N, Kumar A (2023) Biocontrol screening of endophytes: applications and limitations. Plants 12:2480

Kiss L (2001) The role of hyperparasites in host plant-parasitic fungi relationships. In: Biotic interactions in plant-pathogen associations. CABI Publishing, Wallingford, pp 227–236

Knoester M, Pieterse CM, Bol JF, Van Loon LC (1999) Systemic resistance in *Arabidopsis* induced by rhizobacteria requires ethylene-dependent signaling at the site of application. Mol Plant Microbe Interact 12:720–727

Köhl J, Kolnaar R, Ravensberg WJ (2019) Mode of action of microbial biological control agents against plant diseases: relevance beyond efficacy. Front Plant Sci 10:845

Kumar A (2022a) Microbial biocontrol: sustainable agriculture and phytopathogen management: Volume 1. Springer Nature. https://doi.org/10.1007/978-3-030-87512-1

Kumar A (2022b) Microbial biocontrol: food security and post harvest management. Springer, Cham, Switzerland. https://doi.org/10.1007/978-3-030-87289-2

Kumar A, Singh VK, Tripathi V, Singh PP, Singh AK (2018) Plant growth-promoting rhizobacteria (PGPR): perspective in agriculture under biotic and abiotic stress. In: Crop improvement through microbial biotechnology. Elsevier, California, USA, pp 333–342

Kumar A, Solanki MK, Wang Z, Solanki AC, Singh VK, Divvela PK (2024) Revealing the seed microbiome: navigating sequencing tools, microbial assembly, and functions to amplify plant fitness. Microbiol Res 279:127549

Kumari M, Qureshi KA, Jaremko M, White J, Singh SK, Sharma VK, Singh KK, Santoyo G, Puopolo G, Kumar A (2022) Deciphering the role of endophytic microbiome in postharvest diseases management of fruits: opportunity areas in commercial up-scale production. Front Plant Sci 13:1026575

Kurhekar J, Tupas GD, Otero MCB (2019) In-vitro assays for antimicrobial assessment. In: Springer eBooks. pp 279–298

Lahlali R, Ezrari S, Radouane N, Kenfaoui J, Esmaeel Q, El Hamss H, Belabess Z, Barka EA (2022) Biological control of plant pathogens: a global perspective. Microorganisms 10(3):596

Laskowski RA, MacArthur MW, Moss DS, Thornton JM (1993) PROCHECK: a program to check the stereochemical quality of protein structures. J Appl Crystallogr 26(2):283–291

Lecomte C, Alabouvette C, Edel-Hermann V, Robert F, Steinberg C (2016) Biological control of ornamental plant diseases caused by Fusarium oxysporum: a review. Biol Control 101:17–30

Liu P, Luo L, Long CA (2013) Characterization of competition for nutrients in the biocontrol of Penicillium italicum by Kloeckera apiculata. Biol Control 67(2):157–162

Loper JE, Lindow SE (1993) Roles of competition and antibiosis in suppression of plant diseases by bacterial biological control agents. In: Pest management: biologically bases technologies, pp 144–155

Lopes PF (2012) Biocontrol of phytopathogenic fungi relevant for cacao-and olive-derived economy. https://hdl.handle.net/1822/76839

Mageshwaran V, Mondal KK, Kumar U, Annapurna K (2012) Role of antibiosis on suppression of bacterial common blight disease in French bean by Paenibacillus polymyxa strain HKA-15. Afr J Biotechnol 11(60):12389–12395

Manda RR, Addanki VA, Srivastava S (2020) Microbial bio-pesticides and botanicals as an alternative to synthetic pesticides in the sustainable agricultural production. Plant Cell Biotechnol Mol Biol 21(61–62):31–48. https://ikprress.org/index.php/PCBMB/article/view/5640

McPhillie MJ, Cain R, Narramore S, Fishwick CW, Simmons KJ (2015) Computational methods to identify new antibacterial targets. Chem Biol Drug Des 85:22

Mendgen K, Wirsel SG, Jux A, Hoffmann J, Boland W (2006) Volatiles modulate the development of plant pathogenic rust fungi. Planta 224:1353–1361

Mischke S, Wünnemann B (2006) The Holocene salinity history of Bosten Lake (Xinjiang, China) inferred from ostracod species assemblages and shell chemistry: possible palaeoclimatic implications. Quat Int 154:100–112

Moricca S, Ragazzi A, Assante G (2005) Biological control of rust fungi by Cladosporium tenuissimum. In: Rust diseases of willow and poplar. CABI Publishing, Wallingford, pp 213–229

Mukherjee M, Mukherjee PK, Horwitz BA, Zachow C, Berg G, Zeilinger S (2012) Trichoderma–plant–pathogen interactions: advances in genetics of biological control. Indian J Microbiol 52:522–529

Nawaz MA, Mabubu JI, Hua H (2016) Current status and advancement of biopesticides: microbial and botanical pesticides. J Entomol Zool Stud 4:241–246

Nehra S, Gothwal RK, Dhingra S, Varshney AK, Solanki PS, Meena P, Trivedi PC, Ghosh P (2022) Mechanism of antagonism: hyperparasitism and antibiosis. In: Microbial biocontrol: sustainable agriculture and phytopathogen management, vol 1. Springer International Publishing, Cham, pp 257–277

Pathak P, Rai VK, Can H, Singh SK, Kumar D, Bhardwaj N, Roychowdhury R, de Azevedo LC, Kaushalendra VH, Kumar A (2022) Plant-endophyte interaction during biotic stress management. Plants 11(17):2203

Pelech S, Sanghera JS (1992) Mitogen-activated protein kinases: versatile transducers for cell signaling. Trends Biochem Sci 17(6):233–238

Plouvier WN, Wajnberg E (2018) Improving the efficiency of augmentative biological control with arthropod natural enemies: a modeling approach. Biol Control 125:121–130

Puopolo G, Masi M, Raio A, Andolfi A, Zoina A, Cimmino A, Evidente A (2013) Insights on the susceptibility of plant pathogenic fungi to phenazine-1-carboxylic acid and its chemical derivatives. Nat Prod Res 27:956–966

Raaijmakers JM, De Bruijn I, Nybroe O, Ongena M (2010) Natural functions of lipopeptides from Bacillus and Pseudomonas: more than surfactants and antibiotics. FEMS Microbiol Rev 34(6):1037–1062

Raguchander T, Saravanakumar D, Balasurramanian P (2011) Molecular approaches to improvement of biocontrol agents of plant diseases. J Biol Control 25(2):71–84

Rajani G, Latha PC, Sundaram RM, Phule A, Babu KV, Barbadikar KM, Babu MB, Mandal PK, Rani HS (2023) Effect of plant growth promoting endophytic bacteria Gluconacetobacter diazotrophicus, on germination attributes and seedling growth of Rice varieties under in vitro conditions. Int J Plant Soil Sci 35(20):62–71

Rane MR, Sarode P, Chaudhari BL, Chincholkar SB (2007) Detection, isolation and identification of phenazine-1-carboxylic acid produced by biocontrol strains of Pseudomonas aeruginosa. J Sci Indus Res 66:627–631

Romero FM, Rossi FR, Gárriz A, Carrasco P, Ruíz OA (2019) A bacterial endophyte from apoplast fluids protects canola plants from different phytopathogens via antibiosis and induction of host resistance. Phytopathology 109(3):375–383

Ruiu L (2018) Microbial biopesticides in agroecosystems. Agronomy 8(11):235. https://doi.org/10.3390/agronomy8110235
Saraf M, Pandya U, Thakkar A (2014) Role of allelochemicals in plant growth promoting rhizobacteria for biocontrol of phytopathogens. Microbiol Res 169(1):18–29
Sarangi T, Ramakrishnan S (2023) Biocontrol potential and mechanism of Bacillus spp. against phytopathogens: a review. Int J Environ Clim Chang 13(7):512–527
Savchuk S, Dilantha Fernando WG (2004) Effect of timing of application and population dynamics on the degree of biological control of Sclerotinia sclerotiorum by bacterial antagonists. FEMS Microbiol Ecol 49(3):379–388
Schaeffer HJ, Weber MJ (1999) Mitogen-activated protein kinases: specific messages from ubiquitous messengers. Mol Cell Biol 19:2435–2444
Scheuplein NJ, Bzdyl NM, Kibble EA, Lohr T, Holzgrabe U, Sarkar-Tyson M (2020) Targeting protein folding; a novel approach for the treatment of pathogenic bacteria. J Med Chem 63:13355
Shafi J, Tian H, Ji M (2017) Bacillus species as versatile weapons for plant pathogens: a review. Biotechnol Biotechnol Equip 31(3):446–459
Sharma I (2021) Phytopathogenic fungi and their biocontrol applications. In: Fungi bio-prospects in sustainable agriculture, environment and nano-technology. Academic, pp 155–188
Sharma S, Kooner R, Arora R (2017) Insect pests and crop losses. In: Breeding insect resistant crops for sustainable agriculture, pp 45–66
Shehata AS, Mohammed AK, Mosa AA, Ali MK (2018) Evaluation of some fungicides and biocontrol agents for controlling of alternaria rot on citrus fruits. Arab Univ J Agric Sci 26(2):691–699
Shimizu K, Imanishi Y, Toh-e A, Uno J, Chibana H, Hull CM, Kawamoto S (2014) Functional characterization of PMT2, encoding a protein-O-mannosyltransferase, in the human pathogen Cryptococcus neoformans. Fungal Genet Biol 69:13–22
Simionato AS, Navarro MO, de Jesus ML, Barazetti AR, da Silva CS, Simões GC, Balbi-Peña MI, de Mello JC, Panagio LA, de Almeida RS, Andrade G (2017) The effect of phenazine-1-carboxylic acid on mycelial growth of Botrytis cinerea produced by Pseudomonas aeruginosa LV strain. Front Microbiol 8:110
Singh M, Kumar A, Singh R, Pandey KD (2017a) Endophytic bacteria: a new source of bioactive compounds. 3 Biotech 7:1–4
Singh VK, Singh AK, Kumar A (2017b) Disease management of tomato through PGPB: current trends and future perspective. 3 Biotech 7:1–10
Solanki MK, Joshi NC, Singh PK, Singh SK, Santoyo G, de Azevedo LCB, Kumar A (2023) From concept to reality: transforming agriculture through innovative rhizosphere engineering for plant health and productivity. Microbiol Res 279:127553
Solanki MK, Wang Z, Kaushik A, Singh VK, Roychowdhury R, Kumar M, Kumar D, Singh J, Singh SK, Dixit B, Kumar A (2024) From orchard to table: significance of fruit microbiota in postharvest diseases management of citrus fruits. Food Control 165:110698
Sur S, Romo TD, Grossfield A (2018) Selectivity and mechanism of fengycin, an antimicrobial lipopeptide, from molecular dynamics. J Phys Chem B 122(8):2219–2226
Tamizi AA, Mat-Amin N, Weaver JA, Olumakaiye RT, Akbar MA, Jin S, Bunawan H, Alberti F (2022) Genome sequencing and analysis of Trichoderma (Hypocreaceae) isolates exhibiting antagonistic activity against the Papaya Dieback Pathogen, *Erwinia mallotivora*. J Fungi (Basel) 8(3):246
Trott O, Olson AJ (2010) AutoDock Vina: improving the speed and accuracy of docking with a new scoring function, efficient optimization, and multithreading. J Comput Chem 31(2):455–461
Vaz WF, Neves BJ, Custodio JM, Silva LL, D'Oliveira GD, Lemes JA, Lacerda BF, Santos SX, Perez CN, Napolitano HB (2021) In silico-driven identification and structural analysis of nitrodihydroquinolinone pesticide candidates with antifungal activity. J Mol Struct 1226:129326

Verma S, Azevedo LCB, Pandey J, Khusharia S, Kumari M, Kumar D, Kaushalendra, Bhardwaj N, Teotia P, Kumar A (2022) Microbial intervention: an approach to combat the postharvest pathogens of fruits. Plants (Basel) 11:3452

Vilanova L, Teixidó N, Torres R, Usall J, Viñas I, Sánchez-Torres P (2016) Relevance of the transcription factor PdSte12 in Penicillium digitatum conidiation and virulence during citrus fruit infection. Int J Food Microbiol 235:93–102

Vinale F, Sivasithamparam K, Ghisalberti EL, Woo SL, Nigro M, Marra R, Lombardi N, Pascale A, Ruocco M, Lanzuise S, Manganiello G (2014) Trichoderma secondary metabolites active on plants and fungal pathogens. Open Mycol J 8(1)

Volker C, Brown JR (2002) Bioinformatics and the discovery of novel anti-microbial targets. Curr Drug Targets Infect Disord 2(4):279–290

Wang Z, Solanki MK, Kumar A, Solanki AC, Pang F, Ba ZX, Niu JQ, Ren ZX (2023) Promoting plant resilience against stress by engineering root microenvironment with Streptomyces inoculants. Microbiol Res 277:127509

Waterhouse A, Bertoni M, Bienert S, Studer G, Tauriello G, Gumienny R, Heer FT, de Beer TA, Rempfer C, Bordoli L, Lepore R (2018) SWISS-MODEL: homology modelling of protein structures and complexes. Nucleic Acids Res 46(W1):W296–W303

Willger SD, Willger SD, Ernst J, Alspaugh JA, Lengeler KB (2009) Characterization of the PMT gene family in Cryptococcus neoformans. PLoS One 4:e6321

Xiong Z, Niu J, Liu H, Xu Z, Li J, Wu Q (2017) Synthesis and bioactivities of Phenazine-1-carboxylic acid derivatives based on the modification of PCA carboxyl group. Bioorg Med Chem Lett 27(9):2010–2013

Yu Y, Gui Y, Li Z, Jiang C, Guo J, Niu D (2022) Induced systemic resistance for improving plant immunity by beneficial microbes. Plants (Basel) 11(3):386. https://doi.org/10.3390/plants11030386

Zain ME, Awaad AS, Razak AA, Maitland DJ, Khamis NE, Sakhawy MA (2009) Secondary metabolites of Aureobasidium pullulans isolated from Egyptian soil and their biological activity. J Appl Sci Res 5(10):1582–1591

Zhang L, Tian X, Kuang S, Liu G, Zhang C, Sun C (2017) Antagonistic activity and mode of action of Phenazine-1-carboxylic acid, produced by marine bacterium Pseudomonas aeruginosa PA31x, against Vibrio anguillarum in vitro and in a zebrafish in vivo model. Front Microbiol 8:289

Zhang YQ, Zhang S, Sun ML, Su HN, Li HY, Zhang YZ, Chen XL, Cao HY, Song XY (2022) Antibacterial activity of peptaibols from Trichoderma longibrachiatum SMF2 against gram-negative Xanthomonas oryzae pv. oryzae, the causal agent of bacterial leaf blight on rice. Front Microbiol 13:1034779

Zhao X, Mehrabi R, Xu J (2007) Mitogen-activated protein kinase pathways and fungal pathogenesis. Eukaryot Cell 6:1701–1714

Talaromyces spp. Are Promising Biocontrol Agents for Sustainable Agriculture

13

Aqleem Abbas, Sajjad Ali, Mustansar Mubeen, Abid Hussain, Kainat Ahmed Gutumsary, Babar Hussain, and Taixiang Chen

Abstract

Pesticides have long been indispensable in modern agriculture for controlling pests and diseases, yet their widespread use has come at a significant environmental cost. These chemical compounds, while effective in the short term, often persist in ecosystems, leading to adverse impacts on biodiversity, soil health, water quality, and even human health. In response to these challenges, biocontrol agents such as *Talaromyces* spp. have emerged as sustainable alternatives in agricultural settings. Leveraging their natural bioactive metabolites, *Talaromyces*

A. Abbas (✉) · A. Hussain · K. A. Gutumsary
Department of Agriculture and Food Technology, Karakoram International University, Gilgit, Pakistan
e-mail: abid.raka@kiu.edu.pk; azhar.hussain@kiu.edu.pk

S. Ali
Department of Chemistry, Karakoram International University, Gilgit, Pakistan
e-mail: sajjadal@kiu.edu.pk

M. Mubeen
Department of Plant Pathology, College of Agriculture, University of Sargodha, Sargodha, Pakistan
e-mail: manzoor.hussain@kiu.edu.pk

B. Hussain
State Key Laboratory of Geohazard Prevention and Geo Environment Protection, Chengdu University of Technology, Chengdu, People's Republic of China

T. Chen
State Key Laboratory of Herbage Improvement and Grassland Agro-Ecosystems, Gansu Tech Innovation Centre of Western China Grassland Industry, College of Pastoral Agriculture Science and Technology, Lanzhou University, Lanzhou, People's Republic of China
e-mail: chentx@lzu.edu.cn

A. Kumar, M. K. Solanki (eds.), *Microbial Biocontrol Techniques*, Microorganisms for Sustainability 54,
https://doi.org/10.1007/978-981-97-8739-5_13

spp. offers a nuanced approach to pest and disease management. Their antibiotics exhibit potent antifungal, antibacterial, and nematicidal properties, effectively targeting pathogens without the ecological fallout associated with synthetic chemicals. Enzymes like chitinases, glucanases, and proteases produced by *Talaromyces* degrade fungal cell walls, bolstering plant defenses while promoting soil health through organic matter decomposition and nutrient cycling. Furthermore, *Talaromyces* metabolites stimulate plant resistance mechanisms and enhance growth, fostering robust crop development without the need for synthetic growth promoters. Case studies underscore *Talaromyces*' efficacy in managing soilborne and foliar diseases across diverse agricultural contexts. Strategies for mass production and formulation ensure scalability and practical application, while rigorous safety assessments mitigate potential impacts on nontarget organisms and human health. In this chapter, we comprehensively review the multifaceted applications of *Talaromyces* metabolites in agriculture. Specifically, we focus on antibiotics, enzymes, plant resistance induction, growth promotion, mycoparasitism, and present case studies on their efficacy against soilborne and foliar diseases. Additionally, we discuss mass production and formulation strategies, as well as evaluate their potential impact on humans and nontarget organisms.

Keywords

Talaromyces · Biocontrol agents · Antibiotics · Enzymes · Plant resistance induction · Growth promotion · Mycoparasitism · Sustainable agriculture

Abbreviations

CMC	Carboxymethyl cellulose
Fp	*Fusarium pseudograminearum*
GA	Gibberellic acid
IAA	Indole acetic acid
IPM	Integrated pest management
JA	Jasmonic acid
MRSA	Methicillin-resistant *Staphylococcus aureus*
NMR	Nuclear magnetic resonance
PGPF	Plant growth-promoting fungi
PPE	Personal protective equipment
PR	Pathogenesis-related
ROS	Reactive oxygen species
SA	Salicylic acid
TCA	Tricarboxylic acid cycle

13.1 Introduction

Ecologically sustainable and environmentally benign strategies for crop protection have catalyzed the exploration of efficacious biological control agents as alternatives to conventional synthetic chemical pesticides (Abbas et al. 2021). Biological control entails the utilization of living organisms or their naturally derived metabolites and compounds to inhibit the proliferation or activity of phytopathogenic agents, thereby diminishing disease incidence and mitigating severity in affected crops (Bhardwaj et al. 2024; Verma et al. 2022). This approach offers numerous advantages, including reduced environmental impact, lower toxicity to nontarget organisms, and potential for long-term sustainability. The genus *Talaromyces* has emerged as a promising biological agent for combating various plant diseases (Abbas et al. 2021; Houbraken et al. 2012, 2014). The anamorphic state of *Talaromyces* spp. is commonly represented by species belonging to the genus *Penicillium*, particularly the section *Biverticillata–Symmetrica*. *Talaromyces* species are characterized by the production of white, yellow, or red-pigmented, soft, globose ascomata (fruiting bodies). The asci (spore-bearing structures) within the ascomata are evanescent (short-lived), typically containing eight ascospores arranged in chains. The ascospores themselves are globose or ellipsoidal in shape and may exhibit smooth or ornamented surfaces. Several *Talaromyces* spp. are known to produce bioactive compounds with potential applications in various fields. For instance, the initial investigation into the utilization of *Talaromyces* spp. for the biocontrol of fungal pathogens afflicting *Cucumis sativus* commenced in 2010, when Naraghi et al. (2010a, b) isolated a strain of *T. flavus* exhibiting the capacity to suppress the manifestation of the *Verticillium* wilt disease, caused by *Verticillium albo-atrum*, in cucumber crops (Naraghi et al. 2010b). Kazerooni et al. (2019) documented that *T. pinophilus* exhibits promising biocontrol potential against the pre-emergence and postemergence damping-off diseases in *Cucumis sativus* incited by the phytopathogenic oomycete *Pythium aphanidermatum* and the basidiomycete fungus *R. solani*, respectively (Kazerooni et al. 2019). Efficacious biocontrol of the pre- and postemergence damping-off diseases in *Cucumis sativus* and *Solanum lycopersicum*, incited by *Pythium* spp., could be attained through the deployment of the fungal antagonist *Talaromyces variabilis* (Halo et al. 2019). *Talaromyces derxii* is a producer of the secondary metabolites talaroderxines A and B, which exhibit antimicrobial activity against the Gram-positive bacterium *Bacillus subtilis*, while *Talaromyces convolutus*, isolated from *Hordeum vulgare* in Japan, secretes talaroconvolutins, which exert inhibitory effects on the growth and proliferation of phytopathogenic fungi such as *Aspergillus fumigatus*, *Aspergillus niger*, *Candida albicans*, and *Cryptococcus neoformans*. Other notable examples include the biosynthesis of wortmanilactones A–D by *Talaromyces wortmannii* and the production of luteusins A–E, a novel class of azaphilone compounds, by *Talaromyces luteus* (Suzuki et al. 1992; Zhai et al. 2016). Similarly, *T. thailandiasis* demonstrated the highest efficacy in inhibiting the mycelium growth of *Colletotrichum capsici*, *Fusarium oxysporum*, *Phytophthora palmivora*, and *Lasiodiplodia theobromae* while also moderately affecting *Rhizoctonia solani* and *Sclerotium rolfsii* (Manoch and Dethoup 2011). *Talaromyces* spp. strain TpQ2 exhibited strong inhibitory activities against *F. oxysporum* mycelium growth, showcasing its potential as a biocontrol agent (Tian et al. 2021). Similarly, the growth of *F. equiseti* was inhibited by the strain DYM25 through the antibiosis effect (Luo et al. 2021). Among the *Talaromyces* species, *Talaromyces flavus* is one of the most commonly encountered species in soil and has a global distribution (Abbas et al. 2021; Dethoup et al. 2007; Naraghi et al. 2010b, 2012). It is well known for its antagonistic activity against various plant pathogenic fungi, including *Alternaria alternata*, *Fusarium moniliforme*, *Magnaporthe grisea*, *Sclerotium rolfsii*, and *Verticillium albo-atrum*, making it a promising candidate for biological control applications in agriculture (Luo et al. 2021; Manoch and Dethoup 2011; Miyake et al. 2012; Naraghi et al. 2010b; Ali et al. 2024). These fungi possess a remarkable ability to produce a diverse array of bioactive metabolites, exhibit mycoparasitic activities, and promote plant growth and defense responses, making them attractive candidates for biocontrol applications. The multiple modes of action employed by *Talaromyces* species contribute to their efficacy as biocontrol agents. They can produce antimicrobial metabolites, including antibiotics, enzymes, and other bioactive compounds, which directly inhibit or kill plant pathogens. Additionally, some *Talaromyces* strains exhibit mycoparasitic behavior, enabling them to directly attack and parasitize fungal pathogens through the production of cell wall-degrading enzymes and specialized infection structures (Kakvan et al. 2013). Furthermore, *Talaromyces* spp. can induce systemic resistance in host plants, priming their defense mechanisms and enhancing their ability to fend off pathogen attacks (Luo et al. 2021). This is achieved through the activation of various defense-related genes and the production of antimicrobial compounds and pathogenesis-related proteins. Furthermore, some *Talaromyces* strains possess plant growth-promoting traits, such as nutrient solubilization, phytohormone production, and improved uptake of essential minerals, which contribute to overall plant health and vigor. For example *T. flavus* as seed treatment enhances various plant growth parameters of cotton and potato plants (Naraghi et al. 2012). Conventional chemical pesticides, while effective in controlling pests and diseases, have raised concerns about their negative impact on the environment, nontarget organisms, and human health. These concerns, coupled with the increasing prevalence of pesticide resistance in pathogen populations, have highlighted the need for alternative and environmentally friendly approaches to crop protection. Hence, *Talaromyces*-based biocontrol agents have demonstrated remarkable efficacy against a wide range of plant pathogens, including fungi, bacteria, and nematodes, making them valuable tools in integrated pest management strategies. Their ability to combat soilborne and foliar diseases, coupled with their potential for mass production and formulation, has garnered significant interest in the agricultural community. The importance of biological control in sustainable agriculture cannot be overstated. By harnessing the natural abilities of these beneficial microorganisms, farmers and growers can effectively manage plant diseases while minimizing the negative consequences associated with chemical pesticides. The use of biocontrol agents aligns with the principles of sustainable agriculture, which emphasizes the preservation of natural resources, the promotion of biodiversity, and the reduction of environmental degradation. Moreover, the integration of biocontrol agents into crop protection strategies can contribute to long-term sustainability by reducing the reliance on chemical pesticides and promoting a more balanced and resilient agro-ecosystem. This approach not only protects crops from diseases but also supports the overall health and productivity of the agricultural system, ensuring a stable and secure food supply for future generations. This chapter delves into the various aspects of *Talaromyces* spp. as biocontrol agents, exploring their modes of action, efficacy against different plant pathogens, formulation and application strategies, environmental and safety considerations, and future research directions. By harnessing the potential of these beneficial fungi, we can take a significant step towards achieving sustainable and eco-friendly crop protection practices, contributing to enhanced food security and environmental preservation.

13.2 Genus *Talaromyces*

Talaromyces is a genus of ascomycete fungi erected by Benjamin in 1955, with *T. vermiculatus* designated as the type species (Houbraken et al. 2012, 2014). *Talaromyces* is a genus belonging to the class Ascomycetes, order *Eurotiales*, and family *Trichocomaceae* (formerly classified under *Eurotiaceae*). This fungal genus exhibits a cosmopolitan distribution and has been isolated from diverse substrates, including edaphic matrices, decomposing organic matter, animal excreta, agricultural and industrial waste streams, fecal matter, and various food sources. To date, the taxonomic description of the genus *Talaromyces* encompasses 46 distinct species and 6 designated varieties at the global level (Pitt 2014). The etymological origin of the genus name *Talaromyces* can be traced back to the Greek lexeme "taláros," which translates to "basket," an apt descriptor for the ascospore-bearing reproductive structures characteristic of this taxon (Fig. 13.1). Species presently classified under *Talaromyces* were previously subsumed within the genus *Penicillium*; however, subsequent molecular phylogenetic analyses precipitated their taxonomic reassignment to a distinct genus. Historically, species exhibiting sexual reproductive stages concomitant with *Penicillium* anamorphs were classified under the genera *Eupenicillium* and *Talaromyces* (Pitt 2014). However, during the International Botanical Congress convened in Melbourne in July 2011, species formerly classified within the Penicillium subgenus *Biverticillium* were taxonomically reassigned to the genus *Talaromyces*. Species within *Talaromyces* are characterized by their distinctive conidiophores, which are ramified structures bearing chains of asexual spores (conidia) (Figs. 13.1 and 13.2). The conidia typically exhibit an ellipsoidal or cylindrical morphology and can vary in pigmentation, ranging from green and yellow to reddish-brown hues. Certain species demonstrate the capacity to produce both asexual (conidial) and sexual (ascosporic) reproductive propagules, with the latter being formed within specialized fruiting bodies termed ascocarps (Figs. 13.1 and 13.2). Until recently, the presence of yellow or white gymnothecia in association with an asexual state belonging to *Penicillium* subgenus *Biverticillium* was a defining characteristic of *Talaromyces*. However, after the reclassification of

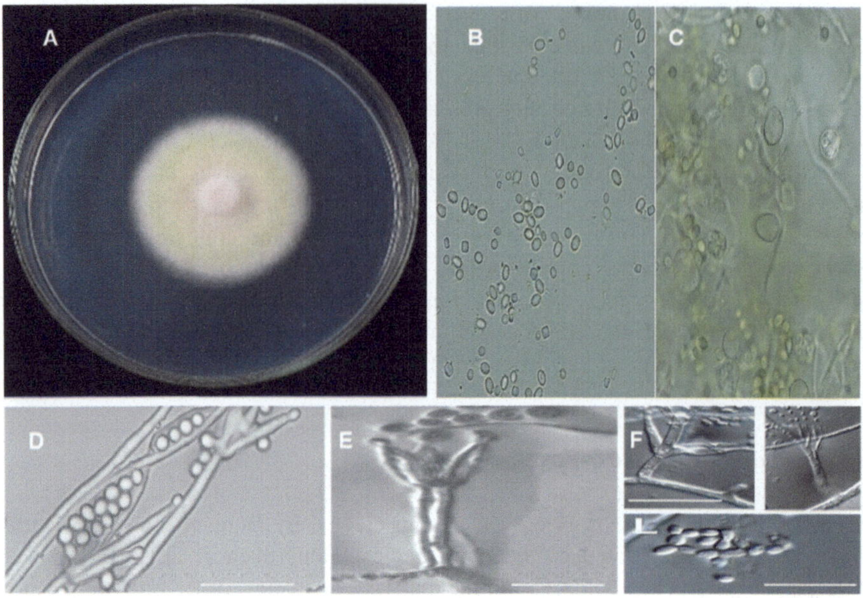

Fig. 13.1 Morphological structures of the *Talaromyces* fungus. (**a**) Colony morphology of *Talaromyces* spp. on a culture plate, showing the characteristic mycelial growth; (**b**) ascospores of *Talaromyces* spp., which are the sexual spores formed in asci; (**c**) conidia (asexual spores) of *Talaromyces* spp., which are produced in chains on conidiophores; (**d**) conidiophores of *Talaromyces* spp., which are specialized structures for producing conidia; (**e**) microscopic view of a branched conidiophore with attached conidia. (Abbas et al. 2021)

species formerly under *Penicillium* subgenus *Biverticillium* into *Talaromyces*, the genus can no longer be solely defined by its ascocarps but also by the production of a conidial state characteristic of the former *Biverticillium* group. While approximately 25 species within *Talaromyces* are known to produce ascospores, the majority of these are edaphic inhabitants. Notably, some ascospore types exhibit thermotolerance, leading to the occasional isolation of *Talaromyces* species from pasteurized fruit juices and fruit-based products, with *T. macrosporus* being the most frequently encountered species in such commodities. In the natural environment, *T. flavus* stands out as the most ubiquitous species within the genus, occasionally isolated from cereals and other food products (Yilmaz et al. 2014). *Talaromyces wortmannii*, though exhibiting many similarities to *T. flavus*, can be readily distinguished by its comparatively slower growth rate and is the only other species likely to be encountered in food matrices. The intricate gymnothecia of *Talaromyces*, once a defining feature, now serves as a unique characteristic alongside the conidial states, reflecting the genus's diverse evolutionary history and ongoing taxonomic refinements guided by modern molecular techniques. The genus *Talaromyces* is renowned for its diverse metabolic capabilities, which enable its members to degrade a wide range of substrates, including cellulose, hemicellulose, lignin, and various aromatic compounds. This metabolic versatility has rendered certain

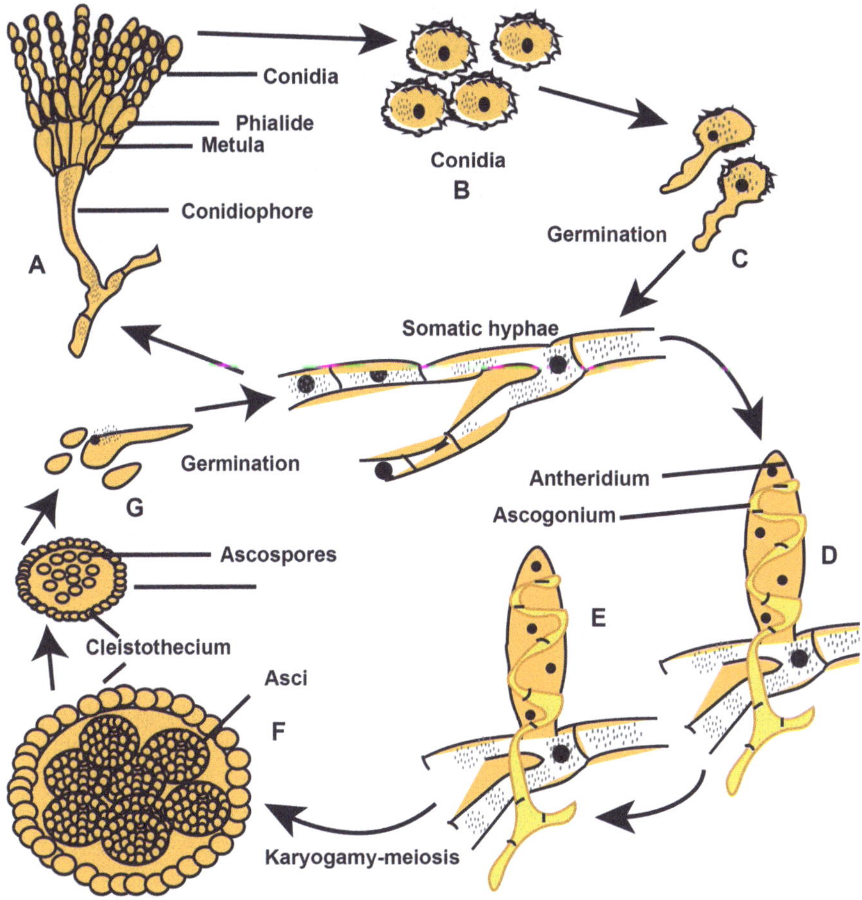

Fig. 13.2 The life cycle of *Talaromyces* spp. (**a**, **b**) Formation of conidiophores, which produce conidia. (**c**) These conidia are then capable of germination, leading to the development of somatic hyphae. (**d**) The sexual reproduction phase involves the formation of antheridium and ascogonium. (**e**) Plasmogamy between these structures. (**f**) This fusion leads to the development of a cleistothecium, within which asci containing ascospores are formed. (**g**) The ascospores are then released and germinate

Talaromyces species valuable in industrial applications, such as enzyme production, biodegradation, and biotransformation processes. Several *Talaromyces* species are known to biosynthesize secondary metabolites exhibiting potential bioactive properties. These metabolites encompass antibiotics, mycotoxins, pigments, and other compounds with prospective applications in the pharmaceutical, agricultural, and food industries. However, it is imperative to note that some of these metabolites can be deleterious to humans, animals, or plants, necessitating the implementation of appropriate safety measures when working with these species. Traditionally, the identification and classification of *Talaromyces* spp. have relied on morphological characteristics, such as colony appearance, conidial ornamentation, and presence or absence of sexual structures. However, contemporary molecular techniques, including DNA sequencing and phylogenetic analyses, have provided a more robust framework for species delimitation and elucidating phylogenetic relationships within the genus. *Talaromyces* spp. is ubiquitous in diverse habitats, including edaphic matrices, atmospheric environments, plant materials, and indoor environments. Certain species are recognized as opportunistic human pathogens, capable of causing infections in immunocompromised individuals or those with underlying health conditions. Proper hygiene and handling practices are essential when working with these fungi. Several *Talaromyces* species have exhibited promising potential as biological control agents against phytopathogens, insect pests, and other deleterious organisms. Their ability to biosynthesize a wide range of secondary metabolites, including antibiotics and other bioactive compounds, renders them attractive candidates for developing environmentally benign pest management strategies.

13.3 Mechanisms of Action of *Talaromyces* spp. Against Plant Pathogens

13.3.1 Production of Antimicrobial Metabolite

One of the primary modes of action employed by *Talaromyces* species in their role as biocontrol agents is the production of a diverse array of antimicrobial metabolites. These metabolites possess the ability to directly inhibit or kill plant pathogens, serving as potent weapons in the battle against various diseases.

13.3.1.1 Antibiotics

Talaromyces species are prolific biosynthetic producers of antibiotics, many of which exhibit antifungal, antibacterial, and nematicidal properties. One notable example is mitorubrin, a polyketide antibiotic biosynthesized by several *Talaromyces* strains. Mitorubrin has demonstrated remarkable antifungal activities against economically significant phytopathogenic fungi such as *Botrytis cinerea* (causative agent of gray mold), *Fusarium oxysporum* (vascular wilt pathogen), and *Rhizoctonia solani* (incitant of root rot diseases). Other antibiotics produced by *Talaromyces* include talaromycins, which have been shown to inhibit the growth and proliferation of various phytopathogenic fungi and bacteria. Numerous studies have elucidated that *Talaromyces* spp. biosynthesizes a diverse array of antibiotics exhibiting potential applications in mitigating plant pathogenic diseases. In another study, it has been demonstrated that proxide, in conjunction with hydrogen peroxide, prevented the formation of microsclerotia by the phytopathogenic fungus *Verticillium dahlia* (Kim et al. 1990). Suzuki et al. (1992) documented the isolation of the secondary metabolites talaroderxines A and B from the fungal species *Talaromyces derxii*, exhibiting antimicrobial activity against the Gram-positive bacterium *Bacillus subtilis* (Suzuki et al. 1992). This discovery underscores the potential of secondary metabolites derived from *T. derxii* as natural antimicrobial agents, providing avenues for further exploration and exploitation in the pharmaceutical and agricultural industries (Suzuki et al. 2000). Moreover, *T. convolutus* isolated from *Hordeum vulgare* in Japan has been demonstrated to biosynthesize talaroconvolutins, which exert inhibitory effects against a diverse array of phytopathogenic and human pathogenic fungi, including *Aspergillus fumigatus*, *A. niger*, *Candida albicans*, and *Cryptococcus neoformans* (Suzuki et al. 2000). This broad-spectrum antifungal activity positions talaroconvolutins as valuable candidates for the development of novel antifungal agents with potential applications in both agricultural crop protection and clinical therapeutics. In addition to these findings, Dong et al. (2006) reported the isolation and characterization of wortmanilactones A–D from the fungal species *T. wortmannii*, while Yoshida et al. (1996) elucidated the chemical structures of luteusins A–E, a novel class of azaphilone compounds, from *T. luteus* (Dong et al. 2006). These newly identified chemical entities contribute to the expanding repertoire of bioactive metabolites sourced from *Talaromyces* spp., highlighting the rich chemical diversity inherent within this genus and its potential for drug discovery and agricultural biotechnology endeavors. Furthermore, *T. flavus* emerges as a prominent species within the genus, renowned for its antagonistic activity against various phytopathogenic fungi, including *Alternaria alternata*, *Rhizoctonia solani*, *Fusarium moniliforme*, *Magnaporthe grisea*, *Sclerotium rolfsii*, and *Verticillium albo-atrum* (Duo-Chuan et al. 2005; Inglis and Kawchuk 2002; Naraghi et al. 2010a). In another study, *Talaromyces* strains indigenous to the Amazon region were isolated with the objective of identifying biologically active secondary metabolites. Thirteen compounds, encompassing polyketides, steroids, anhydrides, and phenolic compounds, were identified through classical chromatographic approaches. Two previously uncharacterized chemical entities, a homodimeric chromanone, and a maleic anhydride methyl ester derivative, were structurally elucidated. The antimicrobial activity of these compounds was evaluated against the Gram-positive bacteria *Bacillus cereus* and *Staphylococcus aureus*, the Gram-negative bacterium *Escherichia coli*, and the phytopathogenic bacterium *Ralstonia solanacearum*. Furthermore, the antifungal potential was assessed against the fungi *Penicillium avellaneum*, *Candida albicans*, and *Candida tropicalis* (Silva et al. 2018). Similarly, endophytic fungi, such as *Talaromyces wortmannii*, are known to produce bioactive compounds that have potential applications against plant pathogens (Bara et al. 2013). Recent research findings have identified several bioactive compounds from *T. wortmannii*, including a newly discovered atropisomer and various derivatives of wortmannin. These compounds have demonstrated significant antibiotic efficacy against a wide spectrum of bacteria, encompassing both Gram-positive and Gram-negative strains, including those that afflict plants. Notably, biemodin, one of these compounds, exhibits potent antibiotic activity against Gram-positive bacteria, notably methicillin-resistant *Staphylococcus aureus* (MRSA). These findings underscore *T. wortmannii*'s promise as a valuable reservoir of novel antimicrobial agents, potentially pivotal in addressing antibiotic-resistant bacterial strains and safeguarding plants from pathogenic threats. However, the current study's focus is limited in certain aspects. It does not delve deeply into identifying the most effective compounds against specific pathogenic microorganisms. Furthermore, the synergistic potential of combining different compounds to enhance antimicrobial activity remains unexplored. Likewise, the mechanisms underlying the action of these isolated compounds against the tested microorganisms warrant further investigation. Moreover, the broad geographic range and antagonistic capabilities of *T. flavus* make it a promising candidate for biocontrol strategies aimed at managing fungal diseases in crops. These findings underscore the importance of bioactive compounds derived from *Talaromyces* in combating microbial pathogens, spanning agricultural and clinical environments alike. Ongoing research into the biosynthetic pathways, mechanisms of action, and potential applications of these compounds show the potential for addressing challenges associated with plant diseases and human health. Additionally, investigating the diversity within *Talaromyces* may reveal novel bioactive molecules with varied biological activities, thereby enriching the arsenal of natural products for pharmaceutical and agricultural purposes. The antibiotics and enzymes produced by *Talaromyces* spp. against plant pathogens are shown in Table 13.1.

13.3.1.2 Enzymes

In addition to their antibiotic properties, *Talaromyces* species are renowned for their production of hydrolytic enzymes capable of breaking down the cell walls of plant pathogens, leading to their lysis and subsequent demise. Notably, chitinases and glucanases play pivotal roles in this process: chitinases degrade chitin, a major component of fungal cell walls, while glucanases target glucan components. Certain strains of *Talaromyces*, such as *T. flavus*, are recognized for their ability to produce substantial quantities of these enzymes, enhancing their efficacy in biocontrol against fungal pathogens. For instance, Kim and Fravel demonstrated that glucose oxidase, produced by *T. flavus*, effectively inhibits the formation of microsclerotia of *V. dahliae* (Kim et al. 1990). Nevertheless, Proksa et al. (1992) documented that 2-methyl sorbic acid, produced by *T. flavus*, exhibited inhibitory effects on the growth of *V. albo-atrum* (Proksa et al. 1992). The glucose oxidase enzyme produced by *T. flavus* functions in the presence of glucose, leading to the production of hydrogen peroxide. Consequently, the addition of glucose to *T. flavus*'s nonvolatile metabolite enhances its inhibitory effect on microsclerotia formation (Murray et al. 1997). Another study evaluated the effectiveness of nonvolatile extracts, specifically chitinase enzymes, obtained from *T. flavus*, in combating soybean stem white rot and bean stem rot diseases. Chitinase enzymes are renowned for their ability to break down chitin, a key constituent of fungal cell walls, thereby disrupting fungal growth and pathogenicity. Application of the extract derived from *T. flavus* on soybean plants effectively controlled bean stem rot caused by *S. rolfsii*. These findings highlight the potential of chitinase enzymes produced by *Talaromyces* spp. as biocontrol agents against these destructive fungal pathogens (Madi et al. 1997). In the realm of effective fungal metabolites, glucanase from *Zygorhynchus moelleri* and glucose oxidase from *T. flavus* stand out for their demonstrated efficacy against a range of fungal pathogens including *R. solani*, *P. intermedium*, *V. albo-atrum*, and *Fusarium oxysporum* f. sp. *lini* (F.o.l.). Glucanase enzymes play a critical role by

Table 13.1 Antibiotics and enzymes of *Talaromyces* spp. used against plant pathogens

S. No.	*Talaromyces* spp.	Plant-pathogen	Diseases	Antibiotics/ enzymes	References
1	*T. flavus*	*Verticillium dahliae*	Verticillium wilt	Glucose oxidase, 2-methyl sorbic acid	Kim et al. (1990), Proksa et al. (1992)
2	*T. wortmannii*	Various	Various	Wortmanilactones A–D	Dong et al. (2006)
3	*T. luteus*	Various	Various	Luteusins A–E	Yoshida et al. (1996)
4	*T. flavus*	*Alternaria alternata*	Leaf spot	Various enzymes	Duo-Chuan et al. (2005), Inglis and Kawchuk (2002)
5	*T. derxii*	*Bacillus subtilis*	Bacterial diseases	Talaroderxines A and B	Suzuki et al. (1992)
6	*T. convolutes*	Various	Various	Talaroconvolutins	Suzuki et al. (2000)
7	*T. flavus*	*Botrytis fabae*	Brown spot disease	Protease	Haggag et al. (2006)
8	*T. flavus*	Various	Various	Protease, chitinase, amylase, cellulase, pectinase	Sahu et al. (2019)
9	*T. muroii*	*Fusarium pseudograminearum*	Fusarium crown rot	Antioxidants, cell wall-degrading enzymes	Yang et al. (2023)
10	*T.* spp. (*Amazon region*)	Various	Various	Polyketides, steroids, anhydrides, phenolic compounds	Silva et al. (2018)
11	*T.* spp.	Various	Various	Unspecified bioactive compounds	Bara et al. (2013)
12	*T. wortmannii*	Various	Various	Wortmanilactones A–D	Dong et al. (2006)
13	*T. luteus*	Various	Various	Luteusins A–E	Yoshida et al. (1996)
14	*T. flavus*	*Alternaria alternata*	Leaf spot	Various enzymes	Duo-Chuan et al. (2005), Inglis and Kawchuk (2002)

(continued)

Table 13.1 (continued)

S. No.	*Talaromyces* spp.	Plant-pathogen	Diseases	Antibiotics/ enzymes	References
15	*T. derxii*	*Bacillus subtilis*	Bacterial diseases	Talaroderxines A and B	Suzuki et al. (1992)
16	*T. convolutes*	*Botrytis cinerea*, *Alternaria solani*	Gray mold, Early blight	Talaroconvolutin A	Lan and Wu (2020)
17	*T. convolutes*	Various	Various	Talaroconvolutins	Suzuki et al. (2000)
18	*T. flavus*	*Botrytis fabae*	Brown spot disease	Protease	Haggag et al. (2006)
19	*T. flavus*	Various	Various	Protease, chitinase, amylase, cellulase, pectinase	Sahu et al. (2019)
20	*T. muroii*	*Fusarium pseudograminearum*	Fusarium crown rot	Antioxidants, cell wall-degrading enzymes	Yang et al. (2023)
21	*T.* spp.	Various	Various	Polyketides, steroids, anhydrides, phenolic compounds	Silva et al. (2018)
22	*T.* spp.	Various	Various	Unspecified bioactive compounds	Bara et al. (2013)
23	*T. meleagris*	Various	Various	Meleagrin	Lan and Wu (2020)

degrading glucans within fungal cell walls, thereby compromising their structural integrity and ultimately leading to cell lysis. On the other hand, glucose oxidase produced by *T. flavus* catalyzes the generation of hydrogen peroxide, a potent antifungal agent that disrupts crucial fungal cellular processes. Beyond its direct fungicidal properties, this enzyme has been noted for its ability to induce systemic resistance in plants. By triggering systemic defense mechanisms, glucose oxidase enhances the overall ability of plants to defend against fungal pathogens, thus offering a dual mechanism of action in disease management strategies (Brown 1987; Singh et al. 2017a, b). The genomes of *Talaromyces* spp. harbor a remarkable diversity of primary and secondary metabolic gene clusters, encoding a wide array of enzymes and bioactive compounds. This genomic richness underlies the fungi's versatility in producing various proteases, carbohydrate-active enzymes, fungal cell wall-degrading enzymes, lectins, and secondary metabolites. Such genetic diversity equips *Talaromyces* spp. with the tools necessary for survival, adaptation, and complex ecological interactions in their environments. The ability to synthesize this broad spectrum of bioactive compounds and enzymes not only contributes to the fungi's ecological success but also presents significant potential for biotechnological applications. Of particular interest is the demonstrated antagonistic activity of *T. muroii* TM28 against *Fusarium pseudograminearum*, the causative agent of

Fusarium crown rot in wheat. In vitro studies have shown that *T. muroii* TM28 can inhibit both conidial and hyphal growth of the pathogen, highlighting its potential as a biocontrol agent in agriculture. This finding exemplifies how the metabolic capabilities of *Talaromyces* spp. can be harnessed for practical applications in crop protection, potentially offering sustainable alternatives to chemical fungicides. The diverse metabolic toolkit of *Talaromyces* spp. opens up numerous avenues for future research and applications. Comparative genomic studies could provide insights into the evolution and distribution of these metabolic capabilities across the genus, while investigations into the specific mechanisms of biocontrol could lead to more targeted and effective strategies for crop protection. Furthermore, the rich enzymatic and metabolic repertoire of these fungi suggests potential applications beyond biocontrol, such as in industrial enzyme production, discovery of novel bioactive compounds, or bioremediation. However, the development of *Talaromyces*-based biocontrol strategies also presents challenges. Translating promising in vitro results to effective field applications requires careful consideration of environmental factors, delivery methods, and potential interactions with other microorganisms. Additionally, as some *Talaromyces* spp. are known to produce mycotoxins, thorough safety assessments would be crucial before widespread agricultural application. Despite these challenges, the genetic and metabolic diversity of *Talaromyces* spp. offers significant opportunities for selecting or engineering strains with enhanced biocontrol properties and minimal safety concerns (Yang et al. 2023). The introduction of *T. muroii* TM28 into wheat ecosystems catalyzed a cascade of biochemical responses, fostering the production of protective antioxidants and an array of cell wall-disintegrating enzymes. This multifaceted intervention culminated in a marked diminution of *Fusarium pseudograminearum* (Fp) proliferation across the rhizospheric milieu, root structures, and basal stem regions of wheat specimens in meticulously orchestrated pot trials. Genomic expression profiling via transcriptome sequencing illuminated a pronounced attenuation of genetic activity linked to cellular architecture synthesis in Fp when subjected to TM28's antagonistic presence. Furthermore, the bioactive exudates from TM28 cultures exerted far-reaching metabolic perturbations in Fp, disrupting crucial anabolic pathways governing lipid biosynthesis, energy metabolism, and central carbon utilization. This metabolic sabotage effectively compromised Fp's capacity for mycotoxin translocation, thus mitigating its virulence potential. The investigation into *T. flavus* and its proteases presents an innovative approach to biocontrol, particularly against the economically significant brown spot disease in faba beans caused by *Botrytis fabae*. This study demonstrates the adaptive capabilities of *T. flavus*, which was induced to produce proteases when cultivated in the presence of casein. The subsequent purification and evaluation of these proteases for their antifungal properties mark a shift from traditional whole-organism biocontrol methods to a more targeted, enzyme-based approach. This research opens up several intriguing avenues for discussion and further investigation. The mechanism by which these proteases inhibit *B. fabae* growth, possibly through the degradation of essential cell wall proteins or extracellular enzymes, warrants deeper exploration. The use of purified proteases, rather than whole *T. flavus* cells, could offer advantages such as more predictable effects, easier standardization, and potentially reduced ecological impact. However, this approach also presents challenges, including ensuring the stability of proteases in field conditions, assessing effects on nontarget organisms, and determining the cost-effectiveness of large-scale production. Comparative studies evaluating the efficacy of these proteases against conventional fungicides and other biocontrol agents would provide valuable insights into their practical potential. Moreover, this protease-based method might offer benefits in resistance management, as it targets fundamental cellular components, potentially making it less prone to the development of resistance compared to conventional fungicides. The success of this approach could inspire similar studies with other fungi and their enzymes against various plant pathogens, potentially leading to a new class of biocontrol agents. To fully realize the potential of this approach, future research should focus on developing appropriate formulations to preserve protease activity and ensure effective delivery to plant surfaces. Additionally, exploring how these proteases could be integrated into existing disease management strategies for faba beans could lead to more comprehensive and sustainable control of brown spot disease. While promising, further investigations are needed to fully understand the efficacy, safety, and practical feasibility of this approach in real-world agricultural settings. Nonetheless, this line of research represents a significant step towards developing more environmentally friendly and sustainable crop protection strategies, potentially reducing reliance on chemical fungicides while maintaining effective disease control (Haggag et al. 2006). The purification of proteases from *T. flavus* involved a sophisticated two-stage isolation protocol, showcasing the researchers' commitment to obtaining enzymes of exceptional purity and potency. Initially, the crude enzyme extract underwent ammonium sulfate precipitation, a classic technique that exploits differences in protein solubility to achieve preliminary separation. This step effectively concentrated the proteases while eliminating a significant portion of extraneous proteins and cellular debris. Following this initial fractionation, the partially purified enzyme preparation was subjected to Sephadex G-200 gel permeation chromatography. This advanced chromatographic method leverages the principle of size exclusion, allowing for precise separation of proteins based on their molecular dimensions. The selection of Sephadex G-200 as the matrix is particularly noteworthy, as it offers an optimal fractionation range for proteins of the size typically associated with fungal proteases. The implementation of this dual-step purification strategy underscores the researchers' meticulous approach to enzyme isolation. By combining these complementary techniques, they were able to achieve a high degree of purification, effectively isolating proteases with superior catalytic activity. This rigorous purification process not only ensures the acquisition of high-quality enzymes for subsequent antifungal efficacy studies but also facilitates more accurate characterization of the proteases' biochemical properties and functional attributes. The emphasis on obtaining highly active proteases through this refined purification method highlights the study's focus on quality and specificity. Such purified enzymes are invaluable for precise evaluation of their antifungal properties against *Botrytis fabae*, as they minimize potential interference from other fungal metabolites or proteins. Furthermore, this level of enzyme purity is crucial for any future applications, whether in agricultural settings or for further biochemical investigations. This methodological detail also opens avenues for future research. For instance, it would be intriguing to explore how variations in the purification process might affect the enzyme's stability, specificity, or antifungal efficacy. Additionally, the successful isolation of these proteases paves the way for structural studies, which could provide insights into their mechanism of action and guide efforts to enhance their antifungal properties through protein engineering. Optimal enzymatic activity was observed at a temperature of 30 °C and pH 6.5, highlighting the stability and efficiency of these proteases under physiological conditions. *T. flavus* exhibited robust extracellular protease production, demonstrating its capability for enzymatic synthesis. The purified proteases showed potent inhibitory effects on *B. fabae*, affecting germination, growth rate, and extracellular enzyme activities at concentrations ranging from 40 to 120 U/mL. Complete suppression of *B. fabae* growth and extracellular enzyme production was achieved at 80 U/mL of protease concentration. Notably, the application of these proteases effectively reduced brown spot disease severity and pathogen sporulation on faba bean leaves inoculated with *B. fabae*. In field trials conducted under natural infestation conditions, foliar application of *T. flavus* protease at 80 U/mL significantly decreased disease incidence on both leaves and pods of faba bean plants. This treatment also led to an increase in pod yield per plant, underscoring the potential of *T. flavus* protease as an eco-friendly biocontrol agent for managing brown spot disease in faba beans. Additionally, *T. trachyspermus* exhibited heightened activity across various hydrolytic enzymes such as protease, chitinase, amylase, cellulase, and pectinase, crucial for their antagonistic properties (Sahu et al. 2019).

13.3.1.3 Other Bioactive Compounds

In addition to antibiotics and enzymes, *Talaromyces* spp. synthesizes a diverse array of bioactive compounds crucial for their biocontrol efficacy. These include organic acids, plant growth regulators, and secondary metabolites with multifaceted biological functions. Organic acids such as gluconic and citric acids modulate pH levels in their environment, creating inhospitable conditions for targeted plant pathogens. Certain *Talaromyces* strains also produce plant growth regulators like indole acetic acid (IAA) and gibberellic acid (GA), which stimulate plant growth and development, bolstering plant resistance against pathogen attacks indirectly. Moreover, *Talaromyces* secondary metabolites exhibit broad-spectrum bioactivities encompassing antifungal, antibacterial, and insecticidal properties. For example, talaroconvolutin A, a polyketide from *T. convolutus*, displays potent antifungal action against notorious plant pathogens such as *Botrytis cinerea* and *Alternaria solani*. Similarly, meleagrin, an alkaloid isolated from *T. meleagris*, demonstrates antibacterial and antifungal efficacy, positioning it as a promising contender for biocontrol strategies. These diverse bioactive compounds underscore the potential of *Talaromyces* spp. as valuable resources in agricultural and pharmaceutical applications (Lan and Wu 2020).

13.3.2 Mycoparasitism

In addition to the production of antimicrobial metabolites, some *Talaromyces* species exhibit mycoparasitic behavior, enabling them to directly attack and parasitize plant pathogenic fungi. This mode of action involves a complex interplay of various mechanisms and specialized structures (Abbas et al. 2021). The mycoparasitic process typically initiates with the recognition of the host fungus by the *Talaromyces* spp. Subsequently, the fungus produces hydrolytic enzymes such as chitinases and glucanases, which play a pivotal role in degrading the cell walls of the host fungus. For instance, chitinases of 41 and 32 kDa have been identified with activity against the cell walls of *Verticillium* spp., *S. sclerotiorum*, and *R. solani*. These enzymes also exhibit inhibitory effects on spore germination and germ tube elongation of *A. alternata*, *F. moniliforme*, and *M. grisea* (Duo-Chuan et al. 2005). Subsequently, the *Talaromyces* strain forms specialized structures like appressoria and haustoria, which facilitate the penetration and colonization of the host fungus. Once inside the host, the *Talaromyces* mycoparasite can utilize the nutrients and resources of the host, ultimately leading to its demise. This process is often accompanied by the production of additional antimicrobial metabolites and enzymes that further contribute to the suppression of the host fungus. The biological potential of *Talaromyces* sp. KNB422 against common rice seedling pathogens was assessed in a research (Miyake et al. 2012). The in vitro experiments provide valuable insights into the mode of action of KNB422, demonstrating its ability to parasitize and inhibit the growth of pathogens like *Rhizopus oryzae*, *Pythium graminicola*, and *Gibberella fujikuroi*. These interactions, observed through scanning electron microscopy and fluorescence microscopy, offer a detailed understanding of how KNB422 effectively combats these detrimental organisms. While the study showcases the efficacy of KNB422 in controlled laboratory conditions, further research is warranted to explore its practical application in real-world agricultural settings. Investigating factors such as long-term effects, field efficacy, scalability, and ecological impacts will be crucial in assessing the viability of KNB422 as a biocontrol agent for managing rice seedling diseases. Another study presents an innovative approach to isolating and evaluating potential biocontrol agents from the genus *Talaromyces* against the notorious plant pathogen *R. solani*. The use of *R. solani* sclerotia as baits in paddy soil demonstrates a clever and targeted isolation technique for *Talaromyces* spp. This method effectively selects *Talaromyces* spp. that naturally interacts with and potentially antagonizes *R. solani*, increasing the likelihood of identifying effective biocontrol candidates. The success in isolating five *Talaromyces* strains (TF-04, TF-03, TF-02, TF-01, and TA-02) through this method underscores the prevalence of potential biocontrol agents in agricultural soils and the efficacy of this selective isolation approach. The subsequent evaluation of these strains against *R. solani* revealed promising results, particularly in their ability to parasitize the pathogen's hyphae and cause decay of its sclerotia. This dual-action mechanism—attacking both the active growth phase (hyphae) and the survival structures (sclerotia) of *R. solani*—suggests a comprehensive biocontrol potential of these *Talaromyces* strains. Notably, the *Talaromyces* strains maintained their efficacy even under conditions of higher water content in both sterilized sand and soil (Abbas et al. 2021). In another study, dual-culture tests demonstrated that *T. pinophilus*'s antagonistic potential against *B. cinerea* relied on mycoparasitism, evidenced by the formation of mycelial overgrowth. Detailed analyses using light microscopy, SEM, and TEM revealed that *T. pinophilus* penetrated and grew intracellularly within *B. cinerea* hyphae, causing significant cytoplasmic disorganization and cell wall disintegration at the penetration sites (Abdel-Rahim and Abo-Elyousr 2018). *Talaromyces flavus* reduced the germinability of *Verticillium dahliae* microsclerotia in soil from 84% to 17% by parasitizing them. The mycoparasite invaded host cells using small penetration pegs, lysing cell walls, and degrading cell contents, thus aiding in the biological control of *Verticillium* wilt disease. Scanning electron micrographs showed heavy fungal colonization on microsclerotia buried in treated soil, indicating ongoing parasitism by *T. flavus* (Fahima et al. 1992). *Talaromyces variabilis* demonstrated strong inhibitory effects on the in vitro growth of *Pythium aphanidermatum* in both solid and liquid media. Abnormalities in the hyphae of *P. aphanidermatum* were observed due to the presence of *T. variabilis*, indicating a direct antagonistic interaction. In a series of groundbreaking studies, various *Talaromyces* spp. have demonstrated remarkable prowess in combating an array of plant pathogens, heralding a new era in biological control strategies. *T. variabilis* emerged as a formidable adversary against *Pythium aphanidermatum*, with its extracts orchestrating cellular disruption and dramatically curtailing oospore formation in the pathogen. This multifaceted assault on *P. aphanidermatum*'s life cycle underscores *T. variabilis*' potential as a robust biocontrol agent. Concurrently, *T. purpurogenus* Q2 exhibited exceptional efficacy in mitigating *Fusarium* wilt in bitter gourd. By strategically reducing the rhizospheric abundance of *Fusarium oxysporum* f. sp. *momordica*, this *Talaromyces* strain significantly diminished disease incidence across both controlled greenhouse environments and real-world field trials. This successful translation from laboratory to field application marks a significant milestone in sustainable disease management. Perhaps most impressively, *T. muroii* strain TM28, originally isolated from the roots of *Panax quinquefolius*, showcased extraordinary antagonistic capabilities against *Fusarium pseudograminearum*. In rigorous in vitro assays, TM28 demonstrated a staggering 87.8% inhibition of *F. pseudograminearum* mycelial growth within just 72 h. Furthermore, the cell-free fermentation filtrate of TM28 exhibited a dual-pronged attack on the pathogen, not only impeding mycelial growth but also thwarting conidial germination. This comprehensive suppression was achieved through the filtrate's ability to compromise the integrity of *F. pseudograminearum*'s cell membranes, effectively neutralizing the pathogen's growth and reproductive capabilities. These findings collectively illuminate the vast potential of *Talaromyces* spp. in revolutionizing plant disease management. By harnessing their diverse mechanisms of action—from inducing cellular leakage to disrupting membrane integrity and inhibiting reproductive structures—these fungi offer a sophisticated, eco-friendly alternative to conventional chemical controls. As research in this field advances, the integration of these *Talaromyces* strains into agricultural practices promises to usher in a new paradigm of sustainable crop protection, potentially reshaping the landscape of global food security.

13.3.3 Induction of Resistance and Promotion of Plant Growth

In addition to direct antagonism, some *Talaromyces* spp. can induce systemic resistance in host plants, enhancing their natural defense mechanisms against pathogens. This mode of action involves the priming of the plant's immune system, leading to the activation of various defense-related genes and the production of antimicrobial compounds and pathogenesis-related proteins. The process of inducing systemic resistance begins with the recognition of *Talaromyces*-derived elicitors by the plant's immune system. These elicitors can be various compounds, such as chitin oligomers, lipopolysaccharides, or other signaling molecules. Upon recognition, the plant's defense signaling pathways are activated, leading to the accumulation of defense-related proteins, such as pathogenesis-related (PR) proteins, and the production of secondary metabolites with antimicrobial properties. *Talaromyces* spp. can prime the plant's defense responses through various mechanisms, including the activation of salicylic acid (SA) and jasmonic acid (JA) signaling pathways. These pathways play crucial roles in regulating the expression of defense-related genes and the production of antimicrobial compounds, such as phytoalexins and reactive oxygen species (ROS). Additionally, some *Talaromyces* strains can induce the production of enzymes involved in cell wall reinforcement, such as peroxidases and polyphenol oxidases. These enzymes contribute to the strengthening of plant cell walls, providing a physical barrier against pathogen invasion and colonization. By priming the plant's defense responses, *Talaromyces* species can enhance the overall resistance of the host plant against a wide range of pathogens, including fungi, bacteria, and nematodes. This mode of action complements the direct antagonistic effects of antimicrobial metabolites and mycoparasitism, making *Talaromyces* an effective and multifaceted biocontrol agent. The potential of *Talaromyces flavus* to promote plant growth has been highlighted in several studies (Abbas et al. 2021; Naraghi et al. 2010a, b, 2012). These studies revealed that different isolates of *T. flavus* can significantly enhance the growth of cotton and potatoes, potentially like the growth promotion observed with *Trichoderma* spp. Several researchers have demonstrated the ability of *T. flavus* to control important soilborne pathogens such as *Verticillium dahliae*, *Rhizoctonia solani*, *Sclerotinia sclerotiorum*, and *Sclerotium rolfsii* in crops like cotton, potato, tomato, eggplant, and bean (Madi et al. 1997; Naraghi and Heydari 2018). Specifically, Naraghi et al. (2010a, b) reported that *T. flavus* could reduce *Verticillium* wilt disease caused by *V. albo-atrum* in potato, tomato, and greenhouse cucumbers (Naraghi et al. 2010a). In addition to its biocontrol capabilities, *T. flavus* has been shown to promote plant growth in various crops. Naraghi et al. (2012) demonstrated that different *T. flavus* isolates could enhance growth parameters like root length, crown length, plant height, fresh weight, and dry weight in cotton and potato plants (Naraghi et al. 2012). This growth promotion effect has been attributed to the production of organic acids by *T. flavus* and other

Penicillium species, which can solubilize phosphate precipitates and chelate phosphate-precipitating cations, thereby increasing phosphorus availability to plants. The ability of *T. flavus* to simultaneously act as a biocontrol agent and a plant growth promoter makes it a promising candidate for sustainable agriculture practices. As plant growth-promoting fungi (PGPF), *T. flavus* and other beneficial fungi can potentially reduce the need for chemical pesticides and fertilizers (Lakshmi et al. 2011). The discovery of *T. wortmannii* FS2 from Japanese agricultural soils marks a significant advancement in the field of plant-microbe interactions and sustainable agriculture. This fungus, classified as a plant growth-promoting fungus (PGPF), demonstrates remarkable capabilities that extend beyond mere growth enhancement, venturing into the realm of plant defense fortification. *T. wortmannii* FS2's most intriguing feature lies in its production of terpenoid like volatile organic compounds (VOCs), with β-caryophyllene emerging as a key player. These airborne chemical messengers act as powerful biostimulants, triggering a cascade of beneficial responses in *Brassica campestris* L. var. *perviridis*, commonly known as Japanese mustard spinach. The dual action of these VOCs—simultaneously boosting plant growth and bolstering resistance mechanisms—showcases the sophisticated nature of this fungal-plant relationship. The ability of *T. wortmannii* FS2 to enhance plant growth through volatile emissions represents a paradigm shift in our understanding of PGPFs. Unlike traditional growth-promoting microorganisms that often require direct contact or root colonization, *T. wortmannii* FS2 can exert its beneficial effects from a distance. This unique mode of action opens up new possibilities for applications in agriculture, potentially allowing for the development of novel biostimulant formulations that do not require direct application to plant roots or foliage. Furthermore, the increased resistance observed in *B. campestris* when exposed to *T. wortmannii* FS2 volatiles suggests a form of fungus-induced systemic resistance. This phenomenon, where plants exhibit heightened defensive capabilities against a broad spectrum of pathogens following exposure to beneficial microorganisms, could prove invaluable in integrated pest management strategies. By priming the plant's own defense mechanisms, *T. wortmannii* FS2 offers a sustainable alternative to chemical pesticides, aligning perfectly with the growing demand for environmentally friendly agricultural practices. The isolation of *T. wortmannii* FS2 from an agricultural field underscores the untapped potential residing in our soils. It serves as a reminder of the vast microbial diversity yet to be explored and the potential solutions to agricultural challenges that may be lying dormant in farmlands worldwide. This discovery not only contributes to our scientific understanding but also paves the way for practical applications in sustainable farming. As agriculture faces mounting pressures from climate change, population growth, and environmental concerns, PGPFs like *T. wortmannii* FS2 offer a beacon of hope. Their ability to enhance crop productivity while simultaneously reducing the need for synthetic inputs aligns perfectly with the principles of sustainable intensification. By harnessing the power of these beneficial fungi, we may be able to develop more resilient, productive, and environmentally friendly agricultural systems. In conclusion, the identification and characterization of *T. wortmannii* FS2 represent a significant leap forward in our quest for sustainable agricultural solutions. As research in this field progresses, we may soon see the development of innovative bioproducts based on fungal volatiles, ushering in a new era of eco-friendly crop enhancement and protection strategies (Yamagiwa et al. 2011). However, the efficacy of plant growth-promoting fungi (PGPF) in agricultural settings is akin to orchestrating a complex symphony, where the fungi serve as conductors, the plants as the orchestra, and the soil as the concert hall—each element must be in perfect harmony, with factors such as plant maturity, soil composition, and intricate web of ecological interactions all playing crucial roles in determining whether the performance will result in a standing ovation or a discordant finale (Abbas et al. 2021). Similarly, the exploration of *T. omanensis* as an endophytic fungus capable of enhancing drought tolerance in tomato plants represents a fascinating frontier in agricultural biotechnology and plant-microbe symbiosis (Anand et al. 2023; Kashyap et al. 2023; Kumari et al. 2022; Pathak et al. 2022; Singh et al. 2017a, b). This discovery illuminates the hidden potential of fungal endophytes in mitigating one of the most pressing challenges facing global agriculture: water scarcity. *T. omanensis*, like many endophytic fungi, establishes a sophisticated relationship with its host plant, in this case, the tomato. Unlike surface-dwelling microorganisms, endophytes like *T. omanensis* colonize the internal tissues of plants, forming an intimate association that allows for direct influence on the plant's physiology and biochemistry. This unique positioning enables the fungus to modulate the plant's responses to environmental stressors, particularly drought (Halo et al. 2020). Similarly, the discovery of *Talaromyces* sp. MR1's ability to enhance both growth and cadmium uptake in *Arabidopsis thaliana* under cadmium stress conditions represents a significant advancement in the field of phytoremediation and plant-microbe interactions. This finding illuminates the complex and beneficial relationships that can exist between plants and their endophytic fungal partners, particularly in the context of environmental stress and pollution management. Cadmium is a highly toxic heavy metal that poses severe risks to human health and ecosystem stability. Its presence in soil, often due to industrial activities, mining, or use of phosphate fertilizers, presents a significant environmental challenge. The ability of plants to take up and accumulate cadmium, a process known as phytoextraction, has been explored as a potential remediation strategy. However, cadmium toxicity often inhibits plant growth, limiting the effectiveness of this approach (Xiao et al. 2023). Recent investigations into the plant growth-promoting capabilities of *Talaromyces* spp. have yielded promising results across various crop systems. Notably, *T. islandicus* has demonstrated remarkable efficacy in enhancing the growth and mineral nutrition of *Phaseolus vulgaris* L. (common bean). This beneficial effect is primarily attributed to the fungus's multifaceted mechanisms, including potassium solubilization, indole-3-acetic acid (IAA) production, and potential biocontrol activity against pathogenic *Fusarium* spp. (Zapata et al. 2024). In a parallel study focusing on maize, a synergistic approach utilizing a consortium of *T. apiculatus* and *Clonostachys rosea* yielded significant improvements in plant performance. Inoculated maize seedlings exhibited markedly increased shoot and root biomass compared to uninoculated controls. Furthermore, the fungal consortium enhanced the uptake and accumulation of essential nutrients, including nitrogen, phosphorus, potassium, calcium, magnesium, and boron, suggesting a comprehensive improvement in nutrient acquisition and utilization efficiency. Expanding the scope to oil palm cultivation, researchers evaluated the impact of *T. apiculatus* and *C. rosea*, both individually and in combination, on rapidly growing seedlings. Intriguingly, the application of these fungal inoculants did not adversely affect the levels of chlorophyll and carotenoids, crucial photosynthetic pigments. This observation is particularly noteworthy as it indicates that the growth-promoting effects of these fungi do not come at the expense of the plant's photosynthetic capacity, a critical factor in overall plant health and productivity. These findings collectively underscore the potential of *Talaromyces* spp., particularly when used in strategic combinations with other beneficial fungi, to significantly enhance crop performance across diverse agricultural systems. The demonstrated ability to improve nutrient uptake, stimulate growth, and potentially offer protection against pathogens, all without compromising photosynthetic efficiency, positions these fungal inoculants as promising tools for sustainable agricultural intensification. Future research directions may include elucidating the molecular mechanisms underlying these beneficial interactions, optimizing inoculation strategies for field-scale applications, and exploring the long-term impacts on soil health and crop productivity in various agroecological zones. Treatment with *T. apiculatus*, *C. rosea*, and their combination resulted in a 4.9–60% reduction in basal stem rot disease caused by *Ganoderma boninense* compared to the untreated control (Goh et al. 2020). In a recent study, the isolate TF-04 of *T. flavus* demonstrated remarkable efficacy in promoting rice growth and combating sheath blight. This isolate not only significantly reduced the severity of the disease but also enhanced the overall yield of rice crops. Furthermore, TF-04 upregulated defense-related genes and boosted enzyme activities within the rice plants, contributing to their enhanced resilience and productivity. This research underscores a microbiome-guided strategy for identifying effective biological control agents, showcasing the substantial potential of *Talaromyces* isolates in both managing rice sheath blight and promoting robust rice growth (Abbas et al. 2021). A separate study identified a range of bioactive compounds produced by *T. assiutensis*, which are believed to be responsible for its growth-promoting and systemic resistance-inducing effects on sunflower plants (Farhat et al. 2023). A greenhouse experiment demonstrated that *T. muroii* strain TM28 significantly boosted both fresh weight and height of wheat plants infected with the *F. pseudograminearum* pathogen. The fungus also enhanced the wheat's antioxidant defense system, mitigating the negative effects of *F. pseudograminearum*. This included reduced disease severity and lower pathogen abundance in the surrounding soil, roots, and stem base of the wheat. A strain of fungus called *T. purpurogenus*, designated Q2, strengthens bitter gourd's defenses against *Fusarium oxysporum* f. sp. *momordicae*, a harmful soilborne pathogen. This beneficial effect is achieved through two mechanisms: activation of the bitter gourd's lignin biosynthesis pathway, which reinforces cell walls, and direct promotion of the plant's growth (Tian et al. 2022). Under greenhouse conditions, treating with crude extract at 5 g/L of *T. tratensis* KUFA 0091 reduced brown spot incidence by 56.74%, and spore suspension of *T. tratensis* at 10^6 spores/mL reduced dirty panicle incidence by 60% (Dethoup et al. 2018). In field conditions, spore suspension treatments reduced dirty panicle incidence by 50–60%. Applying crude extract twice at 10 g/L resulted in the highest yield weight, which was 31.71% greater than the negative control. This study is the first to reveal that *T. albobiverticillius* HNB9, a member of the *Talaromyces* genus, possesses both root colonization and plant growth-promoting capabilities. Plants colonized with *T. albobiverticillius* HNB9 exhibited enhanced growth, evidenced by increased shoot and root lengths compared to untreated controls. This discovery significantly contributes to our understanding of root-colonizing fungi within the *Talaromyces* genus. Moreover, *T. albobiverticillius* HNB9 was found to produce indole acetic acid (IAA), a well-known plant growth hormone, which plays a crucial role in root elongation and differentiation. Additionally, HNB9 demonstrated the ability to solubilize essential nutrients such as phosphorus, zinc, and silica. These nutrient-solubilizing properties are critical for improving plant nutrient uptake and overall plant health. The ability of *T. albobiverticillius* HNB9 to both colonize roots and promote plant growth through IAA production and nutrient solubilization positions it as a valuable agent in sustainable agriculture. This study opens new avenues for exploring the potential of other *Talaromyces* spp. in enhancing crop productivity and soil health. Future research should focus on understanding the mechanisms underlying these interactions and evaluating the field performance of *T. albobiverticillius* HNB9 under various environmental conditions (Kharkwal et al. 2024).

13.4 Case Studies Against Soil and Foliar Plant Pathogens

13.4.1 Soilborne Diseases

Talaromyces spp. have demonstrated remarkable efficacy as biocontrol agents against various soilborne fungal pathogens that can cause devastating diseases and significant yield losses in a wide range of crops. Their ability to suppress these pathogens through multiple modes of action, including antibiosis, mycoparasitism, and induction of plant defense responses, has made them valuable tools in the management of soilborne diseases. *Fusarium oxysporum* is a ubiquitous soilborne fungus responsible for causing vascular wilt in numerous economically important crops, such as tomatoes, cotton, and bananas. Several *Talaromyces* strains have shown promising results in controlling *Fusarium* wilt disease. These biocontrol agents can directly parasitize and inhibit the growth of *Fusarium* through the production of cell wall-degrading enzymes, as well as the release of antifungal metabolites like mitorubrin. Additionally, they can prime the plant's defense responses, enhancing its overall resistance to the pathogen (Tian et al. 2022; Yang et al. 2023). *Rhizoctonia solani*, a notorious soilborne fungal pathogen, wreaks havoc on various crops, causing root rot, damping-off, and other diseases across cereals, vegetables, and ornamentals. *T. flavus shows promise as powerful biocontrol agents against this destructive pathogen* (Abbas et al. 2021; Tian et al. 2022; Yang et al. 2023). Their capacity to generate hydrolytic enzymes, such as chitinases and glucanases, allows them to break down the cell walls of *Rhizoctonia*, thereby suppressing the pathogen. Additionally, the synthesis of antibiotics and various antifungal compounds enhances their overall effectiveness as biocontrol agents. *Verticillium dahliae* is a harmful soilborne fungus responsible for causing vascular wilt in crops such as cotton, olive trees, and potatoes, frequently leading to significant yield reductions. Strains of *Talaromyces*, including *T. flavus*, have been investigated as potential biocontrol agents to combat this pathogen (Abbas et al. 2021; Naraghi et al. 2010a, 2012). The ability of *Talaromyces* spp. to effectively colonize the rhizosphere plays a crucial role in their function as biocontrol agents against soilborne pathogens such as *Verticillium dahliae*. Through mechanisms such as antibiosis and mycoparasitism, these fungi can outcompete and suppress *Verticillium*, thereby mitigating its harmful impact on crops. Antibiosis involves the production of antimicrobial compounds that inhibit the growth of the pathogen, while mycoparasitism refers to the direct parasitism of the pathogen by the biocontrol agent. Additionally, *Talaromyces* spp. can induce systemic resistance in host plants. This means that the presence of *Talaromyces* in the plant's environment can trigger the plant's defense mechanisms, enhancing its ability to resist pathogen attacks. This dual action of direct pathogen suppression and induced systemic resistance makes *Talaromyces* an effective biocontrol agent. Numerous case studies and field trials have demonstrated the efficacy of *Talaromyces*-based biocontrol products in managing soilborne fungal diseases across various cropping systems. For instance, in rice production, the application of *Talaromyces* spp. has been shown to significantly reduce the incidence and severity of *R. solani*, the causal agent of rice sheath blight. This reduction in disease incidence leads to improved yields and quality of the rice crop. These findings highlight the potential of *Talaromyces* spp. as a sustainable and effective alternative to chemical fungicides, offering a promising solution for the management of soilborne diseases in agriculture (Abbas et al. 2021; Naraghi et al. 2010a, 2012). In wheat cultivation, the application of *T. flavus* as a seed treatment has shown significant efficacy in controlling *Fusarium* crown rot diseases caused by soilborne pathogens (Yang et al. 2023). This demonstrates the versatility of *T. flavus* as a biocontrol agent, not only protecting crops but also enhancing their growth and yield. The success of *T. flavus* in wheat highlights its potential application across different crops and agricultural systems. Furthermore, *T. omanensis* has exhibited strong biocontrol properties against *P. aphanidermatum*, a pathogen responsible for damping-off diseases. *T. omanensis* can suppress the growth of *P. aphanidermatum* in both solid and liquid forms, inhibiting oospore production and inducing abnormal hyphal patterns in the pathogen. This comprehensive suppression of *P. aphanidermatum* highlights the multifaceted mechanisms through which *T. omanensis* exerts its biocontrol effects. The ability to inhibit oospore production is particularly significant, as oospores are critical for the pathogen's life cycle and persistence in the soil. The observed efficacy of *Talaromyces* species in various cropping systems underscores their potential as effective biocontrol agents. By reducing the reliance on chemical fungicides, *Talaromyces*-based treatments offer a more sustainable and environmentally friendly approach to managing soilborne diseases. These findings support the broader adoption of *Talaromyces* spp. in integrated pest management strategies, aiming to improve crop health and productivity while minimizing the environmental impact of agricultural practices (Halo et al. 2020).

13.4.2 Foliar Diseases

Talaromyces spp. have shown considerable promise in controlling foliar fungal diseases in addition to their effectiveness against soilborne pathogens, which can lead to substantial economic losses in diverse crop systems. *Botrytis cinerea*, responsible for causing gray mold, is a common fungal pathogen affecting crops like grapes, strawberries, and vegetables. Various *Talaromyces* spp. have been studied for their potential to manage this pathogen by producing antifungal compounds and enzymes, and through their ability to parasitize fungal cells (Abdel-Rahim and Abo-Elyousr 2018). In another experiment, two applications of the dry formulation of *T. tratensis* KUFA 0091 at a concentration of 10% resulted in a significant 50% reduction in *Alternaria padwickii*, *Bipolaris oryzae*, *Curvularia lunata*, and *Fusarium moniliforme*, the pathogens responsible for causing dirty panicle disease in rice. Moreover, this treatment led to a notable 24.06% increase in rice yield compared to untreated controls. The effectiveness of this treatment was comparable to that of the fungicide difenoconazole + propiconazole. Interestingly, the addition of solvents such as dimethyl sulfoxide or carboxymethyl cellulose (CMC) did not improve the formulation's efficacy. These findings underscore the potential of heat-tolerant *T. tratensis* KUFA 0091 as a promising biocontrol agent capable of both disease suppression and enhancing crop productivity in rice cultivation (Dethoup et al. 2018). These studies underscore *Talaromyces* spp. as a viable alternative for sustainable disease control strategies in rice cultivation. Furthermore, the study found that the effectiveness of *Talaromyces* spp. was not improved by the addition of solvents such as dimethyl sulfoxide or carboxymethyl cellulose, affirming the practicality and robustness of its dry formulation for field applications. Overall, these findings highlight the potential of *Talaromyces* spp. to support sustainable agriculture by reducing reliance on chemical inputs while simultaneously improving disease management outcomes and crop yields in rice production systems.

13.4.3 Nematode Diseases

Talaromyces spp. have also demonstrated remarkable nematicidal activities against various plant-parasitic nematodes, which are responsible for significant crop losses worldwide. These fungi produce a diverse array of metabolites that exhibit potent nematicidal effects, either by directly killing or immobilizing the nematodes or by interfering with their life cycle and reproduction. The mode of action of *Talaromyces* involves disrupting the nematode's cuticle and inhibiting egg hatching, thereby reducing the overall nematode population. Soil microorganisms play a crucial role in shaping the population dynamics of plant-parasitic nematodes. Researchers investigated fungi isolated from *Heterodera zeae* cysts to understand the decline in *H. zeae* populations. By analyzing gene sequences (ITS, *BenA*, *CaM*, *RPB2*) and conducting morphological studies, they identified a previously unknown fungus, strain GX1, which demonstrated a strong parasitic activity against *H. zeae* cysts. This fungus, now identified as *T. cystophila*, effectively parasitized and killed *H. zeae* cysts, achieving a parasitism rate of 91.11% within just 3 days posttreatment. The fungus degraded contents within cysts, leading to dense vacuolation, and significantly inhibited egg hatching, demonstrating high nematophagous activity in in vitro and greenhouse trials. Greenhouse experiments further indicated that spore suspensions and fermentation filtrates of *T. cystophila* effectively suppressed *H. zeae* reproduction by 56.17–78.76%, underscoring its potential as a biocontrol agent against plant-parasitic nematodes (Mo et al. 2023). The nematicidal properties of *Talaromyces* spp. offer promising opportunities for integration with other nematode management strategies, contributing to the development of sustainable and environmentally friendly pest management programs.

13.5 Role of *Talaromyces* spp. in Natural and Agricultural Ecosystems

Phosphorus (P) biogeochemical cycling is essential in natural and agricultural environments, supporting plant and microorganism growth. Metabolites such as 2-methylsorbic acid, sorbic acid, bromomethylsorbic acid, and bromosorbic acid, synthesized by *Talaromyces* species, significantly contribute to this cycle by effectively solubilizing and mobilizing inorganic phosphates (Wakelin et al. 2004). Phosphorus is an essential macronutrient for living organisms, but its availability in soil is often limited due to its tendency to form insoluble complexes with various cations, such as calcium, iron, and aluminum. This low solubility of phosphate compounds restricts their uptake by plants and microorganisms, leading to P deficiency and reduced productivity in natural and agricultural systems. The metabolites produced by *Talaromyces* species demonstrate strong chelating abilities, forming stable complexes with metal cations found in insoluble phosphate compounds. This chelation process effectively solubilizes inorganic phosphates, making them accessible for uptake by plants and microorganisms. Consequently, these metabolites play a crucial role in improving the availability of phosphorus in the soil, thereby supporting plant growth and productivity. Additionally, they exhibit antimicrobial and antifungal properties, which can help suppress plant pathogens and promote beneficial microbial communities in the rhizosphere (Solanki et al. 2023). This dual functionality of enhancing nutrient availability and combating pathogens makes these metabolites highly beneficial for sustainable agricultural practices. In natural ecosystems, their ability to solubilize inorganic phosphates facilitates the cycling of phosphorus among various components such as soil, water, and biota. This process ensures the efficient utilization of available phosphorus resources and maintains the balance of nutrient cycles within the ecosystem. However, it is important to note that the production and activity of these metabolites can be influenced by various environmental factors, such as soil pH, temperature, and presence of other microorganisms. Further research is needed to optimize the application of *Talaromyces* species or their metabolites in different agricultural and environmental contexts to maximize their potential in promoting sustainable P cycling and enhancing plant productivity (Harvey et al. 2009). Endophytes enhance phosphorus (P) uptake and produce plant hormones like indole acetic acid (IAA), auxins, abscisins, gibberellins, and ethylene, crucial for plant growth (Singh et al. 2022). These fungi effectively mitigate root-rotting fungi, primary pathogens causing plant diseases, potentially promoting plant growth (Kumar et al. 2018). The biocontrol potential of endophytic fungi, particularly *T. assiutensis* and *T. trachyspermus*, was explored for enhancing systemic resistance in sunflowers. *Talaromyces* spp. exhibited significant efficacy against root-rotting fungi, thereby potentially improving overall plant health and growth. The inhibition of these pathogens was linked to secondary metabolite production by these endophytic fungi. *T. trachyspermus*, known for producing hydrolytic enzymes like chitinase, amylase, cellulose, protease, and pectinase, also synthesizes IAA and siderophores and solubilizes phosphate, contributing to plant growth promotion through antagonistic activities (Sahu et al. 2019). Out of 67 fungal isolates belonging to *T. purpureogenus*, 5 were identified for their exceptional drought tolerance and subsequently evaluated for their abilities to promote plant growth, produce antioxidants, and exhibit antifungal properties. Fungal isolate #8TAKS-3a demonstrated the highest drought tolerance and exhibited significant production of growth-promoting substances and strong antioxidant activity. On the other hand, #6TAKR-1a showed the highest levels of total flavonoids and phenolics, along with effective antifungal properties. These findings highlight the potential of *T. purpureogenus* to enhance various growth parameters under both normal and stressed conditions, suggesting its promising role as a bioinoculant for enhancing plant growth in drought-prone environments (Kaur and Saxena 2023). The utilization of sewage sludge as an organic amendment in agricultural soils presents a dual-edged scenario, offering benefits for soil fertility and plant growth while also posing risks associated with heavy metal contamination. *T. pinophilus* MW695526 has emerged as a promising tool in mitigating these challenges by promoting wheat growth in sewage sludge-amended soils while enhancing phytoremediation potential. The observed enhancement in wheat growth parameters, including plant biomass, photosynthetic pigments, osmolytes, and enzymatic antioxidants, underscores the beneficial impact of *T. pinophilus* on plant physiological processes under challenging soil conditions. These improvements are crucial for sustaining agricultural productivity in environments where sewage sludge application is practiced. Moreover, *T. pinophilus's* ability to reduce the absorption of harmful heavy metals by wheat plants highlights its dual functionality in soil remediation. By promoting plant growth and concurrently restricting heavy metal uptake, *T. pinophilus* demonstrates its efficacy in addressing environmental concerns associated with sewage sludge amendments. This dual role not only enhances soil health and crop productivity but also contributes to minimizing potential risks posed by heavy metal accumulation in agricultural ecosystems (El-Shahir et al. 2021).

13.6 Mass Production Techniques and Formulation Strategies

13.6.1 Solid-State and Liquid Fermentation

The large-scale production of *Talaromyces* spp. is crucial to ensure availability for diverse applications, including biocontrol, bioremediation, and enzyme production. There are two primary methods employed for cultivating *Talaromyces* spp. on a large scale: solid-state fermentation (SSF) and liquid fermentation (submerged fermentation). SSF involves cultivating the fungus on solid substrates like agricultural residues or industrial by-products. This approach offers advantages such as using cost-effective substrates, achieving higher yields, and mimicking natural growth conditions. SSF has successfully produced various enzymes, metabolites, and biocontrol agents derived from *Talaromyces*. In contrast, liquid fermentation involves growing *Talaromyces* spp. in liquid media within bioreactors or fermenters. This method provides better control over conditions like aeration, pH, and nutrient availability, ensuring more consistent and scalable production. Liquid fermentation is particularly suited for producing fungal biomass, enzymes, and specific metabolites that require optimal oxygen transfer rates. Both SSF and liquid fermentation play crucial roles in meeting the demands for *Talaromyces*-derived products, supporting their applications across different sectors of biotechnology and agriculture (Papavizas et al. 1984).

13.6.2 Optimization of Growth Conditions

To achieve optimal yields and product quality, the growth conditions for *Talaromyces* spp. must be carefully optimized. Key factors influencing the growth and metabolite production of these fungi include temperature, pH, nutrient composition, aeration, and moisture content. Temperature plays a crucial role in regulating the growth and metabolic activities of *Talaromyces* spp. Most *Talaromyces* spp. exhibit optimal growth and metabolite production within the mesophilic temperature range of 25–30 °C, although some species may have slightly different temperature preferences (Naraghi et al. 2023; Papavizas et al. 1984). The pH of the growth medium or substrate also significantly impacts the growth and metabolite production of *Talaromyces* spp. While the optimal pH range varies among species, most *Talaromyces* spp. prefer slightly acidic to neutral conditions, with pH values ranging from 4.5 to 7.0. Nutrient composition, including carbon and nitrogen sources, minerals, and growth factors, is another critical factor affecting the growth and productivity of *Talaromyces* cultures. Optimizing the nutrient composition can enhance biomass accumulation, enzyme production, and biosynthesis of desired metabolites. Aeration plays a crucial role in submerged fermentation, as it ensures adequate oxygen supply for the aerobic growth of *Talaromyces* spp. Proper aeration can be achieved through appropriate bioreactor design, agitation rates, and airflow rates. In the case of solid-state fermentation, moisture content is a critical factor influencing the growth and metabolite production of *Talaromyces* spp. Maintaining appropriate moisture levels is essential for ensuring efficient nutrient transfer and preventing water stress or overhydration, which can negatively impact fungal growth and productivity.

13.6.3 Formulation Strategies

To ensure the efficacy and long-term viability of *Talaromyces*-based products, various formulation strategies are employed to enhance the stability and shelf life of these fungi. One key approach is the use of appropriate carriers or additives that provide protection against environmental stresses, such as desiccation, temperature fluctuations, and UV radiation. Inert carriers, such as diatomaceous earth, vermiculite, or clay minerals, are commonly used in the formulation of *Talaromyces*-based biocontrol agents or biofertilizers. These carriers help to maintain the viability of the fungi during storage and application while also facilitating uniform distribution and adherence to plant surfaces or soil. Another strategy involves the use of protective agents, such as osmoprotectants (e.g., glycerol, trehalose) or cell wall-stabilizing compounds (e.g., skimmed milk, sucrose). These additives help to maintain the structural integrity of fungal cells and enhance their tolerance to desiccation and other environmental stresses. Encapsulation techniques, such as gel entrapment or micro/nanoencapsulation, can also improve the stability and shelf life of *Talaromyces* formulations. By encapsulating the fungal propagules or metabolites within protective matrices, these formulations can provide controlled release, enhanced protection against degradation, and targeted delivery to the desired site of action. The application of *Talaromyces*-based biocontrol agents for foliar diseases can be achieved through various methods, including foliar sprays, soil drenches, and seed treatments. Additionally, these biocontrol agents can be integrated into broader integrated pest management (IPM) strategies that combine biological control with other sustainable practices, such as cultural control methods and carefully selected chemical treatments when necessary. This integrated approach ensures a comprehensive and sustainable management of foliar fungal diseases while minimizing the reliance on chemical pesticides. Researchers have developed a convenient way to harness the power of *Talaromyces*. They grow the fungus on sterilized cooked rice for a month. Then, they dry it out in a hot oven for a few hours. Finally, the rice with the fungus is ground into a fine powder, creating a biofungicide and plant growth booster. To see how effective this was against rice dirty panicle disease, researchers tested different concentrations of the powder, with and without additional ingredients. They found that simply using a 10% concentration of the dried *T. tratensis* KUFA 0091 produced the strongest defense against the disease (Dethoup et al. 2018). Recent advancements in biotechnology have facilitated the creation of liquid bioformulations and microcapsules derived from the fungal antagonist *T. flavus*, tailored for commercial greenhouse applications aimed at combating soilborne pathogens. This research focused on optimizing these formulations: a liquid variant incorporating potato dextrose broth and stabilized with dicycloserine or sodium nitrate, and microcapsules composed of combinations such as alginate + myristyl alcohol and guar gum + calcium lignosulfonate. Greenhouse trials demonstrated substantial efficacy, with the microcapsule powder applied via seedling impregnation and the liquid formulation added directly to soil reducing disease severity by 91.6% and 68.4%, respectively. These refined techniques for producing and utilizing *T. flavus* formulations underscore their significant potential for commercial deployment (Naraghi et al. 2023). Recently, nanotechnology has been employed to improve the stability of *T. flavus* through the development of advanced nanoformulations intended for future use in plant disease control. Four distinct nanoformulations were synthesized: two nanocapsules, one nanoemulsion, and one nanopowder. These formulations were created using a polymerization technique that involved blending plant oil and *T. flavus* with hydrophilic polymers such as poly urea formaldehyde, chitosan, and starch. Cross-linking agents like calcium chloride and surfactants were subsequently introduced, followed by homogenization at 35 °C and 10,000 rpm. The resulting nanoformulations (specifically F1 and F3 as nanocapsules, F2 as nanoemulsion, and F4 as nanopowder) exhibit promising potential for stable and effective application in agricultural disease management strategies (Naraghi and Negahban 2018). In a subsequent experiment, laboratory tests identified the most potent antagonistic isolates from each species, which were then utilized to formulate nine bioformulations using carriers such as peat, rice bran, and talc. Greenhouse trials conducted 60 days post-sowing demonstrated that all bioformulations significantly enhanced the number of healthy seedlings compared to the untreated control. The most effective bioformulation observed was talc-*T. harzianum*, followed by peat-*T. flavus*, talc-*T. flavus*, and rice bran-*T. harzianum*. Overall, *T. flavus* emerged as the most efficacious fungal antagonist in both laboratory and greenhouse conditions, with *T. harzianum* and *T. asperellum* also demonstrating notable effectiveness. These findings indicate that *Trichoderma* and *Talaromyces*, utilizing distinct mechanisms, represent promising biocontrol agents against sugar beet damping-off disease induced by *R. solani* (Kakvan et al. 2013).

13.6.4 Delivery Systems and Application Methods

The successful application of *Talaromyces*-based products relies on effective delivery systems and appropriate application methods (Naraghi et al. 2023; Papavizas et al. 1984). Depending on the intended use and target application site, various delivery systems and application techniques can be employed. For soil applications, such as biocontrol agents or biofertilizers, *Talaromyces* formulations can be delivered through seed coatings, soil drenches, or granular formulations. Seed coating involves applying the fungal formulation directly to the seed surface, ensuring early colonization, and protection of the developing root system. Soil drenches involve applying the formulation as a liquid suspension or solution directly to the soil, facilitating the establishment of the fungi in the rhizosphere. For foliar applications, such as biocontrol agents against foliar pathogens or plant growth promoters, *Talaromyces* formulations can be delivered through spraying or misting techniques.

These methods involve suspending the fungal propagules or metabolites in a suitable carrier solution and applying them to the foliage, ensuring adequate coverage and adherence to the plant surfaces. In certain cases, *Talaromyces* formulations can be incorporated into other delivery systems, such as hydrogels, alginate beads, or slow-release matrices. These systems provide controlled release of the fungi or their metabolites, ensuring sustained activity and prolonged efficacy. The choice of delivery system and application method depends on various factors, including the target application site, the intended use (biocontrol, biofertilization, bioremediation, etc.), the physicochemical properties of the formulation, and the compatibility with existing agricultural practices or environmental conditions (Abbas et al. 2021; Naraghi et al. 2023; Papavizas et al. 1984). Effective formulation and application strategies are essential for maximizing the potential of *Talaromyces* spp. in various applications. By optimizing mass production techniques, enhancing stability and shelf life, and employing appropriate delivery systems and application methods, these beneficial fungi can be efficiently utilized for sustainable agriculture, environmental management, and industrial processes (Kumar 2022).

13.7 Environmental Fate and Persistence

13.7.1 Potential Impacts on Nontarget Organisms

While *Talaromyces* spp. are generally beneficial microorganisms with diverse applications in agriculture, bioremediation, and industry, it is crucial to evaluate their potential impacts on nontarget organisms and the environment. These fungi may interact with various components of ecosystems, potentially disrupting indigenous microbial communities in soil (Sharma and Nonzom 2022). This interference could alter nutrient cycling and soil fertility, affecting ecosystem balance. Moreover, the introduction of *Talaromyces* spp. might influence nontarget invertebrates, such as beneficial insects and earthworms, impacting their growth and behavior despite not being pathogenic to them. Furthermore, assessing the persistence and spread of *Talaromyces* spp. in the environment is essential. Some species may establish long-term populations or spread beyond intended areas, posing risks to natural ecosystems and nontarget habitats.

13.7.2 Biosafety Assessments and Regulatory Frameworks

To ensure the safe and responsible use of *Talaromyces* spp., rigorous biosafety assessments and regulatory frameworks are indispensable. These assessments involve comprehensive evaluations of potential risks associated with introducing these microorganisms (Keswani et al. 2016). Regulatory frameworks provide guidelines for handling, containing, and releasing microorganisms, varying across regions but aiming to minimize risks to human and animal health as well as environmental impact. In many countries, regulatory approval or registration is required for using

Talaromyces spp. in agricultural or environmental applications. This process includes submitting detailed data on their characteristics, production methods, and safety profiles derived from laboratory and field studies. Risk assessment protocols guide the evaluation of potential hazards and the formulation of risk management strategies, including confinement procedures and monitoring to mitigate adverse effects.

13.7.3 Human and Animal Safety

13.7.3.1 Toxicological Evaluations

Despite their generally safe status and extensive use, thorough toxicological evaluations are essential to ensure the safety of *Talaromyces* spp. for human and animal exposure. These evaluations assess potential toxicity, allergenicity, and pathogenicity of specific strains or their metabolites. Studies include acute and chronic toxicity assessments using diverse models to evaluate effects on human organ systems via oral, dermal, or inhalation exposure routes. Assessments also explore allergenic potential through in vitro tests and animal models to detect sensitization risks (Cao et al. 2019). Pathogenicity studies assess whether *Talaromyces* spp. pose infection risks, particularly for immunocompromised individuals.

13.7.3.2 Risk Assessment and Management

Based on toxicological data, comprehensive risk assessments determine potential hazards and consequences of *Talaromyces* spp. use. This process considers strain characteristics, production methods, and application strategies, assessing exposure likelihood and severity. Risk management strategies, such as engineering controls, administrative protocols, and personal protective equipment (PPE), aim to mitigate identified risks during production, application, and disposal. Strategies also encompass labeling, usage instructions, monitoring programs, and emergency response plans to safeguard human health and environmental integrity. By adhering to regulatory frameworks and implementing robust risk management measures, *Talaromyces* spp. can be safely utilized while maximizing their agricultural, bioremediative, and industrial benefits.

13.8 Conclusion

The genus *Talaromyces* stands out as a promising reservoir of biocontrol agents, offering sustainable alternatives to conventional chemical pesticides. Throughout this chapter, we have delved into various facets of *Talaromyces* spp. as biocontrol agents, emphasizing their diverse mechanisms of action, efficacy against a wide spectrum of plant pathogens, formulation techniques, environmental considerations, and future research prospects. *Talaromyces* species are renowned for their production of antimicrobial metabolites that encompass antibiotics, enzymes, and other bioactive compounds, which effectively combat fungal, bacterial, and nematodal infections. Their ability to parasitize pathogens through mycoparasitism and induce systemic resistance in plants underscores their multifaceted approach to disease management. Furthermore, advancements in mass production techniques and formulation strategies are pivotal for optimizing their commercial viability and practical application in agricultural settings. The integration of *Talaromyces*-based biocontrol agents into sustainable agriculture practices holds transformative potential. By reducing dependency on chemical pesticides, these agents mitigate environmental impact, safeguard nontarget organisms, and promote biodiversity conservation. Their inclusion in integrated pest management strategies aligns with the principles of sustainable agriculture, offering comprehensive solutions that enhance crop resilience and productivity. Moving forward, leveraging genomic and metabolomic insights, alongside collaborative efforts among researchers, industry stakeholders, and regulatory bodies, will be critical in fully harnessing the ecological and economic benefits of *Talaromyces* spp. for sustainable crop protection and broader biotechnological applications. Overall, the utilization of *Talaromyces* fungi exemplifies a proactive approach towards sustainable agriculture, ensuring agricultural sustainability and food security for future generations.

13.9 Future Perspectives and Research Directions

Advancements in genomics and metabolomics enable deeper exploration of *Talaromyces* spp.'s genetic diversity and metabolic capabilities. These technologies reveal insights into genetic makeup, biochemical pathways, and secondary metabolite production, aiding strain improvement and biotechnological applications. Genomic analyses identify unique genes and pathways, guiding optimization for enhanced productivity and bioactivity. Metabolomics techniques, such as mass spectrometry and NMR spectroscopy, unveil diverse metabolites with potential agricultural and pharmaceutical applications. Integration of genomic, transcriptomic, and proteomic data enhances understanding of metabolite production and regulatory mechanisms, facilitating strategies for optimizing yield and efficacy. Combining *Talaromyces* biocontrol agents with diverse microbial consortia enhances efficacy against pests and diseases, promoting sustainable agriculture. Integrated pest management strategies incorporate *Talaromyces* spp. to mitigate chemical pesticide reliance, optimize crop protection, and foster agroecosystem resilience. Collaboration across disciplines supports innovation, ensuring safe, scalable deployment in agricultural systems. Scaling production of *Talaromyces*-based products demands optimized fermentation, formulation, and delivery strategies. Addressing production challenges and ensuring product quality, stability, and efficacy support market acceptance and adoption. Collaboration between academia, industry, and regulators aligns research with commercial needs, advancing sustainable solutions for global agriculture and biotechnology sectors.

References

Abbas A et al (2021) Isolation and evaluation of the biocontrol potential of Talaromyces spp. against rice sheath blight guided by soil microbiome. Environ Microbiol 23:5946–5961

Abdel-Rahim IR, Abo-Elyousr KAM (2018) Talaromyces pinophilus strain AUN-1 as a novel mycoparasite of Botrytis cinerea, the pathogen of onion scape and umbel blights. Microbiol Res 212–213:1–9. https://doi.org/10.1016/j.micres.2018.04.004

Ali A, Ölmez F, Zeshan MA, Mubeen M, Iftikhar Y, Sajid A, Abid M, Kumar A, Divvela PK, Solanki MK (2024) Yeast-based solutions in controlling plant pathogens. Biocatal Agric Biotechnol 58:103199. https://doi.org/10.1016/j.bcab.2024.103199

Anand U, Pal T, Yadav N, Singh VK, Tripathi V, Choudhary KK, Shukla AK, Sunita K, Kumar A, Bontempi E, Ma Y (2023) Current scenario and future prospects of endophytic microbes: promising candidates for abiotic and biotic stress management for agricultural and environmental sustainability. Microb Ecol 86:1455–1486. https://doi.org/10.1007/s00248-023-02190-1

Bara R, Aly AH, Pretsch A, Wray V, Wang B, Proksch P, Debbab A (2013) Antibiotically active metabolites from Talaromyces wortmannii, an endophyte of Aloe vera. J Antibiot 66:491–493. https://doi.org/10.1038/ja.2013.28

Bhardwaj A, Kaur K, Kumar A, Singh SK, Kharshiing B (2024) Development of microbial biotechnology products for phytopathogen management. In: The potential of microbes for a circular economy. Academic, pp 105–130. https://doi.org/10.1016/B978-0-443-15924-4.00013-8

Brown AE (1987) Activity of glucanases of Zygorrhynchus moelleri in relation to antagonism against some soil-borne plant pathogenic fungi. J Phytopathol 120:298–309

Cao C, Xi L, Chaturvedi V (2019) Talaromycosis (penicilliosis) due to Talaromyces (Penicillium) marneffei: insights into the clinical trends of a major fungal disease 60 years after the discovery of the pathogen. Mycopathologia 184:709–720

Dethoup T, Manoch L, Visarathanonth N, Chamswarng C, Kijjoa A (2007) Morphology and distribution of Talaromyces flavus from soil and potential use as a biological control agent against plant pathogenic fungi. Thai J Agric Sci 40:37–50

Dethoup T, Kaewsalong N, Songkumorn P, Jantasorn A (2018) Potential application of a marine-derived fungus, Talaromyces tratensis KUFA 0091 against rice diseases. Biol Control 119:1–6. https://doi.org/10.1016/j.biocontrol.2017.11.008

Dong Y et al (2006) Wortmannilactones A–D, 22-membered triene macrolides from Talaromyces wortmannii. J Nat Prod 69:128–130

Duo-Chuan LI, Chen S, Jing LU (2005) Purification and partial characterization of two chitinases from the mycoparasitic fungus Talaromyces flavus. Mycopathologia 159:223–229

El-Shahir AA, El-Tayeh NA, Ali OM, Abdel Latef AAH, Loutfy N (2021) The effect of endophytic Talaromyces pinophilus on growth, absorption and accumulation of heavy metals of Triticum aestivum grown on sandy soil amended by sewage sludge. Plants 10:2659

Fahima T, Madi L, Henis Y (1992) Ultrastructure and germinability of Verticillium dahliae microsclerotia parasitized by Talaromyces flavus on agar medium and in treated soil. Biocontrol Sci Tech 2:69–78. https://doi.org/10.1080/09583159209355220

Farhat H, Urooj F, Irfan M, Sohail N, Majeed S, Ullah S, Shafique HA (2023) Biological control potential of endophytic fungi with amelioration of systemic resistance in sunflower and GC–MS metabolic profiling of Talaromyces assiutensis. Curr Microbiol 80:61. https://doi.org/10.1007/s00284-022-03161-4

Goh YK et al (2020) Biocontrol and plant-growth-promoting traits of *Talaromyces apiculatus* and *Clonostachys rosea* consortium against *Ganoderma* basal stem rot disease of oil palm. Microorganisms 8:1138

Haggag WM, Kansoh AL, Aly AM (2006) Proteases from *Talaromyces flavus* and *Trichoderma harzianum*: purification, characterization and antifungal activity against brown spot disease on faba bean. Plant Pathol Bull 15:231–239

Halo BA, Al-Yahyai RA, Maharachchikumbura SSN, Al-Sadi AM (2019) Talaromyces variabilis interferes with Pythium aphanidermatum growth and suppresses Pythium-induced damping-off of cucumbers and tomatoes. Sci Rep 9:11255

Halo BA, Al-Yahyai RA, Al-Sadi AM (2020) An endophytic Talaromyces omanensis enhances reproductive, physiological and anatomical characteristics of drought-stressed tomato. J Plant Physiol 249:153163. https://doi.org/10.1016/j.jplph.2020.153163

Harvey PR, Warren RA, Wakelin S (2009) Potential to improve root access to phosphorus: the role of non-symbiotic microbial inoculants in the rhizosphere. Crop Past Sci 60:144–151

Houbraken J, Spierenburg H, Frisvad JC (2012) *Rasamsonia*, a new genus comprising thermotolerant and thermophilic *Talaromyces* and *Geosmithia* species. Antonie Van Leeuwenhoek 101:403–421

Houbraken J, de Vries RP, Samson RA (2014) Modern taxonomy of biotechnologically important *Aspergillus* and *Penicillium* species. Adv Appl Microbiol 86:199–249

Inglis GD, Kawchuk LM (2002) Comparative degradation of oomycete, ascomycete, and basidiomycete cell walls by mycoparasitic and biocontrol fungi. Can J Microbiol 48:60–70

Kakvan N, Heydari A, Zamanizadeh HR, Rezaee S, Naraghi L (2013) Development of new bioformulations using Trichoderma and Talaromyces fungal antagonists for biological control of sugar beet damping-off disease. Crop Prot 53:80–84. https://doi.org/10.1016/j.cropro.2013.06.009

Kashyap N, Singh SK, Yadav N, Singh VK, Kumari M, Kumar D, Shukla L, Kaushalendra, Bhardwaj N, Kumar A (2023) Biocontrol screening of endophytes: applications and limitations. Plants 12:2480. https://doi.org/10.3390/plants12132480

Kaur R, Saxena S (2023) Evaluation of drought-tolerant endophytic fungus Talaromyces purpureogenus as a bioinoculant for wheat seedlings under normal and drought-stressed circumstances. Folia Microbiol 68:781–799. https://doi.org/10.1007/s12223-023-01051-1

Kazerooni EA, Rethinasamy V, Al-Sadi AM (2019) *Talaromyces pinophilus* inhibits *Pythium* and *Rhizoctonia*-induced damping-off of cucumber. J Plant Pathol 101:377–383

Keswani C, Bisen K, Singh V, Sarma BK, Singh HB (2016) Formulation technology of biocontrol agents: present status and future prospects. In: Arora N, Mehnaz S, Balestrini R (eds) Bioformulations: for sustainable agriculture. Springer, pp 35–52

Kharkwal AC, Joshi H, Shandilya C, Dabral S, Kumar N, Varma A (2024) Isolation and characterization of a newly discovered plant growth-promoting endophytic fungal strain from the genus Talaromyces. Sci Rep 14:6022. https://doi.org/10.1038/s41598-024-54687-5

Kim KK-A, Fravel DR, Papavizas GC (1990) Glucose oxidase as the antifungal principle of talaron from Talaromyces flavus. Can J Microbiol 36:760–764

Kumar A (ed) (2022) Microbial biocontrol: sustainable agriculture and phytopathogen management, vol 1. Springer Nature, Cham. https://doi.org/10.1007/978-3-030-87512-1

Kumar A, Singh VK, Tripathi V, Singh PP, Singh AK (2018) Plant growth-promoting rhizobacteria (PGPR): perspective in agriculture under biotic and abiotic stress. In: Crop improvement through microbial biotechnology. Elsevier, pp 333–342. https://doi.org/10.1016/B978-0-444-63987-5.00016-5

Kumari M, Qureshi KA, Jaremko M, White J, Singh SK, Sharma VK, Singh KK, Santoyo G, Puopolo G, Kumar A (2022) Deciphering the role of endophytic microbiome in postharvest diseases management of fruits: opportunity areas in commercial up-scale production. Front Plant Sci 13:1026575. https://doi.org/10.3389/fpls.2022.1026575

Lakshmi JS, Kuberan T, Anbura J, Sundaravadivelan J, Kumar C, Manorama Dhanaseeli P (2011) Effect of plant growth promoting fungal inoculant on the growth of Arachis hypogea (L.) and it's role on the induction of systemic resistance against Rhizoctonia solani. Asian J Biosci 6:131–139

Lan D, Wu B (2020) Chemistry and bioactivities of secondary metabolites from the genus Talaromyces. Chem Biodivers 17:e2000229

Luo M, Chen Y, He J, Tang X, Wu X, Xu C (2021) Identification of a new Talaromyces strain DYM25 isolated from the Yap Trench as a biocontrol agent against Fusarium wilt of cucumber. Microbiol Res 251:126841. https://doi.org/10.1016/j.micres.2021.126841

Madi L, Katan T, Katan J, Henis Y (1997) Biological control of *Sclerotium rolfsii* and *Verticillium dahliae* by *Talaromyces* flavus is mediated by different mechanisms. Phytopathology 87:1054–1060. https://doi.org/10.1094/phyto.1997.87.10.1054

Manoch L, Dethoup T (2011) A potential use of Talaromyces species as biological agents against plant pathogenic fungi. Thai J Agric Sci 44:81–91

Miyake T, Kato A, Tateishi H, Teraoka T, Arie T (2012) Mode of action of Talaromyces sp. KNB422, a biocontrol agent against rice seedling diseases. J Pestic Sci 37:56–61

Mo YX, Kan YZ, Jia LM, Cao XT, Sikandar A, Wu HY (2023) Characterization and effect of a nematophagous fungus Talaromyces cystophila sp. nov. for the biological control of corn cyst nematode. Phytopathology 114:618–629. https://doi.org/10.1094/PHYTO-02-23-0045-R

Murray FR, Llewellyn DJ, Peacock WJ, Dennis ES (1997) Isolation of the glucose oxidase gene from Talaromyces flavus and characterisation of its role in the biocontrol of Verticillium dahliae. Curr Genet 32:367–375

Naraghi L, Heydari A (2018) A review on biological control of some important plant diseases using fungal antagonist, *Talaromyces flavus*. Int J Agric Sci Res 8:9–34

Naraghi L, Negahban M (2018) The application of nanotechnology in developing Talaromyces flavus formulations. Int J Biotechnol 8:1–10

Naraghi L, Heydari A, Rezaee S, Razavi M, Jahanifar H (2010a) Study on antagonistic effects of Talaromyces flavus on Verticillium albo-atrum, the causal agent of potato wilt disease. Crop Prot 29:658–662

Naraghi L, Heydari A, Rezaee S, Razavi M, Jahanifar H, Khaledi E (2010b) Biological control of tomato Verticillium wilt disease by Talaromyces flavus. J Plant Prot Res 50:360–365

Naraghi L, Heydari A, Rezaee S, Razavi M (2012) Biocontrol agent Talaromyces flavus stimulates the growth of cotton and potato. J Plant Growth Regul 31:471–477

Naraghi L, Fani SR, Jalali S, Negahban M, Naeimi S (2023) Preparation and optimization of liquid and microcapsule formulations of Talaromyces flavus for management of Fusarium wilt disease of tomato. J Biocontrol Plant Prot 11:1–18. https://doi.org/10.22092/bcpp.2023.364217.355

Papavizas GC, Dunn MT, Lewis JA, Beagle-Ristaino J (1984) Liquid fermentation technology for experimental production of biocontrol fungi. Phytopathology 74:1171–1175

Pathak P, Rai VK, Can H, Singh SK, Kumar D, Bhardwaj N, Roychowdhury R, de Azevedo LCB, Kaushalendra, Verma H, Kumar A (2022) Plant-endophyte interaction during biotic stress management. Plants 11:2203. https://doi.org/10.3390/plants11172203

Pitt JI (2014) PENICILLIUM | Penicillium and Talaromyces: introduction. In: Batt CA, Tortorello ML (eds) Encyclopedia of food microbiology, 2nd edn. Academic, Oxford, pp 6–13. https://doi.org/10.1016/B978-0-12-384730-0.00248-2

Proksa B, Adamcová J, Fuska J (1992) 2-Methylsorbic acid, an antifungal metabolite of Penicillium vermiculatum. Appl Microbiol Biotechnol 37:443–445

Sahu S, Prakash A, Shende K (2019) *Talaromyces trachyspermus*, an endophyte from *Withania somnifera* with plant growth promoting attributes. Environ Sustain 2:13–21. https://doi.org/10.1007/s42398-019-00045-5

Sharma B, Nonzom S (2022) Talaromyces stipitatus, a novel agent causing superficial mycosis in a diabetic patient from North India. Microbes Infect 24:104887

Silva PHF et al (2018) Antifungal polyketides and other compounds from amazonian endophytic Talaromyces fungi. J Braz Chem Soc 29:622–630

Singh M, Kumar A, Singh R, Pandey KD (2017a) Endophytic bacteria: a new source of bioactive compounds. 3 Biotech 7:1–14. https://doi.org/10.1007/s13205-017-0942-z

Singh VK, Singh AK, Kumar A (2017b) Disease management of tomato through PGPB: current trends and future perspective. 3 Biotech 7:1–10. https://doi.org/10.1007/s13205-017-0896-1

Singh R, Pandey KD, Singh M, Singh SK, Hashem A, Al-Arjani ABF, Abd Allah EF, Singh PK, Kumar A (2022) Isolation and characterization of endophytes bacterial strains of Momordica charantia L. and their possible approach in stress management. Microorganisms 10:290. https://doi.org/10.3390/microorganisms10020290

Solanki MK, Joshi NC, Singh PK, Singh SK, Santoyo G, de Azevedo LCB, Kumar A (2023) From concept to reality: transforming agriculture through innovative rhizosphere engineering for plant health and productivity. Microbiol Res 279:127553. https://doi.org/10.1016/j.micres.2023.127553

Suzuki K, Nozawa K, Nakajima S, Udagawa S-i, Kawai K-i (1992) Isolation and structures of antibacterial binaphtho-α-pyrones, talaroderxines A and B, from Talaromyces derxii. Chem Pharm Bull 40:1116–1119

Suzuki S, Hosoe T, Nozawa K, Kawai K-i, Yaguchi T, Udagawa S-i (2000) Antifungal substances against pathogenic fungi, talaroconvolutins, from Talaromyces convolutus. J Nat Prod 63:768–772

Tian Y, Zhao Y, Fu X, Yu C, Gao K, Liu H (2021) Isolation and identification of Talaromyces sp. strain Q2 and its biocontrol mechanisms involved in the control of Fusarium wilt. Front Microbiol 12:724842

Tian Y, Fu X, Yan X, Li X, Peng H, Gao K (2022) The control efficacy and mechanism of *Talaromyces purpurogenus* on *Fusarium* wilt of bitter gourd. Biol Control 165:104804. https://doi.org/10.1016/j.biocontrol.2021.104804

Verma S, Azevedo LCB, Pandey J, Khusharia S, Kumari M, Kumar D, Kaushalendra, Bhardwaj N, Teotia P, Kumar A (2022) Microbial intervention: an approach to combat the postharvest pathogens of fruits. Plants 11:3452. https://doi.org/10.3390/plants11243452

Wakelin SA, Warren RA, Harvey PR, Ryder MH (2004) Phosphate solubilization by *Penicillium* spp. closely associated with wheat roots. Biol Fertil Soils 40:36–43

Xiao Y, Chen R, Chen L, Yang B, Jiang L, Fang J (2023) Endophytic fungus Talaromyces sp. MR1 promotes the growth and cadmium uptake of Arabidopsis thaliana L. under cadmium stress. Curr Microbiol 80:346. https://doi.org/10.1007/s00284-023-03453-3

Yamagiwa Y, Inagaki Y, Ichinose Y, Toyoda K, Hyakumachi M, Shiraishi T (2011) *Talaromyces wortmannii* FS2 emits β-caryphyllene, which promotes plant growth and induces resistance. J Gen Plant Pathol 77:336–341. https://doi.org/10.1007/s10327-011-0340-z

Yang H et al (2023) Antagonistic effects of Talaromyces muroii TM28 against Fusarium crown rot of wheat caused by Fusarium pseudograminearum. Front Microbiol 14:1292885. https://doi.org/10.3389/fmicb.2023.1292885

Yilmaz N, Visagie CM, Houbraken J, Frisvad JC, Samson RA (2014) Polyphasic taxonomy of the genus Talaromyces. Stud Mycol 78:175–341. https://doi.org/10.1016/j.simyco.2014.08.001

Yoshida E, Fujinoto H, Yamazaki M (1996) Isolation of three new azaphilones, luteosin C, D, and E, from an ascomycete, Talaromyces luteus. Chemical and Pharmaceutical Bulletin 44: 284–287

Zapata D, López JE, Saldarriaga JF (2024) Plant growth-promoting and biocontrol potential of Aspergillus tubingensis and Talaromyces islandicus. J Soil Sci Plant Nutr 24:2354. https://doi.org/10.1007/s42729-024-01633-z

Zhai M-M, Li J, Jiang C-X, Shi Y-P, Di D-L, Crews P, Wu Q-X (2016) The bioactive secondary metabolites from Talaromyces species. Nat Prod Bioprospect 6:1–24. https://doi.org/10.1007/s13659-015-0081-3

Metagenomics: Unveiling Microbial Communities Theme

14

Ravishankar L. V., Bushra Rasool, R. Nilesh Kumar, Amit Kumar Maurya, Ayushi Sharma, Ranbir Singh, and Mir Zaffar Bashir

Abstract

Soil microbiome plays an important role and coordinates for efficient delivery of nutrients; these mighty creatures are also responsible for the recycling of nutrients in soil by their ability of decomposition and mineralization. The interaction of microbes and plant through host-pathogen interaction regulates the physiology of plants in overcoming the biotic and abiotic stresses. The diversity of microbial population impacts the agriculture ecosystem through various challenges from protective nature of crops to destructive nature as phytopathogenic pathogens. Microbial communication occurs through the signaling molecules which are secreted by pathogens; these chemical substances not only reveal themselves as signaling molecules but can also be exploited in the industrial way for the welfare of the human being. The study of microbial compounds through culture-dependent and culture-independent ways reveals the structural diversity, but the culture-dependent method is strenuous; the second method, i.e., complete genomic study of DNA of microbiome through metagenomic approach, reveals the microbial diversity of the population. This chapter discusses the diversification of microbiome population with possible structural and functional genomic investigation through culture-independent procedure also revealing the

Ravishankar L. V. (✉) · B. Rasool · R. N. Kumar · A. Sharma · R. Singh
Division of Plant Pathology, Faculty of Agriculture, SKUAST-Jammu, Jammu, Jammu and Kashmir, India

A. K. Maurya
Teerthanker Mahaveer University (TMU), College of Agricultural Sciences, Division of Plant Pathology, Moradabad, Uttarpradesh, India

M. Z. Bashir
Division of Basic Sciences and Humanities, FoA, Wadura, SKUAST-Kashmir, Srinagar, Jammu and Kashmir, India

A. Kumar, M. K. Solanki (eds.), *Microbial Biocontrol Techniques*, Microorganisms for Sustainability 54,
https://doi.org/10.1007/978-981-97-8739-5_14

importance of metagenomics in industrial exploitation in revealing microbial theme for sustainable approaches in agriculture.

Keywords

Microbiome · Omics · Pathogenicity · Metagenomics

14.1 Introduction

14.1.1 Microbiome

Microbes are ubiquitous in nature, which are present in diversified forms; the nature of the microbial interactions in ecological niche is vital for plant growth. Soil is the main reservoir to different microbiomes, and inter- and intra-microbial communications enhance the ecology of the soil properties which strengthen the agroecosystems. Sustainable and innovative agriculture practices are used to combat the global challenges of food security due to degradation of quality and quantity of natural resources. The complex interactions among different components of ecosystem suggest the green energy as an innovative idea for biodiversity conservation, which aims for sustainable food production (Kalia et al. 2017). The exponential growth of global population will reach approximately ten billion by 2050; such dramatic growth leads to overexploitation of natural resources and degradation of ecosystems, and there is a need of sustainability to achieve food security (Levy and Lubell 2018). Microbial applications and manipulations are widely used for the well-being of humans; in the same process, agriculture which is the driving force of the human life also explores the microbial communities employed for higher productivity of crops (Iquebal et al. 2022). Traditional knowledge of the natural sciences transcendent with the modern application of biotechnology uncovers the hidden secrets of microbial communities. Metagenomics, a vibrant biotechnology tool used in the microbial analysis to study the heterogenic complexity to find out the presence and actions of microbiomes, helps us to reveal structural complexity and functionality of each microbial member (Scholz et al. 2012).

14.2 Phytomicrobiome: A Reservoir for Sustainable Agriculture

Soil microbes are attracted by the plant photosynthates from a particular ecological niche (Bulgarelli et al. 2012). Microbes are present in majority of the plant parts that support plant growth and development. "Phytomicrobiome" represents biomes which colonize on different plant parts, best explained by symbiotic relationship between rhizobia and leguminous mutual dependence (Vuong et al. 2017). Most of the known phytomicrobiomes promote plant growth by nutrient absorption and hormone production, with resistance to biotic and abiotic stresses (Berendsen et al. 2012; Wallenstein 2017). In contrast to those beneficial phytomicrobiomes, there are some plants pathogenic in nature and responsible for different biotic stresses (Mendes et al. 2013; Burgess et al. 2017). Plant pathogenic fungi and bacteria infect a wide range of agricultural and horticultural crops and are ultimately responsible for high yield losses (Lecomte et al. 2016; Burgess et al. 2017).

14.2.1 Classification of Phytomicrobiomes

Phytomicrobiomes are generally classified into two with regard to their association with the specific plant part, i.e., shoot or root. If the association of microbiomes is with the shoot, then it is phyllosphere microbiome, and if the association is with the root, then it is rhizosphere microbiome. The phyllosphere and rhizosphere harbor both beneficial and harmful microbiomes. Efficient knowledge of phytomicrobiome could help for exploration of microbial population for plant nutrition and stress resistance (Farrar et al. 2014; Smith et al. 2017).

14.2.2 Rhizosphere Microbiome

Plants are considered as autotrophs that derive their energy through photogeochemical process in the presence of sunlight and carbon dioxide. Roots perform other important functions like nutrient as well as water uptake from the soil. During 1904, Lorenz Hiltner, pioneer agronomist and plant physiologist, in Germany gave a term "rhizosphere"; he stated that soil is an energy-rich ecological niche which hosts a diversified microbial population due to the mutualistic behavior of the microbes and root exudates of plant in which 20% of photosynthates were utilized (Hirsch and Mauchline 2012; Lareen et al. 2016).

14.2.3 Beneficial Soil Microbiome

Plants obtain their major source of energy elements from air and water, and nitrogen, phosphorus, and potassium from soil and other sources essential for growth and development. The rhizosphere microbiomes play a very crucial role in the recycling of nutrients by nitrogen fixation and phosphorus and potassium solubilization. Rhizobia group is commonly known for nitrogen fixation in Fabaceae family of leguminous species (Oldroyd 2013; Laranjo et al. 2014; Sprent et al. 2017). Root nodules are the key compartments where nitrogen fixation takes place. Inorganic nitrogen fixation with the help of symbiotic nitrogen-fixing bacteria, i.e., *Azospirillum*, *Klebsiella*, *Burkholderia*, *Bacillus*, and *Pseudomonas*, helps in the acquisition of nitrogen to plants from soil (Arruda et al. 2013; Kuan et al. 2016). The strong association between soil bacteria and arbuscular mycorrhiza (AM) is considered as a widespread symbiotic relationship in which AM fungi are estimated to have 80% of root colonization on terrestrial plants (Williams et al. 2017), which produces specialized structures of mycorrhizal fungi known as arbuscules in cortical cells by exchanging the nutrients (Parniske 2008). Mycorrhizal fungi regulate the growth and development of a plant through symbiotic exchange of plant nutrients, i.e., N, P, Zn, and Cu (Johnson et al. 2013; Bender et al. 2016).

14.2.4 Harmful Rhizosphere Microbiome

Rhizosphere also harbors various plant pathogens which are responsible in causing serious plant diseases. Pathogen *Phytophthora infestans* caused late blight of potato in the nineteenth-century, which resulted in Irish famine (Saville et al. 2016; Larousse and Galiana 2017). Significant crop yield losses were observed under different soilborne pathogens (Latz et al. 2016). Rhizosphere bacteria such as *Pseudomonas*, *Actinobacteria*, and *Bacillus* defend against many facultative saprophytes (Haas and Défago 2005; Mendes et al. 2011; Latz et al. 2016).

14.3 Dwelling Molecular Approaches for Sustainable Microbiome Analysis of Agriculture

Microbial communities' association with agriculture ecology is not well studied with prospects of microbial themes (Bevivino and Dalmastri 2017). Scientists began the study of 16S rRNA isolated from the bacterial samples (Handelsman 2004), and the metagenomic approach came into existence when emphasis of novel bioactive compounds of soil was highlighted; metagenomic approach was exploited by collecting the environmental samples and analyzing samples by extracting the total DNA from DNA extraction kits, through enzymatic fragmentation of DNA by using commercially available library kits. Genomic approaching technologies are paving pathway to know the diversity of microbes (Kalia et al. 2017), particularly metaomics studies of functional genomes with the molecular information of proteomics (proteins) and transcriptomes (all sets of RNA) (Prosser 2015); sequencing information better fits in the metabolic process of microbial niche in agriculture ecosystem, and microbial rhizosphere colonies play an important role that determines the plant health under agroecosystem (Makonde et al. 2015). The application of the metagenomics helps in understanding the taxonomic component and functional redundancy of rhizobiomes, due to their triggering role in host resistance (Gupta et al. 2018). Metagenomics, unlocking the potential of unculturable microbiota under natural environmental conditions that produce novel bioactive compounds with industrial purpose, have versatility with wider applications.

14.3.1 Metagenomics: An Omic Approach

Meta-analysis of microbial population to know the similarities of the unidentical organisms in soil ecosystem is known as metagenomics (Handelsman et al. 1998).

Construction of genomic libraries from the DNA extracted from the environmental related samples is the main step in metagenomics. The genomic libraries were constructed from prokaryotic organisms in sea water with 16S rRNA laid foundation for metagenomic studies (Iquebal et al. 2022). Soil genomic studies were also coming into existence by targeting 16S rRNA, in the soil considered as the most dynamic environment due to its volume and microbial diversity (Daniel 2005). Joshua Lederberg coined the term "microbiome"; diversity and colonization to different plant organs such as phyllosphere, rhizosphere, and endosphere are the major ecological niches of microbial community (Berg et al. 2016); functional and structural information of different microbial populations was determined by sequence-based analysis of a testing environmental sample without employing conventional cultural techniques (Sarangi et al. 2019). DNA extraction and cloning studies of massive microbial populations on terrestrial ecosystem were carried out with the help of 16S rRNA analyses, which would help us to identify the lineage and evolution of microbes (Dhanjal et al. 2020); this study would help us in reviving ecology and physiology of different microbiomes. The traditional culturing of the microorganisms from an ecological niche has a limitation due to difficulty of culturing and their growth complexity conditions, but also noncultivable with active viable cells with only 1% of microbial diversity by using classic cultivation techniques. Metagenomics is broadly classified as unselective random sequencing and targeted direct sequencing (Suenaga 2012). Screening methods of metagenomics are further divided into (1) shotgun analysis, with mass genome sequencing; (2) activity-driven studies for examining microbial functions; and (3) sequence-driven analysis that links phylogenetic information and functionality with the intervention of next-generation sequence (NGS) (Harismendy et al. 2009; Shendure and Ji 2008) technologies that boosted the metagenomic research where large volumes of sequences could be compared between environmental and traditional sequence methods (Edwards et al. 2006; Palenik et al. 2009).

14.3.2 Functional Metagenomics and Their Role in Soil Environment

Soil ecology is the medium for recycling of many nutrients responsible for proper growth and development in plants; the unique chemical reaction under the dominance of particular microbial genomics is responsible for the distribution of nutrients under soil ecology; metagenomic applications have explored microbial genomes and their expressions in biogeochemical cycles, and *Kuenenia stuttgartiensis* fixes anaerobic fixing of ammonia in the soil as documented by van Niftrik et al. (2008). Cytochrome protein required for the iron oxidation formed during acid mine drainage was revealed with shotgun approach through genomics of rhizosphere microbial community (Bevivino and Dalmastri 2017). Paleogenomic applications revealed the distribution of ancient DNA of plants and associated high level of diversity for microbial variations in the soil (Maillard 2018); paleogenomic is alternatively known as genome-wide ancient DNA analysis with high-throughput sequencing which provides information and understanding on the genome of extinct organisms (Gupta et al. 2018).

14.3.3 Exploring Microbial Cultures Through Structural and Functional Metagenomics

Two major approaches of metagenomics are structural metagenomics, which studies about the taxonomic positions of entire microbiomes, and function, which deals with the genomic association of microbiomes. Microbial interactions among different species and genera and their role in biogeochemical reveal their diversity and can be studied through metagenomics. Functional metagenomics studies genomic diversity of microbiome and also studies new genes responsible for the production of bioactive compounds, i.e., proteins and enzymes (Simon and Daniel 2011). Functional metagenomics aims at the identification of new enzymatic protein families (Alvarez et al. 2013). Metagenomics aids in understanding microbiome through amplicon-based metagenomics and shotgun-based metagenomics. 16S ribosomal RNA is used for bacterial identification and ITS 18S region for fungal and eukaryotic characterization, with hypervariable and conserved sequences. 16S sequencing targets V1–V9 regions of 16S ribosomal RNA gene for identifying the diversity in bacterial communities in the microbiome, and this sequencing data of 16S rDNA that has 97% similarity was grouped as operational taxonomic unit. A drawback in amplicon-based metagenomics is best explained by the examples of *Shigella flexneri* and *E. coli* as distant species does not show any variation (Hilton et al. 2016). ITS region is employed to identify the fungal diversity in between 5.8S and 18S rRNA (Bromberg et al. 2015). Taxonomical diversity study analysis is based on ribosomal subunits such as larger subunit 28S, smaller subunit (SSU) 18S, and 5.8S subunit rRNA genes. rRNA sequencing of 16S/18S rRNA gives information about specific communities in microbiome (Franková and Fry 2013). Untargeted microbiome was targeted through shotgun sequencing approach which involves collection of microbiome DNA and enzymatic degradation amplifying sequencing reads with quality filtering assembling of genomic data and analyzing it (Ghosh et al. 2019).

14.4 Applications of Metagenomics for Biotic Stress in Crop Protection

Crop protection management strategies against biotic stress due to plant pathogens are employed in the excessive application of modern conventional practices; symptomology studies of plant disease are very much important in diagnosis, but the identification of strain/species-level target of the pathogen is important to identify the causative agent. Metagenomic approach in pathogen diagnosis has the potential of targeting multiple specific loci of target plant. Conventional farming techniques alter the taxonomic diversity of rhizosphere and phyllosphere microbiome with unique signatures (Khoiri et al. 2021).

14.4.1 Plant Disease Suppressive and Resistance Breeding

Predicting the causative agents in assessing the biotic factors which are responsible for the disease detection can be assessed by the identification of pathogen strain by metagenomic approach; studies have been done on analyzing rhizosphere endophytic bacterial communities' recognition and impact on the disease development including members of Flavobacteriales (Kaushal et al. 2020). Soil microbiome acts as a suppressive factor in agriculture. Polychip-based metagenomic data analysis analyzed 33,000 prokaryotic species (Mendes et al. 2011).

14.4.2 Metagenomic Soothing Tool for Consolidated Bioprocessing

The rise in the concern of eco-friendly strategies makes an alternative resource in the production of green fuels in India with an increase of 9.8% for renewable source of energy (Pabbathi et al. 2021). The efficient production of biofuels in India that increased to 24.9% stands in the second place of renewable energy production followed by Indonesia (37.5%) among Asia Pacific countries (Dudley 2018); the increasing demand for bioenergy will rise to 11.5% from the present utilization by 2040 (Newell et al. 2019). The microbial degradation of the cellular biomass into bioenergy achieved by the microbial enzymes of glycosyl hydrolase (GHase) family aids in complete hydrolysis of the cellulose (Elleuche et al. 2014; Franková and Fry 2013); enzyme degradation with the help of pure cultures is currently utilized in the green energy production, which are known for their biocatalysis. Metagenomics and culture-independent approaches are efficiently exploited for renewable biofuels.

14.4.3 Application of Metagenomics in Identification of Novel Bioactive Compounds

Wide application of novel bioactive compounds which are secreted by the prokaryotic and eukaryotic organism compounds is very much essential in industrial application; these compounds have been categorized as the novel bioactive molecules. Microbial research took a dramatic change with the implementation of genomic tools, which include (1) sequencing of genome, (2) culture-independent protocols, and (3) bioinformatic studies. Microbes have been considered as the active sources of bioactive compounds from the terrestrial environment (Tejesvi et al. 2011). With the advancement of molecular techniques, it became much familiar that the gene which is actively involved in coding of bioactive compounds is identified (Karumuri et al. 2015). Soil metagenomic libraries have been identified with many novel gene products which are industrially important such as agarases, amidase, antibacterials, dehydrogenases, lipases, and β-lactamase (Daniel 2004; Richardson et al. 2002). The details of natural compounds and identified source through metagenomic approach have been described in Table 14.1.

Table 14.1 Natural compounds and identified source through metagenomic approach

S. No.	Novel compounds	Source	References
1.	Alpha-amylase	Cow dung	Pooja et al. (2015)
2.	Asparaginase	Rhizosphere	Arjun et al. (2018)
3.	Biotin production	Rhizosphere	Entcheva et al. (2001)
4.	Chitinase	Suppressive soils with club root of cabbage	Hjort et al. (2014)
5.	Erdacin	Sand dunes	Chuzel et al. (2018), King et al. (2009)
6.	Exosialidase	Freshwater thermal hot spring	Berini et al. (2017)
7.	Fasamycins A, B	Rhizosphere	Feng et al. (2012)
8.	Glycopeptide- and lipopeptide-like antibiotics	Rhizosphere	Owen et al. (2013)
9.	Hemicellulase	Decomposed wheat straw	Maruthamuthu et al. (2016)
10.	Novel isoprene-degrading proteobacteria	Rhizosphere and foliage	Larke-Mejía et al. (2019)
11.	Transaminase	DNA of unculturable archaea and bacteria	Pawar et al. (2018)
12.	β-Galactosidase	Rhizosphere	Liu et al. (2019)

14.5 Conclusion

Microorganisms nourish the plant, which is responsible for plant growth under biotic and abiotic stresses considered as the most essential factor; majority of rhizosphere microbiomes are still uncharacterized under culture-dependent methods targeting microbiome development in plant-microbe communications. Microbial metagenomics will evaluate the functional genes through microbial markers and reveal the complexity of microbial population.

References

Alvarez TM, Paiva JH, Ruiz DM, Cairo JPL, Pereira IO, Paixão DA, de Almeida RF, Tonoli CC, Ruller R, Santos CR (2013) Structure and function of a novel cellulase 5 from sugarcane soil metagenome. PLoS One 8:e83635

Arjun JK, Aneesh BP, Kavitha T, Harikrishnan K (2018) Characterization of a novel asparaginase from soil metagenomic libraries generated from forest soil. Biotechnol Lett 40:343–348

Arruda L, Beneduzi A, Martins A, Lisboa B, Lopes C, Bertolo F, Passaglia LMP, Vargas LK (2013) Screening of rhizobacteria isolated from maize (Zea mays L.) in Rio Grande do Sul State (South Brazil) and analysis of their potential to improve plant growth. Appl Soil Ecol 63:15–22

Bender SF, Wagg C, van der Heijden MG (2016) An underground revolution: biodiversity and soil ecological engineering for agricultural sustainability. Trends Ecol Evol 31(6):440–452

Berendsen RL, Pieterse CM, Bakker PA (2012) The rhizosphere microbiome and plant health. Trends Plant Sci 17(8):478–486

Berg G, Rybakova D, Grube M, Köberl M (2016) The plant microbiome explored: implications for experimental botany. J Exp Bot 67:995–1002

Berini F, Casciello C, Marcone GL, Marinelli F (2017) Metagenomics: novel enzymes from non-culturable microbes. FEMS Microbiol Lett 364:fnx211

Bevivino A, Dalmastri C (2017) Impact of agricultural land management on soil bacterial community: a case study in the Mediterranean area. In: Soil biological communities and ecosystem resilience. Springer, pp 77–95

Bromberg JS, Fricke WF, Brinkman CC, Simon T, Mongodin EF (2015) Microbiota—implications for immunity and transplantation. Nat Rev Nephrol 11:342–353

Bulgarelli D, Rott M, Schlaeppi K, Ver Loren van Themaat E, Ahmadinejad N, Assenza F et al (2012) Revealing structure and assembly cues for Arabidopsis root-inhabiting bacterial microbiota. Nature 488(7409):91–95

Burgess TI, Scott JK, Mcdougall KL, Stukely MJ, Crane C, Dunstan WA et al (2017) Current and projected global distribution of Phytophthora cinnamomi, one of the world's worst plant pathogens. Glob Change Biol 23(4):1661–1674

Chuzel L, Ganatra MB, Rapp E, Henrissat B, Taron CH (2018) Functional metagenomics identifies an exosialidase with an inverting catalytic mechanism that defines a new glycoside hydrolase family (GH156). J Biol Chem 293:18138–18150

Daniel R (2004) The soil metagenome—a rich resource for the discovery of novel natural products. Curr Opin Biotechnol 15:199–204

Daniel R (2005) The metagenomics of soil. Nat Rev Microbiol 3:470–478

Dhanjal DS, Chopra C, Chopra RS (2020) Metagenomic DNA sequencing: technological advances and applications. In: Metagenomics: techniques, applications, challenges and opportunities. Springer, pp 37–53

Dudley B (2018) BP statistical review of world energy 2018. Energy economic, Centre for Energy Economics Research and Policy. British Petroleum. https://www.bp.com/en/global/corporate/energy-economics/statistical-review-of-world-energy/electricity.html

Edwards RA, Rodriguez-Brito B, Wegley L, Haynes M, Breitbart M, Peterson DM, Saar MO, Alexander S, Alexander EC, Rohwer F (2006) Using pyrosequencing to shed light on deep mine microbial ecology. BMC Genomics 7:1–13

Elleuche S, Schröder C, Sahm K, Antranikian G (2014) Extremozymes—biocatalysts with unique properties from extremophilic microorganisms. Curr Opin Biotechnol 29:116–123

Entcheva P, Liebl W, Johann A, Hartsch T, Streit WR (2001) Direct cloning from enrichment cultures, a reliable strategy for isolation of complete operons and genes from microbial consortia. Appl Environ Microbiol 67:89–99

Farrar K, Bryant D, Cope-Selby N (2014) Understanding and engineering beneficial plant–microbe interactions: plant growth promotion in energy crops. Plant Biotechnol J 12(9):1193–1206

Feng Z, Chakraborty D, Dewell SB, Reddy BVB, Brady SF (2012) Environmental DNA-encoded antibiotics fasamycins A and B inhibit FabF in type II fatty acid biosynthesis. J Am Chem Soc 134:2981–2987

Franková L, Fry SC (2013) Biochemistry and physiological roles of enzymes that 'cut and paste' plant cell-wall polysaccharides. J Exp Bot 64:3519–3550

Ghosh A, Mehta A, Khan AM (2019) Metagenomic analysis and its applications. In: Encyclopedia of bioinformatics and computational biology, pp 184–193

Gupta N, Vats S, Bhargava P (2018) Sustainable agriculture: role of metagenomics and metabolomics in exploring the soil microbiota. In: In silico approach for sustainable agriculture. Springer, pp 183–199

Haas D, Défago G (2005) Biological control of soil-borne pathogens by fluorescent pseudomonads. Nat Rev Microbiol 3(4):307–319

Handelsman J (2004) Metagenomics: application of genomics to uncultured microorganisms. Microbiol Mol Biol Rev 68:669–685

Handelsman J, Rondon MR, Brady SF, Clardy J, Goodman RM (1998) Molecular biological access to the chemistry of unknown soil microbes: a new frontier for natural products. Chem Biol 5:R245–R249

Harismendy O, Ng PC, Strausberg RL, Wang X, Stockwell TB, Beeson KY, Schork NJ, Murray SS, Topol EJ, Levy S (2009) Evaluation of next generation sequencing platforms for population targeted sequencing studies. Genome Biol 10:1–13

Hilton SK, Castro-Nallar E, Pérez-Losada M, Toma I, McCaffrey TA, Hoffman EP, Siegel MO, Simon GL, Johnson WE, Crandall KA (2016) Metataxonomic and metagenomic approaches vs. culture-based techniques for clinical pathology. Front Microbiol 7:484

Hirsch PR, Mauchline TH (2012) Who's who in the plant root microbiome? Nat Biotechnol 30(10):961–962

Hjort K, Presti I, Elväng A, Marinelli F, Sjöling S (2014) Bacterial chitinase with phytopathogen control capacity from suppressive soil revealed by functional metagenomics. Appl Microbiol Biotechnol 98:2819–2828

Iquebal MA, Jagannadham J, Jaiswal S, Prabha R, Rai A, Kumar D (2022) Potential use of microbial community genomes in various dimensions of agriculture productivity and its management: a review. Front Microbiol 13:708335

Johnson NC, Angelard C, Sanders IR, Kiers ET (2013) Predicting community and ecosystem outcomes of mycorrhizal responses to global change. Ecol Lett 16:140–153

Kalia VC, Shouche Y, Purohit HJ, Rahi P (2017) Mining of microbial wealth and metagenomics. Springer

Karumuri S, Singh PK, Shukla P (2015) In silico analog design for terbinafine against Trichophyton rubrum: a preliminary study. Indian J Microbiol 55:333–340

Kaushal M, Mahuku G, Swennen R (2020) Metagenomic insights of the root colonizing microbiome associated with symptomatic and non-symptomatic bananas in Fusarium wilt infected fields. Plants 9:263

Khoiri AN, Cheevadhanarak S, Jirakkakul J, Dulsawat S, Prommeenate P, Tachaleat A, Kusonmano K, Wattanachaisaereekul S, Sutheeworapong S (2021) Comparative metagenomics reveals microbial signatures of sugarcane phyllosphere in organic management. Front Microbiol 12:623799

King RW, Bauer JD, Brady SF (2009) An environmental DNA-derived type II polyketide biosynthetic pathway encodes the biosynthesis of the novel pentacyclic polyketide, erdacin. Angew Chem Int Ed Engl 48:6257

Kuan KB, Othman R, Abdul Rahim K, Shamsuddin ZH (2016) Plant growth-promoting rhizobacteria inoculation to enhance vegetative growth, nitrogen fixation and nitrogen remobilisation of maize under greenhouse conditions. PLoS One 11(3):e0152478

Laranjo M, Alexandre A, Oliveira S (2014) Legume growth-promoting rhizobia: an overview on the Mesorhizobium genus. Microbiol Res 169(1):2–17

Lareen A, Burton F, Schäfer P (2016) Plant root-microbe communication in shaping root microbiomes. Plant Mol Biol 90:575–587

Larke-Mejía NL, Crombie AT, Pratscher J, McGenity TJ, Murrell JC (2019) Novel isoprene-degrading Proteobacteria from soil and leaves identified by cultivation and metagenomics analysis of stable isotope probing experiments. Front Microbiol 10:492457

Larousse M, Galiana E (2017) Microbial partnerships of pathogenic oomycetes. PLoS Pathog 13(1):e1006028

Latz E, Eisenhauer N, Rall BC, Scheu S, Jousset A (2016) Unravelling linkages between plant community composition and the pathogen-suppressive potential of soils. Sci Rep 6(1):23584

Lecomte C, Alabouvette C, Edel-Hermann V, Robert F, Steinberg C (2016) Biological control of ornamental plant diseases caused by Fusarium oxysporum: a review. Biol Contr 101:17–30

Levy MA, Lubell MN (2018) Innovation, cooperation, and the structure of three regional sustainable agriculture networks in California. Reg Environ Change 18:1235–1246

Liu P, Wang W, Zhao J, Wei D (2019) Screening novel β-galactosidases from a sequence-based metagenome and characterization of an alkaline β-galactosidase for the enzymatic synthesis of galactooligosaccharides. Protein Expr Purif 155:104–111

Maillard N (2018) Jésus, Lazare, Noé et le mammouth: Résurrection de la chair humaine et déextinction de l'espèce animale. Rev Semest Droit Anim:281–296

Makonde HM, Mwirichia R, Osiemo Z, Boga HI, Klenk H-P (2015) 454 Pyrosequencing-based assessment of bacterial diversity and community structure in termite guts, mounds and surrounding soils. SpringerPlus 4:1–11

Maruthamuthu M, Jiménez DJ, Stevens P, van Elsas JD (2016) A multi-substrate approach for functional metagenomics-based screening for (hemi)cellulases in two wheat straw-degrading microbial consortia unveils novel thermoalkaliphilic enzymes. BMC Genomics 17:1–16

Mendes R, Kruijt M, De Bruijn I, Dekkers E, Van Der Voort M, Schneider JH, Piceno YM, DeSantis TZ, Andersen GL, Bakker PA (2011) Deciphering the rhizosphere microbiome for disease-suppressive bacteria. Science 332:1097–1100

Mendes R, Garbeva P, Raaijmakers JM (2013) The rhizosphere microbiome: significance of plant beneficial, plant pathogenic, and human pathogenic microorganisms. FEMS Microbiol Rev 37(5):634–663

Newell R, Raimi D, Aldana G (2019) Global energy outlook 2019: the next generation of energy. Resour Future 1

Oldroyd GE (2013) Speak, friend, and enter: signalling systems that promote beneficial symbiotic associations in plants. Nat Rev Microbiol 11(4):252–263

Owen JG, Reddy BVB, Ternei MA, Charlop-Powers Z, Calle PY, Kim JH, Brady SF (2013) Mapping gene clusters within arrayed metagenomic libraries to expand the structural diversity of biomedically relevant natural products. Proc Natl Acad Sci 110:11797–11802

Pabbathi NPP, Velidandi A, Tavarna T, Gupta S, Raj RS, Gandam PK, Baadhe RR (2021) Role of metagenomics in prospecting novel endoglucanases, accentuating functional metagenomics approach in second-generation biofuel production: a review. Biomass Convers Bioref 13:1371–1398

Palenik B, Ren Q, Tai V, Paulsen I (2009) Coastal Synechococcus metagenome reveals major roles for horizontal gene transfer and plasmids in population diversity. Environ Microbiol 11:349–359

Parniske M (2008) Arbuscular mycorrhiza: the mother of plant root endosymbioses. Nat Rev Microbiol 6(10):763–775

Pawar SV, Hallam SJ, Yadav VG (2018) Metagenomic discovery of a novel transaminase for valorization of monoaromatic compounds. RSC Adv 8:22490–22497

Pooja S, Pushpanathan M, Jayashree S, Gunasekaran P, Rajendhran J (2015) Identification of periplasmic α-amlyase from cow dung metagenome by product induced gene expression profiling (pigex). Indian J Microbiol 55:57–65

Prosser JI (2015) Dispersing misconceptions and identifying opportunities for the use of 'omics' in soil microbial ecology. Nat Rev Microbiol 13:439–446

Richardson TH, Tan X, Frey G, Callen W, Cabell M, Lam D, Macomber J, Short JM, Robertson DE, Miller C (2002) A novel, high performance enzyme for starch liquefaction: discovery and optimization of a low pH, thermostable α-amylase. J Biol Chem 277:26501–26507

Sarangi M, Chopra C, Usman YA, Dhanjal DS, Chopra RS (2019) Accessing genetic diversity and phylogenetic analysis of microbial population of soil from Hygam Wetland of Kashmir Valley. Res J Pharm Technol 12:2323–2326

Saville AC, Martin MD, Ristaino JB (2016) Historic late blight outbreaks caused by a widespread dominant lineage of Phytophthora infestans (Mont.) de Bary. PLoS One 11(12):e0168381

Scholz MB, Lo C-C, Chain PS (2012) Next generation sequencing and bioinformatic bottlenecks: the current state of metagenomic data analysis. Curr Opin Biotechnol 23:9–15

Shendure J, Ji H (2008) Next-generation DNA sequencing. Nat Biotechnol 26:1135–1145

Simon C, Daniel R (2011) Metagenomic analyses: past and future trends. Appl Environ Microbiol 77:1153–1161

Smith DL, Gravel V, Yergeau E (2017) Signaling in the phytomicrobiome. Front Plant Sci 8:233087

Sprent JI, Ardley J, James EK (2017) Biogeography of nodulated legumes and their nitrogen-fixing symbionts. New Phytol 215(1):40–56

Suenaga H (2012) Targeted metagenomics: a high-resolution metagenomics approach for specific gene clusters in complex microbial communities. Environ Microbiol 14:13–22

Tejesvi MV, Kajula M, Mattila S, Pirttilä AM (2011) Bioactivity and genetic diversity of endophytic fungi in Rhododendron tomentosum Harmaja. Fungal Divers 47:97–107

van Niftrik L, Geerts WJ, van Donselaar EG, Humbel BM, Webb RI, Fuerst JA, Verkleij AJ, Jetten MS, Strous M (2008) Linking ultrastructure and function in four genera of anaerobic ammonium-oxidizing bacteria: cell plan, glycogen storage, and localization of cytochrome c proteins. J Bacteriol 190:708–717

Vuong HB, Thrall PH, Barrett LG (2017) Host species and environmental variation can influence rhizobial community composition. J Ecol 105(2):540–548

Wallenstein MD (2017) Managing and manipulating the rhizosphere microbiome for plant health: a systems approach. Rhizosphere 3:230–232

Williams A, Manoharan L, Rosenstock NP, Olsson PA, Hedlund K (2017) Long-term agricultural fertilization alters arbuscular mycorrhizal fungal community composition and barley (Hordeum vulgare) mycorrhizal carbon and phosphorus exchange. New Phytol 213(2):874–885

15 Antimicrobial Resistance: Introduction and Challenges

Arunima Singh, Yogesh Kumar Vishwakarma, Mayank, Neelmani Bhardwaj, and R. S. Singh

Abstract

The infection-causing microorganisms are a severe issue for the health of humans, plants, and animals. To overcome this issue, various drugs, such as antibiotics, antivirals, antifungals, and antiparasitic, are used to inhibit or kill these infection-causing microorganisms. These pharmaceuticals are collectively called "antimicrobials." But when microorganisms like bacteria, fungi, viruses, and parasites evolve over some time and become capable of flourishing even in the presence of medication which was once used to inhibit them, the phenomenon is known as antimicrobial resistance (AMR). And it occurs because an organism develops resistance to that medication, such as drugs, and it no longer affects it. Misuse and overuse of antimicrobials are significant contributors to the emergence of drug-resistant pathogens. AMR in human pathogens has emerged as the greatest threat to public health in developing countries and worldwide. The World Health Organization (WHO) designated AMR as one of the top ten worldwide public health hazards to humanity. Also, it is one of the hurdles to achieving the Sustainable Development Goals (SDGs). Lack of clean water, poor sanitation, and inadequate infection prevention and control measures encourage AMR pathogens' proliferation. The prolonged illness causes more frequent hospital visits, sometimes more expensive medications, and ultimately financial stress for individuals affected, in addition to mortality and disability.

A. Singh · Y. K. Vishwakarma · R. S. Singh (✉)
Department of Chemical Engineering & Technology, IIT (BHU), Varanasi, India
e-mail: rssingh.che@itbhu.ac.in

Mayank · N. Bhardwaj
Institute of Science, Banaras Hindu University, Varanasi, India

A. Kumar, M. K. Solanki (eds.), *Microbial Biocontrol Techniques*, Microorganisms for Sustainability 54,
https://doi.org/10.1007/978-981-97-8739-5_15

Keywords

Antimicrobial resistance · Infection-causing · Drug-resistant · Pathogens

15.1 Introduction

Earth is the home of countless diverse and ubiquitous microorganisms; the word "microorganisms" includes bacteria and viruses and an enormous number of multicellular organisms such as fungi and algae (Green et al. 2004; Whitman et al. 1998). These minuscule organisms are present in each of the ecological niches on Earth, comprising every surface, cavity, and cellular milieu of every human and from the base of an ecological system. Most microorganisms are harmless to humans; even a few also show a friendly correlation with humans, but some are perilous to human health, causing serious damage and suffering, and can lead to a lethal denouement (Michael et al. 2014).

Refreshing the memories of "the history of human diseases," infections caused by microorganisms have always been the largest portion of human diseases. It was not long ago when microorganisms used to devastate human lives and used to cause plague in human populations in different parts of the world. However, with the development of science, various therapeutic agents aiming against causative organisms have been developed. The antimicrobial agent which was first developed in the world was Salvarsan, discovered by Ehrlich in 1910, and was used as a remedy against syphilis, and by 1935, Domagk developed sulfonamides, although these were synthetic in origin (Saga and Yamaguchi 2009). In 1928, the first naturally produced antimicrobial agent "penicillin" was discovered, and by 1945, it was thoroughly used in World War II to treat soldiers. But just about a few years later of penicillin's extensive use, by the 1950s, the first antibiotic-resistant agent was discovered (Lobanovska and Pilla 2017; Kardos and Demain 2011). In 1961, methicillin-resistant *Staphylococcus aureus* (MRSA) was isolated in the UK. By mid-1960, penicillin-intermediate *Streptococcus pneumoniae* (PISP) strains and β-lactamase-negative ampicillin-resistant (BLNAR) strains were identified. And sooner by 1990, MRSA-causing nosocomial infection became a serious threat to society, and by the end of 1990s, vancomycin-intermediate *S. aureus* (VISA) was also reported in Japan by 2000s. Multidrug-resistant *P. aeruginosa* (MDRP) was isolated (Saga and Yamaguchi 2009). In 2013, the Centers for Disease Control (CDC) in the United States declared that humanity has entered the "post-antibiotic" era. Following suit, in May 2014, the World Health Organization (WHO) proclaimed that the situation surrounding the AMR crisis had escalated to a critical level (Michael et al. 2014). During the COVID-19 pandemic, aside from the direct impact of SARS-CoV-2 infection, an increase in the number of drug-resistant infections was also witnessed. For example, it was observed that there was a rapid rise in bacteria that do not respond well to multiple drugs, such as certain types of *Klebsiella pneumoniae* and *Enterobacterales* that produce enzymes making antibiotics less effective. Similarly, *Acinetobacter baumannii*, methicillin-resistant *Staphylococcus aureus* (MRSA), and fungi like *Candida glabrata* and *Aspergillus fumigatus* had also become resistant to common treatment due to elevated utilization of antimicrobial agents in COVID-19 patients (Lai et al. 2021).

15.2 Mechanisms of Antimicrobial Resistance

A microorganism can develop resistance against a single or different group of drugs. The mechanism of developing resistance can be broadly classified into two types.

1. **Intrinsic Resistance**

   Intrinsic resistance is naturally found in the microorganism due to morphological or anatomical characteristics. For example, as in bacterial communities, some specific genera (or) species have distinctive structural/functional characteristics, which act as tools for providing resistance to certain antibiotics; similarly, fungi are also known to have intrinsic resistance due to idiosyncratic characteristics (Abushaheen et al. 2020).
2. **Acquired Resistance**

   Acquired resistance is not present naturally in microorganisms and is obtained from their resistant member through the process of horizontal gene transfer, as seen in bacteria. However, fungi develop acquired resistance after exposure to antifungal agents, probably through genotypic alterations. However, it is observed much less commonly and even tough to demonstrate. For example, *Candida auris* was first reported in 2009 in Japan isolated from the ear of a patient accountable for the rapid increase in hospital-acquired invasive infections; at present, this fungus has developed resistance against almost all antifungals and represents a menace to critical care units. Other than that, some of the important mechanisms of resistance to antimicrobials are given in Table 15.1.

Table 15.1 Some of the important mechanisms of resistance to antimicrobials

Mechanism of resistance	Affected antibiotics	Example
Inactivation	Beta-lactam antibiotics	Beta-lactamases
	Aminoglycosides	Aminoglycoside-modifying enzyme
Change in the target binding	Beta-lactams/glycopeptides	Changes in penicillin-binding proteins
	Macrolides	Ribosomal methylation
	Fluoroquinolones	DNA gyrase mutation
Efflux and permeability alteration	Pump-specific antibiotics	Efflux pumps
	Most antibiotics (except polymyxins and aminoglycosides)	Porin protein loss

Source: Chandra et al. (2021), Dhawde et al. (2018), Gupta et al. (2014), Kumar et al. (2017), Lübbert et al. (2017), Saha and Bhattacharjya (2019), Saharan et al. (2020)

15.3 Mechanisms of Resistance in Bacteria

Intrinsic resistance can be developed due to the availability or the unavailability of biological "characteristics." For example, it may be the structure of the cell envelope or membrane, as in the gram-negative bacteria, which is intrinsically insusceptible to most antibiotics due to the presence of proportionally thin peptidoglycan layer which is surrounded by outer layer which has much finer molecular mesh (Vaara 1992). This forms a permeability barrier; on the other hand, gram-positive bacteria have a large permeability threshold, which makes them susceptible to most antibiotics. The existence of a drug transport protein such as MDR efflux pumps: many gram-positive and gram-negative bacteria take the help of efflux pumps to actively eliminate the antimicrobial agent from their cellular entity. Efflux pumps can function as substrate specific, or they can be more wide-ranging, i.e., they can efflux either only a single type of molecule or structurally different types of molecules (Piddock 2006). In the shorter term, the intrinsic resistance developed through the following: the absence of a metabolic pathway, the presence of a drug-metabolizing enzyme, the modified structure of the drug target site, and the expression of specific stress response proteins or high repair capacity. Mechanism of antimicrobial resistance in bacteria is shown in Fig. 15.1.

Acquired resistance bacteria can acquire a gene that is responsible for resistance mainly through all the paths by which bacteria can obtain genetic material, i.e., through all paths by which horizontal gene transfer (HGT) can take place, which include transformation, transduction, transposition, and conjugation, and additionally through the mutation which occurs in chromosomal DNA of bacteria. However,

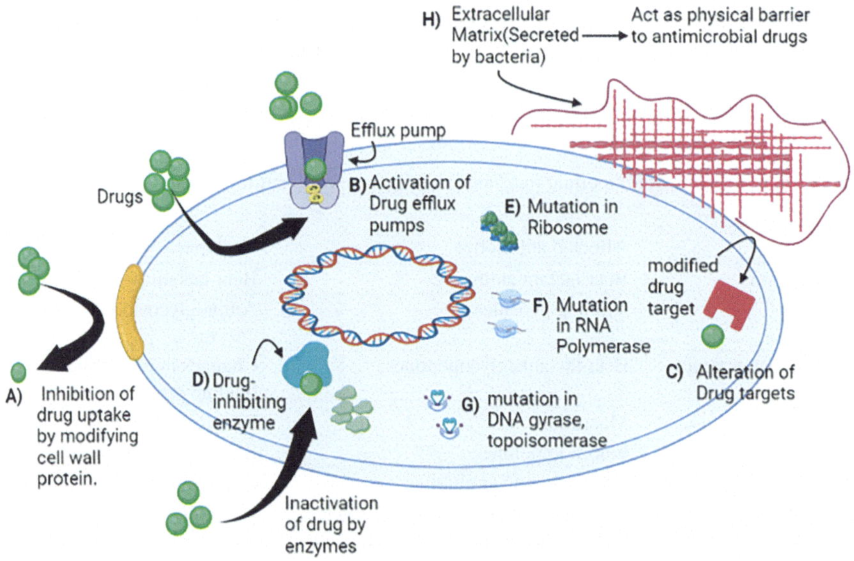

Fig. 15.1 Mechanism of antimicrobial resistance in bacteria

the attainment can be temporary or permanent. Usually, the most common path for attaining external genes is through plasmid-mediated transmission, whereas bacteriophage phage-mediated transmission is the rarest. Some bacteria, such as species of the genus *Acinetobacter*, are competent naturally and hence capable of attaining genetic material straight from the external environment (Reygaert 2018). Internally genetic materials are moved around by the insertion sequences and integrins. Various stressors such as starvation, UV radiation, and chemicals are responsible for inducing mutation, such as substitutions and deletions, in the bacterial chromosome. The average mutation rate in the bacterial chromosome is about 1 at every 10^6 to 10^9 cell divisions, and nearly all of them are injurious to the cell (Coculescu 2009; Davies and Davies 2010). Mutations that are responsible for antimicrobial resistance occur in only some kinds of genes, such as gene encoding for (1) drug targets, (2) drug transporters, (3) regulators that control drug transporters, and (4) antibiotic-modifying enzymes (Martinez 2014). In extension, several mutations are responsible for resistance at the cost of the organism's everyday lifestyle.

15.4 Mechanisms of Resistance in Fungi

A wide variety of mechanisms can be involved in the manifestation of resistance to antifungals. These mechanisms can be either intrinsically present in fungi or acquired after exposure to sublethal drug concentrations (Fig. 15.2). Antifungal drug activity can be divided into three types based on their mode of action: (1) target cytoplasmic membrane (azoles, polyenes), (2) target cell wall (echinocandins), and (3) target DNA and protein synthesis machinery (pyrimidine analogs) (Baddley and Moser 2004).

15.4.1 Resistance Mechanism Against Antifungal Drugs

15.4.1.1 Target Cytoplasmic Membrane

1. Polyene resistance mechanism

   Usually, the most common resistance seen against polyene is natural and organism specific. Acquired resistance to polyenes either involves or limits the capability of AMR to access its target side by altering 1,3-b-D-glucan in the cell wall.

2. Azole resistance mechanisms

   Azole resistance involves several mechanisms: enhancing the efflux by upregulating multidrug efflux transporter genes, substituting Erg11p amino acid, and ergosterol biosynthetic pathway modification.

15.4.1.2 Target Cell Wall by Echinocandin Resistance Mechanisms

Resistance to echinocandin involves two mechanisms:

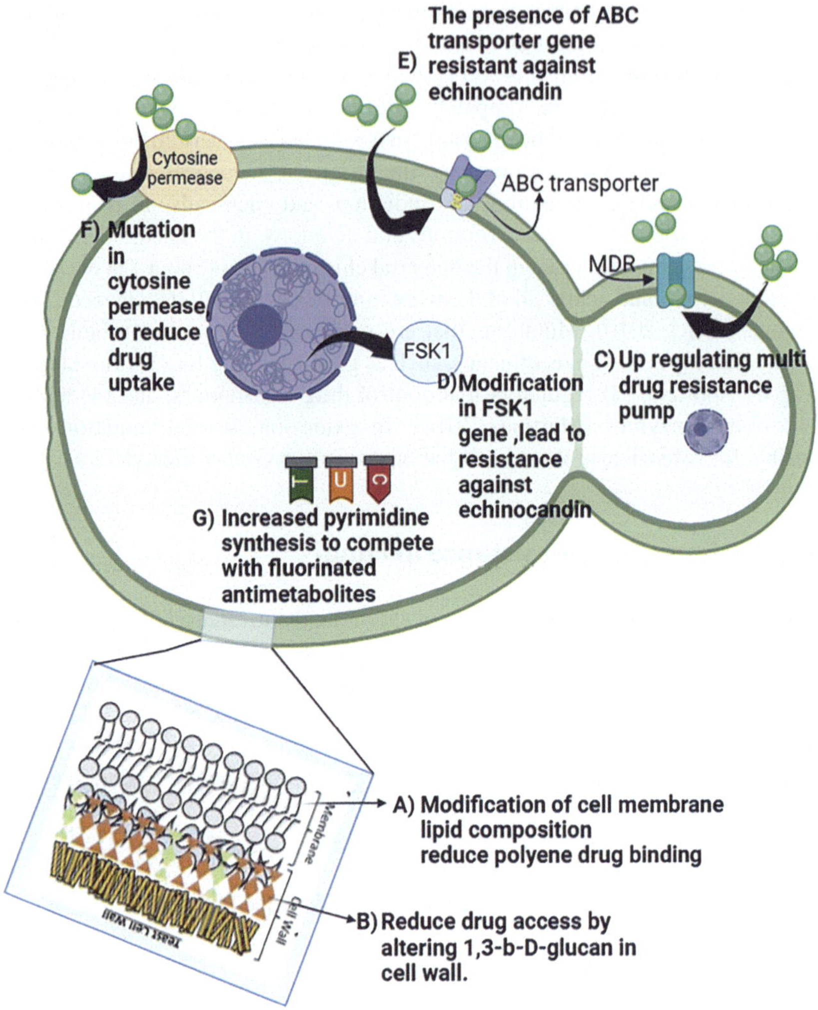

Fig. 15.2 Resistance mechanism against antifungal drugs that target cytoplasmic membrane

1. FKS1 gene (which encodes for subunits of 1,3-b-D-glucan synthase) modification led resistance to *candins*.
2. The second resistant mechanism evolves with the presence of ABC transporter genes.

15.4.1.3 Target DNA or Protein Synthesis

1. Flucytosine resistance mechanisms

   Both intrinsic and acquired resistance are present against flucytosine. It involves various mechanisms, such as a decrease in the uptake or conversion of the drug by mutation in the cytosine permease or deaminase, seen as a case of intrinsic resistance. Also, there is a loss of function of uracil phosphoribosyl transferase, increasing the competitor such as high amount of synthesis of pyrimidines, which compete with the fluorinated antimetabolites.

15.5 Antimicrobial Resistance Scenario in Undeveloped and Developing Countries

Antimicrobial drug resistance is a severe threat to the world's health sector. With the increase in production and consumption of antimicrobial drugs, the adaptation of microbes to and resistance to the drugs are now common phenomena observed throughout the globe (Hart and Kariuki 1998).

Contrary to developed countries where exposure and availability of antimicrobial drugs are more accessible, in developing and underdeveloped countries, access to major antimicrobial drugs is still within the reach of people with the highest population of taxpayers. Instances can be seen where in the majority of cases, only first-line antimicrobial medications (ampicillin, chloramphenicol, co-amoxiclav, erythromycin, etc.) are present in the district hospitals in Kenya; hence, excess use of these drugs expressly and repeatedly has led to the development of resistance of microbes. Second-line antimicrobial medications such as amikacin, cefuroxime, ciprofloxacin, and nalidixic acid are available to a selected section of society who can afford them. In developed countries, however, both first- and second-line antimicrobial medications are available in every section of society.

Also, factors like quality of drugs, production and distribution cost, and prescription without undergoing proper tests or prescriptions based upon previous knowledge are hindering the current situation of antimicrobial drugs in developing countries (e.g., in Africa, a study from Cameroon revealed that out of 284 antimalarials obtained from the country, only 61% were effective against malaria) (Ayukekbong et al. 2017). Several examples support the above points: in countries like Bangladesh, prescribing multiple antimicrobials for a single disease is a common habit. Prescribing antimicrobials based upon experience or exclusively based on symptoms instead of prescription based upon shreds of evidence obtained from conducting tests is practiced there. According to Hoque et al. (2020), antimicrobials were frequently administered for fever, common cold, cough, diarrhea, and acute respiratory infections (ARIs). Similar scenarios prevail in other developing countries, such as Africa, which lack regulating bodies.

Among most developing countries, especially in South Asia, in the past decade, many species and subspecies have caused a threat to the health sector by developing resistance against different antimicrobial drugs.

Among the cases are those of *Salmonella enterica*. Since serotype *typhi* (*S. typhi*) is growing resistant to various antimicrobial medications, including ampicillin, chloramphenicol, and co-trimoxazole (trimethoprim-sulfamethoxazole), the microbe *S. enterica*, serotype *paratyphi* (*S. paratyphi*), is becoming multidrug resistant and may eventually pose a threat to human health in Asia. In regions of Africa and Asia, some cheap and readily available antimicrobials were used intensively, and hence, many microbes have developed resistance against them in the past 20 years. For example, *S. dysenteriae* type I has become resistant to many antibiotics such as ampicillin, co-trimoxazole, chloramphenicol, tetracyclines, and, increasingly, nalidixic acid. Much evidence of *V. cholera* becoming resistant to co-trimoxazole, tetracycline, and other antimicrobials has been found in India, Vietnam, and sub-Saharan Africa, while fluoroquinolone resistance rates rose from 45% to 72% in isolates from Asia and from 17% to 38% in isolates from Africa. According to data, penicillin and erythromycin resistance in community-acquired *S. pneumoniae* is a growing issue in Asia, Mexico, Argentina, Brazil, and some regions of Kenya and Uganda (Okeke et al. 2005). Among different *Candida* species, *C. glabrata* and *C. krusei* have become the most prevalent species in various institutions. *Candida krusei* is consistently resistant to fluconazole (FCZ), and *C. glabrata* may rise as a breakthrough infection, attaining resistance to all triazoles (Walsh and Gamaletsou 2013). Globally, 2.5% and 9% of the *Candida* species isolates were analyzed and were found to resist fluconazole and itraconazole, respectively (Eliopoulos et al. 2002).

15.6 Global Prospect of AMR

A global action plan to combat AMR was approved by the WHO in May 2015. National Strategy for Combating Antibiotic-Resistant Bacteria, published in September 2014 by the US Government, included an exclusive order signed by the US President. The order focused on addressing the expanding threat of antibiotic-resistant bacteria by delineating a comprehensive strategy to address this issue. The Global Antibiotic Resistance Partnership (GARP) was constituted in 2009 to provide a platform for developing practical policy recommendations concerning antibiotic resistance. It includes many African nations and countries with low- and moderate-income levels. However, only countries like India, South Africa, Kenya, Vietnam, Nepal, Mozambique, Tanzania, and Uganda have created national GARP working groups. Furthermore, upon thoughtful inspection of national action plans, it was noticed that the World Health Organization (WHO) has officially identified only one African country, Ethiopia, for having a national action plan particularly addressing AMR (Ayukekbong et al. 2017).

15.7 Antimicrobial Resistance: The Indian Storyline

Globally, India stands first with a 23% increase in sales of antibiotics between 2000 and 2010. As a result of the "scoping report on antimicrobial resistance in India (2017)," research conducted in 2017 shows that more than 70% of *E. coli*, *K. pneumonia*, etc. and 50% of *Pseudomonas* were resistant to fluoroquinolones and third generation. Methicillin resistance rates for the gram-positive bacteria *Staphylococcus aureus* were 42.6%, while vancomycin resistance rates for *Enterococcus faecium* were 10.5%. *Salmonella typhi* and *Shigella* species had resistance rates to

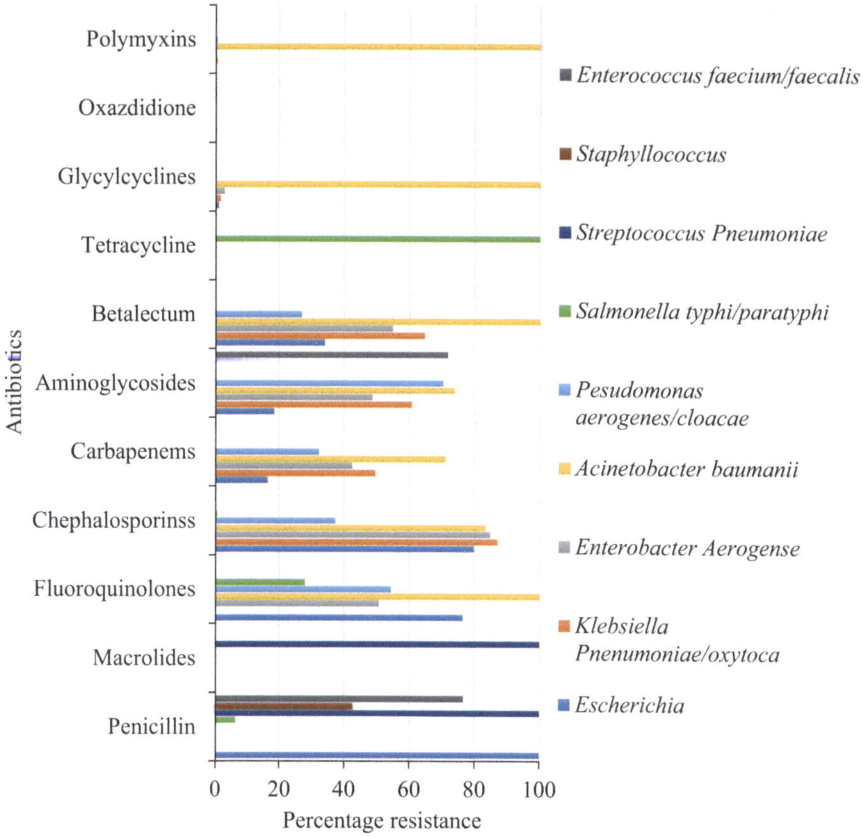

Fig. 15.3 Antimicrobial resistance of different isolates from blood and Cerebrospinal fluid (CSF) samples (ICMR 2015)

ciprofloxacin of 28% and 82%, ceftriaxone of 0.6% and 12%, and co-trimoxazole of 2.3% and 80%, respectively. In various parts of the country, *Vibrio cholera* tetracycline resistance rates ranged from 17% to 75% (Taneja and Sharma 2019). More detailed data for the percentage of antibiotic resistance for the antibiotics are provided in Fig. 15.3.

15.7.1 AMR in Animals

India has become the fourth biggest consumer of antimicrobials for animals after China, the United States, and Brazil. If the conditions prevail, it will emerge as the largest consumer between 2010 and 2030 [Bayesian predictions]. There have been different instances where some exciting results were obtained during AMR testing. One such instance was in 2013–2014, when milk samples of different livestock were sampled for AMR; in West Bengal, milk of cow was reported to have 48% of

Table 15.2 Percentage resistance of different bacterial species against different antibiotics isolated from cow and buffalo milk samples and different aquatic animals from different locations

S. No.	Resistant bacterial species	Percentage of bacteria	Region	Antimicrobial drug	Source
1.	Gram-negative bacilli	47.5	Gujarat	Oxytetracycline	Cow and buffalo
2.	*S. aureus*	2.4	West Bengal	Vancomycin	Cow and buffalo
3.	*S. aureus*	21.4	Karnataka	Methicillin	Cow and buffalo
4.	Coagulase-negative *Staphylococci*	5.6	Karnataka	Methicillin	Cow and buffalo
5.	Enterobacteriaceae	48	Maharashtra	ESBL producers	Tilapia fish
6.	*Vibrio cholera, V. parahaemolyticus*	100	Kerala	Ampicillin	Prawns, shellfish, and crabs
7.	*Vibrio cholera, V. parahaemolyticus*	100	Kerala	Chloramphenicol	Prawns, shellfish, and crabs
8.	*Vibrio cholera, V. parahaemolyticus*	67–96	Kerala	Ceftazidime	Prawns, shellfish, and crabs

gram-negative bacilli, whereas buffalo milk had produced extended-spectrum β-lactamases (ESBL).

India, along with becoming a key hub for the aquaculture industry, producing 9,579,103 tons of fish annually, is also growing in the chicken sector as the country's chicken sector produces 1,916,103 tons of grilled meat annually, with the states of Haryana, West Bengal, and Uttar Pradesh producing the most (18.4%, 17.1%, and 14.1%, respectively). Three research looking at Enterobacteriaceae that produce ESBLs, out of the seven studies on AMR in chicken, show that the rate of ESBL producers ranges from 9.4% in Odisha to 33.5% in MP and 87% in Punjab. Details are described in Table 15.2.

15.7.2 AMR in the Environment

Many rivers of the country are directly dumped, with numerous forms of pollution emerging from different sources. Many of them are also polluted indirectly when different inlets join them, which have various concentrations of multiple antibiotics. Isolates from two of the major rivers from North India, the Ganga and Yamuna, show that gram-negative bacteria were found to have an ESBL rate of 17.4%, with blaNDM-1 and blaOXA48 resistance genes.

15.8 Causes of Antimicrobial Resistance

When antimicrobials were developed, they were termed as "silver bullets" for their quick and effective treatment of diseases and not having any serious observable negative impacts. However, the current rise in AMR rates is mostly because of the wide use of antimicrobials with the assumption that they are universally applicable. A few of the major reasons for rising AMR are discussed below.

15.8.1 Overprescription

1. Due to the practical difficulties in quickly and effectively detecting infectious diseases, antimicrobials are frequently prescribed at random.
2. When a patient presents with symptoms that could end their life, rapid intervention is sometimes necessary.
3. Instead of routinely testing for pathogen and antibiotic susceptibility, the practitioner frequently provides antimicrobial therapy based on prior experience and local epidemiology.

In addition to failing to provide a quick cure, this method repeatedly and intensely applies selective pressure to the patient's microbiota, which promotes and maintains the emergence of AMR among present nonpathogenic organisms. Then, via HGT, this resistance may be spread throughout various microbial communities and, more critically, to pathogenic species that were previously vulnerable. According to studies, the overprescription of antibiotics in low- and middle-income nations caused a considerable hike in antibiotic consumption, which is about more than 65% between the years 2000 and 2015 (Acam et al. 2023).

15.8.2 Public Perception and Behavior

Nonprescription purchases are a significant loophole because manufacturing and distribution are primarily unregulated in some nations.

15.8.3 Agricultural Applications

The constant application of antimicrobials in agriculture to produce higher yields puts pressure on the natural environment, increasing AMR genes.

15.8.4 Commercial Pressures

The current generation is tilted towards many misconceptions, and in the case of microbes, eradicating every microbe from a surface is the best way to survive. This practice, apart from being impossible, is also unnecessary. However, these misconceptions are hyped by business companies to boost their product sales. Successful and varied marketing initiatives persuade consumers to adopt antimicrobials in a range of items, from eye drops to floor cleaners (Michael et al. 2014).

15.8.5 Transmission of AMR Among Human and Animal Populations

Antimicrobial resistance can be interpreted as the potential of microorganisms such as bacteria, viruses, parasites, and fungi to withstand the actions of antimicrobial medications. AMR is primarily caused by selective stress on microbes due to antibiotic exposure. The creation of enzymes, adjustment of target locations, modification of metabolic pathways, alternating the permeability of the external membrane, and efflux pumps are some methods organisms can adapt to develop resistance to antimicrobials (Allcock et al. 2017). Resistant bacteria might exhibit one or more of these pathways, resulting in resistance to more than one antibiotic class. Much research has investigated the relationship between antibiotic usage in animals and human resistance, including direct and indirect transmission mechanisms. The investigations found a link between antibiotic intake and bacterial resistance in people and animals. Antimicrobial usage, misuse, or abuse in medical treatment significantly contributes to AMR's emergence in humans (Allcock et al. 2017).

The enhanced level of antibiotic substances and antibiotic-resistant bacteria in soil and the success of horizontal gene transfer specify that environmental transmission may play a role (Fig. 15.4). Resistant bacteria can adhere to crops or become endophytes, exposing humans to food consumption (Heuer et al. 2011). Plant antibiotic absorption may aid in the formation of resistant endophytes. Agricultural

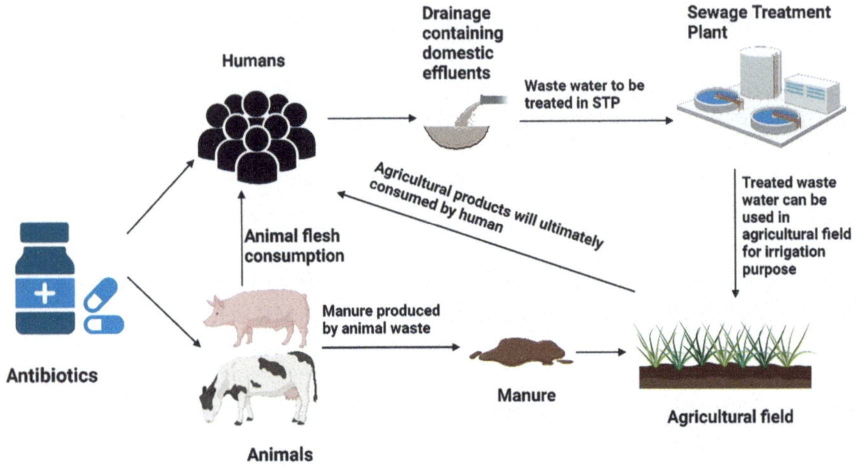

Fig. 15.4 Routes of transmission of antimicrobial resistance

antibiotic use considerably influences the dissemination of resistance in the human population due to the high exposure rate to environmental contamination with transferable antibiotic resistance genes. Furthermore, drainage from the manure storage lagoons assists in disseminating environmental resistance. Recent research conducted in the United States has observed elevated levels of RNA methylase genes and tetracycline resistance in groundwater around sites (Heuer et al. 2011). The wastewater discharged from agriculture and industries contains resistant bacteria; as a result, the treated municipal wastewater contains antibiotic-resistant genes; this treated municipal water is used by the human population, due to which they are exposed to antibiotic-resistant bacteria (LaPara et al. 2011).

15.9 Challenges of AMR

15.9.1 Effect of Antimicrobial Resistance on Human Health

AMR has been associated with increased mortality rates. A study published in The Lancet Infectious Diseases in 2021 analyzed data from 1.7 million patients across 18 countries and found that bloodstream infections caused by drug-resistant bacteria had a high level of mortality rate than infections caused by susceptible bacteria (Ming et al. 2021). About 700,000 fatalities happen worldwide just because of infection caused by drug-resistant bacteria. This figure is anticipated to increase to ten million annually by 2050 (Allcock et al. 2017). Antimicrobial resistance reduces the effectiveness of antimicrobial drugs, leading to treatment failures. A systematic review published in PLOS Medicine in 2020 analyzed various studies and found a high risk of treatment failure in patients infected with drug-resistant bacteria, which usually leads to more extended hospital stays than those patients who are infected with susceptible bacteria. The rise of antimicrobial resistance has led to a decrease in the number of effective antibiotics available for treating infections (Tacconelli et al. 2018).

AMR complicates surgical procedures by increasing the risk of postoperative infections. A study published in JAMA Surgery in 2020 examined the association between antimicrobial resistance and surgical site infections. The findings indicated that infections caused by drug-resistant bacteria were associated with elevated complications and increased healthcare costs. Therefore, to prevent extra costs, antimicrobial sealant use should be avoided after surgical site skin preparation to reduce SSIs (Allegranzi et al. 2016).

15.9.2 Effect of Antimicrobial Resistance on Animal Health

A prominent human population consumes animal flesh as its food resource. To protect their livestock from any diseases, farmers use antibiotics. As meat production has to be enhanced from 200 million tons to 470 million tons by 2050, farmers will rely even more on antibiotics. The inferior quality of veterinary medicine is used for the therapeutic treatment of livestock compounds, which has also paced the problem of antibiotic overuse (Clifford et al. 2018). The worldwide antimicrobial consumption in food animals was evaluated at 63,151 tons in 2010 and 131,109 tons in 2013. India exhibits around 3% of the global usage of antimicrobials in livestock (Mutua et al. 2020). AMR in animals affects them adversely by increasing the frequency of treatment failures and increased severity of infections.

Resistant bacteria from livestock can enter humans through different transmission routes like animal meat consumption, direct exposure to colonized animals, manure dissemination in the surrounding, and antibiotic residues in animal products like milk (Clifford et al. 2018).

15.9.3 Effect of Antimicrobial Resistance on Crop Production

The agriculture field receives antimicrobial resistant bacteria from the manure used in the field. Manure is the "hotspot" of bacteria carrying antimicrobial resistant gene (ARG) and inhabiting on mobile genetic elements. Bacteria carrying ARGs and their metabolites and antimicrobial agents are introduced in soil when the soil is administrated with the manure. Repeated application of manure leads to the accumulation of bacteria with antimicrobial resistance capacity in the soil, and the reduction of bacteria that contribute to soil quality occurs. Plants also receive AMR bacteria through water transport and passive absorption. Since the treated wastewater and lakes also contain ARG, water from these sources for irrigation in crop fields will introduce ARG in plants (Thanner et al. 2016).

15.9.4 The Economic Impact of Antimicrobial Resistance

The economic burden of AMR is substantial. A study published in Health Affairs 2020 estimated that drug-resistant infections in the United States alone cost an additional $1.5 billion in healthcare expenses annually. These costs include extended hospital stays, extra tests, and more costly antibiotics (Klein et al. 2018). The net economic consequence of resistance can be calculated by subtracting the expenses that are involved in the treatment of illnesses that occur due to the resistant isolate ("treatment cost") from the expense involved in preventing such infections ("prevention cost"). The expense analysis must include all resources influenced by disease or intervention. Diagnostic expenditures include the costs of radiologic investigation, laboratory testing, bronchoscopies, and other diagnostic operations, which are mainly of concern to the healthcare facility. Patients bear both direct and indirect healthcare costs (for example, lost productivity resulting in a loss of income). Other indirect costs of antimicrobial drug resistance include pharmaceutical industry expenses due to reduced marketability and enterprise costs due to irrecoverable workers' productive time. All of these elements contribute to the economic consequence of resistance (McGowan Jr 2001).

15.10 Conclusion

Antimicrobial resistance is imminent and escalating rapidly as a formidable challenge to the public health, medicine, and healthcare system. Antimicrobials include antibiotics, antivirals, and antifungals that are predominant in the medicinal world for saving countless lives in the battle against microbial diseases. However, with the passing years, antimicrobial productivity has gradually reduced due to incongruous utilization of this agent, such as overprescription, agriculture misuse, human and animal health care misuse, and inadequate infection control. All these factors pace the evolution of resistance in microorganisms, which has pressured physicians and pharmaceutical companies.

Microorganisms develop resistance through either intrinsic or extrinsic mechanisms. AMR has greatly affected public health, leading to elevated mortality rates, illness duration, and higher health maintenance costs. Many standard medical procedures, such as surgical operations, chemotherapy, and organ transplants, become insecure as the antimicrobials can no longer prevent or treat infections because of AMR. Hence, AMR burdens healthcare systems economically, distorting societal stability by influencing productivity and food safety and augmenting healthcare incongruities. Therefore, we require serious actions to counter AMR, which involves executing antimicrobial stewardship plans to encourage responsible use, investments in research and development programs of novel antimicrobials, better diagnosis, and improved infection prevention and control measures.

References

Abushaheen MA, Fatani AJ, Alosaimi M, Mansy W, George M, Acharya S, Rathod S, Divakar DD, Jhugroo C, Vellappally S, Khan AA (2020) Antimicrobial resistance, mechanisms and its clinical significance. Dis Mon 66(6):100971

Acam J, Kuodi P, Medhin G, Makonnen E (2023) Antimicrobial prescription patterns in East Africa: a systematic review. Syst Rev 12(1):18

Allcock S, Young EH, Holmes M, Gurdasani D, Dougan G, Sandhu MS, Solomon L, Török ME (2017) Antimicrobial resistance in human populations: challenges and opportunities. Glob Health Epidemiol Genom 2:e4. https://doi.org/10.1017/gheg.2017.4

Allegranzi B, Bischoff P, De Jonge S, Kubilay NZ, Zayed B, Gomes SM, Abbas M, Atema JJ, Gans S, Van Rijen M, Boermeester MA, Egger M, Kluytmans J, Pittet D, Solomkin JS (2016) New WHO recommendations on preoperative measures for surgical site infection prevention: an evidence-based global perspective. Lancet Infect Dis 16(12):e276–e287. https://doi.org/10.1016/S1473-3099(16)30398-X

Ayukekbong JA, Ntemgwa M, Atabe AN (2017) The threat of antimicrobial resistance in developing countries: causes and control strategies. Antimicrob Resist Infect Control 6(1):1–8

Baddley JW, Moser SA (2004) Emerging fungal resistance. Clin Lab Med 24(3):721–735

Chandra P, Mk U, Ke V, Mukhopadhyay C, Dinesh Acharya U (2021) Antimicrobial resistance and the post antibiotic era: better late than never effort. Expert Opin Drug Saf 20(11):1375–1390

Clifford K, Desai D, Prazeres Da Costa C, Meyer H, Klohe K, Winkler A, Rahman T, Islam T, Zaman MH (2018) Antimicrobial resistance in livestock and poor quality veterinary medicines. Bull World Health Organ 96(9):662–664. https://doi.org/10.2471/BLT.18.209585

Coculescu BI (2009) Antimicrobial resistance induced by genetic changes. J Med Life 2:114–123

Davies J, Davies D (2010) Origins and evolution of antibiotic resistance. Microbiol Mol Biol Rev 74:417–433

Dhawde R, Macaden R, Saranath D, Nilgiriwala K, Ghadge A, Birdi T (2018) Antibiotic resistance characterization of environmental E. coli isolated from River Mula-Mutha, Pune District, India. Int J Environ Res Public Health 15(6):1247

Eliopoulos GM, Perea S, Patterson TF (2002) Antifungal resistance in pathogenic fungi. Clin Infect Dis 35(9):1073–1080

Green JL, Holmes AJ, Westoby M, Oliver I, Briscoe DA, Dangerfield M (2004) Spatial scaling of microbial eukaryote diversity. Nature 432:747–750. https://doi.org/10.1038/nature03034

Gupta S, Kapur S, Padmavathi DV (2014) Comparative prevalence of antimicrobial resistance in community-acquired urinary tract infection cases from representative states of northern and southern India. J Clin Diagn Res 8(9):DC09

Hart CA, Kariuki S (1998) Antimicrobial resistance in developing countries. BMJ 317(7159):647–650

Heuer H, Schmitt H, Smalla K (2011) Antibiotic resistance gene spread due to manure application on agricultural fields. Curr Opin Microbiol 14(3):236–243. https://doi.org/10.1016/j.mib.2011.04.009

Hoque R, Ahmed SM, Naher N, Islam MA, Rousham EK, Islam BZ, Hassan S (2020) Tackling antimicrobial resistance in Bangladesh: a scoping review of policy and practice in human, animal and environment sectors. PLoS One 15(1):e0227947

Indian Council of Medical Research (2015). India's Antimicrobial Resistance Surveillance & Research Initiatives. https://iamrsn.icmr.org.in/index.php/resources/amr-icmr-data

Kardos N, Demain AL (2011) Penicillin: the medicine with the greatest impact on therapeutic outcomes. Appl Microbiol Biotechnol 92:677–687. https://doi.org/10.1007/s00253-011-3587-6

Klein EY, Van Boeckel TP, Martinez EM, Pant S, Gandra S, Levin SA, Goossens H, Laxminarayan R (2018) Global increase and geographic convergence in antibiotic consumption between 2000 and 2015. Proc Natl Acad Sci 115(15):E3463. https://doi.org/10.1073/pnas.1717295115

Kumar P, Bag S, Ghosh TS, Dey P, Dayal M, Saha B, Das B (2017) Molecular insights into antimicrobial resistance traits of multidrug resistant enteric pathogens isolated from India. Sci Rep 7(1):1–12

Lai CC, Chen SY, Ko WC, Hsueh PR (2021) Increased antimicrobial resistance during the COVID-19 pandemic. Int J Antimicrob Agents 57(4):106324

LaPara TM, Burch TR, McNamara PJ, Tan DT, Yan M, Eichmiller JJ (2011) Tertiary-treated municipal wastewater is a significant point source of antibiotic resistance genes into Duluth-Superior Harbor. Environ Sci Technol 45(22):9543–9549. https://doi.org/10.1021/es202775r

Lobanovska M, Pilla G (2017) Penicillin's discovery and antibiotic resistance: lessons for the future? Yale J Biol Med 90(1):135–145

Lübbert C, Baars C, Dayakar A, Lippmann N, Rodloff AC, Kinzig M, Sörgel F (2017) Environmental pollution with antimicrobial agents from bulk drug manufacturing industries in Hyderabad, South India, is associated with dissemination of extended spectrum beta-lactamase and carbapenemase-producing pathogens. Infection 45(4):479–491

Martinez JL (2014) General principles of antibiotic resistance in bacteria. Drug Discov Today 11:33–39

McGowan JE Jr (2001) Economic impact of antimicrobial resistance. Emerg Infect Dis 7(2):286–292. https://doi.org/10.3201/eid0702.010228

Michael CA, Dominey-Howes D, Labbate M (2014) The antimicrobial resistance crisis: causes, consequences, and management. Front Public Health 2:145

Ming DK, Myall AC, Hernandez B, Weiße AY, Peach RL, Barahona M, Rawson TM, Holmes AH (2021) Informing antimicrobial management in the context of COVID-19: understanding the longitudinal dynamics of C-reactive protein and procalcitonin. BMC Infect Dis 21(1):932. https://doi.org/10.1186/s12879-021-06621-7

Mutua F, Sharma G, Grace D, Bandyopadhyay S, Shome B, Lindahl J (2020) A review of animal health and drug use practices in India, and their possible link to antimicrobial resistance. Antimicrob Resist Infect Control 9(1):103. https://doi.org/10.1186/s13756-020-00760-3

Okeke IN, Laxminarayan R, Bhutta ZA, Duse AG, Jenkins P, O'Brien TF, Klugman KP (2005) Antimicrobial resistance in developing countries. Part I: Recent trends and current status. Lancet Infect Dis 5(8):481–493

Piddock LJ (2006) Multidrug-resistance efflux pumps—not just for resistance. Nat Rev Microbiol 4:629–636

Reygaert WC (2018) An overview of the antimicrobial resistance mechanisms of bacteria. AIMS Microbiol 4(3):482

Saga T, Yamaguchi K (2009) History of antimicrobial agents and resistant bacteria. JMAJ 52(2):103–108

Saha A, Bhattacharjya H (2019) Health-care waste management in public sector Tripura, North-East India: an observational study. Indian J Community Med 44(4):368–372

Saharan VV, Verma P, Singh AP (2020) High prevalence of antimicrobial resistance in Escherichia coli, Salmonella spp. and Staphylococcus aureus isolated from fish samples in India. Aquac Res 51(3):1200–1210

Tacconelli E, Carrara E, Savoldi A, Harbarth S, Mendelson M, Monnet DL, Pulcini C, Kahlmeter G, Kluytmans J, Carmeli Y, Ouellette M, Outterson K, Patel J, Cavaleri M, Cox EM, Houchens CR, Grayson ML, Hansen P, Singh N, Zorzet A (2018) Discovery, research, and development of new antibiotics: the WHO priority list of antibiotic-resistant bacteria and tuberculosis. Lancet Infect Dis 18(3):318–327. https://doi.org/10.1016/S1473-3099(17)30753-3

Taneja N, Sharma M (2019) Antimicrobial resistance in the environment: the Indian scenario. Indian J Med Res 149(2):119–128

Thanner S, Drissner D, Walsh F (2016) Antimicrobial resistance in agriculture. MBio 7(2):e02227–e02215. https://doi.org/10.1128/mBio.02227-15

Vaara M (1992) Agents that increase the permeability of the outer membrane. Microbiol Rev 56:395–411

Walsh TJ, Gamaletsou MN (2013) Treatment of fungal disease in the setting of neutropenia. Hematology Am Soc Hematol Educ Program 2013(1):423–427

Whitman WB, Coleman DC, Wiebe WJ (1998) Prokaryotes: the unseen majority. Proc Natl Acad Sci USA 95:6578–6583. https://doi.org/10.1073/pnas.95.12.6578

Multifunctional Nanocomposites Enhance Biotherapeutics Sensitivity in Replication-Associated Diseases

16

Tanya Kapil, Palak Gupta, Arathi Radhakrishnan, and Rajpal Srivastav

Abstract

Carbon nanomaterials have special qualities and capacities that can be used to achieve better biotherapeutic results, especially in DNA replication-associated diseases including cancer. The remarkable structural and physicochemical properties offer carbon nanocomposites a great potential in improving the sensitivity and efficacy of traditional biotherapeutics. These nanocomposite structures are flexible and find applications as devices for targeted biotherapeutics because of inherent attributes like their broad surface area, biocompatibility, and easier functionalization capability. Various modalities, including drug delivery, photothermal therapy, and imaging, benefit from the use of carbon nanocomposites, making carbon nanocomposites a great promise for biotherapeutics. Their diverse range of applications facilitate new approaches for targeted therapy, therapeutic monitoring, and improved diagnosis with considerations of biosafety and biocompatibility.

Keywords

Nanomaterials · Carbon nanocomposites · Carbon nanotubes · Targeted delivery · DNA defects · Biotherapeutics

T. Kapil · P. Gupta · A. Radhakrishnan · R. Srivastav (✉)
Amity Institute of Biotechnology, Amity University, Noida, Uttar Pradesh, India
e-mail: arathi.krishnan@s.amity.edu; rsrivastav2@amity.edu

A. Kumar, M. K. Solanki (eds.), *Microbial Biocontrol Techniques*, Microorganisms for Sustainability 54,
https://doi.org/10.1007/978-981-97-8739-5_16

16.1 Introduction

Normal cell development and division take place in a highly controlled manner within a living cell. However, the process of replication and cell division might go wrong resulting in cellular defects, diseases, and ultimately cancer and tumorigenic states. Cancer can occur in almost any type of cell in the body and can develop in various organs and tissues (Mohajeri et al. 2018; Debela et al. 2021). The external factors, along with the intrinsic factors, mutations in proto-oncogenes, and expression of tumor suppressor genes, are responsible for DNA replication defects (Srivastav et al. 2019; Chen et al. 2020; Radhakrishnan et al. 2024a). Globally, cancer ranks as the second most prevalent cause of human mortality. Approximately 7.6 million individuals are projected to succumb to cancer annually, accounting for about 13% of total recorded deaths highlighting the need of various biotherapeutics (Gavas et al. 2021). Chemotherapy drugs are designed to impact rapidly dividing cells, not only leading to the elimination of cancer cells but also affecting healthy cells as side effects. Radiation therapy, involving gamma rays, unavoidably results in harmful effects on healthy tissues located along the radiation path, and it also affects gut microbiota, which is a crucial determinant of health status (Radhakrishnan et al. 2023).

Given the deficiencies inherent in existing treatment strategies, a pivotal focus on enhancing cancer therapy revolves around the precise targeting of therapeutic agents to tumor cells while minimizing harm to healthy tissues adjacent to tumors (Peng and Liang 2019). The current formidable state of cancer has compelled researchers to explore diverse techniques for accurate cancer diagnosis and treatment. Among these emerging approaches, nano-oncology stands out as a promising method to address cancer biotherapeutics (Xie et al. 2021), with the remarkable ability of nanoscale system tissues due to the enhanced permeability and retention. These nanoscale structures or nanocomposites offer distinct advantages in the field of cancer biotherapeutics. Tumor tissues have larger blood vessel pore diameters compared to healthy tissues, leading to a preferential accumulation of nanosized anticancer drugs in tumors. This results in increased treatment efficacy and reduces systemic toxicity. Moreover, nanomaterials are being harnessed as thermal nanoscalpels for the purpose of ablating cancer cells, leveraging their impressive optical and electromagnetic properties. Surface functionalization allows for more precise targeting of tumor tissue, enhancing their therapeutic potential while minimizing adverse effects. Carbon nanomaterials, particularly carbon nanotubes (CNTs), find significant applications in both biological and biomedical realms, particularly in cutting-edge biotherapeutics (Xie et al. 2021). These could be utilized to deliver drugs, diagnostic imaging agents, DNA, small interfering RNA, oligonucleotides, and proteins for the improved cancer biotherapeutics (Paunovska et al. 2022; Radhakrishnan et al. 2024b; Ciccone et al. 2023).

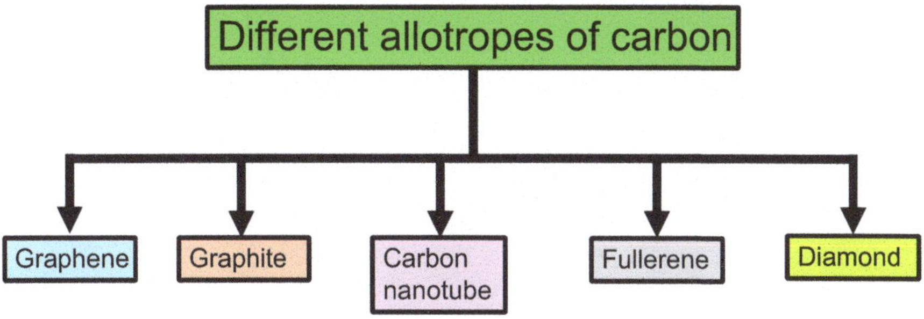

Fig. 16.1 Overview of the different allotropes of the carbon

16.2 Carbon Nanocomposites

Carbon exists in various forms, ranging from zero-dimensional to three-dimensional (3D) structures. Two-dimensional carbon can be shaped into zero-dimensional fullerenes, rolled into one-dimensional nanotubes, or stacked to create three-dimensional graphite (Gulati et al. 2023) (Fig. 16.1). These carbon-based nanomaterials can be categorized as either sp^2 or sp^3 carbon nanomaterials based on their bonding (Azizi-Lalabadi et al. 2020). Within the sp^2 carbon nanomaterial category, we find zero-dimensional fullerenes, one-dimensional nanotubes (CNTs), and two-dimensional graphene (Saleh 2020). On the other hand, sp^3 carbon nanomaterials typically refer to nanodiamonds with crystal sizes in the nano-range. The concept of dimensionality classification is indeed based on dimensions of a nanostructure within the nanoscale range, and this helps us understand the spatial arrangement and properties of nanocomposites. The unique chemical and physical properties exhibited by carbon-based nanocomposites due to their nanoscale dimensions have made them highly attractive for a wide range of applications, which include electrical fields, biosensors, medical treatment, and drug delivery system as important tools of biotherapeutics.

16.3 Carbon Nanocomposites as Carriers of Anticancer Agents

Carbon nanocomposites like carbon nanotubes (CNTs) hold great promise as drug delivery systems in cancer therapy. Their small size and needlelike shapes give them high aspect ratios and extensive specific surface areas, making them adept at binding to or forming conjugates with various therapeutic compounds (Chary et al. 2023). Furthermore, the needlelike structure of CNTs facilitates their uptake into target cells. Extensive research has been conducted on using CNTs for the delivery of cancer biotherapeutic agents such as drugs, genes (including DNA, RNA), and aptamers. Additionally, CNTs can be employed for the delivery of proteins, peptides, and components related to immunotherapy (Fig. 16.2). Different variations of

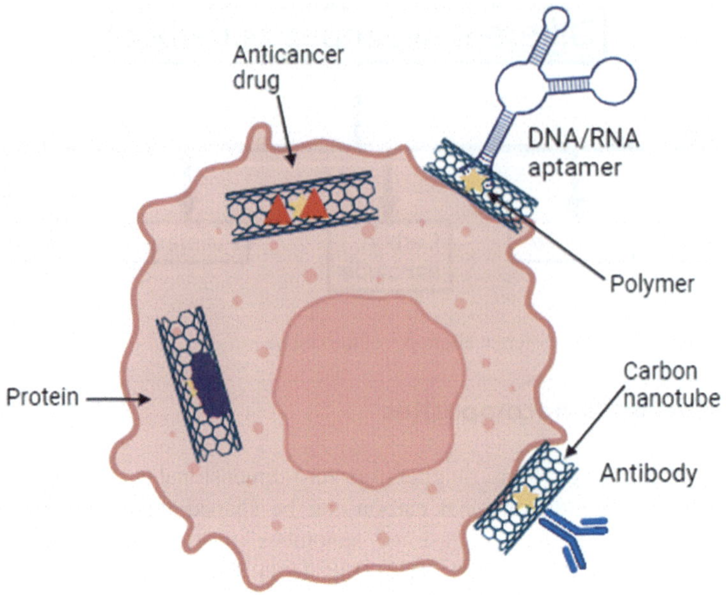

Fig. 16.2 Various nanocomposite conjugations in biotherapeutics

CNT are used as carriers of medicines, nucleic acids, and proteins. These nanocomposites provide cyclic dosage, controlled release of therapeutic agents in constant doses throughout time, and adjustable release of both hydrophilic and hydrophobic medicines and hence play a crucial role in drug delivery (Dias et al. 2021; Madhavi et al. 2017).

The functionalized nanotubes have demonstrated tremendous potential for therapeutic drug delivery. CNTs have been employed as nanocarriers for anticancer medications such as doxorubicin (DOX) and betulinic acid (BA) by conjugate functionalization (Dias et al. 2021). CNT drug delivery is thought to provide excellent biotherapeutic efficacy with minimum doses of the medicine. Due to the EPR impact, nanoparticle retention periods in tumors are longer than in normal tissues. Because solid tumors differ from normal tissues in terms of anatomy and pathophysiology, the EPR effect is a peculiar phenomenon that only affects them. Most solid tumors have greater vascular density (hypervascularity) compared to normal tissues and organs. Angiogenesis is a critical component of tumor progression. Contrary to normal tissues, angiogenic blood arteries in tumor tissues feature gaps between neighboring endothelial cells that range in size of 600–800 nm (Dias et al. 2021). Due to this flawed vascular architecture and inadequate lymphatic drainage, the EPR effect is induced, which causes nanocarriers to accumulate at the site of tissue. The size, surface properties, and degree of angiogenesis of the tumor are some of the variables that determine the accumulation of nanoparticles in tissue (Hoshyar et al. 2016).

Anticancer medications delivered by nanocarriers aggregate in tumor tissue through the EPR effect and exhibit significantly higher activity than free pharmaceuticals with increased sensitivity. Additionally to being designed to target tumor tissue with nanomaterials without a highly selective semipermeable membrane barrier, nanoparticles can also be created to carry medications through a variety of biological barriers. It seems very promising to use nanoparticles to transport across this barrier. This allows for the prolonged delivery of therapeutic doses efficiently (Yu et al. 2020).

16.4 Phototherapy Combined with Nanocomposites

Phototherapy is an emerging and promising therapeutic approach in biotherapeutics, including the detection, localization, and treatment of defective tissue (Jiang et al. 2019). Central to phototherapy are photosensitizers (PSs), which absorb light and release reactive oxygen species within cellular environments. These PSs can be effectively functionalized onto carbon-based nanomaterials, offering several advantages in the context of cancer treatment, and many of these PSs are nontoxic. The field of medical photodynamic therapy (PDT) is still in its early stages, with ongoing research. PDT involves the activation of photosensitizers (PSs) using specific wavelengths of light to eliminate tumors. When exposed to this specific light, a PS molecule or medication generates reactive oxygen species that precisely target and kill the intended defective cell. Compared to conventional chemotherapy and radiotherapy, PDT offers higher therapeutic effectiveness and improved outcomes in biotherapeutics. However, a significant challenge is that most PS molecules are naturally insoluble in water (Dias et al. 2021). To address this issue, recent approaches involve combining PS molecules with various nanomaterials, resulting in significantly increased effectiveness and selectivity in targeting tumor cells when using PDT (Fig. 16.3).

Photothermal therapy (PTT) involves illuminating a photosensitizer with a light source, activating the PS molecule, and causing it to release heat energy that leads

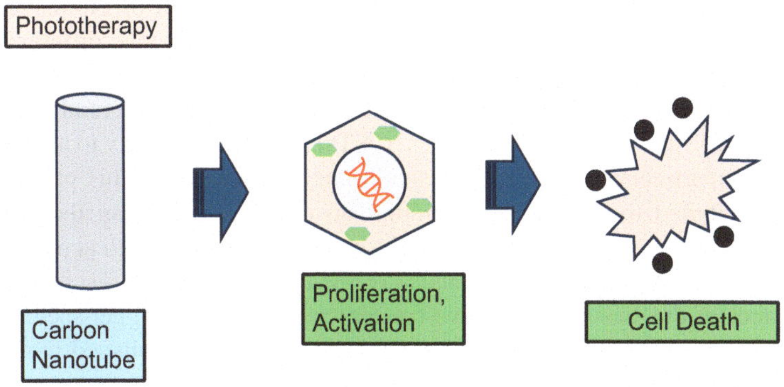

Fig. 16.3 Application and effects of phototherapy in biotherapeutics

to cell death. This increased heat energy, akin to the rise in body temperature during recovery from certain illnesses, can hinder the growth of bacteria, viruses, and cancer cells, ultimately causing their demise in the heated environment.

Recent research has primarily aimed to combat cancer cells by introducing nanoparticles such as gold, silver, fullerene, carbon nanotubes, and graphene into the cells. These nanoparticles generate thermal effects and can be triggered by laser irradiation. In recent years, researchers have shown a growing interest in combining photothermal therapy with commercially available chemotherapy to enhance the overall effectiveness of cancer biotherapeutics (Sundaram and Abrahamse 2020).

A form of alternative medicine known as photobiomodulation (PBM) employs low-level lasers, low-power light-emitting diodes (LEDs), or lasers emitting light within specific wavelength ranges to treat various medical conditions. Photobiomodulation has a positive impact on cell proliferation in a range of cell types, including fibroblasts, endothelial cells, keratinocytes, and lymphocytes. These effects are rooted in the stimulation of mitochondrial signaling pathways and the increased synthesis of growth factors, achieved through transcriptional regulations (Grover et al. 2019; Hassan et al. 2019). The carbon nanocomposites including fullerenes, CNTs, and graphene could be chemically functionalized in diverse ways, which improves their solubility and lowers their toxicity in living things, making them more useful for medication transport and administration (Casais-Molina et al. 2018).

Recent advancements in the surface modification of other nanocomposites like quantum dots (QDs) have opened possibilities for their use in cancer imaging and biotherapeutics. Quantum dots (QDs) are tiny semiconductor particles or nanocrystals, typically ranging from 2 to 10 nm. QD conjugation has proven to be a more efficient and cost-effective method compared to single-color experiments, thanks to advancements in multiplexing capabilities (Nair et al. 2020). This involves their interaction with biomolecules like peptides and antibodies in vivo to target tumors (Namdari et al. 2017). Some studies harness QDs to create biomarkers that expedite processes and offer more stable light intensities compared to traditional fluorescent immunomarkers. Others utilize QDs for labeling cancer, such as using prostate-specific antigen and QDs (Jin et al. 2020). Furthermore, high-sensitivity probes based on quantum dots have been reported for the detection of biomarkers like antigen 125 (CA125) in various specimens, including fixed cells, different tissue sections, and xenografts improving sensitivity of the biotherapeutic analysis.

There are various advantages of these carbon nanocomposites in biotherapeutics. Research suggests that carbon nanotubes (CNTs) have the capability to access intracellular compartments through various cell entry mechanisms. This phenomenon may be attributed to the length-to-width ratio of CNTs, allowing these needle-shaped nanotubes to passively permeate the cell membrane via both active endocytosis and a hypothesized, experimentally verified "nano syringe" mechanism. The significance of this lies in the fact that cancer vaccines aim to trigger specific cytotoxic responses from CD8+ T cells, which can eliminate established tumors and provide long-term protection (Hassan et al. 2019). Enhancing antigen cross-presentation and facilitating the activation of antigen-specific CD8+ T cell responses can be achieved by promoting the translocation of antigens into the cytoplasm of dendritic cells (DCs), where proteasomal processing occurs. CNTs possess the capability to either passively diffuse through cell membranes and reach the cytosol or escape from endosomes into the cytosol after internalization via endocytosis. This makes them suitable candidates for delivering loaded antigens to cytosolic compartments. However, it is crucial to acknowledge the potential issue of endosomal membrane disruption that can occur when CNTs are released from endosomes, potentially causing cellular damage. Membrane disruption is a factor that cannot be overlooked when designing a CNT-based vaccine delivery system, as it could activate the inflammasome and subsequently lead to pyroptotic cell death (Li et al. 2022).

16.5 Enhancements of Sensitivity of Biotherapeutics

Certain types of metallic nanoparticles, such as gold nanoparticles (AuNPs) and silver nanoparticles (AgNPs), can be utilized to enhance carbon nanotubes (CNTs) for the development of nanoscale biosensors. In a study, AuNPs were used to modify multi-walled carbon nanotubes (MWCNTs) to create electrochemical nanosensors for the detection of a specific type of RNA like miR-21. miR-21, which can be found in bodily fluids, plays a crucial role in the early diagnosis of pancreatic cancer (Subramani et al. 2015). However, during the early stages of pancreatic cancer, the levels of miR-21 in the bloodstream are typically quite low, posing a significant challenge to diagnosis. By incorporating AuNPs onto MWCNTs, the sensitivity of miR-21 detection was greatly enhanced (Mazzaglia and Piperno 2022). Their experimental results demonstrated that this nanosensor could offer a wide linear range for the detection of miR-21, with an impressively low detection limit of 3.68 fM, using a source measure unit (SMU). As a result, this biosensor constructed with MWCNTs holds significant promise in biotherapeutics with enhanced sensitivity.

Nanoimmunosensors utilizing antibodies represent another category of nanobiosensors employed in the biotherapeutics. Osteopontin (OPN) serves as a valuable biomarker for the detection of prostate cancer and plays a pivotal role in predicting the survival duration of prostate cancer patients. Conventionally, OPN measurement has relied on the ELISA assay, but this method suffers from drawbacks like limited sensitivity, high costs, and intricate procedures, which significantly impede its effectiveness. Sharma and collaborators took a different approach by covalently attaching monoclonal antibodies specific to OPN onto the surface of carboxylated single-walled carbon nanotubes (SWCNTs), resulting in the creation of a nanoimmunosensor for the electrical detection of OPN in the context of prostate cancer diagnosis (Sharma et al. 2015). This nanoimmunosensor displayed exceptional specificity, exclusively targeting OPN, which ensured remarkable selectivity during the diagnostics. Additionally, the nanoimmunosensor demonstrated a wide range of detectable OPN concentrations and achieved a remarkably low detection limit, enhancing its effectiveness in the diagnosis of prostate cancer.

The potential applications of this nanobiosensor in diagnosis are quite promising. This is because the activity of cell cycle factors like CDK/cyclin, which can be detected using this sensor, holds significance in numerous diseases and is linked to poor prognosis, making it a valuable tool for cancer-related research and diagnosis (Muhammad et al. 2020; Srivastav and Tandon 2022).

Nanoparticles possess a distinct characteristic with their surface-to-volume ratio being 35–45% higher than that of conventional molecules. This specific surface property plays a crucial role in their exceptional value and significantly impacts internal properties, such as their reactivity, which is strongly influenced by size (Pandey et al. 2020). These distinctive attributes of nanoparticles contribute to their multifunctional nature and drive interest in their applications across various fields. Notably, nanoparticles can serve as contrast agents in high-resolution medical imaging, like MRI, and are instrumental as carriers for targeted drug delivery in biotherapeutics. They can be readily modified by controlling and adjusting the preparation parameters. The primary drawback is their susceptibility to changes in microstructure caused by various factors, including unwanted contamination (Kościk et al. 2021). Further, nanocomposites including liposome conjugates exhibit high drug-loading efficiency and can shield encapsulated drugs from premature inactivation, degradation, and dilution. They are easily customizable with various surface modifications and can be formulated in different ways for various routes of administration (Jović et al. 2020). Despite their remarkable potential, they are characterized by low stability and complex preparation procedures. Further, polymer-based nanoparticles exhibit varying characteristics depending on their type, with most being capable of easy modification of their physicochemical properties (Mauro et al. 2021). The nanocomposite substances have extraordinary qualities that can be used to develop novel treatment strategies (Gaur et al. 2021). Reducing collateral damage to healthy tissues, this selectivity could increase the treatment's overall effectiveness. When exposed to light, carbon nanomaterials can effectively turn it into heat or promote the production of reactive oxygen species. By incorporating these materials into biotherapeutics, highly localized and minimally invasive methods might be developed, which would successfully remove tumors while causing the least amount of damage to adjacent tissue (Speranza 2021). Additionally, a new era of diagnostic imaging may be ushered in by the nanocomposites. They are ideal options for boosting imaging processes like MRI and CT scans thanks to their distinctive optical and magnetic features.

Carbon nanoparticles could support early cancer identification and treatment response monitoring by enhancing imaging sensitivity and accuracy, resulting in more proactive and individualized patient care. However, overcoming obstacles is necessary for the path towards the mainstream clinical deployment of carbon nanomaterials. A significant drawback is the potential for increased cytotoxicity, which can arise due to surface modifications. Some of these nanoparticles are also sensitive to environmental factors like pH and temperature. Achieving optimal drug release with these nanoparticles can also be challenging due to variations in pharmacokinetics with the influence of physical parameters.

Luminescent quantum dots, especially carbon quantum dots, are highly suitable for bio-labeling and have demonstrated multiple applications, including biosensing with responses to multiple stimuli and delivery in combination with biomolecules (Kumar et al. 2021; Gavas et al. 2021). Further, it is essential to provide long-term biocompatibility, scalability of synthesis, thorough safety evaluations, and regulatory approvals. A wide range of improvements, for improved therapies and diagnostic tools, are expected to be included into carbon nanocomposites' future roles in biotherapeutics by providing patients with more efficient, minimally invasive, and personalized therapy.

16.6 Conclusive Remarks

The nanocomposites provide a combination of attributes that make them promising candidates for biotherapeutics. The DNA defect diseases including cancer and carcinoma could be targeted and managed using multifunctional nanocomposites with superior sensitivity. These carbon nanocomposites have the capacity to efficiently deliver payloads and chemotherapeutic drugs to tissue locations, minimizing off-target effects and lowering the required dosage, facilitating the precise identification of diseased cells with improvement of sensitivity of tailored biotherapeutics.

Acknowledgements Authors acknowledge Amity University, Noida, Uttar Pradesh, for infrastructural support.

References

Azizi-Lalabadi M, Hashemi H, Feng J, Jafari SM (2020) Carbon nanomaterials against pathogens; the antimicrobial activity of carbon nanotubes, graphene/graphene oxide, fullerenes, and their nanocomposites. Adv Colloid Interf Sci 284:102250

Casais-Molina ML, Cab C, Canto G, Medina J, Tapia A (2018) Carbon nanomaterials for breast cancer treatment. J Nanomater 2018:1–9

Chary PS, Bhawale R, Vasave R, Rajana N, Singh PK, Madan J, Singh SB, Mehra NK (2023) A review on emerging role of multifunctional carbon nanotubes as an armament in cancer therapy, imaging and biosensing. J Drug Deliv Sci Technol 85:104588

Chen L, Liu S, Tao Y (2020) Regulating tumor suppressor genes: post-translational modifications. Signal Transduct Target Ther 5:90

Ciccone G, Ibba ML, Coppola G, Catuogno S, Esposito CL (2023) The small RNA landscape in NSCLC: current therapeutic applications and progresses. Int J Mol Sci 24:6121

Debela DT, Muzazu SG, Heraro KD, Ndalama MT, Mesele BW, Haile DC, Kitui SK, Manyazewal T (2021) New approaches and procedures for cancer treatment: current perspectives. SAGE Open Med 9:205031212110343

Dias LD, Buzzá HH, Stringasci MD, Bagnato VS (2021) Recent advances in combined photothermal and photodynamic therapies against cancer using carbon nanomaterial platforms for in vivo studies. Photochem 1:434–450

Gaur M, Misra C, Yadav AB, Swaroop S, Maolmhuaidh FÓ, Bechelany M, Barhoum A (2021) Biomedical applications of carbon nanomaterials: fullerenes, quantum dots, nanotubes, nanofibers, and graphene. Materials 14:5978

Gavas S, Quazi S, Karpiński TM (2021) Nanoparticles for cancer therapy: current progress and challenges. Nanoscale Res Lett 16:173

Grover R, Burse SA, Shankrit S, Aggarwal A, Kirty K, Narta K, Srivastav R, Ray AK, Malik G, Vats A, Motiani RK, Thukral L, Roy SS, Bhattacharya S, Sharma R, Natarajan K, Mukerji M, Pandey R, Gokhale RS, Natarajan VT (2019) Myg1 exonuclease couples the nuclear and mitochondrial translational programs through RNA processing. Nucleic Acids Res 47:5852–5866

Gulati S, Goyal K, Vijayan S (2023) Carbon nanomaterials: fundamentals, functionalization, and applications. In: Functionalized carbon nanomaterials for theranostic applications. Elsevier, pp 91–105

Hassan HAFM, Diebold SS, Smyth LA, Walters AA, Lombardi G, Al-Jamal KT (2019) Application of carbon nanotubes in cancer vaccines: achievements, challenges and chances. J Control Release 297:79–90

Hoshyar N, Gray S, Han H, Bao G (2016) The effect of nanoparticle size on *in vivo* pharmacokinetics and cellular interaction. Nanomedicine 11:673–692

Jiang B, Zhou B, Lin Z, Liang H, Shen X (2019) Recent advances in carbon nanomaterials for cancer phototherapy. Chemistry 25:3993–4004

Jin C, Wang K, Oppong-Gyebi A, Hu J (2020) Application of nanotechnology in cancer diagnosis and therapy—a mini-review. Int J Med Sci 17:2964–2973

Jović D, Jaćević V, Kuča K, Borišev I, Mrdjanovic J, Petrovic D, Seke M, Djordjevic A (2020) The puzzling potential of carbon nanomaterials: general properties, application, and toxicity. Nanomaterials 10:1508

Kościk I, Jankowski D, Jagusiak A (2021) Carbon nanomaterials for theranostic use. C (Basel) 8:3

Kumar D, Baligar P, Srivastav R, Narad P, Raj S, Tandon C, Tandon S (2021) Stem cell based preclinical drug development and toxicity prediction. Curr Pharm Des 27:2237–2251

Li J, Gao X, Wang Y, Xia T, Zhao Y, Meng H (2022) Precision design of engineered nanomaterials to guide immune systems for disease treatment. Matter 5:1162–1191

Madhavi A, Hingane S, Srivastav R, Joshi N, Subramani C, Muthumohan R, khasa R, Varshney S, Kalia M, Vrati S, Surjit M, Ranjith-Kumar CT (2017). A Screen for novel hepatitis C virus RdRp inhibitor identifies a broad-spectrum anti-viral compound. Scientific Reports. 7(1):5816 https://doi.org/10.1038/s41598-017-04449-3

Mauro N, Utzeri MA, Varvarà P, Cavallaro G (2021) Functionalization of metal and carbon nanoparticles with potential in cancer theranostics. Molecules 26:3085

Mazzaglia A, Piperno A (2022) Carbon nanomaterials for therapy, diagnosis and biosensing. Nanomaterials 12:1597

Mohajeri M, Behnam B, Sahebkar A (2018) Biomedical applications of carbon nanomaterials: drug and gene delivery potentials. J Cell Physiol 234:298–319

Muhammad Q, Jang Y, Kang SH, Moon J, Kim WJ, Park H (2020) Modulation of immune responses with nanoparticles and reduction of their immunotoxicity. Biomater Sci 8:1490–1501

Nair A, Haponiuk JT, Thomas S, Gopi S (2020) Natural carbon-based quantum dots and their applications in drug delivery: a review. Biomed Pharmacother 132:110834

Namdari P, Negahdari B, Eatemadi A (2017) Synthesis, properties and biomedical applications of carbon-based quantum dots: an updated review. Biomed Pharmacother 87:209–222

Pandey H, Khare P, Singh S, Singh SP (2020) Carbon nanomaterials integrated molecularly imprinted polymers for biological sample analysis: a critical review. Mater Chem Phys 239:121966

Paunovska K, Loughrey D, Dahlman JE (2022) Drug delivery systems for RNA therapeutics. Nat Rev Genet 23:265–280

Peng J, Liang X (2019) Progress in research on gold nanoparticles in cancer management. Medicine 98:e15311

Radhakrishnan A, Kapil T, Kapardar RK, Srivastav R (2023) Microbiome additive therapy for the human health. In: Microbiome therapeutics. Elsevier, pp 41–61

Radhakrishnan A, Gangopadhyay R, Sharma C, Kapardar RK, Sharma NK, Srivastav R (2024a) Unwinding helicase MCM functionality for diagnosis and therapeutics of replication abnormalities associated with cancer: a review. Mol Diagn Ther 28:249–264

Radhakrishnan A, Sharma C, Malviya VN, Srivastav R (2024b) Deciphering the molecular functionality of Cdc45 in replisomal complex. Biochem Biophys Rep 37:101643

Saleh TA (2020) Nanomaterials: classification, properties, and environmental toxicities. Environ Technol Innov 20:101067

Sharma A, Hong S, Singh R, Jang J (2015) Single-walled carbon nanotube based transparent immunosensor for detection of a prostate cancer biomarker osteopontin. Anal Chim Acta 869:68–73

Speranza G (2021) Carbon nanomaterials: synthesis, functionalization and sensing applications. Nanomaterials 11:967

Srivastav R, Tandon C (2022) Advances in proteomics approaches and chronic kidney disease. In: Biomedical translational research. Springer Nature, Singapore, pp 201–213

Srivastav R, Sharma R, Tandon S, Tandon C (2019) Role of DHH superfamily proteins in nucleic acids metabolism and stress tolerance in prokaryotes and eukaryotes. Int J Biol Macromol 127:66–75

Subramani R, Gangwani L, Nandy S, Arumugam A, Chattopadhyay M, Lakshmanaswamy R (2015) Emerging roles of microRNAs in pancreatic cancer diagnosis, therapy and prognosis (review). Int J Oncol 47:1203–1210

Sundaram P, Abrahamse H (2020) Phototherapy combined with carbon nanomaterials (1D and 2D) and their applications in cancer therapy. Materials 13:4830

Xie P, Yuan W, Liu X, Peng Y, Yin Y, Li Y, Wu Z (2021) Advanced carbon nanomaterials for state-of-the-art flexible supercapacitors. Energy Storage Mater 36:56–76

Yu W, Liu R, Zhou Y, Gao H (2020) Size-tunable strategies for a tumor targeted drug delivery system. ACS Cent Sci 6:100–116

17 Blockchain Technology for Environmental Conservation

Susmita Shukla, Umme Abiha, Ritambhara Bhutani, Naman Kapoor, and Shiv Kant Shukla

17.1 Introduction

The pursuit of innovative solutions assumes heightened importance in the face of the escalating environmental challenges of our era. Notably, blockchain technology has garnered recent attention and holds the potential to catalyze an upswing in environmental preservation efforts. Blockchain technology, as an emerging breakthrough, exhibits the capacity to potentially revolutionize several sectors, including environmental protection (Tapscott and Tapscott 2016). Its decentralized and transparent nature instills promise for enhancing trust, accountability, and efficacy within environmental programs. Despite its nascent stage, the concept of employing blockchain for environmental safeguarding has garnered considerable attention from scholars, policymakers, and environmental advocates alike (Dinh et al. 2020; Caldeira et al. 2019). Blockchain, as a potent decentralized ledger, facilitates the establishment of immutable records that assure the integrity and traceability of environmental data. This, in turn, addresses concerns surrounding legitimacy and trust, pivotal elements in the development of successful conservation initiatives (Carlson et al. 2018). Our planet faces a monumental environmental crisis, characterized by the devastation of ecosystems due to climate change, biodiversity loss,

S. Shukla (✉) · R. Bhutani · N. Kapoor
Applied Plant Biotechnology Research Lab, Centre for Plant and Environmental Biotechnology, Amity Institute of Biotechnology, Amity University, Noida, India
e-mail: sshukla3@amity.edu

U. Abiha
IDRP, Smart Healthcare, IIT Jodhpur, Rajasthan and All India Institute of Medical Science, Jodhpur, Rajasthan, India

S. K. Shukla
School of Services, Kaushalaya—The Skill University, Ahmedabad, Gujarat, India

A. Kumar, M. K. Solanki (eds.), *Microbial Biocontrol Techniques*, Microorganisms for Sustainability 54,
https://doi.org/10.1007/978-981-97-8739-5_17

deforestation, and pollution, posing grave threats to both the natural world and humanity (Intergovernmental Panel on Climate Change (IPCC) 2021a, b; Intergovernmental Science-Policy Platform on Biodiversity and Ecosystem Services (IPBES) 2019a). Evidently, the scientific community underscores the imperative need for swift, concerted action to mitigate the adverse impacts of these issues. In this context, blockchain technology commands the spotlight.

It harbors a diverse array of potential applications in environmental protection that could confer significant advantages. In one such realm, the market for carbon credits, it can provide a transparent and auditable mechanism for recording and trading these credits. By constructing a trustworthy ledger of carbon emissions and offsets, blockchain can mitigate the risk of fraud and ensure accurate accounting of emission reductions (World Economic Forum 2020). The convergence of blockchain technology, environmental preservation, and renewable energy are the three dynamic fields that represent an exciting juncture. Each of these domains possesses the potential to contribute significantly to transformative advancements independently. For instance, plant biotechnology, through enhancements in genetically modified organisms (GMOs) and environmentally sustainable farming practices, can play a pivotal role in addressing issues concerning food security, reducing the reliance on hazardous pesticides, and promoting climate-resilient crop varieties.

The pivotal objectives of biodiversity preservation, ecosystem maintenance, and climate change mitigation converge under the overarching banner of environmental protection. Furthermore, blockchain technology facilitates sustainable supply chains by enabling product tracking from inception to completion (Economics of Blockchain for Environmental Conservation 2021). This transparency empowers consumers to make well-informed choices by verifying a product's origin, production processes, and environmental impact across sectors like forestry, seafood, and fashion (Global Canopy 2021; Hsu 2018). This approach bolsters the advocacy for sustainable consumption practices and contributes to the fight against illegal logging, overfishing, and other unethical activities (Massachusetts Institute of Technology 2018). Beyond its utility in agriculture, blockchain technology harbors potential applications in tracking the trade of endangered species, monitoring corporate carbon footprints, and validating eco-friendly certifications, among other conservation-related endeavors. For example, blockchain can have a substantial impact on wildlife preservation by combating illegal wildlife trade. A secure and immutable digital record can be established by recording transactions involving animal products, such as ivory or exotic pets, on a blockchain (Hüsemann et al. 2020; Hodler and Karnouskos 2018). This database equips law enforcement agencies with invaluable insights to dismantle organized crime networks engaged in illicit wildlife trafficking. Additionally, blockchain technology holds the potential to revolutionize efforts to combat deforestation and promote sustainable supply chains (World Wide Fund for Nature (WWF) 2021).

Moreover, blockchain technology can construct secure and resilient supply chains. Its decentralized architecture mitigates the risk of data fraud and manipulation, a significant impediment in environmental monitoring. The United Nations Environment Programme (UNEP) underscores block chain's potential in environmental endeavors and furnishes insights and recommendations in this regard (United Nations Educational, Scientific and Cultural Organization (UNESCO) 2022; Prusty and Zhang 2020). While the application of blockchain technology to environmental conservation is undeniably intriguing, a cautious approach to implementation is imperative, considering potential obstacles. Prudent consideration of data privacy, scalability, energy consumption, and interoperability issues is essential to ensure the seamless integration of blockchain into existing environmental systems and frameworks (Huckle et al. 2016). In conclusion, the realm of environmental conservation stands to gain substantially from the integration of blockchain technology. Its autonomous and transparent structure has the potential to augment reliability, accountability, and effectiveness across various domains, encompassing carbon markets, supply chains, wildlife conservation, and renewable energy trade. To fully harness the promise of blockchain in addressing the multifaceted challenges of environmental conservation, further research, collaboration, and investigation are imperative (Drescher 2017).

Blockchain technology has the potential to revolutionize several sectors, including environmental protection. It can provide a transparent and auditable mechanism for recording and trading carbon credits, mitigating the risk of fraud and ensuring accurate accounting of emission reductions. Blockchain technology can also facilitate sustainable supply chains by enabling product tracking from inception to completion, empowering consumers to make well-informed choices by verifying a product's origin, production processes, and environmental impact across sectors like forestry, seafood, and fashion. Blockchain technology can also have a substantial impact on wildlife preservation by combating illegal wildlife trade. A secure and immutable digital record can be established by recording transactions involving animal products, such as ivory or exotic pets, on a blockchain. Additionally, blockchain technology holds the potential to revolutionize efforts to combat deforestation and promote sustainable supply chains. However, a cautious approach to implementation is imperative, considering potential obstacles such as data privacy, scalability, energy consumption, and interoperability issues. Prudent consideration of these issues is essential to ensure the seamless integration of blockchain into existing environmental systems and frameworks.

17.2 Environmental Conservation Challenges

17.2.1 Current Environmental Issues

Numerous pressing environmental issues confront our world today, demanding swift responses. The sustainability of ecosystems and human societies faces significant risks due to climate change, primarily driven by the accumulation of greenhouse gases in the atmosphere. These concerns are exacerbated by diminishing biodiversity, habitat degradation, pollution, improper waste management, deforestation, and excessive exploitation of natural resources (World Wildlife Fund n.d.).

This confluence of problems constitutes a crisis that imperils the delicate ecosystems of our planet.

17.2.1.1 Climate Change and Carbon Footprint

Climate change, a matter of profound concern to policymakers, researchers, and the general populace, takes center stage in these environmental apprehensions. The primary driver of climate change is the escalation of greenhouse gases, particularly carbon dioxide resulting from the combustion of fossil fuels, within Earth's atmosphere (World Wildlife Fund n.d.). Its far-reaching consequences encompass rising global temperatures, heightened frequency and severity of severe weather events, and thawing of polar ice caps and glaciers (World Wildlife Fund n.d.; White et al. 2019). Furthermore, the alteration of established climate patterns disrupts agriculture, leading to food scarcity and potential resource conflicts (White et al. 2019). Blockchain technology can aid carbon offset markets by providing secure and transparent platforms for trading carbon credits. Smart contracts on the blockchain can expedite the verification and transfer of carbon credits, enhancing the effectiveness of emissions offsets and encouraging investments in renewable energy initiatives (World Economic Forum 2019). This empowers individuals and organizations to actively participate in mitigating climate change and assume responsibility for their carbon emissions (World Economic Forum 2019).

17.2.1.2 Biodiversity Loss

Another grave concern is the deterioration of biodiversity. According to the Intergovernmental Science-Policy Platform on Biodiversity and Ecosystem Services (IPBES), biodiversity is declining at an alarming rate, with approximately one million species currently at risk of extinction (World Economic Forum 2019; Intergovernmental Science-Policy Platform on Biodiversity and Ecosystem Services (IPBES) 2019b). The primary culprits behind this loss of biodiversity are overexploitation of natural resources, pollution, invasive species, and habitat destruction. Moreover, biodiversity loss directly impacts human well-being by affecting vital ecosystem services like pollination, clean water provisioning, and climate regulation (World Economic Forum 2019). Blockchain technology can contribute to addressing this issue by creating digital identities and immutable records of wildlife transactions. With blockchain, tracking and tracing the origins of animal products become feasible, making it more challenging for illegal activities to evade detection. This technology enhances supply chain transparency, aids law enforcement efforts, and safeguards endangered species (White et al. 2019; Intergovernmental Science-Policy Platform on Biodiversity and Ecosystem Services (IPBES) 2019b).

17.2.1.3 Deforestation and Resource Depletion

Deforestation, a major threat to global ecosystems, also contributes to climate change. Forests are crucial for the livelihoods of millions and play pivotal roles in conserving biodiversity and sequestering carbon. Their destruction leads to a significant increase in carbon emissions, exacerbating climate change. Additionally, the overexploitation of natural resources such as fisheries and minerals depletes finite reserves and disrupts ecosystems. Unsustainable fishing practices, for instance, have depleted fish stocks, jeopardizing coastal communities' livelihoods and global food security (Pure Earth 2021). Blockchain technology can play a critical role in combating deforestation by digitally recording timber transactions and enhancing transparency throughout the timber supply chain. Tracking the origin and transportation of timber through blockchain makes it easier to detect and halt illicit logging activities, ensuring sustainable and legal procurement processes (Intergovernmental Science-Policy Platform on Biodiversity and Ecosystem Services (IPBES) 2019b; Pure Earth 2021).

17.2.1.4 Pollution and Waste Management

Inefficient waste management and pollution persist as significant global environmental challenges. Blockchain technology can enhance waste management systems by providing a transparent and verifiable record of waste disposal and recycling operations (World Health Organization 2021). Through the integration of sensors and Internet of Things devices with blockchain, stakeholders can monitor waste generation, collection, and recycling, establishing an efficient and sustainable waste management system. This technology encourages responsible waste disposal and supports recycling efforts, thereby mitigating pollution and advancing a circular economy (World Health Organization 2021; Liebenau et al. 2021) (Table 17.1).

17.2.2 Limitations of Traditional Solutions

While traditional methods of environmental conservation have undeniably yielded positive results, they inherently fall short in addressing the scale and complexity of today's environmental problems. These conventional approaches exhibit several limitations:

Table 17.1 Key environmental issues up to 2022

Environmental issues	Data recorded
Climate change	Global average temperature increase of 1.1 °C above pre-industrial levels (World Economic Forum 2019)
Deforestation	Approximately ten million hectares of forest lost annually (Intergovernmental Science-Policy Platform on Biodiversity and Ecosystem Services (IPBES) 2019b)
Wildlife extinction	Up to one million species facing extinction due to human activities (Intergovernmental Science-Policy Platform on Biodiversity and Ecosystem Services (IPBES) 2019b)
Air pollution	Nine out of ten people worldwide breathe polluted air (Pure Earth 2021)
Ocean pollution	An estimated eight million metric tons of plastic waste enter the ocean each year (World Health Organization 2021)
Pollution from industrial activities	Industrial pollution is responsible for 4.4 million premature deaths annually (World Health Organization 2021)
Water scarcity	By 2025, it is projected that nearly two-thirds of the global population could be living under water-stressed conditions (Liebenau et al. 2021)

1. One drawback is the lack of accountability and transparency in traditional conservation methods. When decision-making processes exclude local populations and indigenous communities, which possess deep connections to their ecosystems, their needs and knowledge may be disregarded in policies and practices (United Nations 2018).
2. The challenge of cross-border cooperation is another impediment. Environmental issues such as deforestation, air and water pollution, and climate change transcend international boundaries. Traditional methods often fail to foster international collaboration and coordination, hampering the effectiveness of global conservation efforts. Enhanced international partnerships and agreements are imperative to address these challenges (United Nations 2018, 2021).
3. Additionally, monitoring and enforcing conservation measures in remote or politically unstable regions can be exceptionally challenging. Limited access to these areas, inadequate infrastructure and resources, and potential conflicts hinder effective monitoring and enforcement activities (United Nations 2021). Consequently, conservation efforts may prove less effective, allowing unlawful activities such as poaching, illegal logging, and unsustainable resource extraction to persist.

There is a growing demand for more robust, inclusive, and cutting-edge technological solutions to overcome these barriers. Incorporating diverse perspectives, including those of local communities, and respecting traditional ecological knowledge can enhance the effectiveness and sustainability of conservation initiatives (Azam et al. 2022; Böhme et al. 2015). The utilization of advanced technology such as satellite imagery, remote sensing, and data analytics can significantly augment monitoring and enforcement capabilities even in remote and politically unstable areas (Böhme et al. 2015).

17.2.3 The Need for Technological Innovation

Technological advancements offer a glimmer of hope. Blockchain technology, by providing transparency, traceability, and accountability, has the potential to address many of the shortcomings of conventional conservation efforts. One of the prominent areas where blockchain technology can have a transformative impact is the monitoring and protection of endangered species (Meng et al. 2020). Through blockchain's immutable record of wildlife data, including population sizes, migratory patterns, and poaching incidents, conservation organizations and law enforcement agencies can access accurate information promptly. This enables the timely implementation of conservation measures and effective enforcement of anti-poaching regulations (Meng et al. 2020; Chum et al. 2019). Furthermore, blockchain offers innovative ways to track and verify the origins of environmentally friendly products, such as sustainably harvested timber or organic food (Chum et al. 2019). By empowering consumers to make informed choices, it fosters demand for eco-friendly behaviors and plays a crucial role in promoting sustainability and monitoring carbon emissions. Blockchain-based systems can precisely record and certify carbon credits generated by emission reduction projects, safeguarding the credibility of the carbon market and encouraging businesses and individuals to adopt sustainable practices (Imani et al. 2018). Realizing the full potential of blockchain for environmental protection necessitates collaboration across diverse stakeholders. Governments, environmental organizations, technology firms, and local communities must actively engage in researching and implementing blockchain solutions. Through such cooperative efforts, a robust ecosystem of blockchain applications for conservation can be established, fostering innovation and optimizing resources (Imani et al. 2018; Shin and Lee 2021). The application of blockchain technology to environmental conservation initiatives presents intriguing avenues for overcoming the limitations of conventional methods. By harnessing the transparency, accountability, and reliability inherent in blockchain, we can build trust, promote environmentally responsible behavior, and preserve our natural resources for future generations to inherit.

The article "Current Environmental Issues" provides a detailed exploration of specific challenges and potential solutions within each domain of environmental conservation. For instance, in the context of climate change and carbon footprint, the article discusses the primary driver of climate change, the escalation of greenhouse gases, particularly carbon dioxide resulting from the combustion of fossil fuels, within Earth's atmosphere. It also explains how blockchain technology can aid carbon offset markets by providing secure and transparent platforms for trading carbon credits, enhancing the effectiveness of emission offsets and encouraging investments in renewable energy initiatives.

Similarly, in the context of biodiversity loss, the article explains how blockchain technology can contribute to addressing this issue by creating digital identities and immutable records of wildlife transactions. With blockchain, tracking and tracing the origins of animal products become feasible, making it more challenging for illegal activities to evade detection. The article also discusses how blockchain technology can play a critical role in combating deforestation by digitally recording timber transactions and enhancing transparency throughout the timber supply chain.

Overall, the article provides a comprehensive overview of the potential applications of blockchain technology in environmental protection and a detailed exploration of specific challenges and potential solutions within each domain.

17.3 Applications of Blockchain in Environmental Conservation

17.3.1 Transparent Supply Chain Management

One of the most promising and valuable applications of blockchain technology is transparent supply chain management. Supply chains, vital for the production and distribution of goods, often span global networks. However, they frequently suffer from issues of transparency, inefficiency, and environmentally irresponsible practices. Blockchain technology provides a disruptive solution by revolutionizing how we monitor and manage supply chains in the environmental sector, leveraging its inherent transparency and traceability. Blockchain allows for the tracking of every transaction and exchange of goods, creating a transparent and immutable ledger of the entire supply chain journey (Soni 2018). This aspect is particularly significant for environmental conservation efforts, as it enables the monitoring and validation of sustainable practices such as ethical sourcing, fair trade, and eco-friendly manufacturing methods. Moreover, the real-time tracking and recording of each stage of the production and distribution process are facilitated by blockchain-enabled transparent supply chain management (Soni 2018). Consequently, the path taken by goods from their origin to the end user can be securely recorded on an unalterable database. The transparency afforded by blockchain technology can aid in the fight against illegal environmental activities like wildlife trafficking and unlawful logging. Tracking and verifying every movement of products on the blockchain make it easier to detect and halt illicit operations, contributing to the protection of vulnerable ecosystems and endangered species (IBM n.d.). Furthermore, blockchain-based supply chain management empowers consumers by granting them access to credible and reliable information regarding the environmental impact of the products they purchase. This allows consumers to make more informed decisions based on the ethical standards and environmental commitment of manufacturers and suppliers (Soni 2018; IBM n.d.). Several organizations and projects in the field of environmental conservation are already leveraging blockchain technology for transparent supply chain management. For instance, the IBM (International Business Machines Corporation) Food Trust is a blockchain-powered network that provides traceability for food products, ensuring their safety and authenticity (IBM n.d.).

17.3.2 Carbon Footprint Tracking and Offset

Amid growing concerns about climate change and its associated challenges, blockchain technology emerges as a potent tool for environmental conservation, specifically in the precise tracking and offsetting of carbon emissions. Carbon footprints represent the measurement of greenhouse gas emissions produced by individuals, businesses, or products, contributing to global warming and environmental degradation. Blockchain's transparent, secure, and immutable properties are poised to revolutionize how we measure, reduce, and offset carbon emissions, offering innovative solutions to combat climate change in a world where carbon emissions play a significant role in global warming. Blockchain's distributed transparency and decentralization enable the accurate and unalterable recording of emission data (ClimateTrade n.d.; United Nations Environment Programme (UNEP) 2021). By safeguarding emission data in a distributed database, blockchain ensures the accuracy and reliability of carbon footprint reporting. This transparency addresses issues related to data integrity and accuracy associated with carbon footprint reporting. It now enables the identification of emissions at various stages in a product's life cycle, including during manufacturing, transportation, and consumption. This level of precision allows both organizations and individuals to better understand their carbon footprints and identify opportunities for improvement. Blockchain can simplify the implementation of validated emission reduction initiatives and promote sustainable practices by attributing carbon credits or offsets to specific emission reductions (ClimateTrade n.d.). In the field of environmental conservation, several programs and initiatives are already utilizing blockchain technology for carbon footprint tracking and offsetting. For instance, the blockchain-based initiative Climate Trade facilitates the buying and selling of certified carbon credits, providing complete transparency and traceability throughout the process (ClimateTrade n.d.; United Nations Environment Programme (UNEP) 2021).

17.3.3 Wildlife Conservation and Anti-poaching Efforts

The foremost challenges in environmental protection encompass the preservation of vast biodiversity and the prevention of illegal wildlife trade. In these endeavors, blockchain technology emerges as a potent ally, offering innovative solutions for safeguarding endangered species and bolstering anti-poaching initiatives. Due to its attributes of transparency, authenticity, and security, blockchain technology possesses the potential to revolutionize global biodiversity preservation efforts (World Bank 2021). It can securely store critical wildlife protection data, including statistics on species populations, migratory patterns, and habitat conservation efforts (World Bank 2021). By recording this information on the blockchain, it becomes visible and tamperproof, mitigating the risk of data corruption or manipulation. This transparency facilitates effective collaboration among stakeholders, enabling governments, scientists, and conservation organizations to make informed decisions based on factual and trustworthy information. Blockchain technology empowers local communities and indigenous individuals to play a more active role in wildlife conservation. Such stakeholders can document their conservation activities on an open blockchain, ensuring fair compensation and recognition for their vital contributions to biodiversity conservation (Food and Agriculture Organization (FAO) 2021). This inclusive approach fosters collaboration and grants agency to individuals engaged in conservation efforts. The decentralized nature of blockchain enables the creation of networks and systems that connect various stakeholders involved in wildlife preservation, including park rangers, local residents, law enforcement agencies, and nongovernmental organizations. Organizations can employ blockchain-based systems to record and monitor animal movements, track poaching activities, and share real-time data among stakeholders (Food and Agriculture Organization (FAO) 2021; Food and Agriculture Organization of the United Nations (FAO) 2022). Several projects and companies are already utilizing blockchain technology to support anti-poaching and wildlife conservation efforts. For instance, the blockchain-based platform Wild Chain focuses on funding wildlife conservation programs and involving local communities in these endeavors. Another project, Wild Track, utilizes blockchain to monitor and locate endangered animals by analyzing footprint photographs and storing the data on the blockchain for secure and transparent data management (Wildchain n.d.; WildTrack n.d.).

17.3.4 Sustainable Energy Management

In the era of urgent environmental preservation, the adoption of sustainable energy practices is imperative to reduce carbon emissions and mitigate climate change. Blockchain technology emerges as a powerful tool that can accelerate the transition to sustainable energy management by offering innovative solutions that enhance the generation, distribution, and consumption of renewable energy while promoting transparency and credibility (WildTrack n.d.; Li et al. 2020). It can facilitate the tracking and verification of renewable energy production and consumption, promoting accountability and transparency. By securely recording energy generation data from renewable sources such as solar panels and wind turbines on the blockchain, consumers can easily verify the source and legitimacy of green energy (Li et al. 2020; Energy Web Foundation n.d.). This transparency not only fosters trust among stakeholders but also encourages the adoption of renewable energy sources by ensuring the accuracy of renewable energy claims. Furthermore, blockchain-based platforms can enable peer-to-peer energy trading, eliminating traditional intermediaries and advancing energy decentralization. Individuals or businesses with renewable energy sources, such as solar panels, can generate surplus energy and sell it directly to other consumers within their network through blockchain-enabled smart contracts. Several initiatives and projects are already harnessing blockchain technology for sustainable energy management. For example, the Energy Web Foundation focuses on developing open-source blockchain solutions for the energy industry, with a focus on optimizing grid operations and integrating renewable energy sources (Energy Web Foundation n.d.).

17.3.5 Waste Management and Recycling

The application of blockchain technology in waste management and recycling is a significant facet of environmental conservation. By promoting efficient waste disposal, material reuse, and resource recovery, waste management and recycling efforts aim to mitigate the adverse environmental impacts of waste (Energy Web Foundation n.d.). Blockchain can greatly enhance these endeavors through transparent and verifiable waste tracking systems, incentivizing recycling, and advancing the principles of the circular economy. It can enable the transparent and immutable recording of waste management data, from generation to recycling or final disposal. By securely storing data on waste types, volumes, and origins on the blockchain, consumers gain visibility over the entire waste management process (Energy Web Foundation n.d.). This open and accessible system can help ensure that waste is handled and disposed of in an environmentally friendly manner by tracking and enforcing regulatory compliance. Additionally, blockchain offers substantial benefits for recycling. Digital credentials for recycled materials known as Recycled Content Certificates (RCCs) can be created to increase transparency and confidence in recycling operations by providing verifiable confirmation of the amount of recycled materials in products. Blockchain technology is already being utilized in various initiatives and operations for waste management and recycling. For instance, the blockchain-based platform Plastic Bank allows individuals in underdeveloped countries to exchange their used plastics for digital tokens with various uses (Energy Web Foundation n.d.; Plastic Bank n.d.). Another project, Recycle To Coin, uses blockchain technology to reward individuals for depositing recyclables at specific collection points by offering them Bitcoin (Plastic Bank n.d.; Kumar and Jain 2018) (Fig. 17.1).

Blockchain technology has several potential applications in environmental conservation, including transparent supply chain management and carbon footprint tracking and offset. In transparent supply chain management, blockchain can be used to track every transaction and exchange of goods, creating a transparent and immutable ledger of the entire supply chain journey. This enables the monitoring and validation of sustainable practices such as ethical sourcing, fair trade, and eco-friendly manufacturing methods. Blockchain-based supply chain management can also aid in the fight against illegal environmental activities like wildlife trafficking and unlawful logging. In carbon footprint tracking and offset, blockchain's transparent, secure, and immutable properties can be used to accurately and unalterably record emission data, enabling the identification of emissions at various stages in a product's life cycle. This level of precision allows both organizations and individuals to better understand their carbon footprints and identify opportunities for improvement. Blockchain can also simplify the implementation of validated

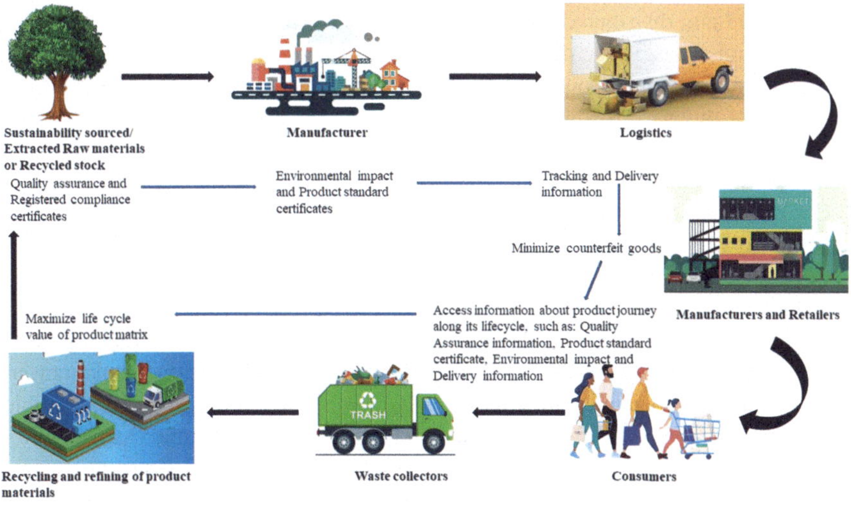

Fig. 17.1 An outline of a supply chain supported by blockchain (Science Division Blockchain Technology and Environmental Sustainability 2020)

emission reduction initiatives and promote sustainable practices by attributing carbon credits or offsets to specific emission reductions. Other potential applications of blockchain in environmental conservation include creating digital identities and immutable records of wildlife transactions to combat biodiversity loss, digitally recording timber transactions to combat deforestation, and enhancing waste management systems by providing a transparent and verifiable record of waste disposal and recycling operations. However, the full potential of blockchain for environmental protection necessitates collaboration across diverse stakeholders, including governments, environmental organizations, technology firms, and local communities.

17.4 Benefits and Advantages of Blockchain Adoption

17.4.1 Data Integrity and Immutability

Blockchain technology offers several advantages, particularly in terms of data integrity and immutability. It has the potential to revolutionize the collection, monitoring, and analysis of environmental data, fostering trust and transparency in conservation efforts. The core promises of immutability and data integrity inherent in blockchain have profound implications for the preservation of the Earth's fragile ecosystems. Due to the fundamental data integrity of blockchain technology, environmental data recorded on the network is guaranteed to be immutable and resistant to fraud (Kissinger et al. 2019). This fundamental attribute instills confidence among all stakeholders, including governments, nonprofits, and concerned individuals who rely on accurate and unalterable information for informed decision-making. One of the primary benefits of using blockchain in environmental protection is the secure storage and tracking of crucial environmental data. By utilizing a decentralized and distributed ledger, blockchain ensures that data related to greenhouse gas emissions, waste disposal, energy consumption, and other environmental metrics remains tamperproof and resistant to fraudulent manipulation. This high level of data integrity fosters stakeholder trust and simplifies accurate reporting. Furthermore, blockchain enhances the immutability of environmental data, making it challenging to alter or manipulate historical records (Kissinger et al. 2019).

Once a transaction or data entry is recorded on the blockchain, it becomes a permanent part of the distributed ledger. This feature enhances the verifiability and flexibility of environmental operations, as historical records cannot be modified over time (Kissinger et al. 2019). For example, data integrity is critical in various domains, including the regulation of endangered species trade and tracking businesses' carbon footprints. Consequently, the risk of fraud is mitigated, and environmental monitoring becomes more transparent thanks to the blockchain's immutable record (Science Division Blockchain Technology and Environmental Sustainability 2020). Deforestation, wildlife trafficking, and habitat preservation are just a few pressing environmental challenges that necessitate such transparency for effective mitigation. Additionally, the use of blockchain technology in environmental conservation enhances data transparency by enabling real-time monitoring and reporting.

"Smart contracts" that run on blockchain technology can automate and enforce compliance with environmental regulations. These self-executing contracts can incorporate predefined criteria and regulations to ensure accountability and adherence to environmental norms. By automating processes and eliminating intermediaries, blockchain streamlines regulation and reduces administrative costs (Kissinger et al. 2019; Witkowska et al. 2019).

17.4.2 Enhanced Transparency and Accountability

Blockchain's capacity to promote transparency and accountability is a significant advancement at a time when the preservation of our planet's ecosystems depends on the accuracy of data and the ability to hold stakeholders accountable. The transparency feature of blockchain is intrinsically embedded in its design. On the blockchain, every transaction and piece of data are instantly accessible to authorized parties. This transparency is vital for environmental conservation activities, as it enables secure and verifiable data sharing related to biodiversity, carbon emissions, and supply chain transparency (Kissinger et al. 2019; Witkowska et al. 2019). Stakeholders, including government entities, conservation organizations, and citizens, have unrestricted access to critical environmental data, enabling them to actively participate in conservation efforts. For example, it can be employed to trace and verify the origin of environmentally friendly goods and materials. The integration of blockchain into supply chain management enables transparency throughout a product's life cycle. Therefore, blockchain's transparency feature serves as a robust deterrent against fraud and unethical environmental conservation practices. Users can monitor and verify the accuracy of claims regarding sustainable practices, eco-labels, and certifications through the immutable ledger. Customers gain the confidence needed to make informed choices and demand eco-friendly products, thereby driving market demand for sustainable business models and ethical environmental responsibility. Transparency and accountability are two more pivotal facets of blockchain technology. Once data is logged, the technology's permanent record ensures its security, tamper resistance, and traceability (Science Division Blockchain Technology and Environmental Sustainability 2020; Kissinger et al. 2019).

17.4.3 Decentralization for Trust

The adoption of blockchain technology offers numerous benefits and advantages that can significantly advance environmental conservation efforts, with decentralization being one of the primary advantages (Kissinger et al. 2019). Decentralization refers to the distribution of this technology across multiple nodes, rather than being under the control of a single central authority (Kissinger et al. 2019). Traditional trust models often rely on centralized entities, which can introduce biases and vulnerabilities into the conservation process. In contrast, blockchain operates on a decentralized network of nodes where no single organization holds complete authority. By distributing decision-making and data management across a network of participants, this decentralization promotes a more democratic and equitable approach to environmental protection. Additionally, due to the absence of intermediaries and centralized points of control, this decentralized system reduces the likelihood of environmental data manipulation and fraud (Kissinger et al. 2019; Witkowska et al. 2019). Decentralization has several advantages in the realm of environmental conservation. First, it enhances trust by eliminating the need to rely on a single trusted authority or institution. Instead, trust is placed in the blockchain network's consensus mechanism, where multiple participants verify the accuracy and quality of environmental data (Witkowska et al. 2019). Decentralization in blockchain also enables a broader range of individuals to participate in environmental conservation efforts. Decentralized applications (DApps) built on blockchain platforms can enable individuals, groups, and organizations from around the world to support conservation activities through monitoring, reporting, or financial contributions (Kissinger et al. 2019; Witkowska et al. 2019). The decentralized nature of blockchains aligns with open-source collaboration. Environmental data is transparent, secure, and collaboratively maintained on a shared platform that conservationists, scientists, and policymakers can use and contribute to. This inclusive approach fosters innovation and a collaborative spirit by encouraging the exchange of ideas and subject matter expertise to address environmental challenges (Witkowska et al. 2019).

17.4.4 Improved Efficiency and Cost Reduction

The utilization of blockchain technology in the field of environmental conservation offers numerous significant advantages, including increased efficiency and reduced costs. Blockchain technology streamlines operations, reduces paperwork, and eliminates intermediaries, resulting in several processes that enhance efficiency and cut costs (Witkowska et al. 2019). Conventional paper-based procedures are often labor intensive, cumbersome, and susceptible to human errors. This technology enables the digitization and automation of these processes, facilitating quicker transaction completion with less waste (López-Pintado et al. 2019). Efficiency is further enhanced through the use of smart contracts, a component of blockchain technology that automates processes. By establishing autonomous contracts with predefined terms and conditions, smart contracts can automate tasks such as verifying emission reductions for carbon offset initiatives within the realm of environmental protection, significantly reducing administrative overhead (Witkowska et al. 2019; López-Pintado et al. 2019). While ensuring the integrity and reliability of transactions, this automation results in significant cost reductions. In addition to expediting processes, the decentralized nature of blockchain technology can help reduce costs associated with environmental conservation efforts. By eliminating the need for intermediaries or transaction fees, blockchain reduces associated expenses. Furthermore, the transparency and auditability of blockchain can improve cost-effectiveness by reducing redundancies and fraud throughout the entire supply chain. By being transparent, blockchain can reduce the cost of resource authentication, quality control, and unethical practices (Witkowska et al. 2019; López-Pintado et al. 2019). Environmental organizations can enhance operational efficiency, reduce operating costs, and improve resource management processes by leveraging blockchain technology. These advancements ultimately lead to cost savings, allowing for greater allocation of resources to actual conservation activities.

Blockchain technology offers several benefits and advantages in terms of data integrity and immutability, particularly in the context of environmental conservation efforts. By utilizing a decentralized and distributed ledger, blockchain ensures that data related to greenhouse gas emissions, waste disposal, energy consumption, and other environmental metrics remains tamperproof and resistant to fraudulent manipulation. This high level of data integrity fosters stakeholder trust and simplifies accurate reporting. Furthermore, blockchain enhances the immutability of environmental data, making it challenging to alter or manipulate historical records. This technology can also aid carbon offset markets by providing secure and transparent platforms for trading carbon credits, contributing to addressing the issue of climate change. Additionally, blockchain can create digital identities and immutable records of wildlife transactions, making it more challenging for illegal activities to evade detection and safeguarding endangered species. Blockchain technology can also enhance waste management systems by providing a transparent and verifiable record of waste disposal and recycling operations, mitigating pollution and advancing a circular economy.

17.5 Case Studies

17.5.1 Carbon Credits on the Blockchain

The incorporation of blockchain technology into carbon trading and offset markets aims to enhance transparency, traceability, and efficiency by integrating carbon credits onto the blockchain. The application of blockchain technology offers numerous advantages in the management of carbon credits (Gebbers and Adamchuk 2010). These benefits encompass immutability, transparency, and decentralization, thereby facilitating the prevention of fraudulent activities and ensuring the reliability of carbon offset initiatives. Through a case study methodology, the research investigates the potential benefits and challenges associated with the integration of blockchain into the carbon credit market (Abbasi et al. 2014; Adil et al. 2015; Bartlett et al. 2015). The analysis highlights the key factors.

Eco-Chain is a conceptualized network that utilizes blockchain technology to streamline the issuance, sale, and retirement of carbon credits. The platform leverages smart contracts to automate various processes, ensuring transparency and reducing administrative burdens within the carbon credit economy. The Eco-Chain portal facilitates the registration of carbon credit projects, including activities such as forestry campaigns, renewable energy installations, and methane capture (Jagannathan and Priyatharshini 2015; Yoshida et al. 2016; Lin et al. 2017; Jiang et al. 2016). The legitimacy of these projects is ensured through third-party auditors' monitoring. Tokenization involves representing each carbon credit as a unique digital asset on the blockchain. These tokens serve as representations of tangible carbon offsets directly linked to projects. Eco-Chain employs a network of auditors to conduct periodic verification of emission reductions achieved by registered projects. The auditors present their findings to the blockchain, documenting them transparently. Smart contracts, deployed on blockchain technology, facilitate automated processes for the issuance, transfer, and retirement of carbon credits. Upon successful verification of emission reductions, a specific quantity of carbon credit tokens is generated and allocated to the project owner's wallet (Yu et al. 2017; Perea et al. 2017; Gouveia and Fonseca 2008; Gipp et al. 2017). The Eco-Chain platform includes a marketplace where individuals can engage in the purchase and sale of carbon credits. The use of blockchain technology ensures the presence of only legitimate and authenticated credits eligible for trading, mitigating the potential for fraudulent activities. Carbon Offset Retirement involves the practice where a firm or an individual seeks to neutralize their carbon emissions by acquiring carbon credits from the marketplace. Upon acquisition, these credits are permanently removed from the blockchain, ensuring their non-duplicability (Collier et al. 2014; Marttunen et al. 2017; NGO Monitor 2015). Company A, a prominent technology corporation, has made a firm commitment to achieving carbon neutrality. They have chosen to participate in Eco-Chain to acquire carbon credits for offsetting their emissions. The term "company" refers to an organization or entity engaged in commercial activities. Upon registration, users must undergo a verification process to authenticate their commitment to achieving carbon neutrality. Users explore the Eco-Chain marketplace, where they acquire a collection of carbon credits earned through a forestry initiative in South America (Dixon and Richards 2016). After acquisition, the carbon credits are promptly transferred to Company A's designated digital storage and rendered inactive on the blockchain, preventing any further trading. Eco-Chain enables the tracking of Company A's carbon offset balance, facilitating its use for marketing purposes and generating reports highlighting its carbon-neutral status (Rejesus et al. 2005). Blockchain technology ensures transparency in all transactions, verifications, and credit retirements, providing both transparency and tamper-proof security. Smart contracts streamline operations by automating various processes, resulting in reduced administrative costs and reduced error likelihood. Traceability is a fundamental aspect of carbon credits, allowing for clear identification and verification of each credit's origin. This traceability feature is crucial for establishing the credibility and authenticity of carbon credits. The implementation of immutable records and auditor control helps mitigate the potential risk of counterfeit carbon credits entering the market (Sun et al. 2016; Lacetera and Zirulia 2011).

17.5.2 Protecting Endangered Species with Blockchain

This case study examines the application of blockchain technology to safeguard endangered species. By leveraging the inherent features of blockchain, such as transparency, immutability, and decentralization, this study explores the potential benefits (Davidson et al. 2016; Wiles 2015). The purpose of this study is to investigate the effects of climate change on biodiversity in tropical rainforests. Wild Chain is a hypothetical blockchain-driven initiative conceptualized with the aim of addressing the issue of illicit wildlife trading and promoting advancements in conservation efforts. The utilization of blockchain technology's inherent characteristics, such as transparency and traceability, enables the establishment of a robust and impervious ecosystem for safeguarding animals. This study employs digital identities to provide unique identification for endangered species, including, but not limited to, rhinoceroses, elephants, and tigers. These digital identities encompass various attributes, such as geographical location, taxonomic classification, age, and overall well-being of the animals. Internet of Things (IoT) devices, such as GPS collars and microchips, are utilized by wildlife rangers and conservationists to equip animals with tracking mechanisms (Environments 2017a; IBM Corporation 2015; Nobre et al. 2016). These devices consistently collect data regarding the animals' movements and behaviors. The Wild Chain platform employs a decentralized ledger system, known as blockchain, to meticulously record relevant information about wildlife, including their digital identities, movement patterns, and historical records. Smart contracts, integrated into the blockchain technology, automate various operations such as data verification, alerts, and fund disbursements. A key focus of this case study was the preservation of rhinoceros populations in South Africa, a species frequently targeted by illegal hunters due to the demand for their horns (Walport 2016; Bloom et al. 2014; Andreopoulou 2012). The process of creating digital identities involves assigning a unique digital identity on the blockchain to each individual rhinoceros within a designated protected reserve. This identity includes distinctive characteristics of the rhinoceros and data obtained from IoT sensors used for tracking purposes (Gouveia et al. 2004; Lin et al. 2015; Banafa 2017). Real-time tracking is facilitated by IoT devices attached to the rhinos, communicating their GPS coordinates and additional sensor data to the blockchain, which is captured and documented in real time. Smart contracts are designed to effectively monitor the movements of rhinos and identify any abnormal patterns, such as entry into high-risk areas or prolonged immobility. If anomalies are detected, the system generates alerts to notify wildlife rangers and law enforcement personnel. Transparency is ensured through a public-facing interface that allows conservation organizations, donors, and the general public to access real-time data related to the rhino population, their movements, and the effectiveness of conservation efforts (MacNeil 2014). Donors can track the allocation of their contributions, which are stored securely within a blockchain-based fund. Smart contracts autonomously allocate funds to various conservation programs based on pre-established conditions, such as the successful safeguarding of rhinoceros populations; the endangered animal species are given in Fig. 17.2.

The implementation of real-time tracking and alert systems plays a pivotal role in mitigating poaching activities and protecting endangered species. Transparency is a crucial element in conservation efforts as it enables the public to monitor and evaluate project progress, fostering accountability and trust. The immutability of

Fig. 17.2 Blockchain technology has the potential to revolutionize wildlife conservation efforts by providing a transparent and secure system for tracking and monitoring endangered species

blockchain ensures data integrity, safeguarding wildlife data from unauthorized alterations or tampering (Lansiti and Lakhani 2017). Implementing smart contracts to automate operations enhances efficiency, allowing conservationists to focus their efforts on species protection. Donors receive transparent information about the allocation and utilization of their contributions, encouraging increased financial support for conservation efforts. In conclusion, the case study of Wild Chain serves as a hypothetical illustration but underscores the genuine and promising potential of employing blockchain technology in wildlife protection (Back et al. 2014). The technology has the capacity to fundamentally transform the approach to safeguarding endangered species and their ecosystems while engaging diverse stakeholders in the conservation endeavor. Nevertheless, practical applications in the real world may require addressing several technical, logistical, and regulatory challenges (DNV GL AS 2017).

The case study on carbon credits on the blockchain highlights several potential benefits and challenges associated with the integration of blockchain technology into the carbon credit market. The benefits include immutability, transparency, and decentralization, which can prevent fraudulent activities and ensure the reliability of carbon offset initiatives. However, there are also challenges, such as the need for third-party auditors to monitor the legitimacy of carbon credit projects and the potential risk of counterfeit carbon credits entering the market.

In terms of specific challenges and potential solutions within each domain, blockchain technology can aid carbon offset markets by providing secure and transparent platforms for trading carbon credits. Smart contracts on the blockchain can expedite the verification and transfer of carbon credits, enhancing the effectiveness of emission offsets and encouraging investments in renewable energy initiatives. This empowers individuals and organizations to actively participate in mitigating climate change and assume responsibility for their carbon emissions.

For biodiversity loss, blockchain technology can contribute to addressing this issue by creating digital identities and immutable records of wildlife transactions. With blockchain, tracking and tracing the origins of animal products become feasible, making it more challenging for illegal activities to evade detection. This technology enhances supply chain transparency, aids law enforcement efforts, and safeguards endangered species.

Regarding deforestation and resource depletion, blockchain technology can play a critical role in combating deforestation by digitally recording timber transactions and enhancing transparency throughout the timber supply chain. Tracking the origin and transportation of timber through blockchain makes it easier to detect and halt illicit logging activities, ensuring sustainable and legal procurement processes.

Finally, for pollution and waste management, blockchain technology can enhance waste management systems by providing a transparent and verifiable record of waste disposal and recycling operations. Through the integration of sensors and Internet of Things devices with blockchain, stakeholders can monitor waste generation, collection, and recycling, establishing an efficient and sustainable waste management system. This technology encourages responsible waste disposal and supports recycling efforts, thereby mitigating pollution and advancing a circular economy.

17.6 Challenges and Considerations

17.6.1 Scalability and Energy Consumption

The development and implementation of various technologies, such as blockchain, cloud computing, and data centers, necessitate meticulous consideration of scalability and energy consumption. These factors play a pivotal role in ascertaining the efficacy and sustainability of these technologies. Scalability, as a concept, pertains to the capacity of a system, process, or organization to manage increased demands. Network congestion emerges when a growing number of users join a network or platform, resulting in a deceleration of transaction processing and data retrieval (Pazaitis et al. 2017). This challenge becomes particularly conspicuous in the realm of public blockchains, exemplified by Bitcoin and Ethereum. Scalability concerns frequently orbit around transaction throughput, signifying the system's ability to process a specific volume of transactions or operations within a defined time frame, typically measured in seconds (Wright and De Filippi 2015). In cases where the network attains its maximum capacity, the potential arises for transactions to experience reduced speed and increased costs. Consensus mechanisms play a pivotal role in numerous blockchain networks, with procedures such as Proof of Work (PoW) or Proof of Stake (PoS) being commonly employed. Notably, Proof of Work (PoW) is renowned for its substantial energy consumption, posing a significant impediment to scalability (Sutherland et al. 2017). Scalability considerations extend to data storage as well. The escalating generation and storage of data on blockchain or cloud settings present substantial challenges in scaling storage solutions. Achieving interoperability between diverse blockchain networks or disparate cloud environments can introduce inherent complexities, thereby hindering scalability across multiple platforms.

Upholding security and decentralization in the context of scalability presents a formidable challenge. Increasing transaction throughput can potentially undermine network security or centralize control. The concept of Proof of Work (PoW) serves as a cryptographic protocol for validating and confirming transactions within a decentralized network. The energy consumption associated with Proof of Work (PoW)-based blockchains, exemplified by Bitcoin, is widely acknowledged for its significant impact on energy consumption (Wood 2014). Mining operations necessitate a substantial amount of processing power, resulting in noteworthy carbon footprints. High energy consumption is not only economically burdensome but also poses significant environmental risks, raising concerns about the long-term viability of blockchain networks. Several blockchain projects are currently exploring energy-efficient solutions, such as the adoption of Proof of Stake (PoS) or hybrid consensus techniques, to mitigate and minimize energy consumption. The procurement of energy for data centers and blockchain mining operations assumes paramount importance, with a particular focus on sustainability (Swan 2015). The utilization of renewable energy sources holds the potential to effectively alleviate adverse environmental consequences. Efficiency enhancements, achieved through continuous improvements in hardware, software, and algorithms, can contribute to reducing energy usage in data centers and blockchain networks. Governments and regulatory bodies may exert pressure through limitations or taxation measures targeting high-energy-consuming technology, with the aim of promoting environmentally friendly alternatives. The relationship between scalability and energy efficiency may entail potential trade-offs, as more efficient systems may not necessarily exhibit optimal performance under substantial workloads (Moreno-Sanchez et al. 2016; Pilkington 2016). Raising awareness among consumers and businesses regarding the environmental impact of technologies can drive demand for greener solutions. Balancing scalability and energy efficiency is a complex endeavor that often involves deliberate design choices, encompassing consensus method selection and hardware decisions aligned with objectives. Continuous innovation stands as a critical factor in effectively addressing scalability and energy consumption challenges. Technological innovation encompasses hardware upgrades, algorithmic optimizations, and software improvements (Meunier 2016). Incorporating diverse methodologies, such as Layer-2 scaling solutions within blockchains and multitiered architecture in cloud computing, can effectively mitigate scalability challenges. Ensuring adherence to regulatory compliance is crucial in integrating energy-saving measures and carbon footprint reporting into technological systems, thereby influencing their development and operational processes.

The adoption of environmental responsibility and the integration of sustainability into technology initiatives are imperative for businesses and organizations. Disseminating knowledge and raising awareness (Scott 2016) are key aspects. Informing and enlightening consumers, developers, and decision-makers about the significance of scalability and energy efficiency can potentially drive positive transformations in the acceptance and utilization of technology. Hence, addressing scalability and energy consumption challenges in technology necessitates a comprehensive strategy encompassing technical advancements, compliance with regulatory standards, commitment to environmental sustainability, and educational initiatives. Achieving an optimal balance between scalability and energy efficiency is of paramount importance for ensuring the sustainable development of technological ecosystems (Environments 2017b; Özsu and Valduriez 2011; Shulman and Co-founder 2006).

The incorporation of blockchain technology into carbon trading and offset markets aims to enhance transparency, traceability, and efficiency by integrating carbon credits onto the blockchain. A case study of Eco-Chain, a conceptualized network that utilizes blockchain technology to streamline the issuance, sale, and retirement of carbon credits, highlights the potential benefits and challenges associated with the integration of blockchain into the carbon credit market. The platform leverages smart contracts to automate various processes, ensuring transparency and reducing administrative burdens within the carbon credit economy. Tokenization involves representing each carbon credit as a unique digital asset on the blockchain, ensuring the legitimacy of registered projects through third-party auditors' monitoring. The auditors present their findings to the blockchain, documenting them transparently. Smart contracts, deployed on blockchain technology, facilitate automated processes for the issuance, transfer, and retirement of carbon credits. The integration of blockchain technology into carbon trading and offset markets offers numerous advantages, including immutability, transparency, and decentralization, thereby facilitating the prevention of fraudulent activities and ensuring the reliability of carbon offset initiatives.

17.6.2 Privacy and Data Security

In the contemporary landscape of digital technology, the preservation of privacy and the assurance of data security have emerged as pivotal concerns. Safeguarding sensitive information and recognizing individuals' privacy rights assume paramount significance for corporations, organizations, and individuals alike. Several noteworthy issues and factors warrant consideration in the domain of privacy and data security. Foremost among these concerns are the challenges that demand attention. The issue of data breaches remains a persistent and substantial risk. Cybercriminal strategies are in a perpetual state of evolution as they endeavor to gain unauthorized access to sensitive data, potentially resulting in severe financial consequences and reputational damage (Schlanger 2017). Complying with dynamic data protection legislation, exemplified by Europe's General Data Protection Regulation (GDPR) and the California Consumer Privacy Act (CCPA) in the United States, poses a formidable challenge for globally operating enterprises (Malakoff and Cornwall 2016; Davis et al. 2007; Rohr and McCoy 2010; March for Science 2017). The practice of data minimization presents a significant hurdle in the collection and maintenance of only essential data. Many organizations tend to amass an excess of data beyond their actual requirements, thereby augmenting vulnerability in the event of a data breach (Davis et al. 2007). Third-party risk is a common phenomenon in business operations, with organizations often relying on external vendors and partners for data processing; data security is shown in Fig. 17.3. Managing data security concerning the sharing of information with external entities presents a substantial challenge. The presence of insider threats introduces substantial risk, as employees or

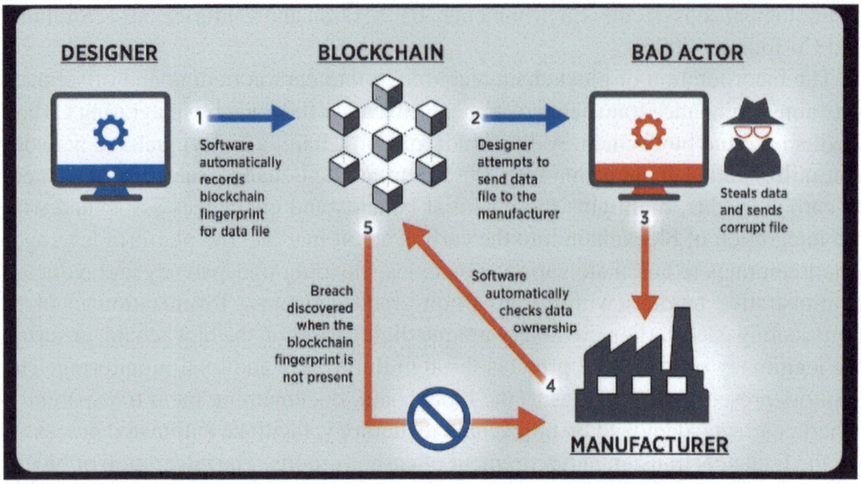

Fig. 17.3 Data security is also a major issue in blockchain technology

individuals with privileged access may engage in either malicious or negligent activities, potentially compromising data security. The advent of emerging technologies, including the Internet of Things (IoT), artificial intelligence, and quantum computing, has introduced novel security issues and vulnerabilities (Rohr and McCoy 2010).

Data localization concerns entail the intricate task of balancing the imperative of storing data in specific geographic locations, ensuring data security, and facilitating global access. Acquiring informed consent from users for data collection and enabling autonomy over their data can pose significant challenges. Data encryption stands as a critical measure for safeguarding sensitive information from unauthorized access during both storage and transmission. To ensure the protection of sensitive data, it is imperative to establish robust access controls and authentication systems, limiting access exclusively to authorized personnel. Data classification involves categorizing data based on sensitivity and significance, enabling the prioritization of security measures and compliance with relevant regulations. Regular auditing and monitoring practices involve ongoing system surveillance and periodic security audits to promptly identify and address security incidents. The concept of privacy by design entails integrating privacy and security measures into the initial stages of product and service design and development (March for Science 2017). Developing a comprehensive incident response plan is essential for efficiently managing data breaches and mitigating their consequences. Providing comprehensive training to employees, users, and stakeholders regarding optimal data security protocols and the importance of safeguarding privacy is crucial. Establishing and implementing data retention policies aim to reduce the accumulation of unnecessary data. It is imperative to thoroughly assess and establish contractual obligations for third-party suppliers and partners, mandating their compliance with specific security and privacy standards. Compliance management involves maintaining current knowledge of relevant data protection regulations and ensuring compliance through the implementation of policies, processes, and documentation (Lewis 2014).

User-friendly privacy controls seek to provide individuals with straightforward and comprehensible mechanisms to manage their personal information effectively. These controls encompass various options that empower users to oversee their data. Data Privacy Impact Assessments (DPIAs) should be conducted to scrutinize and evaluate potential privacy risks associated with new projects or systems. Deliberation regarding the ethical implications of data collection and utilization, especially in emerging technologies such as artificial intelligence (AI) and the Internet of Things (IoT), is imperative. Transparency plays a pivotal role in data practices, involving the provision of comprehensive information to users regarding the collection, use, and sharing of their data (Forbes 2017). Promoting a corporate environment that prioritizes data security and privacy awareness across the organization is advisable. Consulting with legal professionals is essential to ensure compliance with intricate data protection rules and regulations. Hence, addressing privacy and data security concerns requires a comprehensive approach that encompasses technological solutions, compliance efforts, employee education, and a commitment to ethical data practices. This endeavor is ongoing, demanding vigilance and adaptability in response to the ever-evolving landscape of threats and regulatory conditions (Gupta 2017).

17.6.3 Regulatory and Legal Implications

Effectively navigating the regulatory and legal landscape assumes paramount importance in the business sphere, particularly in industries dealing with sensitive information, emerging technologies, or complex global markets. The following elucidates several significant considerations and factors regarding regulatory and legal implications. The present study seeks to delve into the challenges inherent in this subject matter (Hong Kong Monetary Authority 2017). The complexity of regulations is escalating, marked by a proliferation of distinct sets of norms enforced by various jurisdictions. Meeting these diverse requirements can prove challenging, particularly for global corporations. The rapid pace of technological advancement frequently outpaces regulatory updates, leading to legal ambiguities and compliance challenges in nascent domains such as artificial intelligence, blockchain, and biotechnology. The implementation of stringent data privacy and protection legislation, typified by Europe's General Data Protection Regulation (GDPR) and the California Consumer Privacy Act (CCPA), carries profound global ramifications. To mitigate the risk of substantial financial penalties, organizations must possess a comprehensive understanding of how to effectively manage data in compliance with these requirements. The management of data across international borders may introduce complexities, stemming from data localization laws and cross-border data transfer constraints (The Food and Agriculture Organization of the United Nations 2016; Larsen et al. 2009; World Summit on the Information Society 2017).

Industry-specific regulations add a layer of complexity to legal compliance efforts, as sectors like finance, healthcare, and energy maintain distinct regulatory frameworks and compliance obligations. The legal landscape of consumer protection, including advertising and marketing practices, continues to evolve, emphasizing the importance of informed consent and transparent practices. Safeguarding intellectual property, encompassing patents, trademarks, and copyrights, assumes paramount importance, given the potential financial costs and adverse consequences associated with intellectual property disputes (United States Federal Bureau of Investigation 2016). Efficiently managing data breaches and cyberattacks necessitates adherence to breach notification legislation and the formulation of robust incident response strategies. Global antitrust and competition laws significantly impact various aspects of corporate operations, including commercial practices, mergers, acquisitions, and partnerships, with noncompliance in carrying the potential for significant financial penalties. Export controls and sanctions, governing technology and product exports to specific countries, remain susceptible to geopolitical shifts (Acharya and Acharya 2017). It is advisable to designate a compliance officer or team within the organization responsible for actively acquiring and maintaining knowledge about relevant regulations and ensuring compliance. Consultation with legal professionals possessing specialized knowledge in specific legal domains is recommended to address regulatory adherence, contractual matters, and risk mitigation. The process of regulatory monitoring entails continuous observation of regulatory changes, necessitating subsequent adjustments to policies and practices. Subscription services and legal technology tools offer effective means to stay current and informed (Department of Engineering and Irrigation 2017). When conducting global operations, a thorough understanding of the regulatory framework in each country and corresponding practice modifications are imperative. Regular risk assessments are essential to identify and address legal and regulatory issues, and comprehensive compliance programs, encompassing policies, procedures, personnel training, and monitoring techniques, are advisable. The concept of privacy by design entails proactively incorporating privacy and compliance considerations into product and service design to effectively address legal and regulatory obligations. Suitable contractual elements with partners, vendors, and consumers are necessary to manage legal and regulatory compliance effectively. Robust data governance practices are crucial for managing data in compliance with relevant data protection rules (Suichies 2017).

The principle of openness and reporting involves upholding transparency in interactions with consumers, regulators, and stakeholders and promptly notifying instances of noncompliance or data security breaches as required. Active public policy engagement includes interaction with legislators and industry associations to advocate for beneficial rules and contribute to industry standard development (Greenspan 2017a). This proposal advocates for well-defined procedures to settle legal disputes, including alternate dispute resolution methods, to mitigate the financial burden of litigation. Routine internal and external audits are essential for evaluating regulatory adherence and identifying enhancement opportunities. Deliberation on the ethical implications of commercial practices, particularly in the context of evolving technology, and their alignment with legal and societal norms is imperative (Eyal et al. 2016). In essence, adherence to legal and regulatory obligations is pivotal for achieving corporate success and managing reputation effectively. A proactive and continuous commitment is necessary to comprehend and comply with laws and regulations relevant to one's industry and geographical operations. The involvement of legal counsel and compliance professionals is crucial for guiding organizations through the intricacies of this complex environment (Rizzo 2017).

17.6.4 Technological Barriers to Adoption

The implementation and adoption of novel technologies, systems, or innovations can be hindered by technological impediments. These challenges present significant hurdles that both organizations and individuals must overcome. The following discusses several noteworthy issues and factors pertaining to technological barriers to adoption. One of the primary obstacles in this context is the challenge that arises from the initial expenses associated with procuring and integrating new technology (Environments 2017c; Sompolinsky and Zohar 2013; Buterin 2017). These costs can pose substantial barriers, especially for small and medium-sized enterprises (SMEs) and individuals. Insufficient infrastructure, characterized by unreliable Internet access or power supply, can hinder the widespread adoption of advanced technologies in specific regions or businesses. A deficit of skilled individuals capable of effectively operating, maintaining, and resolving issues related to emerging technology can also be a substantial obstacle. Interoperability presents intricate and costly challenges when addressing compatibility issues between existing systems and emerging technologies, particularly within legacy contexts. Concerns regarding the security of emerging technologies, particularly in domains like cloud computing and the Internet of Things (IoT), may impede their widespread adoption due to the inherent risk of data breaches or cyberattacks. The acquisition and utilization of individuals' personal data through emerging technologies, such as artificial intelligence (AI) and machine learning, raise privacy concerns and have the potential to generate resistance from users (Greenspan 2017b; De Filippi 2017; Gcoin 2017; Shafique et al. 2020; Tang et al. 2019).

Integrating new technologies into established corporate operations can disrupt the organization's functioning and necessitate significant workflow and cultural changes. The adoption of novel technology by employees or end users may be impeded by factors such as limited exposure, job security concerns, or discomfort with alterations to established routines. Accessibility poses a significant challenge in ensuring that technology is available to individuals with impairments or in diverse linguistic contexts, making it increasingly crucial to address this issue. Regulatory compliance can also pose a significant obstacle, particularly in sectors with stringent regulations like healthcare or finance, as organizations must adhere to regulatory criteria and obtain essential certifications (Gardašević et al. 2020). An analytical approach involves assessing the potential return on investment (ROI) to determine whether the benefits of adopting new technology outweigh the associated expenses.

Small-scale pilot initiatives can be undertaken to assess the viability and outcomes of novel technologies before extensive deployment. Allocating resources to implement training programs aimed at enhancing the skill sets of staff and end users is crucial for ensuring their proficiency in utilizing emerging technology. Collaboration is essential to address interoperability and security challenges, involving engagement with technology suppliers, industry peers, and subject matter experts. Scalability is a critical consideration when evaluating a technology solution's ability to accommodate an organization's growth and prevent future obstacles. To instill confidence among users, it is imperative to establish and enforce comprehensive data privacy and security protocols while effectively communicating these measures. This study aims to devise change management techniques encompassing comprehensive communication plans, robust leadership support, and active user engagement to effectively address and minimize opposition (Abdollahi et al. 2021).

Regulatory compliance necessitates staying well informed about relevant regulations and compliance obligations and taking proactive measures to address them during technology implementation. The importance of accessibility and inclusivity in technology solutions lies in their ability to cater to a wide range of users. Building solutions with the goal of accommodating various user needs and ensuring equal access is crucial (Mainetti et al. 2011). Prioritizing user experience and designing user-friendly technological solutions are essential strategies to minimize resistance and enhance adoption rates. Gathering feedback from users and stakeholders is crucial to continually improve technology solutions. Risk management involves formulating risk mitigation strategies to address technological, financial, and operational hazards that may arise during adoption. Careful evaluation of the long-term consequences associated with technology adoption and establishing a comprehensive plan for managing the technology life cycle and implementing necessary upgrades are imperative (Xu et al. 2018). The successful resolution of technological barriers in the adoption process often requires the implementation of a well-planned, interdisciplinary strategy. This method involves evaluating unique challenges within a specific context and implementing focused measures to effectively address them. Furthermore, possessing a proactive and adaptive mindset is crucial given the rapid evolution of technology. Continuous efforts are required to ensure the successful adoption and integration of technology (Alam 2022).

17.7 Future Prospects and Trends

17.7.1 Integration with IoT and Sensor Networks

The convergence of the Internet of Things (IoT) with sensor networks holds significant potential across diverse sectors and applications, with rapid development underway. This section provides an analysis of prospects and emerging trends in this domain. The utilization of edge computing to enable real-time processing is expected to gain prominence as IoT implementations expand. With the increasing demand for real-time data processing in IoT scenarios, edge computing, involving data processing at the network edge near the data source, is poised to reduce latency and lessen the necessity of transmitting all data to centralized cloud servers (Alrashidi and Nasri 2021). The deployment of 5G networks, offering high-speed, low-latency connections, is crucial for IoT and sensor networks. This technological advancement will facilitate faster data transmission, enabling applications requiring prompt responsiveness and handling large data volumes. IoT devices and sensors generate substantial data volumes. Incorporating artificial intelligence (AI) and machine learning (ML) into sensor networks enables data analysis and pattern recognition and facilitates predictive maintenance, anomaly detection, and automation (Kumar et al. 2023).

Swarm IoT, integrating swarm intelligence with the IoT paradigm, involves a multitude of cost-effective IoT devices and sensors collaborating as a unified entity. This approach enhances coverage, redundancy, and resilience in IoT deployments. Blockchain technology is expected to witness increased adoption to enhance the security and reliability of data generated and transmitted by IoT devices, ensuring data integrity and provenance, particularly in applications like supply chain tracking and smart contracts. Energy efficiency remains a priority for IoT devices and sensors (Smith et al. 2019). Advancements in low-power hardware, energy harvesting, and battery technology are expected to prolong device lifespans and reduce maintenance needs. The utilization of IoT and sensor networks is set to significantly impact the advancement of smart cities, spanning domains such as traffic control, waste management, environmental monitoring, and resource allocation optimization (Alahi et al. 2023). The healthcare sector is likely to expand its use of IoT-enabled wearable devices and remote monitoring systems to enhance patient care, potentially improving patient outcomes and reducing healthcare costs. In agriculture, IoT sensors and networks have the potential to transform precision farming techniques. Real-time monitoring of soil conditions, weather patterns, and crop health enables efficient resource utilization and maximizes agricultural productivity (Whitmore et al. 2015). Supply chain optimization aims to enhance efficiency and effectiveness. Sensor networks and IoT devices are expected to improve supply chain visibility, tracking commodities throughout their journey, reducing financial losses, enhancing operational efficiency, and increasing transparency. Environmental monitoring utilizing IoT sensors will address concerns such as air and water quality, deforestation, and wildlife preservation, contributing valuable data for sustainable practices and evidence-based policymaking. Customizable sensor networks tailored to industry-specific requirements are expected to become more accessible. Modular and flexible solutions will enable organizations to adapt IoT deployments to their needs (Bilal et al. 2016).

Enhancements in security protocols, encryption techniques, and user consent processes will address rising concerns over security and privacy with the growing popularity of IoT. Interoperability standards will play a crucial role in facilitating seamless communication among diverse IoT devices and platforms, fostering a more interconnected environment. The integration of autonomous systems with IoT and sensor networks is anticipated to advance fields like self-driving vehicles, drones, and robotics, offering benefits to industries such as logistics and transportation (Gashi et al. 2022). The trajectory of IoT and sensor networks is characterized by increased interconnectedness, intelligence, and innovation across various fields. The rapid pace of technological advancement presents extensive opportunities for enhancing efficiency, sustainability, and overall quality of life through IoT and sensor network integration. Organizations adopting these emerging trends are poised to gain a competitive edge and contribute to the sustained growth of this domain (Shukla and Hait 2022).

17.7.2 Interoperability Between Blockchains

The establishment of interoperability among blockchains stands as a critical area of advancement within the blockchain technology and cryptocurrency domain. Ensuring seamless communication and efficient data exchange among diverse blockchain networks is imperative to fully unlock the potential of distributed ledger technology. This discourse provides an analysis of prospective developments and emerging patterns in blockchain interoperability. The central topic of discussion revolves around the establishment of standards for cross-chain communication (Esenogho et al. 2022). The adoption of standardized norms and protocols for facilitating communication between different blockchain networks is poised to have a significant impact on promoting interoperability. Notable initiatives such as Polkadot, Cosmos, and Interledger have already embarked on the mission of developing these standards. Interoperability is expected to greatly enhance the functionality of cross-chain decentralized finance (DeFi) applications, allowing users to seamlessly transfer assets and access financial services across diverse blockchains without reliance on centralized intermediaries (Adeniyi et al. 2021).

Blockchain bridges, serving as technological solutions, aim to bridge different blockchain networks. These bridges enable the transfer of digital assets and data between separate blockchain systems, thus facilitating interoperability. The proliferation of blockchain bridges is expected to become more frequent, enabling seamless asset and data flow between diverse blockchains. Notable examples include Ethereum's Rainbow Bridge and Bitcoin's Lightning Network (Darwish and Hassanien 2018). The concept of interconnected ecosystems refers to the interdependence and interconnectedness of various ecological systems. As interoperability among blockchain networks continues to expand, it is foreseeable that interconnected blockchain ecosystems will emerge, with each blockchain dedicated to specific use cases or functionalities. These ecosystems have the potential to support a wide range of applications (Singh et al. 2014). Layer-2 solutions, operating at the second layer of the OSI model, are expected to increasingly incorporate interoperability elements. This includes solutions like side chains and state channels, which will provide expedited and cost-effective cross-chain transactions while maintaining the security measures of the underlying chains. Cross-chain interoperability is anticipated to offer benefits for non-fungible tokens (NFTs) and blockchain-based games, enabling users to exchange NFTs and various assets across diverse gaming platforms and markets (Rahman et al. 2023). Incorporating interoperability into decentralized identity solutions will facilitate the seamless management of user IDs and secure access to services across various blockchain-based applications and platforms. The intersection of supply chain management and trade finance gains academic and practical significance. Blockchain interoperability can facilitate the seamless flow of data and assets among numerous stakeholders and systems, playing a vital role in supply chain management and trade finance (Kamble et al. 2018).

Interoperability can foster secure data sharing and collaboration among corporations, government agencies, and other organizations by bridging public and private blockchain networks. Cross-chain oracles, mechanisms enabling data and information transfer between different blockchain networks, will witness enhanced interoperability. This will enable smart contracts to securely access reliable data from diverse sources and chains (Sharma et al. 2017).

Addressing regulatory compliance and data privacy concerns is pivotal in the development of interoperability solutions, especially when assets or data traverse diverse jurisdictions and networks. Enhancing security protocols for cross-chain transactions and data transfers will remain a priority to mitigate risks associated with interoperability, including atomic swaps and security audits. Continued research and innovation in blockchain interoperability will drive the advancement of novel technologies and methodologies to overcome existing constraints and challenges. The prevalence of cross-chain decentralized exchanges (DEXs) is expected to rise, allowing users to trade assets across multiple blockchains without relying on centralized exchanges (Zhang et al. 2014). Given the increasing complexity of interoperability, providing user-friendly tools and interfaces is crucial to ensure convenient access and utilization of cross-chain features for both individuals and enterprises. The establishment and widespread acceptance of blockchain technology heavily depend on interoperability. As the blockchain field continues to evolve, collaborative efforts among diverse blockchain networks and innovative approaches to tackle interoperability challenges are expected to emerge. These emerging patterns will contribute to the development of a more interconnected and adaptable blockchain ecosystem (Wang et al. 2016).

17.7.3 The Evolution of Green Cryptocurrencies

Green cryptocurrencies, also referred to as environmentally conscious or sustainable cryptocurrencies, are dedicated to mitigating the ecological impact associated with blockchain technology, with a primary focus on reducing energy consumption within blockchain networks. This discourse provides an analysis of prospective developments and emerging trends in the evolutionary trajectory of environmentally sustainable cryptocurrencies. Cryptocurrencies adopting the Proof-of-Stake (PoS) consensus mechanisms, exemplified by Ethereum 2.0, Cardano, and Polkadot, are gaining momentum in the market. In contrast to Proof-of-Work (PoW) cryptocurrencies like Bitcoin, PoS cryptocurrencies require significantly lower energy consumption (Mohamed 2020). The adoption of the PoS consensus mechanism is expected to reduce the environmental footprint of blockchain networks. Emerging blockchain networks are intentionally designed with a strong emphasis on energy efficiency. These systems employ advanced consensus techniques, including directed acyclic graphs (DAGs) or hybrid Proof of Work/Proof of Stake (PoW/PoS), to mitigate energy consumption. The topic of carbon offsetting and sustainability initiatives plays a crucial role in environmental conservation and sustainable development. An increasing number of cryptocurrency initiatives and organizations are embracing carbon-offsetting measures. Some individuals allocate a portion of their block rewards to support environmental causes or invest in renewable energy projects to offset their carbon emissions (Park and Han 2023).

Green cryptocurrencies are exploring partnerships with renewable energy providers to power their blockchain networks using sustainable energy sources such as solar, wind, and hydroelectric power. Ongoing advancements in energy-efficient mining equipment, notably application-specific integrated circuits (ASICs) and graphics processing units (GPUs) designed for minimal power consumption, will continue to reduce the energy demands of PoW-based cryptocurrencies. Further research into consensus algorithms that balance security and energy efficiency is essential to enhance the environmental sustainability of future blockchain networks (Mihai et al. 2022). Transparency regarding environmental impacts is becoming a trend among cryptocurrency initiatives. Some may seek certification from environmental organizations to validate their commitment to environmentally sustainable practices. Regulatory bodies and government entities have the potential to introduce environmental criteria and standards for cryptocurrencies and blockchain projects, promoting eco-friendly practices. Decentralized energy markets, where energy generation, distribution, and consumption deviate from traditional centralized models, offer Bitcoin miners the opportunity to source energy directly from environmentally sustainable sources through blockchain technology (Elijah et al. 2018).

The concept of green NFTs and sustainable collectibles is emerging, with platforms focusing on sustainability in the creation and exchange of digital art and collectibles. Projects may explore carbon-neutral transaction systems, where a portion of transaction fees compensates for the carbon emissions associated with those transactions. Prominent corporate entities and financial institutions entering the cryptocurrency space may show a preference for environmentally friendly cryptocurrencies and invest in sustainable blockchain technologies (Cao et al. 2022). The development of standardized metrics for assessing the environmental impact of cryptocurrencies and blockchain networks could facilitate the evaluation of users' blockchain activities' sustainability. Research and innovation will drive the advancement of environmentally friendly cryptocurrency solutions, addressing existing challenges and enhancing blockchain networks' sustainability (Cao et al. 2019). The adoption of environmentally friendly cryptocurrencies is driven not only by ecological concerns but also by the growing demand from investors and consumers for sustainable and socially responsible financial opportunities. As blockchain technology matures, its evolution is expected to prioritize both innovation and sustainability, fostering an ecologically conscious cryptocurrency ecosystem (Golpîra et al. 2021).

17.7.4 Collaborative Efforts and Partnerships

Collaborative efforts and partnerships hold paramount significance in the realm of innovation, problem-solving, and attainment of common objectives across diverse industries and sectors. As we gaze forward, several emerging prospects and trends pertaining to collaborative efforts and partnerships warrant consideration. Collaborations between industries that traditionally operated in isolation are poised to proliferate (Alfa et al. 2021). For instance, healthcare and technology enterprises may unite their efforts to advance telemedicine solutions, while automotive and technology firms may engage in joint endeavors aimed at the development of autonomous vehicles. Concurrently, governments and private-sector organizations will forge public-private partnerships (PPPs) to grapple with multifaceted societal challenges, such as infrastructure development, climate change mitigation, and public health crises (Chataut et al. 2023). Businesses are increasingly cognizant of the value inherent in establishing partnerships within their respective ecosystems. This encompasses collaborations with suppliers, customers, and adjacent industries to create value chains and ecosystems that accrue benefits to all stakeholders. Nonprofit organizations and corporations will enter into partnerships to address social and environmental issues, with the potential to yield innovative solutions that positively impact society (Gill et al. 2022). Innovation hubs and clusters will continue their expansion, serving as convergence points for startups, corporations, research institutions, and governmental bodies, fostering innovation, entrepreneurship, and knowledge exchange. Organizations will also initiate data-sharing alliances, pooling resources, and data for research, analytics, and insights, potentially leading to breakthroughs in domains such as healthcare, finance, and urban planning. Partnerships focused on sustainability will gain impetus, driven by environmental, social, and governance (ESG) considerations. Businesses will collaborate intensively to diminish their carbon footprints, endorse ethical practices, and realize sustainability objectives. Moreover, partnerships between educational institutions and businesses will witness a surge, primarily to address skill gaps, offer training programs, and prepare the workforce for rapidly evolving industries (Garrido-Momparler and Peris 2022; Gravina and Fortino 2020; Hart and Martinez 2006; Chen et al. 2014).

Research consortia will convene experts and resources from multiple entities to expedite innovation in sectors such as healthcare, aerospace, and renewable energy. Furthermore, international organizations, governments, and businesses will establish global alliances aimed at confronting worldwide challenges, including climate change, public health, and cybersecurity. Collaboration among technology companies and standard organizations will precipitate the development of interoperable and universally embraced technology standards, facilitating the seamless integration of products and services (Shilpa and Kaur 2022). Cultural institutions, artists, and creative industries will forge alliances to foster cultural exchange, tourism, and preservation of cultural heritage, leveraging the synergy of art and technology. Impact investors, philanthropic organizations, and startups will join forces to channel investments into projects and ventures that yield both financial returns and positive societal or environmental impacts. Additionally, cooperation between organizations and governments to share threat intelligence and enhance cybersecurity measures will be pivotal in countering evolving cyber threats. The imperative of resilient supply chains will spur alliances focused on risk mitigation and disruption amelioration through diversification, digitization, and contingency planning. Blockchain technology and decentralized platforms will facilitate secure and trustless collaborations without necessitating centralized intermediaries, fostering novel modes of cooperation. The thriving open-source communities will persist, affording developers, organizations, and individuals the opportunity to collaborate on software and technological innovations that redound to the benefit of the global community (Da Xu 2011). In summation, collaborative efforts and partnerships shall remain indispensable for the addressing of intricate challenges, the nurturing of innovation, and the cultivation of a more interconnected global ecosystem. Entities and organizations that embrace these trends and cultivate a collaborative mindset are poised for prosperity within the evolving landscape of the future (Furstenau et al. 2023; Ping et al. 2018; Tzounis et al. 2017).

17.8 Conclusion

In conclusion, the potential for blockchain technology to induce profound transformations across numerous industries has been demonstrated through the examination of case studies in carbon credit management and animal conservation. The provision of immutable and transparent ledgers serves to foster accountability, trust, and efficiency within intricate ecosystems. However, as blockchain technology becomes increasingly pervasive, additional noteworthy concerns and considerations come to the fore. Of paramount concern are issues related to scalability and energy consumption. Blockchain networks grapple with the challenge of scalability, particularly in the face of escalating transaction volumes. Additionally, the environmentally problematic nature of certain consensus mechanisms, notably Proof of Work, necessitates immediate attention. Privacy and data security also loom large on the horizon, with the imperative to safeguard sensitive information and adhere to dynamic data protection requirements being pivotal for broader acceptance of blockchain applications. Furthermore, the presence of regulatory and legal ramifications introduces complexity to the landscape, requiring continuous vigilance and adaptability to navigate ever-evolving regulatory frameworks. The adoption of this technology encounters various barriers, including substantial implementation costs, infrastructural limitations, and resistance from users. These hurdles pose substantial challenges to the technology adoption process. Establishing connections across these disparities is essential to facilitate the widespread integration of blockchain technology. Looking toward the future, several promising trends and opportunities come into view. The integration of blockchain technology into Internet of Things (IoT) and sensor networks holds the potential to facilitate innovative avenues for data acquisition, automation, and real-time decision-making across multiple sectors. Interoperability between blockchain networks, facilitated by cross-chain standards and protocols, will promote a more integrated and diverse blockchain ecosystem. The potential for environmentally friendly cryptocurrencies to address the ecological impact of blockchain technology and align it with sustainability goals is driven by the shift toward Proof of Stake, the utilization of renewable energy sources, and the implementation of carbon offset schemes. The blockchain industry stands at a pivotal juncture of development, offering significant opportunities for beneficial advancements while grappling with notable challenges. Successfully navigating the forthcoming hurdles necessitates a conscientious and collaborative approach to harnessing the full potential of blockchain innovations and addressing the intricacies of our evolving digital landscape. Blockchain technology is a secure and transparent platform that can aid in environmental conservation efforts, particularly in mitigating climate change and biodiversity loss. It can provide a secure and transparent platform for trading carbon credits, which can enhance the effectiveness of emission offsets and encourage investments in renewable energy initiatives. Smart contracts on the blockchain can expedite the verification and transfer of carbon credits, empowering individuals and organizations to actively participate in mitigating climate change and assume responsibility for their carbon emissions. Additionally, blockchain technology can help track and verify sustainable practices in supply chains, such as responsible sourcing of materials and ethical labor practices, promoting transparency and accountability in environmental conservation efforts.

References

Abbasi AZ, Islam N, Shaikh ZA (2014) A review of wireless sensors and networks' applications in agriculture. Comput Stand Interfaces 36:263–270

Abdollahi A, Rejeb K, Rejeb A, Mostafa MM, Zailani S (2021) Wireless sensor networks in agriculture: insights from bibliometric analysis. Sustainability 13(21):12011

Acharya AP, Acharya A (2017) Cyberterrorism and biotechnology. Foreign Aff. https://www.foreignaffairs.com/articles/world/2017-06-01/cyberterrorism-and-biotechnology. Accessed 2 Jun 2017

Adeniyi EA, Ogundokun RO, Awotunde JB (2021) IoMT-based wearable body sensors network healthcare monitoring system. In: IoT in healthcare and ambient assisted living, pp 103–121

Adil A, Badarla V, Plappally AK, Bhandari R, Sankhla PC (2015) Development of affordable ICT solutions for water conservation in agriculture. In: Proceedings of the 7th international conference on communication systems and networks (COMSNETS), Bangalore, India, pp 1–6

Alahi MEE, Sukkuea A, Tina FW, Nag A, Kurdthongmee W, Suwannarat K, Mukhopadhyay SC (2023) Integration of IoT-enabled technologies and artificial intelligence (AI) for smart city scenario: recent advancements and future trends. Sensors 23(11):5206

Alam T (2022) Blockchain-based Internet of Things: review, current trends, applications, and future challenges. Computers 12(1):6

Alfa AA, Alhassan JK, Olaniyi OM, Olalere M (2021) Blockchain technology in IoT systems: current trends, methodology, problems, applications, and future directions. J Reliable Intell Environ 7(2):115–143

Alrashidi M, Nasri N (2021) Wireless body area sensor networks for wearable health monitoring: technology trends and future research opportunities. Int J Adv Comput Sci Appl 12(4)

Andreopoulou Z (2012) Green informatics: ICT for green and sustainability. J Agric Inform 3:1–8

Azam H, Mahmud MZ, Vo DB (2022) Blockchain-enabled carbon credits utilization for supply chain sustainability. J Clean Prod 334:147632

Back A, Corallo M, Dashjr L, Friedenbach M, Maxwell G, Miller A, Poelstra A, Timón J, Wuille P (2014) Enabling blockchain innovations with pegged side chains. Blockstream

Banafa A (2017) IoT and blockchain convergence: benefits and challenges. IEEE

Bartlett AC, Andales AA, Arabi M, Bauder TA (2015) A smartphone app to extend use of a cloud-based irrigation scheduling tool. Comput Electron Agric 111:127–130

Bilal M, Oyedele LO, Qadir J, Munir K, Ajayi SO, Akinade OO et al (2016) Big Data in the construction industry: a review of present status, opportunities, and future trends. Adv Eng Inform 30(3):500–521

Bloom N, Garicano L, Sadun R, Van Reenen J (2014) The distinct effects of information technology and communication technology on firm organization. Manag Sci. https://doi.org/10.3386/w14975

Böhme R, Christin N, Edelman B, Moore T (2015) Bitcoin: economics, technology, and governance. J Econ Perspect 29(2):213–238

Buterin V (2017) Problems in cryptoeconomic research. In: Hard problems in cryptoeconomics. Taipei Ethereum Event, Taipei, Taiwan

Caldeira C, Sinderen M, Guerreiro S (2019) The role of blockchain in circular economy: a review. In: IFIP conference on sustainable internet and ICT for sustainability. Springer, pp 229–243

Cao L, Cai Y, Yue Y (2019) Swarm intelligence-based performance optimization for mobile wireless sensor networks: survey, challenges, and future directions. IEEE Access 7:161524–161553

Cao L, Wang Z, Yue Y (2022) Analysis and prospect of the application of wireless sensor networks in ubiquitous power Internet of Things. Comput Intell Neurosci 2022:1

Carlson TN, Arthur ST, Chandler WR, Alavi AS (2018) Remote sensing technologies for mapping and monitoring carbon stocks in tropical forests: a review. Int J For Res 2018:1–16

Chataut R, Phoummalayvane A, Akl R (2023) Unleashing the power of IoT: a comprehensive review of IoT applications and future prospects in healthcare, agriculture, smart homes, smart cities, and industry 4.0. Sensors 23(16):7194

Chen S, Xu H, Liu D, Hu B, Wang H (2014) A vision of IoT: applications, challenges, and opportunities with China perspective. IEEE Internet Things J 1(4):349–359

Chum R, Lens T, Van Fransen H (2019) Blockchain-based wildlife management. arXiv preprint arXiv:1909.07482

ClimateTrade (n.d.) ClimateTrade: climate action economy. https://www.climatetrade.com/

Collier ZA, Bates ME, Wood MD, Linkov I (2014) Stakeholder engagement in dredged material management decisions. Sci Total Environ 496:248–256

Da Xu L (2011) Enterprise systems: state-of-the-art and future trends. IEEE Trans Industr Inform 7(4):630–640

Darwish A, Hassanien AE (2018) Cyber physical systems design, methodology, and integration: the current status and future outlook. J Ambient Intell Humaniz Comput 9:1541–1556

Davidson S, De Filippi P, Potts J (2016) Economics of blockchain. https://doi.org/10.2139/ssrn.2744751

Davis MS, Riske-Morris M, Diaz SR (2007) Causal factors implicated in research misconduct: evidence from ORI case files. Sci Eng Ethics 13:395–414

De Filippi P (2017) It's time to take mesh networks seriously (and not just for the reasons you think). https://www.wired.com/2014/01/its-time-to-take-mesh-networks-seriously-and-not-just-for-thereasons-you-think/. Accessed 22 Apr 2017

Department of Engineering and Irrigation (2017) Council of Agriculture, Executive Yuan Website. http://doie.coa.gov.tw/english/about-us.asp. Accessed 5 May 2017

Dinh TN, Liu R, Zhang M, Chen G, Ooi BC, Wang J (2020) Untangling blockchain: a data processing view of blockchain systems. IEEE Trans Knowl Data Eng 32(12):2348–2365

Dixon J, Richards C (2016) On food security and alternative food networks: understanding and performing food security in the context of urban bias. Agric Hum Values 33:191–202

DNV GL AS (2017) Global opportunity report

Drescher D (2017) Blockchain basics: a non-technical introduction in 25 steps. Apress

Economics of Blockchain for Environmental Conservation (2021) Blockchain technology for agriculture and environmental management. Springer, Cham
Elijah O, Rahman TA, Orikumhi I, Leow CY, Hindia MN (2018) An overview of Internet of Things (IoT) and data analytics in agriculture: benefits and challenges. IEEE Internet Things J 5(5):3758–3773
Energy Web Foundation (n.d.) Energy web: accelerating the decentralized, low-carbon future. https://www.energyweb.org/
Environments (2017a) 4, 50 11 of 13
Environments (2017b) 4, 50 12 of 13
Environments (2017c) 4, 50 13 of 13
Esenogho E, Djouani K, Kurien AM (2022) Integrating artificial intelligence Internet of Things and 5G for next-generation smartgrid: a survey of trends challenges and prospect. IEEE Access 10:4794–4831
Eyal I, Gencer AE, Sirer EG, Van Renesse R (2016) Proceedings of the 13th USENIX Symposium on networked systems design and implementation (NSDI 16), Santa Clara, CA, USA, 16–18 March 2016, vol Bitcoin-ng: a scalable blockchain protocol, pp 45–59
Food and Agriculture Organization (FAO) (2021) Blockchain applications for sustainable agriculture and food systems. Food and Agriculture Organization (FAO)
Food and Agriculture Organization of the United Nations (FAO) (2022) Soil degradation: 33% of global soil resources are degraded. Food and Agriculture Organization of the United Nations (FAO)
Forbes S (2017) How bitcoin will end world poverty. https://www.forbes.com/sites/steveforbes/2015/04/02/how-bitcoin-will-end-world-poverty/2/#211bf31452fc. Accessed 16 Apr 2017
Furstenau LB, Rodrigues YPR, Sott MK, Leivas P, Dohan MS, López-Robles JR et al (2023) Internet of Things: conceptual network structure, main challenges and future directions. Digit Commun Netw 9(3):677–687
Gardašević G, Katzis K, Bajić D, Berbakov L (2020) Emerging wireless sensor networks and Internet of Things technologies—foundations of smart healthcare. Sensors 20(13):3619
Garrido-Momparler V, Peris M (2022) Smart sensors in environmental/water quality monitoring using IoT and cloud services. Trends Environ Anal Chem 35:e00173
Gashi L, Luma A, Januzaj Y (2022) The integration of wireless sensor networks, mobile networks and cloud engineering for a decision support system—a systematic literature review. In: 2022 International congress on human-computer interaction, optimization and robotic applications (HORA). IEEE, pp 1–8
Gcoin (2017). http://g-coin.org/index.html. Accessed 10 May 2017
Gebbers R, Adamchuk VI (2010) Precision agriculture and food security. Science 327:828–831
Gill SS, Xu M, Ottaviani C, Patros P, Bahsoon R, Shaghaghi A et al (2022) AI for next generation computing: emerging trends and future directions. Internet Things 19:100514
Gipp B, Breitinger C, Meuschke N, Beel J (2017) Introducing securely timestamped manuscript submission and peer review feedback using the blockchain. https://www.gipp.com/wpcontent/papercite-data/pdf/gipp2017b.pdf
Global Canopy (2021) ConserveForest: transforming the timber supply chain. https://globalcanopy.org/project/conserveforest
Golpîra H, Khan SAR, Safaeipour S (2021) A review of logistics internet-of-things: current trends and scope for future research. J Ind Inf Integr 22:100194
Gouveia C, Fonseca A (2008) New approaches to environmental monitoring: the use of ICT to explore volunteered geographic information. GeoJournal 72:185–197
Gouveia C, Fonseca A, Camara A, Ferreira F (2004) Promoting the use of environmental data collected by concerned citizens through information and communication technologies. J Environ Manag 71:135–154
Gravina R, Fortino G (2020) Wearable body sensor networks: state-of-the-art and research directions. IEEE Sensors J 21(11):12511–12522

Greenspan G (2017a) Avoiding the Pointless Blockchain Project. http://www.multichain.com/blog/2015/11/avoiding-pointless-blockchain-project/. Accessed 14 Apr 2017

Greenspan G (2017b) Blockchains vs centralized databases. http://www.multichain.com/blog/2016/03/blockchains-vs-centralized-databases/. Accessed 22 Apr 2017

Gupta V (2017) Building the hyperconnected future on blockchains. World Government Summit. http://internetofagreements.com/files/WorldGovernmentSummit-Dubai2017.pdf. Accessed 10 May 2017

Hart JK, Martinez K (2006) Environmental sensor networks: a revolution in the earth system science? Earth Sci Rev 78(3–4):177–191

Hodler R, Karnouskos S (2018) The carbon blockchain: a climate governance innovation. Energy Res Soc Sci 39:15–18

Hong Kong Monetary Authority (2017) Whitepaper on distributed ledger technology. http://www.hkma.gov.hk/media/eng/doc/key-functions/finanical-infrastructure/Whitepaper_On_Distributed_Ledger_Technology.pdf. Accessed 10 May 2017

Hsu C (2018) Blockchain applications for environmental sustainability: a comprehensive survey. Sustainability 10(4):1168

Huckle S, Bhattacharya R, White M, Beloff N (2016) Internet of Things, blockchain, and shared economy applications. Proc Comput Sci 98:461–466

Hüsemann BD, Bhandari KP, Knottnerus OS (2020) Born to be wild: a feasibility study on the use of blockchain technology for wildlife conservation. In: The evolution of digital innovation. Springer, pp 179–187

IBM (n.d.) IBM Food Trust. https://www.ibm.com/food

IBM Corporation (2015) Device democracy: saving the future of the Internet of Things. http://www-01.ibm.com/common/ssi/cgi-bin/ssialias?htmlfid=GBE03620USEN

Imani M, Perera C, Chandrasekaran M (2018) Blockchain applications in supply chain management: a comprehensive study. J Clean Prod 200:139–168

Intergovernmental Panel on Climate Change (IPCC) (2021a) Climate change: global average temperature increase of 1.1 degrees Celsius above pre-industrial levels. Intergovernmental Panel on Climate Change (IPCC)

Intergovernmental Panel on Climate Change (IPCC) (2021b) Summary for policymakers. In: Climate change 2021: the physical science basis. Intergovernmental Panel on Climate Change (IPCC)

Intergovernmental Science-Policy Platform on Biodiversity and Ecosystem Services (IPBES) (2019a) Global assessment report on biodiversity and ecosystem services. Intergovernmental Science-Policy Platform on Biodiversity and Ecosystem Services (IPBES)

Intergovernmental Science-Policy Platform on Biodiversity and Ecosystem Services (IPBES) (2019b) Wildlife extinction: up to 1 million species facing extinction due to human activities. Intergovernmental Science-Policy Platform on Biodiversity and Ecosystem Services (IPBES)

Jagannathan S, Priyatharshini R (2015) Smart farming system using sensors for agricultural task automation. In: Proceedings of the IEEE technological innovation in ICT for agriculture and rural development (TIAR), Chennai, India, pp 49–53

Jiang JA, Wang CH, Liao MS, Zheng XY, Liu JH, Chuang CL, Hung CL, Chen CP (2016) A wireless sensor network-based monitoring system with dynamic convergecast tree algorithm for precision cultivation management in orchid greenhouses. Precis Agric 17:766–785

Kamble SS, Gunasekaran A, Gawankar SA (2018) Sustainable industry 4.0 framework: a systematic literature review identifying the current trends and future perspectives. Process Saf Environ Prot 117:408–425

Kissinger G, Herold M, De Sy V (2019) Drivers of deforestation and forest degradation: a synthesis report for REDD+ policymakers. Lexeme Consulting

Kumar R, Jain S (2018) RecycleToCoin: a framework for incentivising recycling using blockchain and cryptocurrency. In: 2018 IEEE 9th annual information technology, electronics and mobile communication conference (IEMCON), pp 912–918

Kumar M, Kumar A, Verma S, Bhattacharya P, Ghimire D, Kim SH, Hosen AS (2023) Healthcare Internet of Things (H-IoT): current trends, future prospects, applications, challenges, and security issues. Electronics 12(9):2050

Lacetera N, Zirulia L (2011) The economics of scientific misconduct. J Law Econ Organ 27:568–603

Lansiti M, Lakhani KR (2017) The truth about blockchain. Harv Bus Rev 95:119–127

Larsen K, Kim R, Theus F (2009) Agribusiness and innovation systems in Africa. World Bank Publications. https://openknowledge.worldbank.org/handle/10986/2643. Accessed 10 May 2017

Lewis DL (2014) Science for sale: how the US government uses powerful corporations and leading universities to support government policies. In: Silence top scientists, jeopardize our health, and protect corporate profiles. Skyhorse Publishing Inc., New York, NY

Li X, Liang X, Jin Q, Huang X, Chen H, Malik TQ (2020) A comprehensive blockchain primer: development insights, challenges & solutions. IEEE Access 8:218671–218696

Liebenau J, Quintas P, Ferreira P, Huysman M (2021) Blockchain technology for environmental conservation. AI Soc:1–16

Lin YP, Deng DP, Lin WC, Lemmens R, Crossman ND, Henle K, Schmeller DS (2015) Uncertainty analysis of crowd-sourced and professionally collected field data used in species distribution models of Taiwanese moths. Biol Conserv 181:102–110

Lin YP, Chang TK, Fan C, Anthony J, Petway JR, Lien WY, Liang CP, Ho YF (2017) Applications of information and communication technology for improvements of water and soil monitoring and assessments in agricultural areas—a case study in the Taoyuan irrigation district. Environments 4:6

López-Pintado F, Fernández LD, Martín C (2019) Blockchain for sustainable supply chains. Socio Econ Plan Sci 68:100661

MacNeil R (2014) Canadian environmental policy under conservative majority rule. Environ Polit 23:174–178

Mainetti L, Patrono L, Vilei A (2011) Evolution of wireless sensor networks towards the Internet of Things: a survey. In: SoftCOM 2011, 19th international conference on software, telecommunications and computer networks. IEEE, pp 1–6

Malakoff D, Cornwall W (2016) Trump team targets key climate metric. Science 354:1364–1365

March for Science (2017). https://www.marchforscience.com. Accessed 16 Apr 2017

Marttunen M, Belton V, Lienert J (2017) Are objectives hierarchy related biases observed in practice. A meta-analysis of environmental and energy applications of multi-criteria decision analysis. Eur J Oper Res 265:178–194

Massachusetts Institute of Technology (2018) Blockchain applications for the energy sector. https://jwel.mit.edu/sites/default/files/MIT_J-WEL_Blockchain_Applications_for_the_Energy_Sector_2018.pdf

Meng X, Malone C, Shi L (2020) Blockchain technology in the energy sector: a systematic review of challenges and opportunities. Appl Energy 266:114848

Meunier S (2016) Blockchain technology—a very special kind of distributed database

Mihai S, Yaqoob M, Hung DV, Davis W, Towakel P, Raza M et al (2022) Digital twins: a survey on enabling technologies, challenges, trends and future prospects, vol 24. IEEE Commun Surv Tutorials, p 2255

Mohamed M (2020) A comparative study on Internet of Things (IoT): frameworks, tools, applications and future directions. J Intell Syst Internet Things 1(1):13–39

Moreno-Sanchez P, Zafar MB, Kate A (2016) Listening to whispers of ripple: linking wallets and deanonymizing transactions in the ripple network. Proc Priv Enhanc Technol 2016:436–453

NGO Monitor (2015) Amnesty International: failed methodology, corruption, and anti-Israel bias. http://www.ngo-monitor.org/books/amnesty_international_failed_methodology_corruption_and_anti_israel_bias/

Nobre CA, Sampaio G, Borma LS, Castilla-Rubio JC, Silva JS, Cardoso M (2016) Land-use and climate change risks in the Amazon and the need of a novel sustainable development paradigm. Proc Natl Acad Sci USA 113:10759–10768

Özsu MT, Valduriez P (2011) Principles of distributed database systems. Springer Science & Business Media, London, UK
Park JH, Han MH (2023) Enhancing livestock management with IoT-based wireless sensor networks: a comprehensive approach for health monitoring, location tracking, behavior analysis, and environmental optimization. J Sustain Urban Futur 13(6):34–46
Pazaitis A, De Filippi P, Kostakis V (2017) Blockchain and value systems in the sharing economy: the illustrative case of backfeed. Technol Forecast Soc Change 125:105
Perea RG, Garcia IF, Arroyo MM, Diaz JAR, Poyato EC, Montesinos P (2017) Multiplatform application for precision irrigation scheduling in strawberries. Agric Water Manag 183:194–201
Pilkington M (2016) Blockchain technology: principles and applications. In: Research handbook on digital transformations. Edward Elgar, Cheltenham, UK
Ping H, Wang J, Ma Z, Du Y (2018) Mini-review of application of IoT technology in monitoring agricultural products quality and safety. Int J Agric Biol Eng 11(5):35–45
Plastic Bank (n.d.) Plastic bank: social plastic & recycling rewards. https://plasticbank.com/
Prusty BG, Zhang X (2020) The role of blockchain technology in sustainability: a systematic literature review, framework development, and case study. J Clean Prod 271:122572
Pure Earth (2021) Pollution from industrial activities: industrial pollution is responsible for 4.4 million premature deaths annually. Pure Earth
Rahman A, Hasan K, Kundu D, Islam MJ, Debnath T, Band SS, Kumar N (2023) On the ICN-IoT with federated learning integration of communication: concepts, security-privacy issues, applications, and future perspectives. Futur Gener Comput Syst 138:61–88
Rejesus RM, Little BB, Lovell AC (2005) Using data mining to detect crop insurance fraud: is there a role for social scientists? J Financ Crime 12:24–32
Rizzo P (2017) Mike Hearn: bitcoin farewell post was no 'Banker Conspiracy'. http://www.coindesk.com/mike-hearn-bitcoin-post-banker-conspiracy/. Accessed 9 May 2017
Rohr JR, McCoy KA (2010) Preserving environmental health and scientific credibility: a practical guide to reducing conflicts of interest. Conserv Lett 3:143–150
Schlanger Z (2017) Rogue scientists race to save climate data from Trump
Science Division Blockchain Technology and Environmental Sustainability (2020). https://wedocs.unep.org/bitstream/handle/20.500.11822/34226/FB019.pdf
Scott B (2016) How can cryptocurrency and blockchain technology play a role in building social and solidarity finance? UNRISD Working Paper
Shafique K, Khawaja BA, Sabir F, Qazi S, Mustaqim M (2020) Internet of Things (IoT) for next-generation smart systems: a review of current challenges, future trends and prospects for emerging 5G-IoT scenarios. IEEE Access 8:23022–23040
Sharma L, Javali A, Nyamangoudar R, Priya R, Mishra P, Routray SK (2017) An update on location based services: current state and future prospects. In: 2017 International conference on computing methodologies and communication (ICCMC). IEEE, pp 220–224
Shilpa, Kaur T (2022) Digital healthcare: current trends, challenges and future perspectives. In: Proceedings of the future technologies conference (FTC) 2021, vol 2. Springer International Publishing, pp 645–661
Shin D, Lee N (2021) Blockchain technology for circular economy and economic and social sustainability. Resour Conserv Recycl 169:105456
Shukla S, Hait S (2022) Smart waste management practices in smart cities: current trends and future perspectives. In: Advanced organic waste management. Elsevier, pp 407–424
Shulman A, Co-founder C (2006) How to mitigate the most significant database vulnerabilities. Top ten database security threats
Singh D, Tripathi G, Jara AJ (2014) A survey of Internet-of-Things: future vision, architecture, challenges and services. In: In 2014 IEEE world forum on Internet of Things (WF-IoT). IEEE, pp 287–292
Smith AD, Li Q, Vyas A, Haque MM, Wang K, Velasco A et al (2019) Carbon-based electrode materials for microsupercapacitors in self-powering sensor networks: present and future development. Sensors 19(19):4231

Sompolinsky Y, Zohar A (2013) Accelerating bitcoin's transaction processing: fast money grows on trees, not chains. https://www.smithandcrown.com/open-research/accelerating-bitcoinstransaction-processing-fast-money-grows-on-trees-not-chains/. Accessed 10 May 2017

Soni S (2018) Blockchain technology and its applications in environmental sustainability. In: Proceedings of the 3rd international conference on computational systems and information technology for sustainable solutions. Springer, pp 36–41

Suichies B (2017) Why blockchain must die in 2016. https://medium.com/@bsuichies/whyblockchain-must-die-in-2016-e992774c03b4. Accessed 13 Apr 2017

Sun J, Yan J, Zhang KZ (2016) Blockchain-based sharing services: what blockchain technology can contribute to smart cities. Financ Innov 2:26

Sutherland WJ, Barnard P, Broad S, Clout M, Connor B, Cote IM, Dicks LV, Doran H, Entwistle AC, Fleishman E et al (2017) A horizon scan of emerging issues for global conservation and biological diversity. Trends Ecol Evol 2017(32):31–40

Swan M (2015) Blockchain thinking: the brain as a DAC (decentralized autonomous organization). In: Proceedings of the Texas bitcoin conference, Austin, TX, USA, 27–29 March, pp 27–29

Tang S, Shelden DR, Eastman CM, Pishdad-Bozorgi P, Gao X (2019) A review of building information modeling (BIM) and the Internet of Things (IoT) devices integration: present status and future trends. Autom Constr 101:127–139

Tapscott D, Tapscott A (2016) Blockchain revolution: how the technology behind bitcoin is changing money, business, and the world. Penguin

The Food and Agriculture Organization of the United Nations (2016) International Telecommunication Union.2E-Agriculture Strategy guide piloted in Asia-Pacific countries. The Food and Agriculture Organization of the United Nations, International Telecommunication Union, Bangkok, Thailand. http://www.fao.org/3/a-i5564e.pdf. Accessed 10 May 2017

Tzounis A, Katsoulas N, Bartzanas T, Kittas C (2017) Internet of Things in agriculture, recent advances and future challenges. Biosyst Eng 164:31–48

United Nations (2018) Global Environmental Outlook—GEO 6: summary for policymakers. United Nations

United Nations (2021) Blockchain for sustainable development. https://sdgs.un.org/sdgs/goal-9-industry-innovation-and-infrastructure/techfacial-technologies/blockchain-sustainable

United Nations Educational, Scientific and Cultural Organization (UNESCO) (2022) Water scarcity: by 2025, it is projected that nearly two-thirds of the global population could be living under water-stressed conditions. United Nations Educational, Scientific and Cultural Organization (UNESCO)

United Nations Environment Programme (UNEP) (2021) Ocean pollution: an estimated 8 million metric tons of plastic waste enter the ocean each year. United Nations Environment Programme (UNEP)

United States Federal Bureau of Investigation (2016). Smart farming may increase cyber targeting against US Food and Agriculture Sector. https://publicintelligence.net/fbi-smart-farm-hacking/. Accessed 2 Jun 2017

Walport M (2016) Distributed ledger technology. Blackett Rev. https://www.gov.uk/government/publications/distributed-ledger-technology-blackett-review

Wang S, Wan J, Li D, Zhang C (2016) Implementing smart factory of industrie 4.0: an outlook. Int J Distrib Sensor Netw 12(1):3159805

White T, Wallensteen P, Särd H, Cunningham M (2019) Blockchain for biodiversity protection–legal insights and recommendations. Environ Sci Pol 93:101–110

Whitmore A, Agarwal A, Da Xu L (2015) The Internet of Things—a survey of topics and trends. Inf Syst Front 17:261–274

Wildchain (n.d.) Wildchain—blockchain for wildlife conservation. https://wildchain.io/

WildTrack (n.d.) Combining footprint identification technology (FIT) and blockchain. https://www.wildtrack.org/

Wiles N (2015) The radical potential of blockchain technology. https://www.meetup.com/London-Futurists/events/221734513/

Witkowska D, Białończyk D, Sowa A (2019) Blockchain technology and its impact on green logistics. J Clean Prod 232:417–430
Wood G (2014) Ethereum: a secure decentralised generalised transaction ledger. Ethereum Project: Yellow Paper
World Bank (2021) Deforestation: approximately 10 million hectares of forest lost annually. World Bank
World Economic Forum (2019) Deep dive into blockchain for supply chains. https://www.weforum.org/whitepapers/deep-dive-into-blockchain-for-supply-chains
World Economic Forum (2020) Harnessing AI and blockchain for environmental challenges. https://www.weforum.org/whitepapers/harnessing-artificial-intelligence-and-blockchain-for-environmental-challenges
World Health Organization (2021) Air pollution: 9 out of 10 people worldwide breathe polluted air. World Health Organization
World Summit on the Information Society (2017). http://www.itu.int/net/wsis/index.html. Accessed 16 Apr 2017
World Wide Fund for Nature (WWF) (2021) Blockchain: a game changer in tackling deforestation and promoting sustainable supply chains. World Wide Fund for Nature (WWF)
World Wildlife Fund (n.d.) Blockchain for nature. https://www.worldwildlife.org/innovation-technology/blockchain-for-nature
Wright A, De Filippi P (2015) Decentralized blockchain technology and the rise of lex cryptographia. SSRN Electron J. https://doi.org/10.2139/ssrn.2580664
Xu LD, Xu EL, Li L (2018) Industry 4.0: state of the art and future trends. Int J Prod Res 56(8):2941–2962
Yoshida K, Tanaka K, Hariya R, Azechi I, Iida T, Maeda S, Kuroda H (2016) Contribution of ICT monitoring system in agricultural water management and environmental conservation. In: Serviceology for designing the future. Springer, Tokyo, Japan, pp 359–369
Yu QY, Shi Y, Tang HJ, Yang P, Xie AK, Liu B, Wu WB (2017) eFarm: a tool for better observing agricultural land systems. Sensors 17:453
Zhang Y, Sun L, Song H, Cao X (2014) Ubiquitous WSN for healthcare: recent advances and future prospects. IEEE Internet Things J 1(4):311–318

Exploring Host Immunity and Virulence in Mucorales

18

Aastha Chauhan, Pooja Sen, Aman Singh,
Krishna Prajapati, Vanshika Jain, and Pooja Vijayaraghavan

Abstract

Fungal spores, ubiquitous in the environment, are significant allergens and pathogens. Mucormycosis is a severe infection primarily affecting immunocompromised individuals, such as those with uncontrolled diabetes, bone marrow or organ transplants, corticosteroid treatment, hematological cancers, or traumatic injuries and COVID-19 patients, highlighting the urgent need to understand its pathogenesis. Mucorales fungi are known to be innately resistant to common antifungals such as azoles and Amp-B, making the treatment difficult and challenging to diagnose. The interaction between Mucorales and host immune responses—spanning innate and adaptive immunity—plays a crucial role in disease progression. Iron availability and the dimorphic nature of fungi also influence virulence. Recent research has focused on the role of iron uptake mechanisms and the calcineurin pathway in fungal pathogenicity. This comprehensive review underscores the complex interplay between Mucorales, host immune defenses, and emerging therapeutic targets to combat mucormycosis effectively.

Keywords

Mucorales · Host immune system · Iron acquisition · Spore coat protein · Dimorphism

A. Chauhan · P. Sen · A. Singh · K. Prajapati · V. Jain · P. Vijayaraghavan (✉)
Amity Institute of Biotechnology, Amity University Uttar Pradesh,
Noida, Uttar Pradesh, India
e-mail: vrpooja@amity.edu

A. Kumar, M. K. Solanki (eds.), *Microbial Biocontrol Techniques*,
Microorganisms for Sustainability 54,
https://doi.org/10.1007/978-981-97-8739-5_18

18.1 Introduction

Fungal spores are ubiquitous environmental contaminants and a cause of human health problems from allergy to life-threatening infections. In fact, these spores are so extensive that their airborne levels may be as much as 1000 times more than other typical allergens (Nicolás et al. 2020) and can exist both indoors and outdoors. The increasing prevalence of mucormycosis, a severe fungal infection, is being reported across the world and notably in India and China among a high percentage of patients with uncontrolled diabetes mellitus (Prakash et al. 2019).

It is one of the three most common fungal diseases to affect hematological patients other than aspergillosis and candidiasis (Skiada et al. 2013). This infection primarily endangers immunocompromised individuals, such as those with uncontrolled diabetes, bone marrow or organ transplants, corticosteroid treatment, hematological cancers, or traumatic injuries. Both innate and adaptive immune responses are crucial in defending against the infection in these vulnerable patients. While the immune responses to fungi like *Candida* and *Aspergillus* have been extensively studied, much less is known about how the immune system responds to fungi in the order Mucorales (Obar et al. 2016).

The recent increase in the cases of mucormycosis, particularly among COVID-19 patients, has brought to light a reality that this disease can be more severe, and treatment is often not effective (Xu et al. 2014). However, some research in the past years has approached to unveil molecular machineries related to virulence of these fungi (Hassan and Voigt 2019).

Mucorales are highly resistant to the most common antifungal drugs, e.g., voriconazole, and their capability of escaping host immune response may contribute to this resistance (Katragkou et al. 2014). Invasive mucormycosis is a challenging disease to manage with amphotericin B (Roilides and Simitsopoulou 2017), due to various obstacles that exist such as difficultly in diagnosing early because it has nonspecific symptoms, low efficiency of the serological test culture, and molecular diagnostic method delaying antifungal therapy initiation (Roilides and Simitsopoulou 2017). One of the chief targets in recent research is the iron uptake systems among Mucorales since most living organisms rely on this element, and it has also been recognized as a secondary virulence factor for some other pathogens. Together, these studies have uncovered an intricate relationship between iron acquisition and fungal dimorphism that likely involves novel control mechanisms for potential future drug development efforts (Navarro-Mendoza et al. 2018).

18.2 Host Immune Response to Mucorales Infections

The first line of defense against sporangiospores involves physical barriers: the skin, respiratory tract, and intestinal mucosa. Among important innate immune cells that play a pivotal role in host defense from these organisms are neutrophils, monocytes/macrophages, and dendritic cells (Roilides and Simitsopoulou 2017). However, the Mucorales can breach physical barriers by percutaneous inoculation, ingestion, or inhalation. The innate immune cells that Mucorales face once across these physical barriers include macrophages, neutrophils, and dendritic cells (Ghuman and Voelz 2017). The process of mucormycosis initiation in humans normally begins when the innate immune system that is naturally able to destroy spores and inhibit the germination of fungal spores fails to function protectively. Once a pathogen breaks through nonspecific barriers like skin and mucosal layers, the first responders are from the innate immune system, and macrophages and neutrophils act as key cellular defenders against the foreign invader (Ghuman and Voelz 2017). Mucorales penetrate physical barriers, probably by entry through skin wounds or by ingestion. Its infective particles are asexual sporangiospores, as with most other Mucorales, which are ubiquitous in the environment. Such spores can swell and become active (germinate) creating fast-growing hyphae in their life cycle. In the host, this germination and filamentous growth can eventually lead to angioinvasion resulting in vessel thrombosis and tissue necrosis (Ghuman and Voelz 2017).

The activation of innate and adaptive immunity along with platelets is key to protecting against infections by the host (Ghuman and Voelz 2017). Epidermal growth factor (EGF) pathways also play an important role in this protective effect since they do not allow cell to be extruded and thus are able to prevent mucormycosis as well, which helps skin and other barriers (Wurster et al. 2021; Gullì et al. 2024). The innate immunity involves nonspecific defense mechanisms that include the physical barriers of skin and respiratory mucosa, which stop microorganisms from entry into the host; complement proteins mediate antigen recognition by immune cells (antibody independent), a network of immune system components to mount a defense against the pathogen (Ben-Ami et al. 2009). Extensive tissue invasion and destruction are the hallmarks of the process of mucormycosis, and it has been described as the most angioinvasive of all diseases caused by fungi, which usually results in disseminated diseases.

Mucorales fungi infect nasal and alveolar epithelial cells through their respective interactions with the host glucose-regulated protein 78 kDa and integrin $\alpha 3\beta 1$, respectively (Alqarihi et al. 2020). The process initiates through contact of asexual resting spores with skin or mucosa. While innate physical barriers to the latter generally obstruct tissue invasion, a disruption in the former or inability of the immune cell effectors can result in mucormycosis (Hussell and Bell 2014) (Fig. 18.1).

Numerous studies have examined how Mucorales species interact with various immune cells, revealing key insights into the pathogenesis of mucormycosis.

18.2.1 Epithelial Cell Interaction with Mucorales

Epithelial cells are the first line of defense at infection sites, including alveoli and skin (Liu et al. 2010). Damage to these barriers is therefore considered among the major risk factors for Mucorales infections. Damage exposes basement membrane proteins, which then may support the first step of a *Rhizopus oryzae* infection: spore adhesion, leading to spore germination and host infection (Liu et al. 2010). Further, endothelial cells can phagocytose and damage Mucorales spores (Liu et al. 2010);

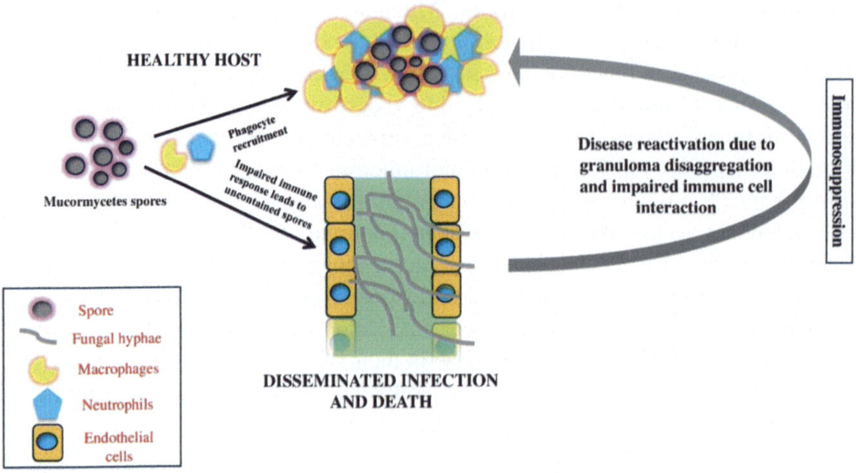

Fig. 18.1 The interface between mucormycetes and the host

the glucose-regulated protein 78 (GRP78) serves as a receptor, recognizing specifically *Mucor* species but no other fungal pathogens like *A. fumigatus*. GRP78 also mediates the invasion and damage of endothelial cells by mucoralean fungi. Similar to levels observed in diabetic keto acidosis (DKA), elevated iron and glucose enhance GRP78 cell surface expression on human endothelial cells within the brain, lungs, and sinuses (Hassan and Voigt 2019).

18.2.2 Endothelial Cell Interaction with Mucorales

Angioinvasion is one of the hallmarks of mucormycosis (Ghuman and Voelz 2017) that involves the penetration of endothelial cells lining the inner layer of blood vessels. Besides structure, these cells take a lead role in pathogen recognition and are critically required in physiological functions (Aird 2012). *R. oryzae* adheres to endothelial cells directly, leading to damage by internalization. This entry was mediated through GRP78, whose expression was markedly upregulated by hyperglycemic conditions in DKA (Ibrahim et al. 2012). GRP78 does not only favor the invasion of endothelial cells by Mucorales; it is also implicated in the endothelial cell damage (Ghuman and Voelz 2017). Immunohistochemical studies of ethmoidal sinus tissue infected with Mucorales have revealed a high expression of GRP78 on endothelial cells, especially in necrotic tissue, where it is also found on macrophage surfaces (Shumilov et al. 2018).

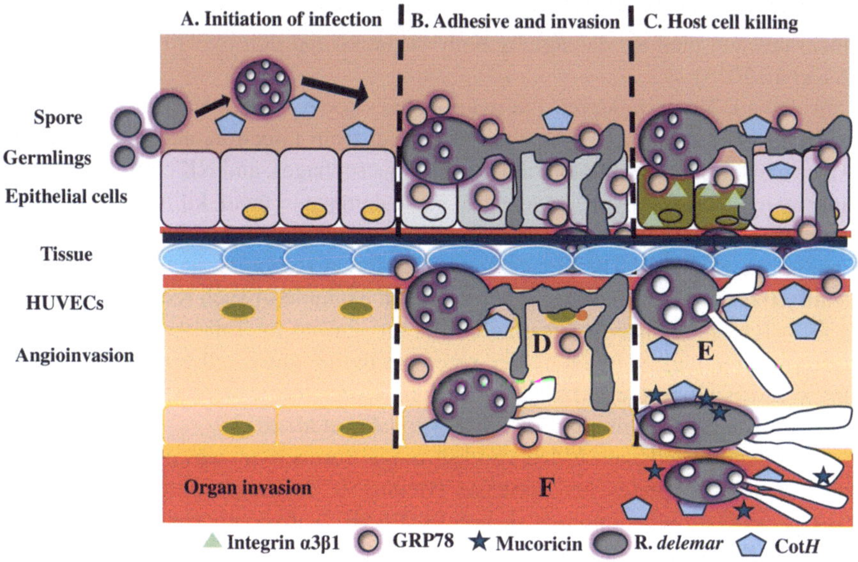

Fig. 18.2 The hypothesized sequence of events leading to adhesion, invasion, and host cell death is as follows: (A) Inhaled spores attach to epithelial cells through CotH/integrin α3β1 interaction, initiating germination. The resulting germlings secrete mucoricin. (B) Spore invasion causes substantial disruption of the epithelial layer within hours of infection (C) Within 48 h, mucoricin triggers host cell death (D). Hyphae and sporulated cells invade blood vessels by interacting with CotH and endothelial GRP78. (E) Infection with *R. delemar* spores results in endothelial cell damage and subsequent tissue necrosis. (F) The spread through the bloodstream leads to organ seeding, with tissue swelling and organ failure caused by excessive vascular leakage

18.2.3 T-Cell Responses to Mucorales

It is at this mucormycetes-host T-cell—a component of the adaptive immune system—interface that pathogenic fungi are eradicated via the excretion of certain cytokines necessary for hyphal damage. Such Mucorales-specific T cells that secreted IL-4 and IL-10 and produced IFNγ were detected only in patients with diagnosed mucormycosis (Potenza et al. 2011). This identified a potential antigen-specific T cell-based diagnostic test for the clinical management of mucormycosis (Potenza et al. 2011). Also, the reactivation of resting T cells by IL7 or even better with a combination of both cytokines will potentiate Mucorales-specific responses (T cell/fungitoxic macrophages) in vitro. It should generate Mucorales-specific CD4+ T-cell help and provide an adjunct to adaptive immunity aimed at combating this invasive fungal infection (Potenza et al. 2011) (Fig. 18.2).

18.2.3.1 Innate Immunity

Alveolar macrophages are the first line of defense against inhaled Mucorales spores. These macrophages (Ibrahim and Voelz 2017), unlike those from immunocompetent individuals, which easily kill *Aspergillus fumigatus* conidia, have poor killing ability against phagocytized Mucorales spores. They can attach to the hyphae of Mucorales and mediate damage by both oxidative and non-oxidative mechanisms (Lee et al. 2015).

In diabetic mice, however, the macrophages seem incapable of preventing the germination of spores and hence fail to protect from infection, and such mice die (Alqarihi et al. 2023). Although monocytes, macrophages, and NK cells do have the capacity to recognize hyphae and cause some damage to them, killing has not been observed. PMNs, on the other hand, kill the filamentous forms of the fungi in vitro (Schmidt et al. 2016). The monocytes use different receptors, including PRRs, to identify the PAMPs in the fungal cell wall. An example of such receptors includes CLRs, which recognize β-glucan; chitin is recognized by Toll-like receptors, and mannose is recognized by NOD-like receptors in the fungal cell wall. These receptors would mediate the induction of different signaling cascades that eventually orchestrate the innate immune response (Lionakis et al. 2017).

Mononuclear cells and PMNs can kill fungal forms by means of toxic metabolites, fungicidal peptides, and proteases (Heung 2020; Roilides and Simitsopoulou 2017). Peripheral blood mononuclear cells (PBMCs) are composed of ~80% T and B, 10% NK, and 10% monocytes that play a key role in both innate and adaptive immunity (Montaño and Voigt 2020). Activated phagocytes also secrete chemokines and cytokines that can support T-cell activation (pro-inflammatory activitics), helping to link innate immunity with adaptive immune response. Human PBMCs in response to mucoralean fungi produce TNF-α, IL-1β, IL6, IL8, MCP01, and GM-CSF following an in vitro incubation period of 9 h with killed spores and germ tubes from *Rhizopus arrhizus*, *Cunninghamella bertholletiae*, *Mucor circinelloides*, etc. Production of IL-1β by these Mucorales was significantly greater than the quantity produced by the challenge with *A. fumigatus* (Wurster et al. 2017).

Opportunistic Mucorales are resistant to innate immune cells. Their chemotaxis and killing activities could be significantly impaired in hyperglycemia, especially in settings of DKA (Alqarihi et al. 2023). At the adult stage, the bone marrow differentiating monocytes are distinguishable from resident macrophages, such as alveolar macrophages or glial cells, which are originally derived in part either from fetal yolk sac progenitor cells or from fatal liver monocytes (Hussell and Bell 2014).

Mucormycetes are highly proliferative fungi that outgrow the suppressed immune response. It has already been well documented that the hyphal growth of the *Mucor circinelloides* mutant locked in the yeast form is required for its pathogenicity. Known chemical inhibitors of the calcineurin pathway, such as FK506, or knockout of the calcineurin regulatory subunit cnbR greatly impairing hyphal growth, rendered the *M. circinelloides* avirulent in the infected wax moth larvae. Species of Mucorales with high germination rates, such as *Cunninghamella bertholletiae*, are much more virulent than other species showing slower rates of germination, like *Rhizopus oryzae* and *Mucor circinelloides*, in a neutropenic rabbit model of pulmonary mucormycosis. Increased virulence was directly related to the burden in the lungs, with amplified angioinvasion and reduced survival rates.

Platelets also contribute to antimicrobial host defense through platelet microbicidal protein release (Perkhofer et al. 2009). The protection from mucormycosis is thus achieved by the innate immune system through control of spores within the phagocytes and granulomatous clusters, and inhibition of spore germination. In susceptible individuals, there is a loss of control over spores and subsequent filamentous fungal growth. New findings now show that Mucorales may act as intracellular pathogens within granulomas, opening up latent infections' possibility. This could provide a new avenue for therapies directed at resting spores prior to the emergence of full-blown hyphae, which might represent an effective prophylactic treatment (Alqarihi et al. 2023).

18.2.3.2 Adaptive Immunity

The adaptive immune system plays a supportive role in combating invasive fungal infections caused by Mucorales. Initially, T cells were thought to have minimal importance in this process because T-cell deficiencies did not appear to significantly increase the risk of Mucorales infections (Ibrahim and Kontoyiannis 2013). However, further research has highlighted that while T cells might not be the primary defense, they do contribute to the immune response, especially through the interplay between innate and adaptive immunity (Wurster et al. 2017).

Activated phagocytes, through antigen presentation and induction of T-helper cell responses, act as a bridge between innate and adaptive immunity. Dectin-1 is required for the induction of adaptive immunity in response to Mucorales fungi by human dendritic cells. This signaling pathway, similar to *Aspergillus* infection immune responses, also effectively induces IL-23 and Th17 response (Chamilos et al. 2010). The Th-17 response is particularly relevant as it gives rise to immune cytokines like IL-17 that draw the abundant number of neutrophils to the site of infection and broaden still more our capacity for severe combat against this fungus (Chamilos et al. 2010).

Even though B cells and the production of these same known (anamnestic) specific antibodies have not been as extensively studied in Mucorales infection, one may argue that they could exert an important immune role neutralizing fungal elements or contributing to opsonization, thereby easing phagocytosis. If mucormycosis can induce memory in the adaptive immune system, it is possible that these persons who recover from disease may be less prone to recurrent infections.

Overall, the adaptive immune system is not inherently protective against Mucorales but substantially reinforces components of innate immunity by redirecting T-helper cell-mediated activity and regulating cytokine signal pathways that recruit activated additional immune cells (Ibrahim and Kontoyiannis 2013).

18.3 Immunocompromised Condition Interactions with Mucorales

Diabetes mellitus, a leading cause of kidney dysfunction, blindness, strokes, lower limb amputation, and heart attacks, is a significant predisposing factor for mucormycosis. The World Health Organization (WHO) reports that diabetes caused 1.5

million deaths worldwide in 2019, placing it among the top ten leading causes of death globally (Skyler et al. 2017).

Individuals with uncontrolled diabetes experience alterations in their immune system, making them highly susceptible to infections by Mucorales. Many reasons are responsible for tilting us towards becoming vulnerable. Diabetes impairs phagocytosis and the chemotaxis of polymorphonuclear (PMN) cells, hindering the ability of macrophages to engulf spores (Fathima et al. 2021). This failure allows spores to remain within cells, where they enlarge, bud, and eventually develop into hyphae. The oxidative burst failure in PMN cells exacerbates vascular invasion and leads to the hematogenous dissemination of fungal growth, ultimately resulting in mucormycosis (Fathima et al. 2021). Persistent hyperglycemia also disrupts mitochondrial function, leading to the production of reactive oxygen species (ROS) (Berlanga-Acosta et al. 2013), which damage pancreatic beta cells and impair insulin production. This damage activates stress-responsive intracellular signaling molecules, promoting cellular destruction. Additionally, diabetes accelerates the formation of atherosclerotic plaques, contributing to cardiovascular complications (Khanna et al. 2021).

Iron plays a crucial role in both human cellular function and growth of infectious agents (Khanna et al. 2021). Pathogens, including Mucorales, employ various strategies to acquire iron from the host (Panigrahi et al. 2014). Limiting the availability of free iron is essential for preventing the proliferation of infectious agents, particularly Mucorales, which grow poorly without iron (Raizada et al. 2018). In diabetic patients, the combination of free iron and acidic pH increases their susceptibility to mucormycosis (Raizada et al. 2018).

18.4 Virulence Factors of the Mucorales

The virulence factors of pathogens play a crucial role in causing damage to the host (Brunke et al. 2016). The pathogenesis of mucormycosis is less understood compared to other fungal infections (Rogers 2008). Mucormycosis primarily affects immunocompromised individuals, such as those with severe and prolonged neutropenia, hematological malignancies, poorly controlled diabetes, and other conditions requiring immunosuppressive treatments (Ibrahim et al. 2012). Although advances have been made in understanding the mechanisms by which *Mucorales* spp. invade endothelial cells, little is known about how they initiate infection by binding to, entering, and damaging the airway epithelium (Chibucos et al. 2016).

Mucormycetes are fast-growing fungi, which allows them to outcompete the immune response when it is suppressed. In a neutropenic rabbit model of pulmonary mucormycosis, *Mucorales* species with fast germination rates (e.g., *Cunninghamella bertholletiae*) are significantly more virulent than those with slower germination rates (e.g., *Rhizopus oryzae*, *R. microsporus*, *Mucor circinelloides*). The increased virulence is characterized by a higher lung burden, amplified angioinvasion, and lower survival rates. A protective immune response may require spore clearance before the onset of filamentous growth (Petraitis et al. 2013).

18.4.1 High-Affinity Iron Permease in Mucorales

Fungi cannot survive without iron, a necessary nutrient that exists in free form as ferrous (+2) or ferric (+3) ions but is sequestered by host mechanisms (e.g., ferritin and lactoferrin) (Bullen et al. 2006), preventing it from being available to fungi for uptake in physiological tissue fluids like plasma. In DKA and hyperglycemia itself, the low pH of blood interferes with plasma chelators which then become less stable; as a consequence, concentration may rise above its critical level for basal glycemic values, and susceptibility to mucormycosis is increased.

Mucorales fungi have evolved iron acquisition systems that allow them to capture and internalize Fe^{2+} from the environment particularly during low-iron environments, those which include genes such as FTR1 whose product is involved in iron uptake and transport (Gebremariam et al. 2016). This consists of a three-protein strategy: iron reductase (Fre), ferroxidase (Fet3), and the permease for transferring from outside to inside cell, Ftr1 (Navarro-Mendoza et al. 2018).

Proteins like Fob1 and Fob2 on the Mucorales surface mediate iron transfer through this reductase/permease system (Liu et al. 2015).

The results clearly demonstrated that under iron depletion and infection with *M. circinelloides*, three ferroxidase genes, fet3a, fet3b, as well as fet3c (Navarro-Mendoza et al. 2018), were essential providers of favorable environment for the fungal pathogen inner part survival. Dimorphism in *M. circinelloides* regulates the expression of these ferroxidase genes, and for this fungus, it is a necessary step to become virulent by changing from yeast to mycelium (Lax et al. 2020).

Rhizopus species secrete rhizoferrin, a siderophore belonging to the polycarboxylate family, which supplies iron to the fungus through a receptor-mediated, energy-dependent process (Lax et al. 2020) (Fig. 18.3).

In the host, iron is typically bound to proteins like hemoglobin, ferritin, and transferrin, preventing the toxic effects of free iron. However, Mucorales have evolved mechanisms to acquire iron, including the reductive pathway, hem and hemoglobin utilization, and ferric iron acquisition from siderophores. The reductive iron uptake system involves the transport of reduced iron by high-affinity ferric transporters, a process dependent on the dual-protein complex of ferroxidase (Fet3) and permease (Ftr1) (Gerwien et al. 2018). In *R. delemar*, reduced virulence is observed due to the deletion of the iron permease gene (*FTR1*) (Andrianaki et al. 2018).

Fungi also use a non-reductive pathway of iron acquisition called siderophore uptake (Wilson et al. 2016). Siderophores are small molecules with an exceptionally high affinity for ferric iron. Mucorales fungi, including *Rhizopus*, produce siderophores like rhizoferrin with a dissociation constant much higher than other biologically relevant iron-binding ligands (Haas 2014).

18.4.1.1 Spore Coat Protein (CotH Gene)

In the past, investigators showed that spore coat (CotH) proteins are a universal virulence factor on the surface of all *Mucorales* spp. but not other fungal species such as *A. fumigatus* (Hassan and Voigt 2019). By destroying and breaching immune

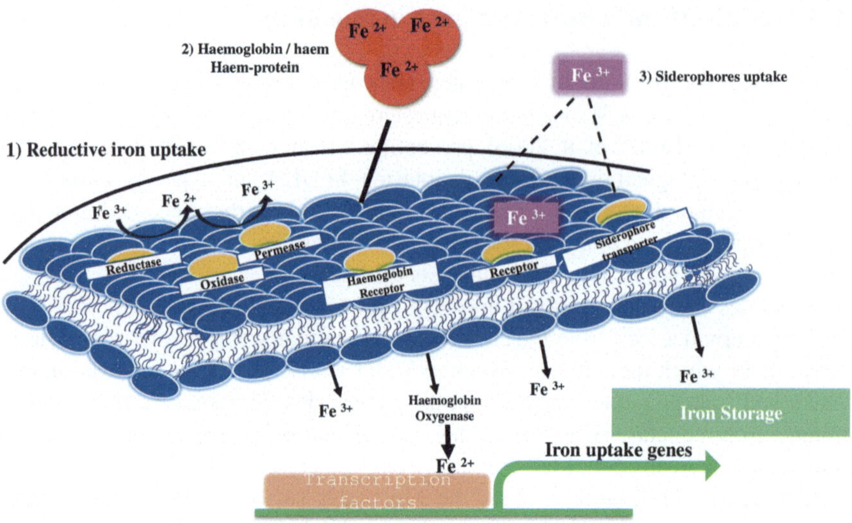

Fig. 18.3 The general iron acquisition strategies of pathogenic fungi include (1) the reductive pathway, involving reduction and uptake processes to subsequently transport ferrous or ferric iron into cytoplasm by specific Fe permeases; (2) heme-iron assimilation system enabling the extraction of intracellular/host-derived heme/hemin from hemoglobin/heme-containing proteins followed by its breakdown; (3) siderophore-assisted process enhancing utilization of various types of endogenous siderophores as well as xenosiderophores increased in response to host immune effectors (Ibrahim 2014)

cells, CotH proteins are crucial for the invasion and pathogenesis of mucormycosis (Gebremariam et al. 2014).

In *R. delemar* CotH proteins such as CotH1 and CotH2, one of the eight known homologous encrypted cotineal trivalent gourd family (COTH4), and pleased" to "In R. delemar CotH proteins, such as CotH1 and CotH2, are part of the encrypted cotineal trivalent gourd family (COTH4), which includes eight known homologous members. CotH2 and CotH3 interact with endothelial glucose-regulated protein 78 (GRP78), whereas CotH7 interacts with integrin α3β1 for invasion of alveolar epithelial cells (Chibucos et al. 2016). This interaction is critical for fungal invasion since levels of GRP78 expression were dramatically induced in the sinuses and lungs during DKA, rendering these anatomic compartments vulnerable to infection (Alqarihi et al. 2020).

The mechanism of tissue invasion relies on CotH2 and CotH3 interacting with GRP78 on epithelial cells (Lax et al. 2020). GRP78 is usually located in the endoplasmic reticulum (ER) (Ibrahim et al. 2019). But it is released to the cell surface under stress conditions, where it interacts with CotH proteins. This interaction is essential for the development of mucormycosis, as demonstrated in diabetic mouse models (Liu et al. 2010).

Dimorphism: Tangible and Apparent Shape Shifting as a Mechanism to Command Pathogenic Outcomes of *Mucor circinelloides*

Dimorphism in certain fungi, of which yeast and mycelium forms exist simultaneously during vegetative growth, is seen in some species of Mucorales. Yeast, for example in some *Mucor* species, is grown under anaerobic conditions and fermentable hexose is used as a carbon source, while wild-type hyphae are formed aerobically on nutrient-poor medium (Lax et al. 2020). Nevertheless, this is not the case for all *Mucor* dimorphic species regarding their response to environmental conditions. To provide examples, *Mucor rouxii* only grow yeast forms under anaerobic conditions in the presence of hexose, whereas *Mucor genevensis* is capable to form yeasts when growing aerobically with a high amount of hexoses available around it. *Mucor circinelloides* is one example of a second species in the Mucorales order with dimorphic processes where this fungus uses budding yeast mode for growth anaerobically and as filamentous fungus aerobically.

However, recent studies were able to identify a relationship between dimorphism and virulence in Mucorales, thus presenting it as an interesting target for potential antifungal therapy against mucormycosis. Among Mucorales, the zygomycete fungus occurs as it is especially noteworthy for its genetic amenability and availability of numerous molecular tools to investigate different aspects in this organism (Navarro-Mendoza et al. 2018). Studies on calcineurin pathway in *M. circinelloides* depicted its essential role, controlling the evolution from yeast to mycelium and virulence (Lee et al. 2015). Calcineurin is a Ca2+/calmodulin-dependent serine-threonine phosphatase composed of two subunits: the catalytic subunit A, which possesses phosphatase activity, and the regulatory subunit B. The regulatory subunit binds to calcium-bound calmodulin to activate the enzyme complex (Fig. 18.4).

Interruption of the CnbR gene or exposure to calcineurin inhibitors yields strains that are locked in yeast phase and have attenuated virulence, suggesting a role for dimorphic switching in fungal virulence. In contrast, null cnaA mutants had

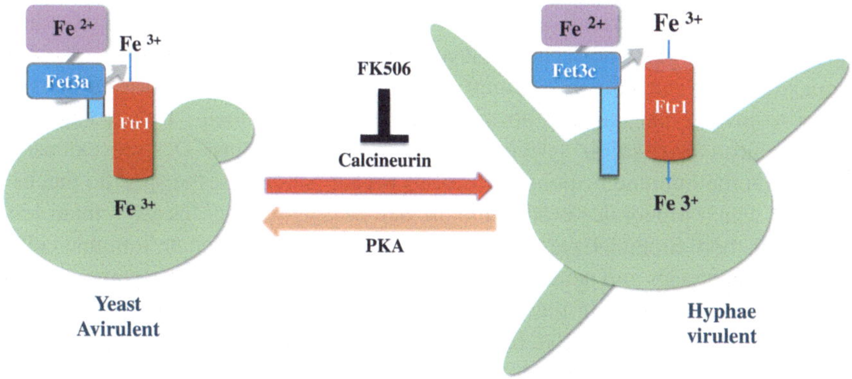

Fig. 18.4 Calcineurin mediates dimorphism and virulence in *M. circinelloides* via coordinating with cyclic AMP-dependent (cAMP) protein kinase, a signal transduction pathway during iron uptake: similar to other dimorphic fungi, it dephosphorylates transcription factors followed by nuclear entrance for activating target gene expression. One regulatory B subunit (CnbR) and three catalytic A subunits (CnaA, CnaB, and CnaC) present in fungus (Boyce and Andrianopoulos 2015)

filamentous growth phase, reduced polarity maintenance capacity, increased susceptibility to calcineurin inhibitors, as well as cell wall abnormalities and enlarged spores. These larger spores are more virulent, as this size inhibits phagosomal maturation when macrophage is engulfed, much like yeast cells that cannot block the process of phagolysosomal maturation (Lee et al. 2015). In *M. circinelloides* pathogenesis, calcineurin interacts closely with protein kinase A (PKA), which too is indispensable in this species. Deletion of the gene encoding calcineurin yielded spores that scored larger, suggesting a role of this phosphatase in size control as well as possibly decoupling virulence (Hassan and Voigt 2019).

In addition to the calcineurin pathway, other pathways—such as the cyclic AMP (cAMP) pathway and its target enzyme, cAMP-dependent protein kinase A (PKA)—also control dimorphism in *M. circinelloides* (Lax et al. 2020). Two catalytic (PKAC) and two regulatory (PKAR) subunits make up the tetrameric holoenzyme PKA. Three of the four pkaR genes in *M. circinelloides* are involved in the dimorphic transition. While pkaR1 deletion results in abnormalities in the yeast-mycelium transition, overexpression of the protein encourages mycelial development and growth, whereas its absence results in abnormalities in the yeast-mycelium transition (Lax et al. 2020).

18.5 Conclusion

Mucormycosis, a severe fungal infection primarily affecting immunocompromised individuals, poses significant challenges in diagnosis and treatment due to its aggressive nature and resistance to common antifungal drugs. The infection's rising prevalence, particularly among patients with uncontrolled diabetes and those recovering from COVID-19, underscores the urgent need for improved understanding and management strategies. The pathogenesis of mucormycosis involves complex interactions between Mucorales fungi and host immune systems, with the fungi's ability to invade and damage host tissues facilitated by their virulence factors, including iron acquisition systems and spore coat proteins. Despite extensive research on the immune response to various fungal pathogens, the specific mechanisms by which Mucorales evade immune defenses and cause disease remain less well understood. Studies have highlighted the crucial role of innate immune cells, such as macrophages and neutrophils, in combating fungal invasion, yet their effectiveness is compromised in the context of diabetes and other predisposing conditions. Additionally, the adaptive immune response, while not inherently protective, plays a supportive role in reinforcing innate defenses and promoting T-cell-mediated responses against the infection. Advances in understanding the molecular mechanisms of fungal virulence, particularly through the study of iron acquisition and fungal dimorphism, offer potential pathways for novel therapeutic interventions. Addressing the gaps in knowledge about mucormycosis, especially in relation to its interaction with host immune responses and the development of resistance, is critical for developing more effective treatments and preventive measures. The complexity of mucormycosis highlights the need for continued research and innovation to better manage and ultimately mitigate the impact of this challenging and potentially life-threatening infection.

References

Aird WC (2012) Endothelial cell heterogeneity. Cold Spring Harb Perspect Med 2:a006429. https://doi.org/10.1101/cshperspect.a006429

Alqarihi A, Gebremariam T, Gu Y, Swidergall M, Alkhazraji S, Soliman SSM, Bruno VM, Edwards JE, Filler SG, Uppuluri P, Ibrahim AS (2020) GRP78 and integrins play different roles in host cell invasion during mucormycosis. mBio 11:e01087–e01020. https://doi.org/10.1128/mBio.01087-20

Alqarihi A, Kontoyiannis DP, Ibrahim AS (2023) Mucormycosis in 2023: an update on pathogenesis and management. Front Cell Infect Microbiol 13:1254919. https://doi.org/10.3389/fcimb.2023.1254919

Andrianaki AM, Kyrmizi I, Thanopoulou K, Baldin C, Drakos E, Soliman SSM, Shetty AC, McCracken C, Akoumianaki T, Stylianou K, Ioannou P, Pontikoglou C, Papadaki HA, Tzardi M, Belle V, Etienne E, Beauvais A, Samonis G, Kontoyiannis DP, Andreakos E, Bruno VM, Ibrahim AS, Chamilos G (2018) Iron restriction inside macrophages regulates pulmonary host defense against Rhizopus species. Nat Commun 9:3333. https://doi.org/10.1038/s41467-018-05820-2

Ben-Ami R, Luna M, Lewis RE, Walsh TJ, Kontoyiannis DP (2009) A clinicopathological study of pulmonary mucormycosis in cancer patients: extensive angioinvasion but limited inflammatory response. J Infect 59:134–138. https://doi.org/10.1016/j.jinf.2009.06.002

Berlanga-Acosta J, Schultz GS, López-Mola E, Guillen-Nieto G, García-Siverio M, Herrera-Martínez L (2013) Glucose toxic effects on granulation tissue productive cells: the diabetics' impaired healing. Biomed Res Int 2013:256043. https://doi.org/10.1155/2013/256043

Boyce KJ, Andrianopoulos A (2015) Fungal dimorphism: the switch from hyphae to yeast is a specialized morphogenetic adaptation allowing colonization of a host. FEMS Microbiol Rev 39:797–811. https://doi.org/10.1093/femsre/fuv035

Brunke S, Mogavero S, Kasper L, Hube B (2016) Virulence factors in fungal pathogens of man. Curr Opin Microbiol 32:89–95. https://doi.org/10.1016/j.mib.2016.05.010

Bullen JJ, Rogers HJ, Spalding PB, Ward CG (2006) Natural resistance, iron and infection: a challenge for clinical medicine. J Med Microbiol 55:251–258. https://doi.org/10.1099/jmm.0.46386-0

Chamilos G, Ganguly D, Lande R, Gregorio J, Meller S, Goldman WE, Gilliet M, Kontoyiannis DP (2010) Generation of IL-23 producing dendritic cells (DCs) by airborne fungi regulates fungal pathogenicity via the induction of TH-17 responses. PLoS One 5:e12955. https://doi.org/10.1371/journal.pone.0012955

Chibucos MC, Soliman S, Gebremariam T, Lee H, Daugherty S, Orvis J, Shetty AC, Crabtree J, Hazen TH, Etienne KA, Kumari P, O'Connor TD, Rasko DA, Filler SG, Fraser CM, Lockhart SR, Skory CD, Ibrahim AS, Bruno VM (2016) An integrated genomic and transcriptomic survey of mucormycosis-causing fungi. Nat Commun 7:12218. https://doi.org/10.1038/ncomms12218

Fathima AS, Mounika VL, Kumar VU, Gupta AK, Garapati P, Ravichandiran V, Dhingra S, Murti K (2021) Mucormycosis: a triple burden in patients with diabetes during COVID-19 pandemic. Health Sci Rev (Oxf, Engl) 1:100005. https://doi.org/10.1016/j.hsr.2021.100005

Gebremariam T, Liu M, Luo G, Bruno V, Phan QT, Waring AJ, Edwards JE, Filler SG, Yeaman MR, Ibrahim AS (2014) CotH3 mediates fungal invasion of host cells during mucormycosis. J Clin Invest 124:237–250. https://doi.org/10.1172/JCI71349

Gebremariam T, Lin L, Liu M, Kontoyiannis DP, French S, Edwards JE, Filler SG, Ibrahim AS (2016) Bicarbonate correction of ketoacidosis alters host-pathogen interactions and alleviates mucormycosis. J Clin Invest 126:2280–2294. https://doi.org/10.1172/JCI82744

Gerwien F, Skrahina V, Kasper L, Hube B, Brunke S (2018) Metals in fungal virulence. FEMS Microbiol Rev 42:fux050. https://doi.org/10.1093/femsre/fux050

Ghuman H, Voelz K (2017) Innate and adaptive immunity to Mucorales. J Fungi 3:48. https://doi.org/10.3390/jof3030048

Gullì SP, Hallur V, Kale P, Menezes GA, Russo A, Singla N (2024) From spores to solutions: a comprehensive narrative review on mucormycosis. Diagnostics 14:314. https://doi.org/10.3390/diagnostics14030314

Haas H (2014) Fungal siderophore metabolism with a focus on Aspergillus fumigatus. Nat Prod Rep 31:1266–1276. https://doi.org/10.1039/c4np00071d

Hassan MIA, Voigt K (2019) Pathogenicity patterns of mucormycosis: epidemiology, interaction with immune cells and virulence factors. Med Mycol 57:S245–S256. https://doi.org/10.1093/mmy/myz011

Heung LJ (2020) Monocytes and the host response to fungal pathogens. Front Cell Infect Microbiol 10:34. https://doi.org/10.3389/fcimb.2020.00034

Hussell T, Bell TJ (2014) Alveolar macrophages: plasticity in a tissue-specific context. Nat Rev Immunol 14:81–93. https://doi.org/10.1038/nri3600

Ibrahim AS (2014) Host-iron assimilation: pathogenesis and novel therapies of mucormycosis. Mycoses 57(Suppl 3):13–17. https://doi.org/10.1111/myc.12232

Ibrahim AS, Kontoyiannis DP (2013) Update on mucormycosis pathogenesis. Curr Opin Infect Dis 26:508–515. https://doi.org/10.1097/QCO.0000000000000008

Ibrahim AS, Voelz K (2017) The mucormycete-host interface. Curr Opin Microbiol 40:40–45. https://doi.org/10.1016/j.mib.2017.10.010

Ibrahim AS, Spellberg B, Walsh TJ, Kontoyiannis DP (2012) Pathogenesis of mucormycosis. Clin Infect Dis 54(Suppl 1):S16–S22. https://doi.org/10.1093/cid/cir865

Ibrahim IM, Abdelmalek DH, Elfiky AA (2019) GRP78: a cell's response to stress. Life Sci 226:156–163. https://doi.org/10.1016/j.lfs.2019.04.022

Katragkou A, Walsh TJ, Roilides E (2014) Why is mucormycosis more difficult to cure than more common mycoses? Clin Microbiol Infect 20(Suppl 6):74–81. https://doi.org/10.1111/1469-0691.12466

Khanna M, Challa S, Kabeil AS, Inyang B, Gondal FJ, Abah GA, Minnal Dhandapani M, Manne M, Mohammed L (2021) Risk of mucormycosis in diabetes mellitus: a systematic review. Cureus 13:e18827. https://doi.org/10.7759/cureus.18827

Lax C, Pérez-Arques C, Navarro-Mendoza MI, Cánovas-Márquez JT, Tahiri G, Pérez-Ruiz JA, Osorio-Concepción M, Murcia-Flores L, Navarro E, Garre V, Nicolás FE (2020) Genes, pathways, and mechanisms involved in the virulence of Mucorales. Genes 11:317. https://doi.org/10.3390/genes11030317

Lee SC, Li A, Calo S, Inoue M, Tonthat NK, Bain JM, Louw J, Shinohara ML, Erwig LP, Schumacher MA, Ko DC, Heitman J (2015) Calcineurin orchestrates dimorphic transitions, antifungal drug responses and host-pathogen interactions of the pathogenic mucoralean fungus Mucor circinelloides. Mol Microbiol 97:844–865. https://doi.org/10.1111/mmi.13071

Lionakis MS, Iliev ID, Hohl TM (2017) Immunity against fungi. JCI Insight 2:e93156. https://doi.org/10.1172/jci.insight.93156

Liu M, Spellberg B, Phan QT, Fu Y, Fu Y, Lee AS, Edwards JE, Filler SG, Ibrahim AS (2010) The endothelial cell receptor GRP78 is required for mucormycosis pathogenesis in diabetic mice. J Clin Invest 120:1914–1924. https://doi.org/10.1172/JCI42164

Liu M, Lin L, Gebremariam T, Luo G, Skory CD, French SW, Chou T-F, Edwards JE, Ibrahim AS (2015) Fob1 and Fob2 proteins are virulence determinants of Rhizopus oryzae via facilitating iron uptake from ferrioxamine. PLoS Pathog 11:e1004842. https://doi.org/10.1371/journal.ppat.1004842

Montaño DE, Voigt K (2020) Host immune defense upon fungal infections with Mucorales: pathogen-immune cell interactions as drivers of inflammatory responses. J Fungi 6:173. https://doi.org/10.3390/jof6030173

Navarro-Mendoza MI, Pérez-Arques C, Murcia L, Martínez-García P, Lax C, Sanchis M, Capilla J, Nicolás FE, Garre V (2018) Components of a new gene family of ferroxidases involved in virulence are functionally specialized in fungal dimorphism. Sci Rep 8:7660. https://doi.org/10.1038/s41598-018-26051-x

Nicolás FE, Murcia L, Navarro E, Navarro-Mendoza MI, Pérez-Arques C, Garre V (2020) Mucorales species and macrophages. J Fungi 6:94. https://doi.org/10.3390/jof6020094

Obar JJ, Hohl TM, Cramer RA (2016) New advances in invasive aspergillosis immunobiology leading the way towards personalized therapeutic approaches. Cytokine 84:63–73. https://doi.org/10.1016/j.cyto.2016.05.015

Panigrahi MK, Manju R, Kumar SV, Toi PC (2014) Pulmonary mucormycosis presenting as non-resolving pneumonia in a patient with diabetes mellitus. Respir Care 59:e201–e205. https://doi.org/10.4187/respcare.03205

Perkhofer S, Kainzner B, Kehrel BE, Dierich MP, Nussbaumer W, Lass-Flörl C (2009) Potential antifungal effects of human platelets against zygomycetes in vitro. J Infect Dis 200:1176–1179. https://doi.org/10.1086/605607

Petraitis V, Petraitiene R, Antachopoulos C, Hughes JE, Cotton MP, Kasai M, Harrington S, Gamaletsou MN, Bacher JD, Kontoyiannis DP, Roilides E, Walsh TJ (2013) Increased virulence of Cunninghamella bertholletiae in experimental pulmonary mucormycosis: correlation with circulating molecular biomarkers, sporangiospore germination and hyphal metabolism. Med Mycol 51:72–82. https://doi.org/10.3109/13693786.2012.690107

Potenza L, Vallerini D, Barozzi P, Riva G, Forghieri F, Zanetti E, Quadrelli C, Candoni A, Maertens J, Rossi G, Morselli M, Codeluppi M, Paolini A, Maccaferri M, Del Giovane C, D'Amico R, Rumpianesi F, Pecorari M, Cavalleri F, Marasca R, Narni F, Luppi M (2011) Mucorales-specific T cells emerge in the course of invasive mucormycosis and may be used as a surrogate diagnostic marker in high-risk patients. Blood 118:5416–5419. https://doi.org/10.1182/blood-2011-07-366526

Prakash H, Ghosh AK, Rudramurthy SM, Singh P, Xess I, Savio J, Pamidimukkala U, Jillwin J, Varma S, Das A, Panda NK, Singh S, Bal A, Chakrabarti A (2019) A prospective multi-center study on mucormycosis in India: epidemiology, diagnosis, and treatment. Med Mycol 57:395–402. https://doi.org/10.1093/mmy/myy060

Raizada N, Jyotsna VP, Kandasamy D, Xess I, Thakar A, Tandon N (2018) Invasive fungal rhinosinusitis in patients with diabetes. J Infect Dev Ctries 12:787–793. https://doi.org/10.3855/jidc.9699

Rogers TR (2008) Treatment of zygomycosis: current and new options. J Antimicrob Chemother 61:i35–i40. https://doi.org/10.1093/jac/dkm429

Roilides E, Simitsopoulou M (2017) Immune responses to Mucorales growth forms: do we know everything? Virulence 8:1489–1491. https://doi.org/10.1080/21505594.2017.1368942

Schmidt S, Schneider A, Demir A, Lass-Flörl C, Lehrnbecher T (2016) Natural killer cell-mediated damage of clinical isolates of mucormycetes. Mycoses 59:34–38. https://doi.org/10.1111/myc.12431

Shumilov E, Bacher U, Perske C, Mohr A, Eiffert H, Hasenkamp J, Trümper L, Wulf GG, Ströbel P, Ibrahim AS, Venkataramani V (2018) In situ validation of the endothelial cell receptor GRP78 in a case of rhinocerebral mucormycosis. Antimicrob Agents Chemother 62:e00172–e00118. https://doi.org/10.1128/AAC.00172-18

Skiada A, Lanternier F, Groll AH, Pagano L, Zimmerli S, Herbrecht R, Lortholary O, Petrikkos GL, European Conference on Infections in Leukemia (2013) Diagnosis and treatment of mucormycosis in patients with hematological malignancies: guidelines from the 3rd European Conference on Infections in Leukemia (ECIL 3). Haematologica 98:492–504. https://doi.org/10.3324/haematol.2012.065110

Skyler JS, Bakris GL, Bonifacio E, Darsow T, Eckel RH, Groop L, Groop P-H, Handelsman Y, Insel RA, Mathieu C, McElvaine AT, Palmer JP, Pugliese A, Schatz DA, Sosenko JM, Wilding JPH, Ratner RE (2017) Differentiation of diabetes by pathophysiology, natural history, and prognosis. Diabetes 66:241–255. https://doi.org/10.2337/db16-0806

Wilson BR, Bogdan AR, Miyazawa M, Hashimoto K, Tsuji Y (2016) Siderophores in iron metabolism: from mechanism to therapy potential. Trends Mol Med 22:1077–1090. https://doi.org/10.1016/j.molmed.2016.10.005

Wurster S, Thielen V, Weis P, Walther P, Elias J, Waaga-Gasser AM, Dragan M, Dandekar T, Einsele H, Löffler J, Ullmann AJ (2017) Mucorales spores induce a proinflammatory cytokine response in human mononuclear phagocytes and harbor no rodlet hydrophobins. Virulence 8:1708–1718. https://doi.org/10.1080/21505594.2017.1342920

Wurster S, Ruiz OE, Samms KM, Tatara AM, Albert ND, Kahan PH, Nguyen AT, Mikos AG, Kontoyiannis DP, Eisenhoffer GT (2021) EGF-mediated suppression of cell extrusion during mucosal damage attenuates opportunistic fungal invasion. Cell Rep 34:108896. https://doi.org/10.1016/j.celrep.2021.108896

Xu S, Zhou Z, Du G, Zhou J, Chen J (2014) Efficient transformation of Rhizopus delemar by electroporation of germinated spores. J Microbiol Methods 103:58–63. https://doi.org/10.1016/j.mimet.2014.05.016

The manufacturer's authorised representative in the EU is Springer Nature Customer Service Centre GmbH, Europaplatz 3, 69115 Heidelberg, Germany. If you have any concerns regarding our products, please contact ProductSafety@springernature.com

Printed and bound by CPI Group (UK) Ltd, Croydon, CR0 4YY

01/10/2026

02225923-0017